ECOLOGY OF PESTICIDES

ECOLOGY OF PESTICIDES

A. W. A. BROWN

Department of Entomology
Michigan State University
East Lansing, Michigan

A Wiley-Interscience Publication
JOHN WILEY & SONS
New York • Chichester • Brisbane • Toronto

094827

Library of Congress Cataloging in Publication Data:

Brown, Anthony William Aldridge, 1911–
   Ecology of pesticides.

  "A Wiley-Interscience publication."
  Includes bibliographical references and index.
  1. Pesticides—Environmental aspects. I. Title.
QH545.P4B74   574.5′222   77-11730
ISBN 0-471-10790-5

Printed in the United States of America

10 9 8 7 6 5 4 3

PREFACE

The public anxiety about environmental pollution has made the side effects of pesticides one of the principal research activities of biologists since 1962. A compilation of the ecological effects of pesticides on nontarget organisms was issued by the Executive Office of the President (United States) in 1971, but it was out of print a mere 2 years later. The present volume attempts to make up for this lack of an information source available for students in the field of pesticide ecology. Initially designed to be a concise text, it has metamorphosed into a detailed review of the extensive research work performed on this controversial subject.

The organochlorine insecticides had given rise to the greatest amount of investigative work, and now several of them have been suspended from general application in the United States and some other countries. Nevertheless, this work is reviewed from a historical standpoint, as providing a framework for present studies on the current pesticides; it is also relevant for countries where organochlorines are still widely applied. It is hoped that the detailed subject-matter here collected on these and other pesticides, and the way in which it is organized, will be useful to those wishing to make their own syntheses for teaching or research purposes.

My gratitude is expressed to Michigan State University for providing the opportunity, facilities, and time to produce this book. Special thanks are due to Marian Mahler for preparing the illustrations, to my secretary Jane Fortman for coping so efficiently with a volume of work at once so extensive and so detailed, and to Dr. David Pimentel for his kindness in giving me one of the last available copies of his valuable compilation on the effects of pesticides on nontarget organisms.

A. W. A. Brown

East Lansing, Michigan
March 1977

CONTENTS

ECOLOGY OF PESTICIDES

INTRODUCTION TO INSECTICIDES, HERBICIDES, AND FUNGICIDES

A. DISCOVERY AND MODE OF ACTION

In order to obtain a full understanding of the material in this section, the reader is referred to Kenaga and End (1974) for the structural formulae of insecticides, and to the *Farm Chemicals Handbook* (1976) for the nature and uses of all pesticides. The very full chapter of Metcalf (1971) describes the chemistry and mode of action of insecticides, herbicides, and fungicides, while the pioneering book of Corbett (1974) provides a synthesis of the mode of action of pesticides as a whole. For insecticides, mode of action (and chemistry) is covered by the books by O'Brien (1967) and by Matsumura (1975). For herbicides, a general description with chemical formulae is given by Klingman and Ashton (1975), and the mode of action is described by Ashton and Crafts (1973) and by Audus (1976). A general description of fungicides is to be found in the book by Sharvelle (1969).

Insecticides

Arsenicals were the first insecticides to be employed to protect foliage against the attack of insects. Paris green was adapted in 1865 to combat the Colorado potato beetle, lead arsenate appeared in 1892 as an orchard spray, and calcium arsenate followed as a crop dust in 1907. These chemicals were stomach poisons, characteristically entering the insect's body through the gut wall, and causing diarrhoea and a flaccid paralysis. The pentavalent As in these arsenates uncouples the dephosphorylation of diphosphoglycerate in the glycolysis of glucose so that the phosphorus is not available for the production of ATP, and thus the muscle is paralyzed. The trivalent As eventually arising from these arsenicals also combines with SH groups, and prevents the further oxidation of glucose by inhibiting not only the pyruvic dehydrogenase that fuels the Krebs cycle with acetate, but also the keto-glutarate dehydrogenase in the tricarboxylic cycle itself (Fig. 1.1).

Hydrogen cyanide was first adapted for agricultural use in 1886, and subsequently citrus trees were fumigated under tarpaulin tents in California. The first synthetic insecticide to appear was dithiocyanodiethyl ether, when it was introduced in 1929 as a fly spray. HCN, and probably the thiocyanate, act by inhibiting the cytochrome oxidase necessary for the final transport of H to O_2 in the terminal oxidation process of the electron-transport system. In 1927 the fish-poison rotenone was introduced as an insecticide; it causes a flaccid paralysis by preventing NAD from passing on H to flavoprotein in the electron-transport system, thus causing the inhibition of ketoglutarate dehydrogenase which requires NAD as an H acceptor. DNOC (3,5-dinitro-*o*-cresol) was the second synthetic insecticide, introduced in 1936 but now discontinued; it uncoupled the dehydrogenation of flavoprotein from the ATP production process which normally occurs also at this point.

Thus the principal insecticides utilized before World War II were inorganic compounds and/or inhibitors of the carbohydrate oxidation that produces the ATP necessary for muscular and other activity. The exceptions were two neurotoxic insecticides, one being pyrethrum introduced from Persia to Europe in 1818, and thence into the United States in 1858, and the other being nicotine, from the tobacco juice recommended by Erasmus Darwin in 1763 and marketed as nicotine sulfate (Blackleaf 40) in 1909.

The synthetic organochlorine insecticide DDT (dichlorodiphenyltri-chloroethane) was discovered to be a remarkable residual insecticide at

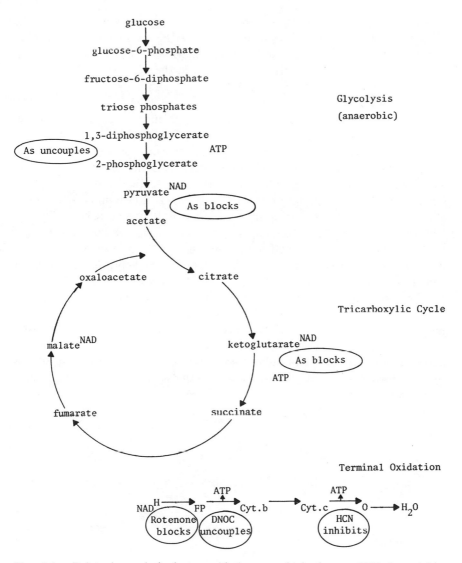

Fig. 1.1. Points in carbohydrate oxidation at which the pre-1939 insecticides inhibit ATP synthesis.

Basle, Switzerland in 1939. Houseflies had only to alight or walk on a deposit of DDT for a few seconds to be destined for death, and these deposits remained insecticidal for months. The compound has an affinity for the lipoidal membrane-sheaths of the nerve axons, causing repetitive discharges in the nerves, which throw the insect into tremors and eventually prostrate it (Fig. 1.2). The instability of the nerve axon is due to the presence of DDT in the axon membrane altering its permeability to Na^+ and K^+ ions, possibly because of the formation of charge-transfer complexes between DDT and certain molecules in the membrane. Relatives of DDT, such as methoxychlor and DDD, act in the same way on nerve axons. The pyrethrins, and synthetic pyrethroids such as allethrin, also act in this manner, but their action on the insect is much more rapid and often not irreversible. Some more remote relatives of DDT, which are more polar (e.g., dicofol, ovex) have been developed as acaricides for orchard mites.

BHC (benzene hexachloride or more correctly hexachlorocyclohexane) was discovered to be a simple and effective insecticide in 1942 in both French and English laboratories. Of its isomers, the gamma is the most effective, and an almost pure grade of gamma-BHC is marketed as lindane. Chlordane, a mixture of terpenoid compounds, was discovered to be a highly effective residual insecticide in the United States in 1945, and 3 years later the most active principle of chlordane, called heptachlor, was made available, along with two other cyclodiene derivatives, aldrin and dieldrin. At the same time a product obtained by the chlorination of turpentine, containing a large number of chlorinated camphenes, was marketed under the name of toxaphene. All of the preceding group of cyclodiene derivatives, including the comparative newcomer endosulfan (Fig. 1.3), are neurotoxic, acting at the nerve ganglion rather than along the nerves of insects, probably by the formation of charge-transfer complexes within the presynaptic membranes.

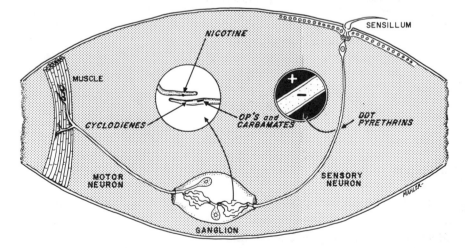

Fig. 1.2. Points in the insect nervous system at which the post-1945 insecticides derange or inhibit action-potential transmission.

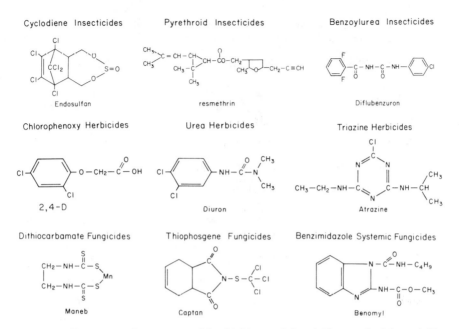

Cyclodiene Insecticides

Endosulfan

Pyrethroid Insecticides

resmethrin

Benzoylurea Insecticides

Diflubenzuron

Chlorophenoxy Herbicides

2,4-D

Urea Herbicides

Diuron

Triazine Herbicides

Atrazine

Dithiocarbamate Fungicides

Maneb

Thiophosgene Fungicides

Captan

Benzimidazole Systemic Fungicides

Benomyl

Fig. 1.3. Representative groups of herbicides and fungicides, and of insecticides not represented in subsequent text-figures.

The organophosphorus (OP) compounds first appeared in 1945 as the result of the success of German industry in finding modifications of chemical warfare agents useful for insect control; the first to appear were tepp and parathion, followed by malathion and many others some 4 yr later; the Swiss product diazinon appeared soon afterwards. Now there are scores of OP compounds, some liquids and some solids, mainly in the molecular forms known as phosphorothioates, phosphorodithioates, and phosphates (Table 1.1). Some of them (e.g., demeton, schradan, phorate) are systemic insecticides, and spread through the plant after application to the foliage or roots. In insects as in mammals, they act by inhibiting the enzyme cholinesterase (ChE) that normally breaks down the neurotransmitter acetylcholine (ACh) at the synapse immediately its work is done. Thus they are ganglionic rather than axonic poisons, causing first the facilitation and then the blocking of the reflex arc, so that the initial hyperactivity and the subsequent convulsions are followed by tetanic paralysis.

Organophosphorus compounds inhibit ChE by combining with (phosphorylating) the active site of the enzyme. The inhibiting potency (measured as the I_{50}) of the OP compound is decided not only by its two alkyl substituents (usually methyl or ethyl), but also by the much larger third sub-

Table 1.1. **The principal insecticides in the various groups of organophosphorus compounds.**

Phosphorothioates	Phosphorodithioates	Phosphates

$$\begin{array}{c} ^-O \\ {}_-O \end{array} P{\stackrel{S}{-}}O{-}$$ $$\begin{array}{c} ^-O \\ {}_-O \end{array} P{\stackrel{S}{=}}S{-}$$ $$\begin{array}{c} ^-O \\ {}_-O \end{array} P{\stackrel{O}{=}}O{-}$$

Phosphorothioates	Phosphorodithioates	Phosphates
Parathion	Malathion	Dichlorvos
Methyl parathion	Azinphosmethyl	Naled
Fenitrothion	Dimethoate	Phosphamidon
Ronnel	Phorate	Monocrotophos
Temephos	Disulfoton	Dicrotophos
Diazinon	Ethion	Stirofos
Chlorpyrifos	Carbophenothion	Schradan
Demeton	Phosmet	Mevinphos

Phosphonothioates	Phosphonodithioates	Phosphonates

Phosphonothioates	Phosphonodithioates	Phosphonates
Trichloronat	Fonofos	Trichlorfon
EPN		
Leptophos		

stituent (phenyl, heterocyclic, or aliphatic); this constitutes the "leaving group" which in due course is hydrolyzed away after the phosphorylation.

Carbamate insecticides, developed as relatives of the ChE inhibitor eserine (physostigmine), first appeared in 1953, when United States scientists introduced carbaryl (Sevin). They have an analogous action, carbamylating rather than phosphorylating the enzyme, and the ChE recovers more readily from carbamates than from OP compounds. All of them are N-methyl carbamates, except for the N-dimethyl carbamates isolan and dimetilan. Carbamates may also block the receptors through which ACh performs its normal function of transmitting the nerve impulses across the synapse. This is a nicotinic effect, and indeed nicotine acts as an insecticide by blocking the ACh receptors in the insect ganglion.

Since 1969 three other groups of insect control chemicals have appeared. The first were the formamidines, of which the Swiss product Galecron (chlordimeform) is an example; they are sympathomimetic agents, inhibiting the monoamine oxidase that normally removes the neurotoxic amines such as serotonin. The second were the juvenile hormone mimics, of which the American product Altosid (methoprene) is an example; they act either by mimicking the natural juvenile hormone or by inhibiting its disappearance, and thus delay and disrupt the metamorphosis process. The third were the chitin synthesis inhibitors, of which the Dutch urea-derivative Dimilin (diflubenzuron, Fig. 1.3) is an example, showing great promise in the control of forest defoliators, mosquito larvae, and other insect pests.

A separate range of compounds was developed as acaricides for the control of Tetranychid mites on orchard trees and field crops. The first group (dicofol, chlorobenzilate, DMC) has —C— between the two benzene rings, as in DDT. The second group (tetradifon, sulphenone, tetrasul) had —S— between the rings, and the third (ovex, Genite, fenson) were benzenesulfonate derivatives with —S—O— between the benzene rings. The mites developed resistance to them all including propargite (Omite), which had —O— between the rings. Entirely new molecules now being applied are oxythioquinox (Morestan), which is a quinoxaline derivative, and cyhexatin (Plictran), the hydroxide of tricyclohexyl-tin.

Herbicides

The first herbicide was sodium arsenite, introduced as a soil sterilant in 1900. It was followed by sodium chlorate, diesel oil, and the volatile but highly phytotoxic Stoddard's solvent.[44] Still later MSMA (monosodium methanearsonate), as well as DSMA and cacodylic acid, appeared for control of undesirable grasses. The arsenical herbicides cause rapid contact injury due to membrane degradation, their mode of action being the inhibition of SH enzymes and uncoupling of oxidative phosphorylation as in insects. DNOC appeared in 1935 as the first synthetic organic herbicide, rapidly destroying the roots and conducting vessels, because of its uncoupling carbohydrate oxidation from the phosphorylation of ADP.

Work in Britain and the United States discovered the phenoxy herbicides in 1944, resulting in MCPA and 2,4-D for selective control of broad-leaved weeds, followed by 2,4,5-T and silvex for control of woody perennials.[6] They act in the same way as auxins, the natural growth hormones, except that they are translocated into all cells and cause general anarchy in the plant growth. Cell proliferation is uncontrolled, axis growth is abnormal, and apical growth is inhibited. Sometimes tumor tissue develops, connected with the stimulating effect of the phenoxy compounds on the synthesis of DNA and proteins. The consequence is a reduction in water uptake, leaf expansion and chloroplast production, followed by a softening of the root cortex and necrosis in all tissues.

The urea herbicides, whose activity was discovered by research during World War II, became commercial in 1951 with monuron, followed by diuron, fenuron, linuron, and so on. They are essentially nonselective soil sterilants, and cause chlorosis followed by the collapse of the parenchyma vessels. Their basic mode of action (Fig. 1.4) is the inhibition of Photosystem II in the chloroplast, in which electrons are removed from water to leave hydrogen ions and O_2; this process is known as the Hill reaction, since it can be demonstrated in cellfree extracts when ferricyanide is added as an electron acceptor. The result is that no ATP is synthesized, and no NADH ultimately appears. Probably these urea-based Hill-reaction inhibitors also release free radicals that would be toxic to the plant.

The triazine herbicides were a Swiss development; the first to appear was simazine in 1952, to be soon followed by atrazine and many others. They are general herbicides which are particularly adapted to remove weeds from corn (*Zea mays*), which is resistant to them. They cause chlorosis, growth inhibition and necrosis, and like the ureas they are Hill-reaction inhibitors and also release a secondary toxic agent.

The bipyridylium (quaternary ammonium) compounds such as diquat and paraquat appeared on the market in 1958. The rapidity with which they cause wilting and desiccation of the foliage to which they are applied make

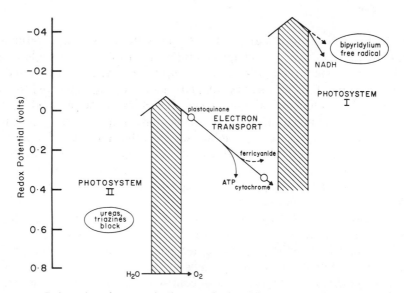

Fig. 1.4. Points in photosynthesis at which Hill-reaction-inhibiting herbicides inhibit ATP and NADH production (from Corbett, 1974).

them especially suitable as preharvest desiccants for root crops such as potatoes, besides being useful as herbicides. The molecule is converted into a free radical by Photosystem I in the photosynthetic process in the leaves, and this produces H_2O_2 as it is oxidized back to the ion, and the peroxide free radicals proceed to disrupt the plant membranes. An immense variety of herbicide groups have appeared in recent years, including amides and acylanilides, carbamates and thiocarbamates, phenols and dinitroanilines, aliphatics, and others, each with their characteristic uses and mode of action.[28]

Fungicides

Lime–sulfur was the first formulation to be applied to foliage; developed in 1851 to protect the ornamental shrubs and trees at Versailles, France against fungal disease, as well as aphids, scales and mites, it consisted of 2 parts of sulfur to 1 part of slaked lime ($Ca(OH)_2$). A second fungicide effective also against arthropod pests was Bordeaux mixture, consisting of copper sulfate with 1–2 parts slaked lime, developed in France in 1882 to combat the downy mildew of the grape. Organic mercurials were developed in Germany in 1913 under the trade name of Ceresan to protect seed against pathogenic soil fungi. By 1923 micronized sulfur, and by 1936 copper silicate, had been developed for foliage sprays.

The first organic fungicides appeared in 1940 with the quinones chloranil and dichlone, relatives of pentachlorophenol (PCP) sufficiently safe to be used for seed treatment.[46] Whereas PCP acts as an uncoupler of oxidative phosphorylation, these quinones inhibit a variety of enzymes by their keto groups combining with the SH groups and free amine groups of the enzyme molecules. The following year saw the introduction of the dithiocarbamate fungicides (e.g., ferbam, nabam, zineb) which could be applied to foliage, for example to control apple scab. Combined with kations such as Na, Mn, Fe, or Zn, they inhibit the glucose 6-phosphatase at the start of the gly-

colytic chain by their $NH-C\overset{\displaystyle C}{\underset{\displaystyle \parallel}{}}-$ group becoming free to combine with the SH group of this enzyme. They also chelate Cu^{2+} ions, thus inhibiting the pyruvic dehydrogenase complex.

Captan appeared in 1951, and soon became the most important foliar fungicide. It combines with SH enzymes, in the process producing and

liberating thiophosgene $Cl-C\overset{\displaystyle C}{\underset{\displaystyle \parallel}{}}-Cl$, which proceeds to react with the free amine and hydroxyl groups of enzymes. Dodine and dinocap also came on the scene in the 1950s, the first as a foliar spray, the second as a dormant spray. The first systemic fungicides that would spread from the sprayed leaves throughout the plant appeared in 1966 with the oxathiin compounds carboxin and oxycarboxin. They were soon superseded in 1968 with the benzimidazole systemic fungicides benomyl and thiophanate. These are converted to the methyl or ethyl esters (respectively) of benzimidazole carbamic acid (MBC or EBC). Their effect appears to be on the spindle fibers determining the movements of the chromosomes in mitosis, and so the DNA synthesis of the plant is inhibited. Subsequently, the synthesis of RNA and protein also ceases, and a second metabolite (butyl isocyanate) inhibits the respiration of subcellular particles. Benomyl, and to some extent thiophanate, was adopted for the control of a wide range of diseases on fruit, vegetables, field crops, and turf, but quite rapidly came to encounter the development of resistance in the target fungi.

B. THE USE OF PESTICIDES

Economic Necessity

Pesticides are needed in agriculture and forestry because, as judged by the United States figures over the decade 1951–1960, roughly one-third (34%) of the production of food and fibre is lost to pests. Of this total, the losses to insects amounted to an average of 13% for the United States in 1970,[41] to be compared to Fletcher's estimate of 10% for the United States in 1891[38] and the estimate of 10% by Neave (1930) for the British Commonwealth. The value of the loss in United States production in 1970 due to insects was $5.5 billion, while the loss due to nematodes was $0.4 billion, that due to diseases of plants and stock was $2.7 billion, and that due to weeds was estimated at $2.5 billion.[53] Thus the total loss in 1970 was $11.1 billion, as compared to the $10 billion average annual loss from 1951 to 1960 for United States crop production alone. Annual losses due to insects in the United States in 1957, which amounted to a total value of $3.5 billion at that time, were made up of the following (in $ billions):[34]

Staple crops	1.46	Fruits and vegetables	0.35
Stored products	0.65	Health and home	0.30
Livestock	0.47	Forests	0.25

In the world, the annual loss of human health due to insect-borne disease mainly due to malaria, which amounted to some 300 million cases and 3 million deaths, has been reduced (despite a doubling of the population since that time) to some 120 million cases and 1 million deaths.[10]

It has been estimated by Pimentel (1973) that a $10 billion average loss in the United States in 1960 would have been $12 billion if no pesticides had been applied. The price of such pesticides, in 1966 for example, was $0.56 billion, so that the total cost including that of their application would involve about $0.75 billion annually, thus representing nearly $3 saved for every $1 spent. A previous estimate[24] had concluded that the marginal value of a $1 expenditure for chemical pesticides in the United States was approximately $4, while an English estimate put the ratio higher at 1:6 for world production.[47] It is not really worth inaugurating chemical control operations on a crop in a developing country until this cost–benefit ratio reaches 1:3, whereas in highly developed agriculture a ratio of 1:1.15 could provide economically valid grounds for chemical control.[39] The annual net profit of using insecticides in Illinois over the decade 1963–1972 was about $25 million for the corn crop alone, while the use of herbicides on corn saved about $25 per acre.[13]

The need for chemical pesticides still persists, and indeed increases, as the human population pressure continues to rise. Extensive contiguous areas of monoculture demand the use of herbicides, insecticides, and fungicides. New species introduced into a country, thus escaping their autochthonous control agencies, demand chemical pesticides to cope with them. Crops liable to virus damage transmitted by aphids or leafhoppers demand the complete control achievable through insecticides. The canning industry calls for insectfree, and the export trade blemishfree, crop products for which only pesticides can at present give the answer. The methods of pest management by the integration of biological control with pest-resistant varieties and proper husbandry will reduce, it is hoped, the need for chemicals; since they will not eliminate that need, it is necessary to understand the side effects of chemicals and to be well informed on the ecology of the pesticides in the broader sense.

Volumes Employed and Areas Covered

In 1963, the annual sales of herbicides and fungicides in the United States were each of the order of 100 million pounds, as compared to 435 million pounds of insecticides, including fumigants (Fig. 1.5). The insecticides included 155 million pounds of DDT and 100 million pounds of the aldrin–toxaphene group (Fig. 1.6); of this annual consumption, 60 million pounds of DDT and 80 million pounds of the cyclodienes were used in the continental United States (President's Scientific Advisory Committee, *Use of Pesticides,* May 15, 1973), and more than 50 million pounds of DDT were employed for the world malaria eradication program. By 1968 the United States' sales of organophosphorus compounds were coming to

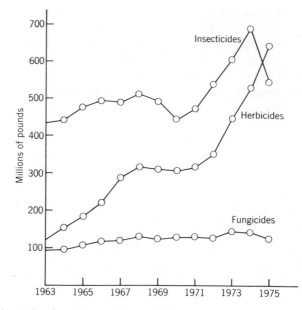

Fig. 1.5. Annual sales of organic pesticides, domestic and foreign, from United States producers (U.S. Tariff Commission and International Trade Commission).

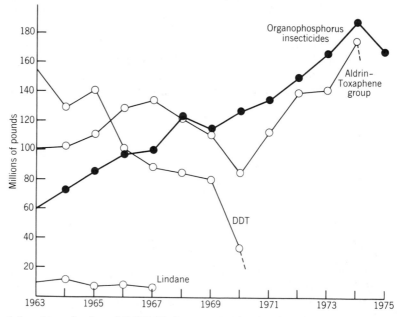

Fig. 1.6. Annual sales of DDT, lindane, cyclodiene, and OP insecticides, domestic and foreign, from United States producers (Tariff Commission).

exceed those of either category of the organochlorines; in the United Kingdom also, the annual use of the OPs was 240 tons as compared to 220 tons DDT and 80 tons of the cyclodiene group, besides 10 tons of carbamate insecticides.[7] This trend was partly due to the greater effectiveness of some OP compounds, and partly to the spread of insect resistance to the organochlorines; the change has been made mandatory by the undesirable side effects found to be caused by the organochlorines on birds and other life forms in the long term. It usually involved a greater cost for the insecticide in its application (Table 1.2), any reductions in the concentration of compound required for effectiveness being on the average offset by the necessity for more frequent applications.

In 1975, the total annual sales, domestic and export, of the herbicides produced in the United States had come to exceed those of insecticides, the totals being, respectively, 645 and 546 million pounds. These totals do not include the inorganic insecticides such as lead and calcium arsenate (6 million pounds in 1971) and the inorganic herbicide sodium chlorate (26 million pounds in 1971), but they do include 35 million pounds of fumigants such as methyl bromide. The total figure of 127 million pounds for fungicides does not include the sulfur and copper sulfate used for that purpose (amounting, respectively, to 100 million and 28 million pounds in 1971), but they do include 36 million pounds of pentachlorophenol, the use of which is becoming restricted to preserving wood. In considering the total load of 1.3 billion pounds of pesticides being applied annually (1975) to the United States outdoor environment, one must include the oil solvents used as carriers, which amounted to nearly 200 million pounds in 1971. Already by 1964, nearly twice as much of the crop area in the United States was treated with herbicides as with insecticides or fungicides, being 65 million acres (100 thousand square miles) as compared to 35 million acres. Between 1966 and 1971, according to the USDA Economic Research Service, the amount of herbicides annually applied by US farmers doubled from 51 to 102 million pounds, fungicides from 40 to 68 million, while insecticides increased from 66 up to 76 million pounds. In 1970, fully 57% of the corn crop with its 66 million acres was treated with herbicides, and 28% of the 54

Table 1.2.   Prices of insecticides in the United States in 1971: dollars per pound (Metcalf, 1975).

DDT	0.22	Methyl parathion	0.46
Toxaphene	0.24	Malathion	0.79
Chlordane	0.59	Dichlorvos	3.75
Aldrin	1.37	Carbaryl	0.80

million acres of wheat (Table 1.3). Well over 90% of the 1.3 million acres of
apple and citrus orchards were treated with insecticides, and over 70% with
fungicides. More than 50% of the 10 million acres of cotton received herbicides, and slightly more received insecticides. Thus, just as nearly half of
the herbicides consumed in the United States goes onto the corn crop,
nearly half of the insecticide tonnage is destined for cotton fields (Table
1.4).

For the same reason, toxaphene was the organochlorine insecticide
applied in greatest volume in the United States in 1971, being used primarily in the southeast against cotton insects, more than twice as much as
the chlordane and aldrin applied in the midwest against corn rootworms
(Table 1.5). By reason of its effectiveness against cotton insects, methyl
parathion was the OP insecticide applied in greatest volume. Malathion and
carbaryl, wide-spectrum insecticides suitable for a variety of crops, including fruit trees, take a secondary place. Among the herbicides, atrazine
exceeded the 2,4-D formulations in volume; and among the fungicides,
captan surpassed all the dithiocarbamate compounds combined.

The amount of insecticide received by an acre of agricultural land
depends on the potency of the insecticide and the stubbornness of the
infestation complex. By and large, insecticides are applied at an average
dosage of 1 lb per acre (1 lb/a) at each application, equivalent to 1.12 kg per

Table 1.3.   **Percent of the acreage of each United States crop that is
treated with insecticides, herbicides, or fungicides: 1968 and
1970 (Pimentel, 1973).**

Crop	Acreage	Insecticides	Herbicides	Fungicides
	(thousands)			
Citrus	713	97	29	73
Apples	623	92	16	72
Potatoes	1,425	89	59	24
Tobacco	980	81	2	7
Peanuts	1,425	70	63	35
Cotton	10,244	54	52	2
Corn	66,186	33	57	2
Rice	1,959	10	52	0
Soybeans	37,324	4	37	1
Wheat	54,428	2	28	1

Table 1.4. **Crops that received the bulk of the insecticides, herbicides, and fungicides applied in the United States in 1968: percent of the total of each type (Pimentel, 1973).**

Insecticides		Herbicides		Fungicides	
Cotton	47	Corn	41	Apples	28
Corn	17	Soybeans	9	Vegetables	25
Vegetables	8	Hay-crop	9	Field Crops	19
Apple	6	Wheat	7	Citrus	13
Tobacco	3	Cotton	6	Peanuts	4
Soybeans	2	Vegetables	5	Tobacco	1
Citrus	2	Peanuts	3	Others	10
Others	15	Others	20		
	100		100		100

hectare (1.12 kg/ha). On many crops, a single annual application suffices, but some insect problems on cotton have required as many as 25 applications in the growing season. Orchards usually receive a total of 10 lb/a during the summer, accumulated in the five or more applications required. Forest spraying, generally conducted with DDT at 0.5 lb/a, requires 1 lb/a with the less residual substitute insectides. Dosages of 0.1 lb or less per acre are sufficient for mosquito control, while at the other extreme the control of beetle vectors of Dutch elm disease require application of 1–5 lb per tree.

Dosages of herbicides applied to cropland are of the order of 1 lb/a, while weed-sterilant applications on nonagricultural industrial land and rights-of-way are of the order of 10 lb/a. Aquatic herbicides, principally phenoxy compounds, employed to treat about 1 million acres of ponds and estuaries each year in the U.S.A. to eliminate water weeds, are usually applied at approximately 4 lb/a (Table 1.6).

C. CONSTRAINTS ON THE CHOICE AND USE OF PESTICIDES

Toxicity and Hazard to Man

The toxicity of pesticides to mammals is primarily assessed as the acute oral LD_{50} on laboratory rats (Table 1.7). Lists have been published of the oral and dermal LD_{50} figures for rats[5,19] and for rabbits, guinea pigs, and

Table 1.5. United States consumption for use of certain important pesticides in 1971: millions of pounds (v. Rumker, Lawless, and Meiners, 1975).

Insecticides				Herbicides		Fungicides	
Organophosphorus		Other					
Methyl parathion	39.7	Carbaryl	25.0	Atrazine	75.0	Captan	16.0
Malathion	16.2	Carbofuran	5.0	2,4-D	48.0	Maneb	7.6
Parathion	10.0	Toxaphene	58.0	Alachlor	21.0	Zineb	2.5*
Diazinon	7.0	Chlordane	15.0	Trifluralin	17.0	Nabam	2.0*
Disulfoton	5.0	Aldrin	12.7	Diuron	6.7	Ferbam	1.5*
				Bromacil	3.0	Hg C'p'ds	0.9

*1969 figures

Approximate annual use of aquatic herbicides in the United States during the early 1970s (v. Rumker, Lawless, and Meiners, 1975).

	Acres treated (thousands)	Pounds expended (millions)		Acres treated (thousands)	Pounds expended (millions)
4-D	400	1.20	2,4,5-T*	20	0.08
MA	80	0.32	Diuron	9	0.09
lvex	60	0.24	Bromacil	8	0.04
raquat	40	0.02	Atrazine	5	0.05
lapon	30	0.60	Ammate	5	0.50

dmixed with 2,4-D

Table 1.7. **Acute oral toxicity of insecticides and other pesticides to the laboratory rat, *R. norvegicus:* LD_{50} in mg/kg (Pimentel, 1971).**

Endrin	20	Parathion	13	Mexacarbate	20
Aldrin	50	Azinphosmethyl	16	Carbaryl	560
Lindane	90	Monocrotophos	21	DNOC	45
Dieldrin	100	Phosphamidon	28	Pb arsenate	800
Endosulfan	110	Methyl parathion	42	Pyrethrins	1300
Heptachlor	115	Chlorpyrifos	150	Na arsenite	30
Toxaphene	160	Dimethoate	215	2,4-D	370
DDT	250	Fenthion	310	2,4,5-T	500
Dicofol	850	Naled	430	2,4-D ester	750
Chlorobenzilate	1950	Trichlorfon	500	Dalapon	970
Ovex	2025	Fenitrothion	680	Atrazine	3080
TDE	3400	Malathion	1650	Diuron	3400
Methoxychlor	6000	Temephos	5000	Captan	9000

mice.[26] The World Health Organization has issued a classification of pesticides categorized according to their general hazard to man.[55] While organic herbicides and fungicides usually have a very low toxicity, the insecticides range from highly toxic to virtually nontoxic. There is a premium on the discovery and use of compounds that are highly insecticidal but of low mammalian toxicity (e.g., malathion, temephos), while instructions for safe use of pesticides aim at eliminating the hazard of the toxic insecticides to applicators and harvesters, and to the general public. At present, however, pesticide toxicity is a serious matter in the world at large, the cases of accidental poisoning amounting to some 500,000 men, women and children each year, with a mortality rate exceeding 1%.[54]

Governmental regulation goes to great pains to prevent secondary poisoning among the public who purchase food stuffs that might have been treated. The mechanism is a set tolerance level for residue content above which the food product is liable to seizure and legal action, and no pesticide may be applied until tolerances have been set for it. Feeding experiments on the laboratory rat (among others) maintained for its entire lifetime establish the maximum level at which no effects can be observed; by applying a 100-fold safety factor to these chronic toxicity figures, which are available in monographs,[18] a figure is decided for the acceptable daily intake (ADI) in mg/kg body weight.[30,35] By applying a food factor indicating the proportion of the food product in the diet of that country, a tolerance figure is set for the pesticide in that food product or crop. In agricultural practice the last application of the pesticide is put on a number of days before harvest sufficient to allow the residues to decrease below the tolerance levels. For example, for a 2 lb/a application:

Insecticide	ADI, mg/kg	Crop	Tolerance, ppm	Preharvest Period, days
Disulfoton	0.001	Potatoes	0.5	75
Carbaryl	0.01	Rice	5	14

There have been no cases in the United States of the public having been poisoned from crop foods in international commerce. The only incidents reported were nonfatal and involved the consumption of home produce, one among 11 persons who consumed mustard greens sprayed the day before with nicotine at twice the recommended dosage, and the other being two families who ate collard greens and chard sprayed with toxaphene contrary to the directions.[22]

Toxicity is one thing, but carcinogenicity is another. It is not the place of this book to discuss such matters, which are medical. Governmental bills (e.g., the Delaney amendment in the United States) have ensured that no crops contaminated with a carcinogen are sold; for example, the entire cranberry crop of 1959, originating from five states, was seized and destroyed because it contained residues (maximum, 0.05 ppm) of the herbicide amitrole which when fed to rats at 200 ppm for 4 months had caused a thyroid enlargement, which was reversible.[3] The acaricide Aramite was found to be carcinogenic to dogs, rats, and mice,[14] and the product was discontinued.

Lifetime feeding experiments on an especially tumor-prone strain of laboratory mice showed that DDT increased the incidence of liver-cell tumors, that some of these tumors metastasized to the lung,[50] and that DDE was no less active in this regard.[49] In another strain of mice, tumors were found only at 250 ppm DDT and there was no metastasis.[48] No increase in tumors was induced by feeding DDT to golden hamsters,[1] and no tumors were induced in a small number of dogs and monkeys.[16] Feeding DDT to men for nearly 2 yr did not result in tumors,[23] and no tumors were found in men whose occupation was the manufacture, formulation, or application of DDT.[16] This insecticide was suspended in the United States in 1972, but meanwhile 40,000 tons are being applied annually to the interiors of houses to combat malaria throughout the world, DDT having failed to have caused, of itself, a single case of fatal poisoning in 20 yr of the closest contact with the human race.[2] This use is within the walls of houses, and involves little contamination of the outdoor environment. The Public Enquiry of 1972 came to the conclusion, after hearing the environmental evidence, that "there is a present need for the essential uses of DDT,"[15] and the present United States uses of DDT to control the flea vectors of plague and bats which carry rabies virus fall into that category. On the evidence of tumors induced in mice only, the cyclodiene insecticides aldrin and dieldrin were suspended in 1974, while chlordane and heptachlor came under notice of suspension in 1975. The evidence obtained from laboratory animals up to 1972 had been that DDT could be tumorigenic rather than carcinogenic, and that aldrin and dieldrin were neither carcinogenic nor tumorigenic.[27]

Resistance to Insecticides

An important factor in the ecology of pesticides is the susceptibility of the target pest species as compared to that of the nontarget species. During the present century, it has frequently been the experience that successive applications of an insecticide over the years have achieved less and less complete control of the insect or acarine pest, due to the target populations having achieved first a greater tolerance and eventually a decisive resistance to it. Resistance of fungi to fungicides has also recently become important, while resistance of weeds to herbicides has appeared in a few instances; these are discussed in Chapters 11 and 15. Resistance to rodenticides has also developed, to warfarin, in the populations of the Norway rat (*Rattus norvegicus*) and more recently the black rat (*R. rattus*), and resistance to sodium pentachlorophenate as a molluscicide has been developed in *Australorbis* and *Oncomelania* snails.[56]

Characteristically, the resistance first appears in a certain population within the distribution of a pest species, usually where the insecticide has received its most intensive use. The first case, the resistance of the San Jose scale insect to lime–sulfur, appeared in 1908 in the Clarkston Valley of Washington, contrasting with the good control still being obtained in apple orchards elsewhere in that state. Resistance of other scale insects to HCN used in citrus-tree fumigation later appeared in certain California orchards and often remained restricted to them. Resistance of the codling moth to lead arsenate developed during the 1920s in the Grand Valley of Colorado. The resistance to aldrin and other cyclodiene insecticides in the western corn rootworm first appeared in southeastern Nebraska and later spread eastward in the species' distribution in the American midwest, while the

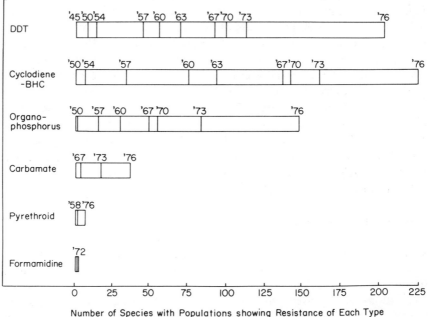

Fig. 1.7. Numbers of arthropod species in which populations have developed resistance to insecticides of the six principal types available.

Table 1.8. Number of species with various types of resistance (Georghiou and Taylor, 1976).

	DDT	Cyclod.	OP	Carb.	Other	Total
Diptera	91	100	40	6	4	133
Coleoptera	26	48	18	7	8	56
Hemiptera	14	23	31	4	4	55
Lepidoptera	31	32	22	12	4	52
Acarina	21	10	32	6	13	43
Other Orders	20	12	4	1	2	25
Total	203	225	147	36	35	364

resistance to organophosphorus compounds in the irrigation-water mosquito *Aedes nigromaculis* developed particularly in the area around Tulare, California where these compounds had been most intensively used as larvicides. On the other hand, the housefly and Tetranychid mites have developed resistance to virtually any insecticide in any part of the world where they have been employed for some time.[8]

Before the advent of DDT in 1945, it was realized that in the 12 cases already encountered the resistance was inherited, and that it was a matter of resistant races, or more correctly resistant populations within a species normally susceptible to that insecticide. By 1955 it had become evident from experience with the housefly and two-spotted mite that resistance was specific to the selecting insecticide and its chemical relatives, so that for example among the organochlorines DDT resistance was separate from cyclodiene and BHC resistance, and OP resistance was something else again.

By 1976, resistance to DDT and its relatives had developed in populations of 203 species of insects and mites, cyclodiene resistance in 225 species, OP resistance in 147 species, and carbamate resistance in 36 species (Fig. 1.7); of the other insecticide groups, pyrethrin resistance had developed in 6 species and formamidine resistance in 2 species, while it has proved possible to develop experimentally resistance to the juvenile-hormone mimics and the insect growth inhibitor diflubenzuron. Among the 364 species then known to have developed resistance,[21] most of them were in the Diptera, mainly of medical or veterinary importance, followed by the Coleoptera, Hemiptera, Lepidoptera, and Acarina, mainly pests of foliage or roots of crop plants (Table 1.8).

A large number of species, such as the housefly, *Heliothis virescens* caterpillars, *Tetranychus urticae* mites, and *Boophilus microplus* ticks, have

developed these different resistance types in succession. As each new insecticide type was introduced to combat the preceding resistance (Table 1.9), it
went down to failure in its turn. Thus insecticide resistance has been in the
driver's seat in the succession of pesticide management from one decade to
the next. It will be noted that its major effect has been to replace the
persistent organochlorines with the less persistent OPs and carbamates,
which on environmental grounds is a movement in the right direction.
However, where the original insecticide was not replaced, but simply
applied in higher concentration and more frequently in the hope of
somehow obtaining control, the effect was simply to intensify the resistance
by eliminating all the natural enemies of the pest;[43] this is what happened on
the cotton crop of Nicaragua, where attempts to combat resistant *Heliothis*
and *Spodoptera* caterpillars with methyl parathion and endrin (*inter alia*)
went as far as 50 applications per crop season.[52]

Insecticide resistance, being inherited, is not postadaptive and cannot be
induced by lifetime exposure to nonkilling doses; its development depends
on the presence in the arthropod population of the preadaptations, that is,
the genes, for the resistance to the insecticides applied.[9] Thus the development of resistance is due to Darwinian selection causing in each successive

**Table 1.9. The succession of insecticides employed to control the
irrigation-water mosquito, the tobacco budworm on cotton, and
the two-spotted spider mite, 1946–1976.**

Aedes nigromaculis	Heliothis virescens	Tetranychus urticae
DDT	DDT	parathion
HCH, aldrin	toxaphene, endrin	azinphosmethyl
parathion, malathion	malathion	carbophenothion
methyl parathion	methyl parathion	ovex, fenson
fenthion	monocrotophos	chlorobenzilate
chlorpyrifos	carbaryl	dicofol
diflubenzuron	chlordimeform	tetradifon
	synthetic pyrethroids	propargite
	Bacillus thuringiensis	chlordimeform
	Heliothis NP virus	Pentac
		oxythioquinox
		cyhexatin

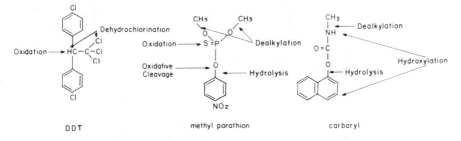

Fig. 1.8. Detoxicative resistance mechanisms developed in arthropods to the DDT group, OP compounds, and carbamate insecticides.

generation an increase in the proportion of resistant genotypes, first mainly heterozygotes and later mainly homozygotes, in the population under selection pressure.[20] The speed of its development depends on the intensity of selection, the number of generations per year, and the degree of isolation of the population from dilution with surrounding untreated populations.

The mechanism of resistance, what it is about one genotype that makes it resistant to the insecticide in contrast to another, the normal type, has proved in most cases to be an ability to detoxify that insecticide. DDT resistance has usually proved to be due to enzymic dehydrochlorination (Fig. 1.8) of DDT to DDE, although oxidative degradation, nerve insensitivity, and decreased uptake through the cuticle occasionally play a part. Cyclodiene resistance is apparently due to the cyclodiene insecticide being sequestered by protective proteins forming charge-transfer complexes with it. Organophosphorus resistance is principally due to detoxication of the insecticide itself of its initial bio-oxidation product, by esterases hydrolyzing it, by microsomal oxidases cleaving it, or by transferases desalkylating it; in some cases the OP resistance has been due to a less-sensitive cholinesterase, the enzyme that is the target site. Resistance to carbamates and pyrethroids is due almost entirely to oxidative degradation.

Crossing and backcrossing experiments between resistant and susceptible strains proved time and time again that the resistance was principally due to a single gene allele; this was always dominant in the case of OP resistance, intermediate in cyclodiene resistance, and often recessive in DDT resistance. Detailed work on DDT resistance in the housefly (Fig. 1.9) has shown that the gene *Deh* determines the production of the dehydrochlorinase enzyme that detoxifies DDT to DDE; this was on chromosome 2 of the housefly's karyotype, while two other genes on chromosome 3 determine nerve insensitivity (*kdr*), and reduced cuticular penetration (*tin*). The genes for OP resistance consist of *ox* for oxidative cleavage, *a* for hydrolysis, and *g* for desalkylation, all on chromosome 2.

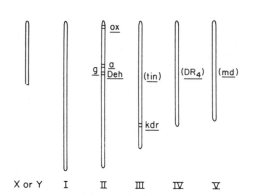

X or Y   I   II   III   IV   V

Fig. 1.9. Approximate (brackets) or proximate locations of genes for insecticide resistance on the chromosomes of the housefly.

The gene *md* on chromosome 5 is responsible for oxidation of DDT and carbamates, and of OP compounds and pyrethrins in some strains.[42]

Species differ with respect to their developing resistance; for example the European corn borer did not develop resistance to DDT, nor the boll weevil to malathion. Such differences can result in a change of species in the ecosystem; for example in tobacco fields in Ontario treated with aldrin the root maggot *Hylemya platura* became replaced by *H. florilega,* which had developed cyclodiene resistance faster; likewise the cocoa capsid *Sahlbergiella* in Ghana became replaced by *Distantiella* where BHC was being employed as the control agent. In apple orchards in the United States treated with azinphosmethyl for codling moth control the tetranychid spider mites infesting the foliage became OP resistant and multiplied excessively until their principal predators, *Amblyseius fallacis* in the northeast and *Typhlodromus occidentalis* in the northwest, in their turn developed enough OP resistance to survive the azinphosmethyl sprays.[12]

REFERENCES CITED

1. C. Agthe, H. Garcia, P. Shubik, L. Tomatis, and E. Wenyon. 1970. Study of the potential carcinogenicity of DDT in Syrian golden hamsters. *Proc. Soc. Exp. Biol. Med.* **134:**113–116.

2. Anonymous. 1963. Ingestion of DDT (an editorial). *Br. Med. J.* 1963. Vol. 1, pp. 205–206.

3. Anonymous. 1965. Technology notes: The cranberry incident. *World Rev. Pest Control* **4:**101.

4. F. M. Ashton and A. S. Crafts. 1973. *Mode of Action of Herbicides.* John Wiley. 504 pp.

4a. L. J. Audus (Ed.). 1976. Herbicides: *Physiology, Biochemistry, Ecology* 2nd Ed'n. Academic. Vol. 1, 608 pp. Vol. 2, 564 pp.

5. R. Bey-Dyke, D. M. Sanderson, and D. N. Noakes. 1970. Acute toxicity data for pesticides. *World Rev. Pest Control* **9:**119–127.

6. R. C. Brian. 1964. The classification of herbicides and types of toxicity. In *The Physiology and Biochemistry of Herbicides*. Ed. L. J. Audus. Academic. pp. 1-37.

7. G. T. Brooks, 1972. Pesticides in Britian. In *Environmental Toxicology of Pesticides*. Ed. F. Matsumura, G. M. Boush, and T. Misato. Academic. pp. 66-114.

8. A. W. A. Brown. 1971. Pest resistance to pesticides. In *Pesticides in the Environment*. Ed. R. White-Stevens. Marcel Dekker. Vol. 1, Part II, pp. 457-552.

9. A. W. A. Brown and R. Pal. 1971. *Insecticide Resistance in Arthropods*. World Health Org. Monogr. Ser. No. 38. 491 pp.

10. A. W. A. Brown, J. Haworth, and A. R. Zahar. 1976. Malaria eradication and control from a global standpoint. *J. Med. Entomol.* **13**:1-25.

11. J. R. Corbett. 1974. *The Biochemical Mode of Action of Pesticides*. Academic. 330 pp.

12. B. A. Croft and A. W. A. Brown. 1975. Responses of arthropod natural enemies to insecticides. *Annu. Rev. Entomol.* **19**:285-335.

13. G. C. Decker. 1974. Costs and benefits of pesticides. In *Survival in Toxic Environments*. Ed. M. A. Q. Khan and J. P. Bederka. Academic. pp. 53-81.

14. W. F. Durham and C. H. Williams. 1972. Mutagenic, teratogenic, and carcinogenic properties of pesticides. *Annu. Rev. Entomol.* **17**:123-148.

15. Environmental Protection Agency. 1972. *Consolidated DDT Hearings: Recommended Findings, Conclusions, and Orders*. Document 40 CFR 164.32. Washington, D.C., 25 April. Edmund M. Sweeney, Hearing Examiner.

16. Environmental Protection Agency. 1975. *DDT: A Review of Scientific and Economic Aspects of the Decision to Ban Its Use as a Pesticide*. Document EPA-540/1-75-022. 300 pp.

17. *Farm Chemicals Handbook*. 1976. Meister Publishing Co., 37841 Euclid Ave., Willoughby, Ohio 44094. 466 pp.

18. Food and Agriculture Organization. 1975. *1973 Evaluations of Some Pesticide Residues in Foods: The Monographs*. FAO, Rome. Processed Publ. AGP/1973/M/9/1. 491 pp.

19. T. B. Gaines. 1960. The acute toxicities of pesticides to rats. *Toxicol. Appl. Pharmacol.* **2**:88-99.

20. G. P. Georghiou. 1972. The evolution of resistance to pesticides. *Annu. Rev. Ecol. Syst.* **3**:133-168.

21. G. P. Georghiou and C. E. Taylor. 1976. Pesticide resistance as an evolutionary phenomenon. *Trans. XVth Int. Congr. Entomol.* pp. 759-785.

22. W. J. Hayes. 1960. Pesticides in relation to public health. *Annu. Rev. Entomol.* **5**:379-404.

23. W. J. Hayes, E. D. Dale, and C. I. Pirkle. 1971. Evidence of safety of long-term high oral doses of DDT for man. *Arch. Environ. Health* **22**:119-135.

24. J. C. Headley. 1968. Estimating the productivity of agricultural pesticides. *Am. J. Agric. Econ.* **50**:13-23.

25. E. Kenaga and C. S. End. 1974. *Commercial and Experimental Organic Insecticides*. Special Publ. 74-1. Entomol. Soc. Am., 4603 Calvert Road, College Park, MD 20740. 77 pp. (A previous edition appeared in *Bull. Entomol. Soc. Am.* **12**:151-217. 1966).

26. S. H. Kerr and J. E. Brogdon. 1959. Relative toxicity to mammals of 40 pesticides. *Agric. Chem.* **14**(9):44-45, 135.

27. W. W. Kilgore and M. Y. Li. 1973. The carcinogenicity of pesticides. *Residue Rev.* **48**:141-161.

28. G. C. Klingman. 1961. *Weed Control; As a Science*. Wiley. 421 pp.

29. G. C. Klingman and F. M. Ashton. 1975. *Weed Science: Principles and Practice*. Wiley. 431 pp.

30. C. F. Lu. 1973. Toxicological evaluation of food additives and pesticide residues and their acceptable daily intake for man. *Residue Rev.* **45**:81–93.

31. F. Matsumura. 1975. *Toxicology of Insecticides*. Plenum. 503 pp.

32. R. L. Metcalf. 1971. The chemistry and biology of pesticides. In *Pesticides in the Environment*. Ed. R. White-Stevens. Dekker. Vol. I, Part I, pp. 1–144.

33. R. L. Metcalf. 1975. Insecticides in pest management. In *Introduction to Pest Management*. Ed. R. L. Metcalf and W. H. Luckmann. Wiley. pp. 235–273.

34. C. L. Metcalf, W. P. Flint, and R. L. Metcalf. 1962. *Destructive and Useful Insects*. McGraw-Hill. 4th edition. pp. 41–43.

35. P. H. Mollenhauer. 1967. The acceptable daily intake value as a base for legislative measures regarding food additives. *Residue Rev.* **19**:1–10.

36. S. A. Neave. 1930. *A Summary of Data Relative to Economic Entomology in the British Empire*. Imperial Bureau of Entomology, London.

37. R. D. O'Brien. 1967. *Insecticides: Action and Metabolism*. Academic. 332 pp.

38. G. Ordish. 1952. *Untaken Harvest*. Constable and Co., London. 171 pp.

39. G. Ordish. 1962. Economics and pest control. *World Rev. Pest Control* **1**(4):31–38.

40. D. Pimentel. 1971. *Ecological Effects of Pesticides on Non-Target Species*. Exec. Off. President, Off. Sci. Technol. Sup't. Doc. Washington. 220 pp. (Sup't. Doc'ts Stock No 4106-0029).

41. D. Pimentel. 1973. Extent of pesticide use, food supply, and pollution. *J. N.Y. Entomol. Soc.* **81**:13–33.

42. F. W. Plapp. 1976. Biochemical genetics of insecticide resistance. *Annu. Rev. Entomol.* **21**:179–197.

43. H. T. Reynolds, P. L. Adkisson, and R. F. Smith. 1975. Cotton insect pest management. In *Introduction to Insect Pest Management*. Ed. R. L. Metcalf and W. H. Luckmann. Wiley. pp. 379–443.

44. W. W. Robbins, A. S. Crafts, and R. N. Raynor. 1942. *Weed Control*. McGraw-Hill. 543 pp.

45. R. von Rumker, E. W. Lawless, and A. F. Meiners. 1975. Production, Distribution, Use, and Environmental Impact Potential of Selected Pesticides. U.S. Environmental Protection Agency, Washington, D. C. Mim. Publ. EPA 540/1-74-001. 439 pp.

46. E. G. Sharvelle. 1969. *Chemical Control of Plant Diseases*. University Publishing, P. O. Box 856, College Station, TX 77840. Processed publ. 340 pp.

47. A. H. Strickland. 1970. Some economic principles of pest management. In *Concepts of Pest Management*. North Carolina State Univ. pp. 30–44.

48. B. Terracini, M. C. Testa, J. R. Cabral, and N. Day. 1973. The effects of long-term feeding of DDT to BALB/c mice. *Int. J. Cancer* **11**:747–764.

49. L. Tomatis, V. Turusov, R. T. Charles, and M. Boicchi. 1974. Effect of long-term exposure to DDE and DDD on CF-1 mice. *J. Natl. Cancer Inst.* **52**:883–891.

50. V. S. Turusov, N. E. Day, L. Tomatis, E. Gati, and R. T. Charles. 1973. Tumors in CF-1 mice exposed for 6 consecutive generations to DDT. *J. Natl. Cancer Inst.* **51**:983–997.

51. U.S. Government. 1963–1975. *Synthetic Organic Chemicals: United States Production and Sales*. Tariff Commission until 1971, then International Trade Commission.

52. M. A. Vaughan and G. Leon. 1976. Pesticide management on a crop with severe resistance problems. *Trans. XVth Int. Congr. Entomol.* pp. 812–815.

53. K. Walker. 1970. Benefits of pesticides in food production. In *The Biological Impact of Pesticides in the Environment.* Oregon State University, Environmental Health Science Series No. 1, pp. 149–152.

54. World Health Organization. 1973. *Safe Use of Pesticides.* World Health Org. Tech. Rep. Ser. No. 513. p. 42.

55. World Health Organization. 1975. Recommended classification of pesticides by hazard. *WHO Chronicle* **29**:397–401.

56. World Health Organization. 1976. *Resistance of Vectors and Reservoirs of Disease to Pesticides.* World Health Org. Tech. Rep. Ser. 585. 88 pp.

2

INSECTICIDES AND THE ARTHROPOD FAUNA OF PLANT COMMUNITIES

A. GENERAL ECOLOGICAL CONSIDERATIONS

Pest Resurgence and New Pests

It has frequently happened that the application of insecticides in agriculture and forestry has given rise to (*i*) resurgence of the target pests against which the chemicals are applied, and (*ii*) outbreaks of arthropod species hitherto unimportant. This is due essentially to the reduction of the natural enemies, namely the predators and parasites nearly all of which are also insects or acarines. It is not so much that the insecticides are more toxic to the natural enemies, for often they are less so, especially to predator larvae and nymphs. It is rather that, first, the insecticide if successful reduces the pests which are the food supply, so that the population of the natural enemies virtually disappears, and second, the pest rapidly recovers its numbers before the reinvasion of natural enemies can catch up and control it. Some pests have amazing powers of multiplication; for example, parthenogenetic aphids each producing 40 offspring would in a season of 12 generations produce 250 million tons of aphids if every one of them survived.[77] Thus the experience with DDT when it was introduced to kill the codling moth, the principal pest in arthropods, was that it resulted in increased numbers and finally outbreaks of (*i*) aphids, (*ii*) scale insects and mealybugs, (*iii*) Tortricid leafrollers, and (*iv*) Tetranychid mites. The situation is often aggravated by the pest developing insecticide resistance, while the natural enemies do not, or if they do it is later and at a slower rate.

Reduction of Species Diversity

The effect of the insecticide is not only to reduce the numbers in any one kind of natural enemy, but to reduce the number of species of natural enemies in the ecosystem, since some of them are driven out for lack of prey or starved out for lack of hosts. Application of the spray schedule recommended at the time for cabbage pests to 0.17-acre cole plots in New York State, namely endrin and parathion mixtures, resulted in the disappearance of 22 out of the 27 species of predators and parasites (Table 2.1) while the numbers of the plant-feeding species were reduced only slightly. Surprisingly, the treatments with DDT caused little loss of natural enemy species, even less than with rotenone.[72]

However in grassland, where the phytophagous species are mainly Homoptera and Hemiptera, treatments with carbaryl at 2 lb/a which first reduced the biomass by 95% finally resulted in the return of the normal biomass 7 wk later, the spiders and the predaceous insects being the least to

Table 2.1.   **Number of species (or species groups) in cole plots in New York
submitted to spray schedules of insecticides or insecticide mixtures
(Pimentel, 1961a).**

	Herbivores	Parasites	Predators
Control Untreated	23	13	14
6 sprays DDT at 0.25 lb/a	20	11	14
7 sprays Rotenone at 0.07 lb/a	18	10	12
4 sprays Endrin-Parathion at 0.38 lb/a	19	2	3

suffer and the first to recover.[4] Cornfields in a 28-sq mi area at Sheldon,
Illinois treated for 5 successive years with soil applications of dieldrin at 2
or 3 lb/a did not suffer loss or even reduction of their Chrysopid, Nabid,
Syrphid, and Coccinellid predators, nor in the Tachinid *Lydella grisescens*
that parasitizes the corn borer. Nevertheless the corn borer population
became about twice as numerous, a phenomenon evidently associated with
the great size of the area, since there was no corn borer increase when
isolated fields were treated with dieldrin.[55] When an abandoned cornfield in
New Jersey was treated with soil applications of diazinon at 14 lb/a in early
May, the depression in the soil microarthropod and herbivorous insect
population was followed in September by a total insect density greater than
that in the untreated half of the field. This was due to the rapid resurgence
of some species, and was perhaps associated with the greater luxuriance and
diversity of the vegetation developing on the treated as compared to the
untreated section.[59]

Stable ecosystems, where the checks and balances due to maximum
species diversity militate against outbreaks, are ideal situations that may be
found in some forests, are rare in orchards, and are nonexistent in annual
crops whether chemicals are employed or not.[83] In agroecosystems, the goal
of maximum species diversity is usually somewhat irrelevant or not
practically useful, and biological control can often be achieved by a simplified system of pest and natural enemy.[22] Structural and spatial diversity
may be more of a stabilizing influence than species diversity, in the sense
that alternative hosts of useful parasites may be found in certain vegetation
in the vicinity of the crop field, such as hawthorn to support the wild host of
the *Horogenes* parasite of the diamond-back moth.[29] A supply of the
parasites of the *Heliothis* pests of cotton may be ensured by planting some
other crops in the cotton area, such as corn, beans, and flax in the Canete
Valley of Peru, to support a reasonable population of caterpillars which
support such parasites.[83] The control of hornworms and budworms on

tobacco may be achieved with less expenditure of insecticide and with greater conservation of natural enemies by treating only the tops of the tobacco plants where the healthy caterpillars are and leaving the bottom leaves where the parasitized larvae survive.[38] The percentage parasitism of the alfalfa caterpillar by *Apanteles* may be increased by restricting treatments to places where the natural enemy ratio is inadequate.[82] The proper timing of applications is important, for example to apply the diazinon or demeton necessary to control aphids and scales in orchards early in the season while the Phytoseiid predators have not yet climbed up into the trees,[21] or to apply methyl parathion or malathion necessary to control the alfalfa weevil in late March well before its parasite *Bathyplectes curculionis* has emerged from the cocoon.[60]

Choice of Insecticide

Obviously one insecticide is more deleterious to natural enemies than another, and it is important to know how they compare with each other in this respect in making a wise choice of the insecticide to recommend. With adults or larvae of Coccinellids, no less than 20 investigations have compared a number of insecticides and ranked them according to their toxicity. Taking them all together, it is possible to derive an overall ranking (Table 2.2), from which it may be seen that the OP compounds are more lethal than the organochlorines. Among the OP compounds the nonpersistent ones (e.g., trichlorfon) and the systemics (e.g., demeton, schradan) are the least toxic, and among the organochlorines the chlorinated acaricides (e.g., tetradifon, chlorobenzilate) are virtually nontoxic. It should be noted that *Bacillus thuringiensis* spores, a biological insecticide, affects only leaf-eat-

Table 2.2. Relative toxicity of insecticides to predaceous lady beetles: generalization from 20 investigations on 10 species (Croft and Brown, 1975).

Highly Toxic		Mod'ly Toxic		Rel. Non-Toxic
1	2	3	4	5
Parathion	Phosdrin	Demeton	Lindane	Schradan
Methyl parathion	Phosphamidon	Methyl demeton	Toxaphene	Chlorobenzilate
Malathion	Diazinon	Trichlorfon	Endrin	Tetradifon
Azinphosmethyl	Dimethoate	Carbophenothion	DDT	Dicofol
Carbaryl	Ethion	Thiometon	Endosulfan	Binapacryl

ing insects and is thus harmless to the predators and parasites. Similarly, the investigations made on adults of 17 species of hymenopterous parasites and 1 tachinid (Table 2.3) clearly show that here the organochlorines were more lethal than the OP compounds, and trichlorfon and methyl demeton are again the least toxic of the latter group.

B. FOREST ECOSYSTEMS

Effects of DDT

The most inclusive studies were made on DDT applied to deciduous forest in eastern North America, the results being presented in terms of the total numbers in the various orders or families 1 wk after, and again 1 month after, a single application.[43] When a bottomland forest in Maryland characterized by tulip tree and black gum was sprayed at 2 lb/a, all the orders showed a considerable reduction in numbers in the first week. Although the Diptera as a whole were not reduced, the calyptrate flies suffered an 85% reduction and did not recover their numbers until 8 wk after the application[42]; 1 month after the treatment the Hymenoptera and Lepidoptera had attained numbers greater than those before it (Table 2.4). Reductions in the spiders and the adult beetles were still evident after 1 month, the latter being largely confined to the Chrysomelids and two abundant but obscure species, *Ptilodactylus serricollis* and *Cyphon obscurus*. Samples taken in the ground cover showed that the Carabid

Table 2.3. **Relative toxicity of insecticides to adult hymenopterous parasites: generalization from results obtained with 18 species (Croft and Brown, 1975).**

Highly Toxic	Moderately Toxic	Relatively Non-Toxic
Aldrin	Parathion	Trichlorfon
Dieldrin	Methyl parathion	Carbophenothion
Endrin	Malathion	Methyl demeton
Lindane	Azinphosmethyl	Thiometon
DDT	Phosphamidon	Endosulfan
	Toxaphene	
	Carbaryl	

Table 2.4. **Percent reductions in total numbers (corrected for change in control untreated areas) in a Maryland deciduous forest sprayed with DDT at 2 lb/a (Hoffmann et al. 1949).**

	Sticky Boards after 1 wk	Light Traps after 1 wk	Light Traps after 3 wks	Sticky Boards after 4 wks
Mecoptera	100	-	-	87
Lepidoptera	84	69	0	0
Homoptera	81	-	-	44
Hymenoptera	74	0	0	0
Araneida	69	-	-	56
Coleoptera	52	71	12	61
Chrysomelids	73	-	-	92
Mordellids	73	-	-	10
Cantharids	70	-	-	0
Staphylinids	39	-	-	25
Coccinellids	0	-	-	57
Lampyrids	0	-	-	6
Elaterids	0	-	-	0
Trichoptera	-	46	0	-
Diptera	0	0	0	0

Chlaenius aestivus and the Silphid *Necrophorus orbicollis* were greatly reduced, and the ants *Tapinoma sessile* and *Prenolepis imparis* were moderately reduced.[43] Otherwise the remaining Carabids and Silphids, and most of the Staphylinids, were unaffected.[42]

When a second-growth stand of red maple and white oak in Pennsylvania was sprayed with DDT at 4.5 lb/a, *Carabus limbatus* was completely eliminated, while *Tapinoma sessile* and the ant species *Aphaenogaster treatae* were much reduced.[43] There was a great and persistent reduction in the numbers of Cicadellids, Membracids, and Chrysomelids.[42] On the forest floor the total numbers of Acarina were unaffected, while those of Collembola increased by 50%. In the trees the *Phytocoris* Mirids were eliminated, and there were reductions in some species of spiders, for example, *Theridion murarium* and the orb-weaver *Leucauge venusta,* but not in other even closely related species. Six weeks after the spraying, the percent reductions in the parasitic Hymenoptera were as follows:

Ichneumonidae	91	Diapriidae	78
Platygasteridae	90	Chalcidoidea	74
Braconidae	82		

However, the numbers had recovered a year later (in the application at 1 lb/a the total numbers in these groups were back to normal 1 wk after spraying). The decimation of predators in the families Miridae, Lygaeidae, Nabidae, Anthocoridae, Syrphidae, Cantharidae, Coccinellidae, and Chrysopidae by this high dosage of DDT resulted about a month later in a general infestation of at least 14 species of Aphidae, which had survived the spray on the undersides of the leaves and then multiplied faster than the surviving predators. Moreover, an infestation of the spider mite *Paratetranychus ununguis* developed on oak and maple, because of the destruction of its neuropteran predator *Coniopteryx vicina,* sufficient to cause damage on red maple in the following spring.[42] Similarly, antimosquito spraying with a 0.1% suspension of DDT in a wooded area in Connecticut had led to harmful infestations of *Paratetranychus bicolor* on oaks and maples and of *Tetranychus urticae* on planted Hydrangea.[14] An outbreak of the spruce spider mite followed a year after the DDT spraying of a Douglas fir forest in Montana.

Scale insects also survived forest spraying with DDT at 2 lb/a.[42] Antimalarial spraying of Manila, Philippines with DDT resulted in outbreaks of 10 species of scales and mealybugs on the shade trees and ornamental shrubs; they could be controlled by direct application of a strong DDT emulsion, but not surprisingly a far worse infestation appeared 2–4 wk later.[62]

A dosage of DDT at 7.5 lb/a applied against the mountain pine barkbeetle in Wyoming reduced the entire insect population by about 90%. At the dosage of 1 lb/a usually employed to control the gypsy moth in deciduous forest, the DDT killed caterpillars, Psyllids, small moths, and a variety of small flies, but 3 wk later the total amount of insects collected was greater than before the spray.[42]

During the period 1952–1966, coniferous forests in New Brunswick, Canada were extensively airsprayed to combat the spruce budworm with DDT applied annually to about 2 million acres, first at 1 lb/a and later at 0.5 lb/a, any given area being treated one to three times. These applications did not have an overall deleterious effect on the natural enemies and did not prolong an outbreak period, although at the peak of an outbreak a failure to respray in one or two subsequent years would result in full defoliation by the budworm. Of the predators, the Coccinellids were always found to be higher in sprayed than in unsprayed areas, while the Chrysopids and Syrphids were lower only in the year of the spray but were higher in subsequent postspray years.[57] Of the hymenopterous parasites, *Apanteles fumiferanae* was higher in sprayed areas than in unsprayed,[56] *Itoplectis conquisitor* was higher in the year of the spray but not thereafter, and *Phaeogenes hariolus* was lower in the year of the spray but higher thereafter. During the years of the study *Glypta fumiferanae* showed the same downward trend for year to year in sprayed areas as it did in unsprayed, while *Meteorus trachynotus* showed the same increase at the end of an outbreak in unsprayed as in sprayed areas.[57] In an area in Oregon sprayed with DDT at 1 lb/a against the western spruce budworm, both *Apanteles fumiferanae* and *Glypta fumiferanae* remained at much the same level, partly because of their survival on alternative Tortricid hosts such as *Argyrotaenia dorsalana*.[16] In these western forests, hymenopterous parasites such as *Phytodietus fumiferanae* survived better than the dipterous parasites; overall, the parasitism was roughly unchanged, in contrast to New Brunswick where on balance it showed an increase in sprayed areas.

Effects of OP Insecticides

Many organophosphorus compounds have been tested or employed as substitutes for DDT. Malathion sprayed at 2 lb/a over 20 acres of deciduous forest in Ohio was found to greatly reduce the numbers of insects and other arthropods, but the recovery of the numbers *in toto* was rapid and made itself felt 1 wk later. Judging by tree-band traps, the initial reduction was of the order of 40%, and from the weight of the insects that fell on catch cloths the spray killed 2.5 lb of insects over the area in the first day. Light-trap catches indicated that Phalaenid moths, Microlepidoptera, Tipulid flies, Cecidomyiid wasps, and Cantharid beetles were among the groups that particularly suffered. The population of Collembola in the leaf litter was about 75% reduced, but had almost completely recovered its numbers 2 wk later; earthworms and snails were unaffected.[70]

Fenitrothion applied in two sprays each of 0.4 lb/a in close succession to northwestern Ontario forest against the spruce budworm had an adverse effect on predators, reducing the numbers of the Lycosid spider *Trochosa terricola* by one-half, the Carabid *Pterostichus pennsylvanicus* by two-thirds, and the Carabid *Agonum retractum* by 95%.[30] In the following year these species, and the Lycosid *Tarentula aculeata,* were about half as numerous in the treated areas as in untreated control areas; this effect, since fenitrothion did not persist from one year to the next, must have resulted from a persistent disturbance of the ecosystem.[31] It should be noted that fenitrothion applied to English agricultural soil at 1.5 lb/a did not reduce the numbers of four out of the five Carabid species investigated.[36]

C. ORCHARD ECOSYSTEMS

Apple Orchards

Species Composition.   In North America, where the codling moth is the most important single pest species, the insect fauna to be found above ground in orchards is of great complexity. In two surveys, one made in two orchards sprayed with ryania and lead arsenate in southern Indiana,[19] and the other in a 3-acre neglected orchard in Door county, Wisconsin,[68] the following were found:

Indiana: 421 species in 315 genera in 106 families in 15 orders
Wisconsin: 763 species in 515 genera in 158 families in 14 orders.

The families with most species were in the parasitic Hymenoptera, besides three families characteristically predaceous, followed by the Aphids and Heleids, viz:

	Ind.	Wis.		Ind.	Wis.
Eulophids	24	55	Mirids	14	24
Pteromalids	16	37	Coccinellids	10	18
Encyrtids	17	14	Chrysopids	3	9
Braconids	30	36	Aphids	6	24
Ichneumonids	15	34	Heleids	4	15 species

The genera richest in species again were among the parasitic Hymenoptera, as follows:

		Ind.	Wis.
Tetrastichus	(Eulophidae)	7	10
Sympiesis	(Eulophidae)	3	7
Habrocytus	(Pteromalidae)	2	6
Apanteles	(Braconidae)	6	6 species

In southern Indiana there were yet fewer species (226 cf 421) and total numbers of insects in orchards treated with DDT and OP compounds than in those treated with ryania and lead arsenate.

Λ detailed inventory of the predators of the codling moth was made from 1938 to 1940 in Virginia in two 20-acre blocks, one treated with lead arsenate and nicotine cover-sprays, the other untreated by cover-sprays but the trees were banded with tanglefoot adhesive.[48] The total numbers of ant colonies, ground beetles, and other predaceous beetles found or collected in the 3 yr were only some 15% less in the treated orchard than in the untreated one (Table 2.5).

Codling Moth Parasites. In the Virginia orchard treated with lead arsenate and nicotine, there was a slight decrease in the thrips predator *Leptothrips mali* that attacks the eggs of the codling moth. But there was a great difference in the percentage parasitism by hymenopterous parasites, viz.:

	Untreated	Treated
Eggs parasitized by		
Trichogramma minutum	13.9	5.7
Larvae parasitized by		
Ascogaster quadridentata	4.5	1.4

This difference was partly due to the fact that the parasites from larvae trapped on the tanglefoot were returned to the untreated orchard. In New Jersey orchards heavily sprayed with lead arsenate[27] the percentage parasitism as compared with unsprayed orchards was as follows:

	Untreated	Treated
Codling moth *Carpocapsa pomonella*		
Eggs parasitized by *Trichogramma* sp.	60	4
Larvae parasitized by *Ascogaster*		
carpocapsae	71	8
Apple leafhoppers, mainly *Empoasca mali,*		
parasitized by *Aphelopus* sp.	34	1

Most of the OP compounds developed as sprays against the codling moth and other apple pests, including parathion, diazinon, and malathion, are deleterious to the thrips and Mirids that pray on the codling moth (Table 2.6); azinphosmethyl is the least harmful of this insecticide group to the Mirids, although it is just as harmful to the thrips.[58]

Leafroller Pests. The introduction of DDT in 1946 as the cover-spray insecticide against codling moth resulted in an upsurge of other pests of

Table 2.5. Predaceous beetles, ants and spiders in orchards untreated or treated with cover sprays of lead arsenate and nicotine: West Virginia (Jaynes and Marucci, 1947).

	Untreated	Treated
Spiders	177	174
Calathus opaculus	358	314
C. gregarius	41	30
Harpalus pennsylvanicus	288	301
H. faunus	46	102
H. caliginosus	14	57
Amara muscula	19	68
Am. sp.	427	197
Anisodactylus rusticus	75	63
An. agricola	30	26
Total Carabids (ca. 30 spp.)	1541	1295

	Untreated	Treated
Staphylinids	108	58
Conoderus lividus	35	33
Cantharids	93	79
Monomorium minimum	93	10
Solenopsis molesta	376	303
Aphaenogaster fulva	215	7
Tetramorium caespitum	49	131
Formica fusca	107	42
Lasius claviger	108	13
Total Ant Colonies (>9 spp.)	1096	624

Table 2.6. Influence of organophosphorus insecticides on the mite predator fauna of Nova Scotia orchards (MacPhee and Sanford, 1961).

	Parathion 0.010%	Diazinon 0.0125%	Malathion 0.025%	Trithion 0.025%	Azinphosmethyl 0.065%
Acarina					
Typhlodromus rhenanus	-	++	++	-	-
T. pyri	+	+	++	-	+0
Phytoseiulus macropilis	-	-	-	-	-
Anystis agilis	++	0	++	+	-
Mediolata novaescotiae	+	0	+	-	-
Thysanoptera					
*Haplothrips faurei	++	++	+	++	++
*Leptothrips mali	++	-	-	-	-
Hemiptera					
Campylomma verbasci	++	++	++	-	-
*Hyaliodes harti	++	+	++	+	++
Diaphnidia pellucida	++	-	-	++	++
*Deraeocoris fasciolus	+	+0	-	+	0
*Plagiognathus obscurus	++	++	-	-	0
Anthocoris musculus	+	++	++	+0	0

* Also egg and/or larvae predatory of the codling moth

apple orchards, namely (*i*) the red-banded leafroller, (*ii*) the woolly apple aphid, and (*iii*) the European red spider and other Tetranychid mites. Leafrollers form an important part of the orchard ecosystem, eight species of Tortricids and six species of Olethreutids having been found in a 3-acre Wisconsin orchard.[68] In orchards in southern Ontario where 18 species of leafrollers had been found during a long monitoring period which started before World War II, the introduction of DDT resulted in a marked increase by 1950–1952 of the red-banded leafroller *Argyrotaenia velutinana,* and the elimination of the fruit-tree leafroller *Archips argyrospilus* which had formerly been a pest species (Table 2.7); by 1962 sprayed orchards showed an almost total absence of the other species of leafrollers to be

Table 2.7.   Upsurge of the red-banded leafroller *Argyrotaenia velutinana* in orchards
sprayed with DDT from 1947 onwards: Norfolk county, Ontario
(Hikichi, 1964).

	Sprayed[a]		1962[b]	
	1941-43	1950-52	Unsprayed	Sprayed
Archips semiferanus	459	8	5	0
A. argyrospilus	1581	0	8	0
Choristoneura rosaceana	575	142	119	2
C. fractivittana	347	0	5	0
Argyrotaenia velutinana	36	395	5	297
A. quadrifasciana	6	0	27	0
Pandemis limitata	391	10	32	0

[a] Numbers of adults taken in bait pails at 4 sites

[b] Numbers of larvae taken in collections in 3 orchards

found in unsprayed orchards, while *A. velutinana* was 60 times as abundant
in the sprayed as in the unsprayed orchards.[40] This increase in the red-
banded leafroller was observed as early as 1947 in New York State and
thought to be associated with the effect of DDT on the codling moth
parasites,[33] while in West Virginia its rise to pest status was associated with
the inability of DDT to penetrate the leaf-rolls, which could be corrected by
substituting DDD which did penetrate.[35] The effect of DDT on a Braconid
parasite was quantitated in an Ohio peach orchard, where the parasitism of
the oriental peach moth by *Macrocentrus ancylivorus* fell from the normal
58% down to 37% in 1946 and 23% in 1947.[76] The abrupt rise in the red-
banded leafroller in North America was paralleled by the rise of *Tortrix
postvittana* when DDT was introduced to orchards in Australia.[69]

Aphids and Scales.   The woolly apple aphid *Eriosoma lanigerum* was
observed at Yakima, Washington to develop very heavy infestations on
trees which had been sprayed with DDT in 1946, that is, for the first time;
the resurgence was attributed to a decrease in its parasitization by the
chalcidoid wasp *Aphelinus mali,* which on trees sprayed with lead arsenate
was about 45%, but on the DDT-sprayed trees had been reduced to about
30%.[66] By 1952, the percent parasitism in the DDT-sprayed orchards was
one-fifth to one-tenth lower, and the aphid infestation was three to six times

higher, than in orchards sprayed with lead arsenate. DDT sprays caused a similar decrease in the encyrtid parasite *Pseudaphycus,* the adults of which were very susceptible to poisoning by DDT residues; the percentage parasitism of the Comstock mealybug by this species, which was 11% in unsprayed blocks, was scarcely more than 1% on DDT-sprayed trees. Although the parasitism recovered to 86% at the end of the season, it was far short of the 100% parasitism found at this time in unsprayed blocks which normally keeps this Coccid at bay.[44]

Tetranychid Mites. Plant-feeding acarines had almost always greatly increased in numbers towards the end of the season in orchards which had been sprayed with 0.25% suspensions of DDT to control the codling moth which is the principal apple pest. The increase in European red spider *Panonychus ulmi* and the two-spotted mite *Tetranychus urticae* to pest proportions was first connected, in Indiana experiments, with the DDT eliminating the Coccinellid predator *Stethorus punctum.*[84] In Virginia, where there was an increase in *T schoenei* also, the elimination of the predaceous phytoseiid mite *Amblyseius fallacis,* and the scarcity of *Leptothrips mali* and *Scolothrips sexmaculata,* were causes additional to the disappearance of *Stethorus.*[18] In southern Ontario, the disappearance of *Haplotrips faurei* was additive to that of *Amblyseius* and *Stethorus,* while in Nova Scotia, the decimation of the Mirid predators *Diaphnidia* and *Hyaliodes* as well as *Amblyseius fallacis* (described as *Iphidulus tiliae*) was an important factor in the upsurge of Tetranychid mites after DDT applications.[54] In Washington and British Columbia, the destruction of *Stethorus picipes* and other predators resulted in a surge of *Tetranychus pacificus* after DDT sprays.[65] In the warmer districts of California and South Australia, orchard applications of DDT have resulted in infestations of the clover mite *Bryobia praetiosa.*[15]

In England, where 42 species of predators were found attacking the red spider *Panonychus ulmi* (Table 2.8), it was found that the main difference between sprayed and unsprayed orchards was that the predator population in a sprayed orchard was low in early summer and thus the red spider population could build up by July.[20] DDT sprays greatly reduced the principal Mirid predator *Blepharidopterus* as well as *Psallus ambiguus*; lead arsenate had a much less pronounced effect,[63] while nicotine and BHC did not change the numbers of predators or their spider-mite prey.[64]

These serious effects on mite predators were not shown by lead arsenate or ryania, either in Nova Scotia (Table 2.9) or in England.[17] Thus in Nova Scotia, where apple growers are confronted with only one generation of the codling moth each year, ryania with lead arsenate could establish satisfactory control. Often ryania alone was adequate, and in a few orchards no

**Table 2.8. Species of arthropods predaceous on *Panonychus ulmi* in
an apple orchard in Essex, England (Collyer, 1953).**

Hemiptera

  Anthocoris nemorum

  A.   nemoralis

  A.   confusus

  Orius minutus

  O.   majusculus

  Himacerus apterus

  H.   lativentris

  Phytocoris tiliae

  P.   reuteri

  P.   ulmi

  Camptobrochis lutescens

  Deraeocoris ruber

  Campyloneura virgula

  Pilophorus perplexus

  Blepharidopterus angulatus

  Orthotylus nassatus

  O.   marginalis

  Capsus meriopterus

  Malacocoris chlorizans

  Psallus ambiguus

  Atractonemus mali

  Plagiognathus arbustorum

  Campylomma verbasci

Acarina

  Typhlodromus tiliae

  T.   finlandicus

  Phytoseius spoofi

  Anystis agilis

  Cheyletus spp.

Neuroptera

  Hemerobius humulinus

  H.   lutescens

  Chrysopa carnea

  Conwentzia psociformis

  C.   pineticola

  Coniopteryx tineiformis

  Semidalis aleyrodiformis

Coleoptera

  Oligota flavicornis

  Adalia bipunctata

  Coccinella septempunctata

  Stethorus punctillum

  Scymnus auritus

Thysanoptera

  Thrips tabaci

  Aeolothrips melaleucus

sprays were required because the natural control was sufficient[71]; but this is
almost a unique situation. When attempted in an orchard in upstate New
York, the suspension of all insecticide sprays for a 10-yr period was disas-
trous. Even though the orchard was tended with all the other regular pomi-
cultural practices including fungicide sprays, the crop was worthless by the

second year and thereafter. It was true that Tetranychid mites were not troublesome, but the codling moth, apple maggot, and other insect pests were simply too abundant. There was even an example of a new pest appearing, when a little-known species, *Grapholitha prunivora,* became a major problem in the final 2 yr.[34]

Once DDT and other wide-spectrum sprays have been used in orchard pest control, it takes about 5 yr to restore normal predator–prey relationships.[71] In Washington State, this period of recovery takes about 4 yr.[46] From experiments in British Columbia, it is clear that wide-spectrum DDT and OP sprays have eliminated the predators *Typhlodronus caudiglans* and *T. pyri,* leaving only the *T. occidentalis* to rely upon.[26] A detailed study in New Jersey (Table 2.10) showed that the predaceous mite fauna in inten-

Table 2.9. **Influence of insecticides on the arthropod predators of phytophagous mites in Nova Scotia orchards (MacPhee and Sanford, 1961).**

	Pb Arsenate	DDT	Nicotine SO$_4$	Ryania	Carbaryl
	0.4%	0.025%	0.125%	0.5%	0.05%
Acarina					
Typhlodromus rhenanus	+	+	0	0	–
T. pyri	0	+	0	0	++
Phytoseiulus macropilis	++	++	–	0	–
Anystis agilis	++	++	0	0	+
Mediolata novaescotiae	0	0	0	0	–
Thysanoptera					
*Haplothrips faurei	0	++	+	0	–
*Leptothrips mali	0	++	+	0	–
Hemiptera					
Campylomma verbasci	–	+0	0	0	0
*Hyaliodes harti	0	+	0	0	++
Diaphnidia pellucida	0	–	+	+	–
*Deraeocoris fasciolus	0	0	0	0	+
*Plagiognathus obscurus	+0	0	0	0	–
Anthocoris musculus	0	0	+0	0	+

* Also egg and/or larval predator of the codling moth

Table 2.10. Occurrence of predaceous and plant-feeding mite species in 15 orchards in New Jersey (Knisley and Swift, 1972).

	Abandoned out of 7 orchards	Intensively Treated out of 3 orchards	Treatments Discontinued out of 5 orchards	Total Occurring
Typhlodromus pomi	6	0	4	10
T. conspicuus	1	0	2	3
T. longipilus	0	2	0	2
T. putmani	0	0	1	1
Amblyseius fallacis	0	3	4	7
A. andersoni	1	0	2	3
A. morgani	1	0	0	1
Phytoseiulus macropilis	5	2	4	11
Tydeus kochi	3	1	3	7
Pronematus sp.	2	0	4	6
Cunaxa croceus	0	0	1	1
Panonychus ulmi	1	3	3	7
Tetranychus urticae	0	1	2	3
T. schoenei	3	0	1	4
Bryobia sp.	5	0	1	6
Aculus schlechtendali	3	1	2	6
Brevipalpus lewisi	1	0	0	1

sively treated orchards had become restricted to *Amblyseius fallacis* and two other species of Phytoseiids as compared to eight Phytoseiid species in orchards where treatments had been discontinued for at least 1 yr, or which had been abandoned for several years. Among the Tetranychid plant-feeding mites, the *Bryobia* that characterized abandoned orchards was replaced in the sprayed orchards by *Panonychus* and *Tetranychus urticae,* which then might persist in some of these orchards where the sprays had been subsequently discontinued. The Stigmaeid predators *Agistemus fleschneri* and *Zetzellia mali,* however, were present in all of these three types of orchards.[51]

Although the carbamate insecticide carbaryl (Sevin) is markedly toxic to predaceous mites (Table 2.11) and has caused resurgences of Tetranychid mites,[17] most of the OP compounds offend less in this respect. Allowing for differences between Nova Scotia, where both groups of mites and OP-susceptible (Table 2.12) and Michigan, where the Tetranychids and some predaceous mites are OP resistant (Table 2.11) it is seen that the OPs least

Table 2.11. Effect of insecticides and acaricides on plant-feeding mites, predaceous mites, and lady beetles in Michigan orchards, 1974 (Croft, 1975).

	Dosage lb/a	*Panonychus* & *Tetranychus*	*Aculus schlechtendali*	*Amblyseius fallacis*	*Agistemus* & *Zetzellia*	*Stethorus punctum*
Azinphosmethyl	2	0	+0	0	0	+0
Phosphamidon	1	0	+	+	+0	+
Dimethoate	8	+0	+0	+0	0	+0
Phosmet	4	0	+0	0	0	+
Phosalone	5	+	+	++	+	+
Demeton	1.5	+	+0	+	0	+
Carbaryl	8	0	++	++	++	++
Endosulfan	4	0	++	+0	++	+
Chlordimeform	2	++	++	++	+	+
Oxythioquinox	2	++	++	+	++	0
Propargite	5	+	++	0	++	0
Cyhexatin	0.3	++	++	+0	++	+0

0 Non-toxic or slightly toxic   +0 Moderately toxic   + Highly toxic

++ Highly toxic, long residual action

harmful to predaceous mites are azinphosmethyl,[21,58] phosmet,[21] and menazon.[79]

Under modern systems of orchard pest management, populations of the apple rust mite *Aculus schlechtendali* are so managed that there are sufficient number of this species to serve as food for *Typhlodromus occidentalis* in orchards in western United States[45] or for *Amblyseius fallacis* in the eastern orchards[21]; if *Aculus* gets out of hand, it is knocked back with the organochlorine endosulfan, which is relatively kind to *Amblyseius.*

Of the acaricides which would be necessary to control the plant-feeding Tetranychid mites, tetradifon (Tedion) and ovex were kind to Phytoseiids (Table 2.12), but the Tetranychids have developed resistance to them. Oxythioquinox (Morestan) is effective but is deleterious to the Phytoseiid predators, and also to *Aphelinus* and *Mormoniella* parasites in the orchard.[11] The propynyl sulfite Omite (propargite) or the cyclohexyltin hydroxide Plic-

Table 2.12.   Effect of insecticides and acaricides on predaceous mites in Nova Scotia orchards, 1966 (Sanford, 1967; Sanford and Herbert, 1967).

	Amblyseius *fallacis*	*Typhlodromus* *rhenanus*	*T.* *pyri*	*Phytoseiulus* *macropilis*	*Anystis* *agilis*	*Mediolata* *novaescotiae*
Dimethoate	++	++	++	-	-	+
Carbophenothion	++	++	++	-	-	+
Methyl-demeton	++	++	++	-	-	0
Menazon	0	+	+	-	-	0
Dicofol	++	++	+	-	0	++
Dimite	+	+	+0	-	-	++
Tetradifon	+	0	0	-	+	0
Chlorbenside	+	0	++	-	-	+
Chloropropylate	-	+	++	-	-	++
Ovex	-	0	0	+	0	0
Tetrasul	0	+	0	-	-	+0
Oxythioquinox	+	++	++	-	-	++
Propargite	+	++	-	-	-	+

- Not determined   0 No effect   +0 Possible reduction

+ Reduction   ++ Practical Elimination

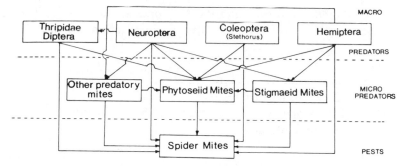

Fig. 2.1. Food-web relationships between spider-mites and their principal arthropod predators (from Croft and Brown, 1975).

tran (cyhexatin) are at present the best materials for suppressing Tetranychids and the apple rust mite and thus maintaining a favorable predator–prey ratio[21] of 1:20, or 1:10 in difficult cases; tetrasul has also been employed for this purpose.[47]

In controlling Tetranychid spider mites, the coccinellid *Stethorus* is as important as the Phytoseiids, while those of second rank are the Stigmaeid mites, the predaceous thrips, and the spiders; then follow in order of importance the Hemiptera (Anthocorids, Mirids, Lygaeids and Nabids), and finally the Chrysopid and Hemerobiid Neuropterans (Fig. 2.1).[22] The effect of azinphosmethyl (Guthion) and of the formamidine insecticide chlordimeform (Fundal) on the other predators was determined in California pear orchards[9] to be as follows, in counts per 40 beats:

	Guthion 2 oz/100 gal	Fundal 8 oz/100 gal	Untreated
Spiders	2.5	2.4	6.9
Anthocorids	0.3	0.7	1.5
Neuroptera	2.2	0.8	5.4

The acaricide oxythioquinox was found in the Netherlands to be definitely dangerous to *Anthocoris nemorum*, as well as to the Mirid *Orius*.[85]

Citrus Orchards

The application of DDT wettable-powder suspensions to citrus in California increased the numbers of the red mite *Panonychus citri* and decreased those of its predators, which were the Coccinellid *Stethorus picipes*, the

Staphylinid *Oligota oviformis,* and the Neuropteran *Conwentzia hageni.*[23] The numbers per 100 leaves 2 months after the application were as follows:

	Predators	Mites
12.5% DDT in Talc	9.7	2302
Talc Alone	32.8	1896
Control Untreated	28.8	377

It will be noted that the dust diluent alone increased the mite population without decreasing the predators, probably because the inert deposits make for better adherence for the mites on the leaves. In Florida, the destruction of predators by DDT, notably the ladybeetle *Chilocorus stigma,* resulted in infestations of the citrus rust mite *Phyllocoptrutes oleivora* and the citrus mealybug *Pseudococcus citri.*[37] DDT sprays also stimulated the development of infestations of the citrus rust mite on oranges in New South Wales.[39]

Application of DDT to citrus orchards also increased the numbers of scale insects. This was in marked contrast to the old method of fumigation with HCN, which caused less than 50% mortality of the Coccinellids,[89] while parasites such as *Aphytis* and *Cryptochaetum* survived inside the scales they parasitized.[32] In Florida, the red scale *Chrysomphalus aonidum* was increased by DDT treatments because they killed its principal parasite *Pseudohomalopoda prima.*[37] In California, DDT treatments led to the increase of the yellow scale *Aonidiella citrina* by killing its parasite *Comperiella bifasciata.*[89] Parathion applications led to serious increases in the brown scale *Coccus hesperidum,* since the sprayed foliage remains lethal to the principal parasite *Metaphycus luteolus* for 2–3 months.[7] Increase in the cottony-cushion scale *Icerya purchasi* due to destruction of the vedalia ladybeetle *Rodolia cardinalis* which is cultured and liberated to control it has been induced by experimental sprays of DDT[89] and subsequently this occurred over many thousands of acres to which DDT had been routinely applied to control the citricola scale.[78] Ten years later *Icerya* became a pest in Imperial and Coachella Valley orchards due to malathion drifting into the orchards from surrounding fields treated from aircraft and killing the vedalia.[8] Comparison of various organophosphorus compounds for their lethality to Coccinellids and parasites (Table 2.13) shows that the safest are nonpersistent or systemic insecticides such as tepp and demeton, respectively, whereas azinphosmethyl and the rest of the regular OP compounds are highly deleterious. Carbaryl and endosulfan were also very deleterious, but residues of lindane were not.[6]

Table 2.13. Contact toxicity of organophosphorus insecticide residues to Hymenopterous parasites and Coccinellid predators of citrus scales and mealybugs (Bartlett, 1963).

	Parathion 0.05%	Diazinon 0.05%	Malathion 0.05%	Dimethoate 0.05%	Azinphosmethyl 0.05%	Tepp 0.025%	Demeton 0.025%
Aphytis lingnanensis	H	H	H	H	H	M-H	H
Metaphycus luteolus	H	H	H	H	H	L-M	M-H
M. helvolus	H	H	H	H	H	L-M	M
Leptomastix dactylopii	H	M	H	H	H	O-L	H
Cryptolaemus montrouzieri	H	L-M	H	L-M	H	O-L	O-L
Lindorus lophanthae	M-H	M	H	M	H	L	L-M
Rodolia cardinalis	H	H	H	H	H	L	M-H
Stethorus picipes	H	H	H	H	H	O	O-L
Hippodamia convergens	H	H	H	H	H	O	O-M
H. quinquefasciatus	H	H	H	M	M	O-L	L-M

H high, M medium, L low, O no toxicity

D. ANNUAL CROP ECOSYSTEMS

Cotton Agroecosystem

In the southeastern states, control of the boll weevil with calcium arsenate resulted in the cotton aphid *Aphis gossypii*) becoming a serious pest because of destruction of its natural enemies.[75] The introduction of lindane to control the aphids, as well as the boll weevil and the *Heliothis zea* and *H. virescens* caterpillars, led to outbreaks of *Tetranychus* mites for the same reason.[10] The addition of OP compounds such as malathion and methyl parathion to organochlorine insecticides, which controlled the cotton leafworms along with the boll weevil, resulted in outbreaks of the *Heliothis* caterpillars as well as *Tetranychus*.[67] Even the systemic insecticide demeton applied against the cotton aphid could kill larvae of Coccinellids and of Syrphids when they fed on the poisoned aphids.[2] By the late 1960s, the treated cottonfields in the southeastern United States had been virtually sterilized of natural enemies.[1] Complete reliance on OP compounds consequent on the suspension of the organochlorines resulted in unusually severe outbreaks of *Heliothis* species, especially *H. virescens,* due to the development of highly OP-resistant populations in cottonfields denuded of their natural enemies.[75] The systemic carbamate insecticide aldicarb (Temik), even when applied as a seed dressing, was also fatal to the natural enemies of *Heliothis* caterpillars.[60]

In California, where the boll weevil was absent, *Lygus* bugs were the key pest demanding control; when DDT was applied to control them it led to outbreaks of the cotton leaf perforator, *Bucculatrix thurberiella,* and other pest species. Control of Lygus could however be achieved by carefully timed applications of nonpersistent OP compounds such as trichlorfon, or systemic insecticides such as methyl-demeton, without the upsurge of *Bucculatrix*. Although the natural enemies were thus conserved, hemipteran predators such as *Geocoris, Orius,* and *Nabis* suffered to some degree from the systemic insecticides through sucking the treated sap.[74] With the advent of the pink bollworm (*Pectinophora gossypiella*) in the late 1960s, the applications of carbaryl necessary to control it resulted in severe outbreaks of the cotton leaf perforator and the salt marsh caterpillar, as well as spider mites and beet armyworms (*Spodoptera exigua*) not only on the cotton but also neighboring crops.[82] The use of monocrotophos (Azodrin) against the cotton bollworm had to be abandoned because it was so hard on the natural enemies, and so persistent.[12] It usually took 2–4 yr to restore a shattered natural-enemy complex.[82]

The introduction of DDT and toxaphene to control the cotton leafworm

Spodoptera littoralis in Egypt resulted in problems not only with spider mites and aphids, but also with *S. exigua* and the spring bollworm *Earias insulana.*[75] Excessive use of these two insecticides, as well as endrin and methyl parathion, on the Nicaragua cotton crop resulted in the cabbage looper *Trichoplusia ni* and *Creontiades* plant-bugs, additional to *S. exigua,* becoming abundant and causing high levels of damage.[86]

Truck-Crop Agroecosystems

Cabbage. The main pests of cole crops in North America are the cabbage worms *Pieris* and *Plurella,* the cabbage aphid *Brevicoryne brassicae,* and two other species of aphids, each with their own parasites and predators among the 177 species of insects present (Fig. 2.2). Two-thirds of the species of the cabbage ecosystem in upstate New York were found to be unaccompanied by any congener, and the parasites and predators tended to prevent any one species exerting undue competition.[73] A cabbage ecosystem studied in Minnesota contained much the same pest species, and the important species among the 199 different kinds of insects could be organized by dividing it according to habitat (Fig. 2.3).

The effect of DDT sprays, as judged by a Brussels sprouts field in England, was to eliminate the cabbageworm parasite *Apanteles rubecula* each year; however, *A. glomeratus* and two Tachinid parasites were less affected, and the populations of *Harpalus* and other Carabid predators were helped by immigration and an increased supply of their Collembolan food.[24] Nevertheless the build-up of the cabbageworm (*Pieris rapae*) was more rapid on a sprayed than an unsprayed plot, the resurgence resulting from the fact that the pest lives on the growing plant where the effect of the chemical is soon lost, whereas its principal enemies live in the soil where the effect of the insecticide persists the longest.[25] On various cole crops in 0.17-acre plots in New York State, the application of DDT so reduced the predator–prey ratio for aphids (from 4.0 down to 0.6) that the cabbage aphid, and to a lesser extent *Myzus persicae,* reached outbreak proportions; curiously enough, rotenone had an even greater effect.[72] Organophosphorus insecticides such as parathion applied against cabbage aphids in England were highly deleterious to the Braconid parasite *Aphidius* (*Diaeretus*) *brassicae,* so that serious aphid infestations developed 10–14 days after application. The use of paraoxon resulted, for this reason, in the most enormous cabbage aphid infestation ever seen in that country.[78] By contrast, the systemic insecticide schradan, which did not affect the natural enemies either directly or by secondary poisoning through ingesting aphids, achieved control for this reason.

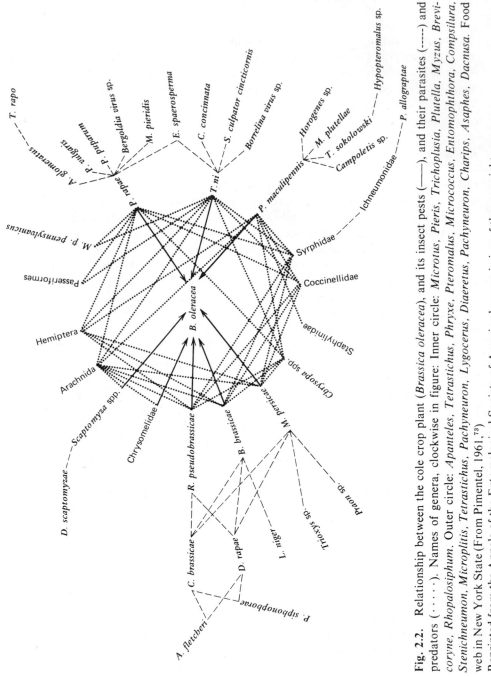

Fig. 2.2. Relationship between the cole crop plant (*Brassica oleracea*), and its insect pests (——), and their parasites (- - - -) and predators (· · · ·). Names of genera, clockwise in figure: Inner circle: *Microtus, Pieris, Trichoplusia, Plutella, Myzus, Brevicoryne, Rhopalosiphum.* Outer circle: *Apanteles, Tetrastichus, Phryxe, Pteromalus, Micrococcus, Entomophthora, Compsilura, Stenichneumon, Microplitis, Tetrastichus, Pachyneuron, Lygocerus, Diaeretus, Pachyneuron, Charips, Asaphes, Dacnusa.* Food web in New York State (From Pimentel, 1961,[73])

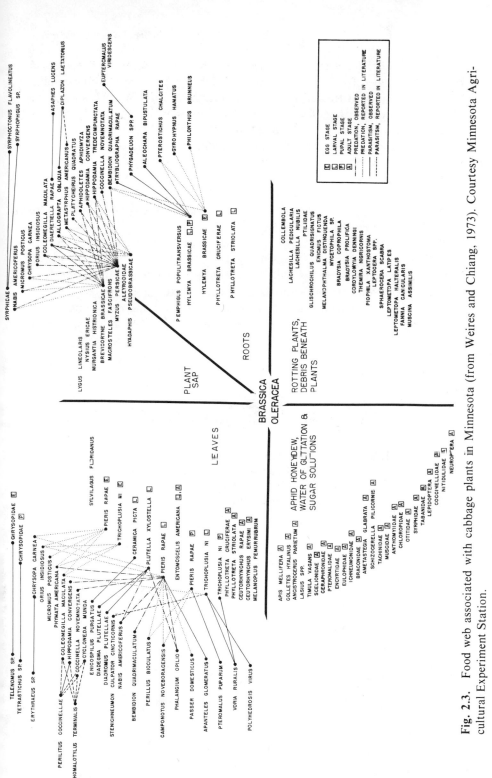

Fig. 2.3. Food web associated with cabbage plants in Minnesota (from Weires and Chiang, 1973). Courtesy Minnesota Agricultural Experiment Station.

Nicotine sulphate was an outstanding aphicide because parasites could emerge unharmed from nicotine-killed aphids. Thus when a cabbage aphid infestation had been treated under a trailing sheet, the few aphids left alive were eliminated by the parasite *A. (D.) brassicae* which had preferentially survived. But the success in the battle lost the war, because now the lack of their aphid food starved out the parasite while the predators naturally left the area, which became recolonized by immigrant aphids that multiplied without check.[77]

Other Truck Crops. Increased populations of the two-spotted mite *Tetranychus urticae* have been observed on potatoes and beans grown in soil treated with DDT, BHC, chlordane, or aldrin at dosages above 250 lb/a in Washington State.[50] Since natural enemies of mites are unimportant on beans, it was considered that the effect of these very heavy applications was through the plants being made more receptive to them. In New South Wales, DDT was incapable of controlling *T. urticae* on beans, while applications to tomato gave rise to severe infestations of the tomato mite *Phyllocoptes lycopersici.*[53]

Alfalfa and Clover Ecosystems

Organophosphorus insecticides have long been necessary for forage crops to avoid the residue problems that organochlorines pose for cattle feed.

Table 2.14.   **Effect of dimethoate on the insect fauna of an Indiana clover field: numbers per 100 sweeps 2–4 wk after treatments (Barrett and Darnell, 1967).**

	0.25 lb/a	0.50 lb/a	Untreated
Orthoptera	8	2	53
Hemiptera	112	73	250
Homoptera	106	62	138
Lepidoptera	8	0	2
Coleoptera	140	130	155
Diptera	70	55	55
Hymenoptera	2	10	12
Total	446	332	665

Parathion or malathion applied to control the spotted alfalfa aphid in California resulted in severe infestations of spider mites, the beet armyworm, the pea aphid, and the hitherto unimportant caterpillar *Platynotia stultana*.[13] The situation was corrected by substituting the systemic insecticide demeton, or the short-lived insecticides trichlorfon or mevinphos, which do not kill the parasites inside the alfalfa aphids. The effect of a moderately persistent OP compound, namely dimethoate, applied to a 22-acre clover field in southern Indiana was to cause a reduction in Hemiptera, Homoptera, and Orthoptera that lasted for at least 4 wk (Table 2.14); the reduction of the total insect population by one-third at 0.25 lb/a and by one-half at 0.5 lb/a significantly reduced the population of the house mouse *Mus musculus* which is omnivorous and normally uses insects for one-half of its food; on the other hand, the treatments increased the numbers of the prairie vole *Microtus ochrogaster*, which is a purely herbivorous species.[5]

E. SIDE EFFECTS ON BEES

The activities of *Apis mellifera* are necessary to ensure pollination and hence the growth of the fruit, especially in apple orchards. The danger of insecticides to bees comes not only from direct contact poisoning, but also from the taking of poisoned nectar and water, and the transport of poisoned nectar or pollen into the hive. Insecticidal dusts are more hazardous than sprays, and oil solutions or concentrates are more hazardous than emulsions or suspensions in water.

A half-century ago, arsenicals were particularly deleterious to the honeybee populations, the worst being the calcium arsenate then employed for aerial dusting. At that time, up to 1500 hives a year were poisoned in the Imperial Valley, California by off-target drift of dust applied to tomatoes and corn (which are not worked by bees) despite the fact that few of the hives were closer than 1 mile to the dusted fields.[28] The substitution of DDT for lead arsenate made orchards much safer for honeybees, the deposits being less toxic and also excitorepellent, so that the bees move on. It proved possible to treat alfalfa, and even apple, in full bloom with only slight mortality of bees. Forest spraying could be conducted with DDT at no risk to honeybees, and there was no hive poisoning with this organochlorine.

From assessments made in California (Table 2.15) it is seen that BHC and the cyclodiene derivatives are highly hazardous. So are most of the organophosphorus compounds,[3] including the azinphosmethyl (Guthion) widely used in orchards; even the nonpersistent tepp and mevinphos, and the

Table 2.15. Relative toxicity of insecticides to honeybees as determined by laboratory and field tests (Anderson and Atkins, 1968).

Highly Toxic	Moderately Toxic	Relatively Non-Toxic
Aldrin, BHC, dieldrin, heptachlor	Chlordane, DDT, endosulfan, endrin	Methoxychlor, TDE, toxaphene
_____	_____	Chlorbenside, chlorobenzilate, dicofol, fenson, ovex, tetradifon
Azinphosmethyl, chlorpyrifos, diazinon, dicapthon, dichlorvos, dicrotophos, dimethoate, famphur, fensulfothion, fenthion, malathion, methyl parathion, mevinphos, monocrotophos, phosmet, phosphamidon, tepp, thionazin	Carbophenothion, crotoxyphos, demeton, disulfoton, methyl demeton, oxydemetonmethyl, phorate, phosalone, ronnel, trichloronat	Dioxathion, ethion, menazon, trichlorfon, schradan
Aldicarb, aminocarb, carbaryl, isolan, methocarb, mexacarbate, propoxur	_____	_____
DNOC	_____	Allethrin, nicotine, pyrethrins, rotenone, ryania
Arsenicals	_____	Binapacryl, dinex
		Cryolite

systemics phosphamidon and dimethoate, are dangerous to bees. All of the carbamate insecticides involve high risk; carbaryl when applied on corn has been responsible for direct-contact and hive poisoning[61]; and when used on cotton in California to combat the invasion of pink bollworm into that state, carbaryl poisoned some 30,000 hives.[60] Malathion is safer than most OP compounds, but when applied undiluted in ULV (ultra low volume) applications on alfalfa it killed large numbers of honey bees.[52] The only low-risk OP compounds are trichlorfon, which is nonpersistent, and the systemic insecticide schradan. All the botanical insecticides offer little risk to bees, and all the chlorinated and other acaricides developed for orchard mite control are virtually nonhazardous. The hazards to honeybees have been included as one of four criteria in a list of pest management ratings for the common insecticides.[60]

The use of azinphosmethyl or carbaryl in orchards, or of carbaryl or methyl parathion on field crops, necessitates a delay in making applications until the crop has completed blooming. Toxaphene and DDT are the only insecticides safe to apply against the rape blossom beetle in Europe up to the time of blooming.[81] When aerial application are to be made, the hives in the vicinity should be covered with burlap at the time; by this method, the high mortality incurred in antimosquito aerial spraying with ULV malathion at 3 oz/a can be reduced to negligible proportions.[41] Spraying must be avoided within a quarter-mile of hives, unless the day temperature is below 15°C; if night spraying is resorted to, it must not be done on warm nights above 22°C.[3]

The alfalfa leafcutting bee, *Megachile rotundata*, responsible for pollination of seed alfalfa in western North America, can be poisoned by the leaves it collects for its nest as well as through residues in pollen and nectar; for this species carbaryl is less toxic than DDT.[87] A categorization of the insecticides according to hazard to wild bees[49] shows that *Megachile*, and also the alkali bee *Nomia melanderi*, are generally more susceptible than the honeybee. Bumblebees are apparently more resistant.

REFERENCES CITED

1. P. L. Adkisson. 1969. *How Insects Damage Crops*. Conn. Agric. Exp. Sta. Bull. 708. pp. 155–164.
2. M. K. Ahmed, L. D. Newsom, R. B. Emerson, and J. S. Roussel. 1954. The effects of Systox on some common predators of the cotton aphid. *J. Econ. Entomol.* **47**:445–449.
3. L. D. Anderson and E. L. Atkins. 1968. Pesticide usage in relation to beekeeping. *Ann. Rev. Entomol.* **13**:213–238.

4. G. W. Barrett. 1968. The effects of an acute insecticide stress on a semi-enclosed grass-land ecosystem. *Ecology* **49**:1019–1035.

5. G. W. Barrett and R. M. Darnell. 1967. Effects of dimethoate on small mammal populations. *Am. Midl. Nat.* **77**:164–175.

6. B. R. Bartlett. 1963. The contact toxicity of some pesticide residues to hymenopterous parasites and coccinellid predators. *J. Econ. Entomol.* **56**:694–698.

7. B. R. Bartlett and W. H. Ewart. 1951. Effect of parathion on parasites of *Coccus hesperidum. J. Econ. Entomol.* **44**:344–347.

8. B. R. Bartlett and C. F. Lagace. 1960. Interference with the biological control of cottony-cushion scale by insecticides. *J. Econ. Entomol.* **53**:1055–1058.

9. W. C. Batiste. 1972. Integrated control of codling moth on pears in California. *Environ. Entomol.* **1**:213–218.

10. I. J. Becnel, H. S. Mayeux, and J. S. Roussel. 1947. Insecticide tests for the control of cotton boll weevil and cotton aphids in 1946. *J. Econ. Entomol.* **40**:508–513.

11. A. F. H. Besemer. 1964. The available data on the effect of spray chemicals on useful arthropods in orchards. *Entomophaga* **9**:263–269.

12. R. van den Bosch. 1969. The toxicity problem: Comments by an applied insect ecologist. In *Chemical Fallout.* Ed. M. W. Miller and G. G. Berg. Charles C Thomas. pp. 97–109.

13. R. van den Bosch and V. M. Stern. 1962. The integration of chemical and biological control of arthropod pests. *Annu. Rev. Entomol.* **7**:367–400.

14. S. W. Bromley. 1948. Mosquito spraying with DDT and the two-spotted mite. *J. Econ. Entomol.* **41**:508.

15. A. W. A. Brown. 1951. Insecticides and the balance of animal populations. In *Insect Control by Chemicals.* Wiley. pp. 720–780.

16. V. M. Carolin and W. K. Coulter. 1971. Trends of western spruce budworm and associated insects in Pacific Northwest forests sprayed with DDT. *J. Econ. Entomol.* **64**:291–297.

17. J. R. Chiswell. 1962. Field comparisons of insecticides for control of the codling moth. *J. Hortic. Sci.* **37**:313–325.

18. D. W. Clancy and H. N. Pollard. 1948. Effect of DDT on several apple pests and their natural enemies. *J. Econ. Entomol.* **41**:507–508.

19. M. L. Cleveland and D. W. Hamilton. 1958. The insect fauna of apple trees in southern Indiana. *Proc. Indiana Acad. Sci.* **68**:205–217.

20. E. Collyer. 1953. Biology of some predatory insects and mites associated with *Metatetranychus ulmi* in south-eastern England. II and IV. *J. Hortic. Sci.* **28**:85–97, 246–259.

21. B. A. Croft. 1975. *Integrated Control of Apple Mites.* Mich. State Univ. Ext. Bull. E-825. 12 pp.

22. B. A. Croft and A. W. A. Brown. 1975. Responses of arthropod natural enemies to insecticides. *Annu. Rev. Entomol.* **20**:285–335.

23. P. DeBach. 1947. Predators, DDT, and citrus red mite populations. *J. Econ. Entomol.* **40**:598–599.

24. J. P. Dempster. 1967. A study of the effects of DDT applications against *Pieris rapae* on the crop fauna. *Proc. 4th Br. Insectic. Fungic. Conf.* **1**:19–25.

25. J. P. Dempster. 1975. Effects of organochlorine insecticides on animal populations. In *Organochlorine Insecticides: Persistent Organic Pollutants.* Ed. F. Moriarty. Academic. pp. 231–248.

26. R. S. Downing and R. K. Moilliet. 1972. Replacement of *Typhlodromus occidentalis* by *T. caudiglans* and *T. pyri* after cessation of sprays on apple trees. *Can. Entomol.* **104**:937–940.

27. B. F. Driggers and B. B. Pepper. 1936. Effect of orchard practices on codling moth and leafhopper parasitism. *J. Econ. Entomol.* **29**:477–480.

28. J. E. Eckert. 1944. The poisoning of bees, with methods of prevention. *J. Econ. Entomol.* **37**:551–552.

29. H. F. van Emden and G. F. Williams. 1974. Insect stability and diversity in agroecosystems. *Annu. Rev. Entomol.* **19**:455–475.

30. R. Freitag, G. W. Ozburn, and R. E. Leech. 1969. Effects of Sumithion and phosphamidon on populations of Carabid beetles and a spider. *Can. Entomol.* **101**:1328–1333.

31. R. Freitag and F. Poulter. 1970. Effects of Sumithion and phosphamidon on populations of Carabid beetles and Lycosid spiders. *Can. Entomol.* **102**:1307–1311.

32. W. W. Froggatt. 1905. The effects of fumigation with hydrocyanic gas upon ladybird beetle larvae and other parasites. *Agric. Gaz. N.S.W.* **16**:1088–1089.

33. E. H. Glass and P. J. Chapman. 1948. Red-banded leaf roller problem in New York. *J. Econ. Entomol.* **42**:29–35.

34. E. H. Glass and S. E. Lienk. 1971. Apple insect and mite populations developing after discontinuance of insecticides: 10-year record. *J. Econ. Entomol.* **64**:23–26.

35. E. Gould and E. O. Hamstead. 1948. Control of the red-banded leaf roller. *J. Econ. Entomol.* **41**:887–890.

36. D. C. Griffiths, F. Raw, and J. R. Lofty. 1967. Effect on soil fauna of insecticides tested against wireworms in wheat. *Ann. Appl. Biol.* **60**:479–490.

37. J. T. Griffiths and W. L. Thompson. 1947. The use of DDT on citrus trees in Florida. *J. Econ. Entomol.* **40**:386–388.

38. F. E. Guthrie, R. L. Rabb, T. G. Bowery, F. R. Lawson, and R. L. Baron. 1959. Control of hornworms and budworms on tobacco with reduced insecticide dosage. *Tob. Sci.* **3**:65–68.

39. P. C. Hely. 1945. DDT as an insecticide: Results of preliminary trials. *Agric. Gaz. N.S.W.* **56**:397–400.

40. A. Hikichi. 1964. The status of apple leaf rollers in Norfolk county, Ontario. *Proc. Entomol. Soc. Ont.* **94**:38–40.

41. E. F. Hill, D. A. Eliason, and J. W. Kilpatrick. 1971. Effects of ULV application of malathion on non-target animals. *J. Med. Entomol.* **8**:173–179.

42. C. H. Hoffman and E. P. Merkel. 1948. Fluctuations in insect populations associated with aerial applications of DDT to forests. *J. Econ. Entomol.* **41**:464–473.

43. C. H. Hoffmann, H. K. Townes, H. H. Swift, and R. I. Sailer. 1949. Field studies on the effects of airplane applications of DDT on forest invertebrates. *Ecol. Monogr.* **19**:1–46.

44. W. S. Hough, D. W. Clancy, and H. N. Pollard. 1945. DDT and its effect on the Comstock mealybug and its parasites. *J. Econ. Entomol.* **38**:422–425.

45. S. C. Hoyt. 1969. Integrated chemical control of insects and biological control of mites on apple in Washington. *J. Econ. Entomol.* **62**:74–86.

46. S. C. Hoyt. 1970. The developing program of integrated control of pests of apple in

Washington. *A.A.A.S. Symposium on Theory and Practice of Biological Control.*
Boston, Dec. 30–31, 1969.

47. S. C. Hoyt and E. C. Burts. 1974. Integrated control of fruit pests. *Annu. Rev.
Entomol.* **19:**231–252.

48. H. A. Jaynes and P. E. Marucci. 1947. Effect of artificial control practices on the
parasites and predators of the codling moth. *J. Econ. Entomol.* **40:**9–25.

49. C. A. Johansen and J. Eves. 1967. *Toxicity of Insecticides to the Alkali Bee and the
Alfalfa Leafcutting Bee.* Wash. Agric. Exp. Sta. Circ. 475. 15 pp.

50. E. C. Klostermeyer and W. B. Rasmussen. 1953. The effect of soil insecticide treatment
on mite population and damage. *J. Econ. Entomol.* **46:**910–912.

51. C. B. Knisley and F. C. Swift. 1972. Qualitative study of mite fauna associated with
apple foliage in New Jersey. *J. Econ. Entomol.* **65:**445–448.

52. M. D. Levin, W. B. Forsyth, G. L. Fairbrother, and F. B. Skinner. 1968. Impact on
colonies of honey bees of ultra-low-volume (undiluted) malathion applied for control of
grasshoppers. *J. Econ. Entomol.* **61:**58–62.

53. N. C. Lloyd, P. C. Hely, A. H. Friend, and G. Pasfield. 1945. DDT as an insecticide:
Results of preliminary trials. *Agric. Gaz. N.S.W.* **56:**347–348, 397–400, 455–456, 467.

54. F. T. Lord. 1949. The influence of spray programs on the fauna of apple orchards in
Nova Scotia. III. Mites and their predators. *Can Entomol.* **81:**202–214, 217–230.

55. W. H. Luckmann. 1960. Increase of European corn borers following soil application of
large amounts of dieldrin. *J. Econ. Entomol.* **53:**582–584.

56. D. R. Macdonald. 1959. Biological assessment of aerial forest spraying against spruce
budworm in New Brunswick. III. Effects on two overwintering parasites. *Can Entomol.*
91:330–336.

57. D. R. Macdonald and F. E. Webb. 1963. Insecticides and the spruce budworm. In *The
Dynamics of Epidemic Spruce Budworm Populations.* Ed. R. F. Morris. Mem.
Entomol. Soc. Can. No 31. pp. 288–310.

58. A. W. MacPhee and K. H. Sanford. 1954–1961. The influence of spray programs on the
fauna of apple orchards in Nova Scotia. *Can Entomol.* **86:**128–135; **88:**631–639;
93:671–673.

59. C. R. Malone, A. G. Winnett, and K. Helrich. 1967. Insecticide-induced responses in an
old field ecosystem: Persistence of diazinon in the soil. *Bull. Environ. Contam. Toxicol.*
2:83–89.

60. R. L. Metcalf. 1975. Insecticides in pest management. In *Introduction to Insect Pest
Management.* Ed. R. L. Metcalf and W. Luckmann. Wiley. pp. 235–273.

61. J. O. Moffett, R. H. Macdonald, and M. D. Levin. 1970. Toxicity of carbaryl-
contaminated pollen to adult honey bees. *J. Econ. Entomol.* **63:**475–476.

62. A. W. Morrill and F. Q. Otanes. 1947. DDT emulsion to control mealy-bugs and scale.
J. Econ. Entomol. **40:**599–600.

63. M. G. Morris. 1968. The effect of sprays on the fauna of apple trees. V. DDT/BHC and
lead arsenate/nicotine applied at the green cluster stage. *J. Appl. Ecol.* **5:**409–429.

64. R. C. Muri. 1964. *The Influence of Certain Fungicides and Insecticides on* Panonychus
ulmi *and Its Predator.* Rep. East Malling Res. Sta. 1964. pp. 167–170.

65. E. J. Newcomer and F. P. Dean. 1953. Control of woolly apple aphid in orchards
sprayed with DDT. *J. Econ. Entomol.* **46:**54–56.

66. E. J. Newcomer, F. P. Dean, and F. W. Carlson. 1946. Effect of DDT, xanthone, and nicotine bentonite on the woolly apple aphid. *J. Econ. Entomol.* **39**:674–676.

67. L. D. Newsom. 1967. Consequences of insecticide use on nontarget organisms. *Annu. Rev. Entomol.* **12**:257–286.

68. E. R. Oatman, E. F. Legner, and R. F. Brooks. 1964. An ecological study of arthropod populations on apple in northeastern Wisconsin. *J. Econ. Entomol.* **57**:978–983.

69. G. T. O'Loughlin. 1948. Codling moth control. *J. Agric. Victoria (Melbourne).* **46**:442–444.

70. T. J. Peterle and R. H. Giles. 1964. *New Tracer Techniques for Evaluating the Effects of an Insecticide on the Ecology of a Forest Fauna.* Ohio State Univ. Res. Found. Rep. 1207 (to USAEC). 435 pp.

71. A. D. Pickett. 1962. Pesticides and the biological control of arthropod pests. *World Rev. Pest Control* **1**:19–25.

72. D. Pimentel. 1961. An ecological approach to the insecticide problem. *J. Econ. Entomol.* **54**:108–114.

73. D. Pimentel. 1961. Competition and the species-per-genus structure of communities. *Ann. Entomol. Soc. Am.* **54**:323–333.

74. H. T. Reynolds. 1971. Recent developments with systemic insecticides for insect control on cotton. *Summary Proc. of the West. Cotton Prod. Conf.* pp. 18–20.

75. H. T. Reynolds, P. L. Adkisson and R. F. Smith. 1975. Cotton insect pest management. In *Introduction to Insect Pest Management.* Ed. R. L. Metcalf and W. Luckmann. Wiley. pp. 379–443.

76. R. W. Rings and C. R. Weaver. 1948. Effects of benzene hexachloride and DDT upon parasitization of the oriental fruit moth. *J. Econ. Entomol.* **41**:566–569.

77. W. E. Ripper. 1944. Biological control as a supplement to chemical control of insect pests. *Nature* **153**:448.

78. W. E. Ripper. 1956. Effect of pesticides on balance of arthropod populations. *Annu. Rev. Entomol.* **1**:403–438.

79. K. H. Sanford. 1967. The influence of spray programs on the fauna of apple orchards in Nova Scotia. XVII. Effects on some predacious mites. *Can. Entomol.* **99**:197–201.

80. K. H. Sanford and H. J. Herbert. 1967. Ibid. XVIII. Predator and prey populations in relation to miticides. *Can. Entomol.* **99**:689–696.

81. F. Schneider. 1966. Some pesticide–wildlife problems in Switzerland. *J. Appl. Ecol.* **3**(Suppl.):15–20.

82. R. F. Smith. 1970. Pesticides: Their use and limitations in pest management. In *Concepts of Pest Management.* N.C. State Univ. pp. 103–118.'

83. F. W. Stehr. 1975. Parasitoids and predators in pest management. In *Introduction to Insect Pest Management.* Ed. R. L. Metcalf and W. Luckmann. Wiley. pp. 147–188.

84. A. Steiner and S. A. Summerland. 1944. Laboratory and field tests of DDT for control of the codling moth. *J. Econ. Entomol.* **37**:156–157.

85. M. van de Vrie. 1967. The effect of some pesticides on the predatory bugs *Anthocoris* and *Orius,* and the woolly aphid parasite *Aphelinus mali. Entomophaga, Mem. hors Ser.* **3**:95–101.

86. M. A. Vaughan and G. Leon. 1976. Pesticide management on a major crop with severe resistance problems. *Trans. XVth Internat. Congr. Entomol.* pp. 812–815.

87. G. D. Waller. 1969. Susceptibility of an alfalfa leaf-cutting bee to residues of insecticides on foliage. *J. Econ. Entomol.* **62**:189–192.

88. R. W. Weires and H. C. Chiang. 1973. *Integrated Control Prospects of Major Cabbage Insect Pests in Minnesota.* Univ. Minn. Agric. Exp. Sta. Tech. Bull. 291. 42 pp.

89. R. S. Woglum, J. R. LaFollette, W. E. Landon, and H. C. Lewis. 1947. The effect of field applied insecticides on beneficial insects of citrus in California. *J. Econ. Entomol.* **40**:818–820.

3

INSECTICIDES AND SOIL INVERTEBRATES

A. EFFECTS OF INSECTICIDES ON SOIL MICROARTHROPODS

Introductory

Although the microflora is responsible for about 90% of the biological respiration activity in the soil,[69] the microarthropods along with earthworms play an important role in comminuting dead plant material and making it the more available for bacterial and fungal breakdown. When the microarthropods were eliminated by naphthalene fumigation, the breakdown of leaf litter was reduced by 25%.[101] Even nonpersistent soil insecticides such as the OP compound phorate[95] and the carbamate compound carbaryl[4] have slowed litter breakdown by as much as 15%, through their suppressive effect on Collembola and saprophagous Acarina. An application of the organochlorine isobenzan (Telodrin) to a New Zealand pasture at 2 lb/a for grass-grub control reduced the microarthropod fauna so severely that the undecomposed plant debris formed a mat on the surface, while the underlying soil became nonporous and dense; this effect lasted for at least 3 yr, by which time 90% of the pasture grass was rooted only in the mat of surface debris.[53]

In addition to insecticide applications aimed directly at soil pests, the treatments applied to annual and perennial crops are also sources of soil contamination. For DDT sprays, 33% of that applied to a field of Brussels sprouts[15] and 30% of that applied to a pea crop[14] is deposited on the ground. Fully 80% of the insecticide applied to apple trees in English orchards is deposited on the ground cover.[84] When lead arsenate was applied to pome orchards, the soil arthropod population remained high even where the average of four sprays per year involved a total input of 20–120 lb/a. The substitution of the organochlorine insecticides for lead arsenate resulted in decreases in the soil microarthropod population, as the following posttreatment census numbers[37] demonstrate:

Pb Arsenate, 120 lb/a	4171	Chlordane, 100 lb/a	37
DDT, 80 lb/a	778	Parathion, 20 lb/a	4403
Technical BHC, 80 lb/a	156	Untreated	1100

Thus the substitution of DDT was slightly deleterious, but with chlordane and BHC (13% gamma) the situation was considerably worse, apart from the control of subterranean woolly apple aphids by the BHC.

The microarthropods have their own balance of populations, the predaceous mites in the order Mesostigmata (especially the Rhodacaridae) feeding on the eggs of Collembola and the early stages of the Oribatid mites which break down the plant debris.[54] A typical inhabitant of northern temperate soils is *Folsomia candida,* a springtail in the family Isotomidae. In contrast to its high susceptibility to OP compounds such as thionazin (Zinophos), dyfonate and phorate, and to carbamates such as methomyl and aldicarb (Table 3.1), *F. candida* is completedly resistant to DDT.[88] This *Folsomia* can grow on a yeast powder containing 15% DDT—in fact the greater the DDT contamination the more eggs it lays[9]—while it detoxifies the insecticide to DDE.[10]

Organochlorine Insecticides

DDT.   Populations of soil Collembola in general are not reduced by DDT until the application level is raised to 50 lb/a,[42] when it is the Sminthurids and Podurids that suffer.[56] While the saprophagous mites are also tolerant of DDT, the predaceous Mesostigmata, of which *Rhodacarellus silesiacus* is an important example, are susceptible. Thus when DDT was tilled into

Table 3.1.   **Toxicity of insecticides in sandy soil to the Isotomid springtail *Folsomia candida*: grouped, and ranked within each group, in the order of descending toxicity (Thompson and Gore, 1972).**

< 0.04 ppm	0.04 - 0.3 ppm	0.3 - 2.0 ppm	> 2.0 ppm
Thionazin	Disulfoton	Lindane	Carbaryl
Isobenzan	Diazinon	Methyl parathion	Heptachlor
Dyfonate	Carbofuran	Dieldrin	Mexacarbate
Phorate	Methomyl	Fenitrothion	Chlordane
Endrin	Aldicarb	Aldrin	Trichloronat
Chlorpyrifos	Fensulfothion	Chlorfenvinphos	DDT
Parathion	Bux	Azinphosmethyl	

arable soil at 10 lb/a there was an almost twofold increase in the population of Isotomid springtails (Table 3.2) consequent on a threefold reduction in the Mesostigmatid mites,[25] the effect being greatest 7 months after the application.[20] A threefold increase was shown in the Arthropleona (which include the Isotomids) in grassland treated with DDT at 3 lb/a (Table 3.3), and this augmented level persisted into the following year.[34] When a deciduous forest was sprayed with DDT at 4 lb/a, there was an approximately 50% increase in the Collembola population in the leaf litter and humus, all the other forest-floor groups (including the mites) remaining apparently unaffected.[45]

Technical BHC.   Where DDT at the same dosage induced large increases of *Onychiurus* and *Tullbergia,* BHC (13% gamma isomer) applied to pasture at 13 kg/ha reduced the Collembola population by two-thirds; the reduction in the predaceous Mesostigmata was one-third while with DDT it was one-half.[77] Applied at 4 kg/ha in Russia, technical BHC decreased the phytophagous and predaceous microfauna but increased the saprophagous species.[39] In Australia, BHC (13% gamma) at 1.8 lb/a provided the most complete control of the lucerne flea *Sminthurus viridis,* and did not reduce its mite predator *Biscirus* by more than one-third.[93] Lindane, the gamma isomer of BHC (HCH), applied to potato fields in northern Germany at 0.09 kg/ha, reduced springtail populations where DDT at 0.6 kg/ha had no effect on them.[79] The reductions in certain species of Collembola and Acarina induced by lindane at 2.5 kg/ha still persisted 3 yr later.[51] The extensive literature on the effects of DDT and HCH products in the soil has been reviewed by Satchell (1955), Bauer (1964), Edwards and Thompson (1973), and Thompson and Edwards (1974).

Table 3.2.   Effects of DDT and aldrin applied to arable land in Somerset, U.K.: total numbers found in core samples (Edwards, Dennis, and Empson, 1967).

		DDT, 10 lb/a	Aldrin, 4 lb/a	Untreated
Collembola	Onychiuridae	1307	1131	1338
	Isotomidae	571	134	331
	Entomobryidae	174	17	154
	Sminthuridae	34	13	40
Acarina	Mesostigmata	85	156	236
	Astigmata	19	15	48
	Prostigmata	96	32	173

Table 3.3. Effects of DDT and aldrin applied to Nova Scotia grassland in
May, 1957: post-application numbers obtained in eight cores
2.5 inches diameter, 3 inches deep (Fox, 1967).

	DDT, 3 lb/a		Aldrin, 6 lb/a		Control	
	1957	1958	1957	1958	1957	1958
Arthropleona*	147	158	52	32	58	61
Symphypleona†	3.0	4.9	2.0	4.3	n.d.	1.8
Mites	102	90	80	53	n.d.	33

* *Entomobrya, Isotoma* (2 spp.), *Lepidocyrtus* (3 spp.),

  Onychiurus, Orchesella (3 spp.)

† *Sminthurus* (2 supp.), *Bourletiella* (2 spp.)

Cyclodiene Insecticides. Organochlorines in this group have been widely
employed for soil applications against wireworms, rootworms, and root
maggots. In contrast to DDT, the cyclodiene insecticides reduce Collembola and predaceous mites alike.[19] Aldrin tilled into the soil at 4 or 6 lb/a
reduced the populations of all microarthropods (Tables 3.2, 3.3), although
the species to be killed in greatest biomasses were the pests such as
wireworms and *Tipula* larvae.[25] Heptachlor applied to arable land at 2–4
lb/a caused similar reductions among the springtails and mites.[17] After
grassland plots had been treated with aldrin, dieldrin, or heptachlor at 6 lb/
a, the populations of Arthropleona springtails were reduced during the year
following the application, but not thereafter.[34] Endrin at 4–8 lb/a had a
similar reducing effect on springtails and mites.[24] Chlordane, which in
orchard soil had caused such large reductions at an aggregate dosage of 100
lb/a,[37] did not reduce the populations of Collembola or mites in sugarcane
fields treated at 2 lb/a, the uncommon Pauropoda and Diplura being the
only groups to suffer.[64]

Organophosphorus Insecticides

Applications of the nonpersistent organophosphorus and carbamate compounds to soil have the same selective effect as DDT of reducing predaceous mites and encouraging Arthropleona springtails.[17] While the
parasitic mites are somewhat more susceptible to OP compounds than to
DDT, the surface-living springtails are even less susceptible to OPs than to
DDT.[29]

Parathion applied to a Virginia apple orchard in amounts aggregating 20
lb/a over the season induced a fourfold increase in the microarthropods in
the soil.[37] Parathion treatment of citrus orchards in Transvaal resulted in

the disappearance of 10 of the 28 species of mites previously present; all the predaceous Trombidiform mites disappeared, while the springtail *Isotoma thermophila* became four times more abundant than before.[67]

Diazinon was not only the first OP compound to be widely used as a soil insecticide, but also had the strongest effect on soil arthropods,[18] parathion excepted (Table 3.4); a consequence of its reducing the Trombidiform mites was that it doubled or tripled the populations of Oribatid mites and deep-living Isotomid springtails.[29]

Of the systemic OP insecticides, disulfoton (Di-Syston) applied from granules at 2 lb/a in cottonfield furrows to control aphids resulted in a 95% reduction of Collembola and mites, which returned almost to the normal population level in the third month after application.[1] Phorate and demeton reduced springtail populations when applied at 1 lb/a, in contrast to diazinon or malathion where the reduction occurred only when dosages reached 25 lb/a. Phorate applied broadcast at 1 lb/a from granules suppressed soil Collembola for at least 4 months, by which time the multiplication of Enchytraeid worms has more than made up for the resultant shortfall in plant debris breakdown.[95] Menazon, a less toxic aphicide, caused a transient 2-wk reduction in Onychiurid springtails when applied at 2 lb/a from granules, and Podurids were not affected until the dosage was raised

Table 3.4.   Effects of OP insecticides applied at 8 lb/a to heavy clay loam in Kent, U.K.; numbers of arthropods in percent of the control plot at postapplication peak population time (Edwards, Thompson and Lofty, 1967).

	Parathion	Diazinon	Disulfoton	Phorate
Surface-dwelling Springtails*	49	53	34	43
Deep-living Springtails†	34	194	171	29
Parasitic Mites	32	42	60	51
Trombidiform Mites	13	19	65	50
Oribatid Mites	42	262	137	31
Pauropods	2	4	7	37
Symphylids	50	50	30	40
Fly larvae	8	7	33	31
Beetles	5	36	44	40

* e.g. Entomobrya, Orchesella, Tomocerus   †Folsomia, Tullbergia

Table 3.5.   Effects of OP insecticides applied to English soil at 1.5 lb/a: numbers per 4-inch diameter core, 8 inches deep (Griffiths, Raw, and Lofty, 1967).

		Control	Fenitrothion	Thionazin	Trichloronat	Fonofos	Aldrin*
M	Oribatidae	0.59	0.59	0.62	0.61	0.60	0.55
	Rhodacaridae	0.57	0.51	0.50	0.40	0.44	0.48
	Parasitidae	1.24	1.15	1.15	0.82	1.25	1.11
C	Onychiuridae	1.78	1.84	1.76	1.91	1.69	1.84
	Poduridae	0.51	0.41	0.48	0.73	0.51	0.47
	Isotomidae	0.73	0.67	0.81	0.63	0.65	0.49

* 2.25 lb/a   M Mites   C Collembola

to 8 lb/a.[72] Fensulfothion (Dasanit) applied at 3 lb/a caused a reduction in Collembola that persisted for at least 4 months.[41]

Among other OP soil insecticides, thionazin (Zinophos), which was the most toxic of the compounds tested against *Folsomia* springtails (see Table 3.1), left even higher populations of Isotomids 1 yr after a 1.5-lb/a application than trichloronat, one of the least toxic to *Folsomia*. At this application rate, thionazin stimulated the population of the predaceous·Rhodacarid mites while decreasing the Parasitid mites.[38] Fenitrothion and fonofos did not change the numbers of Collembola or mites, while trichloronat only reduced the Parasitid mites (Table 3.5). However, one of the applications of chlorfenvinphos (Birlane) applied to field crops at 4 lb/a reduced both predaceous and saprophagous mites, and stimulated the Collembolan populations.[28] Dyfonate and Tetrachlorvinphos (Gardona) at 8 kg/ha did not suppress soil microarthropods, with the exception of Prostigmata mites.[27] Fenthion applied at normal dosages against soil arthropod pests was particularly toxic to *Folsomia* and *Isotoma*.[100]

Malathion, which degrades too fast to be useful as a soil insecticide, proved harmless to Collembola when applied at 1.5 kg/ha.[92] Trichlorfon (Dipterex), another labile OP compound, is harmless to soil mites and actually stimulates springtail populations.[52]

Carbamate Insecticides

Carbaryl applied at 5 kg/ha to control *Ixodes* ticks in cleared forest reduced Oribatid mites by 25%.[92] Applied to hardwood stands at 10 lb/a, carbaryl reduced forest-floor populations of Collembola and mites by 90%, although the mites rapidly recovered their numbers; the reduction was 50% at 1.25 lb/a, the usual dosage for gypsy moth control.[78] When carbaryl had been applied to grassland at 2 lb/a, the rate of plant debris degradation was reduced by about one-sixth.[4] Aldicarb at 10 lb/a reduced Collembola and mites by 60%, being particularly hard on the mesostigmatid predaceous Gamasina as well as the Oribatids; methomyl at this dosage was less destructive, and did not affect the Gamasina mites.[22]·The effects of OP compounds and carbaryl on soil arthropods have been reviewed in detail by Edwards and Thompson (1973) and Thompson and Edwards (1974).

B. EFFECTS OF INSECTICIDES ON SOIL MACROARTHROPODS

Insecta

The principal predators in the soil are the Carabid ground beetles, carnivorous in both adult and larval stages; the larvae are found well in the soil, while the adults range over the surface and the vegetation. Staphylinid rove beetles are also important carnivores, while ants and Cicindelid tiger beetles play a part in soil predation.

Organochlorine Insecticides. The overall effect of organochlorine insecticides (e.g., DDT, lindane, dieldrin) applied to cropland is a net reduction in the Carabid population and a change in the species composition. Sites that had received applications of organochlorines for two or more consecutive years had about one-fifth as many ground beetles as sites treated once or untreated, and *Harpalus rufipes* was replaced by *Feronia melanaria* as the predominant species in the more heavily treated sites.[13]

DDT applied at the excessively high level of 25 lb/a to pasture land in North Carolina practically eliminated the Carabids, while exerting little permanent effect on Staphylinids and ants.[32] Applications of DDT at 2 lb/a decreased the numbers of Carabids and Staphylinids in cropfields[24] and almost completely eliminated several species of Carabids in deciduous forest.[45] Repeated applications (2–6 per annum) of DDT at 0.7 lb/a to a Brussels sprouts crop in England over a period of 3 yr induced a reduction in the number of *Bembidion,* and in the growth rate of *Harpalus* ground beetles, that was partly compensated by immigration of adults into the quarter-acre field; *Trechus quadristriatus* and *Nebria brevicollis* actually increased their numbers, probably because the springtail populations on which they feed had been increased sixfold by the DDT treatments.[15] In Ontario peach orchards receiving DDT sprays amounting to 5 lb/a each year between 1950 and 1955, the Carabid populations were reduced only in the case of *Pterostichus* (Table 3.6); the other species maintained their numbers, while *Anadaptus* and *Anisodactylus* actually increased as compared to orchards to which DDT had not been applied during the 6-yr period.[43] In deciduous forest, although *Pterostichus* ground beetles were unaffected, *Chlaenius aestivus* was greatly reduced by DDT at 2 lb/a in bottomland, and *Carabus limbatus* was virtually eliminated by 4 lb/a applied to oak-maple stands. Two species of ants, *Tapinoma sessile* and *Prenolepis imparis,* and the cricket *Ceuthophilus,* also suffered population reductions.[45]

Technical BHC (10% gamma content) applied from a dust at 5 lb/a to control grasshoppers in alfalfa fields eliminated Carabids, as well as ants and Cicindelids.[7] Treatment of soil plots with BHC at 5 kg/ha reduced the Carabid population to 40% of that in an untreated control; Staphylinids did not suffer until the dosage was raised to 30 kg/ha.[39] At the dosages required for wireworm control, lindane is less destructive to Carabids and Staphylinids than aldrin or chlordane,[33] nor is it as toxic as parathion.[96]

Aldrin or dieldrin applied at 1 lb/a reduced populations of Carabids in a cole crop, but the activating effect of the insecticide stimulated the numbers trapped; once the residues had decayed to below 0.5 ppm an actual increase in the census of *Bembidion* and *Trechus* was noted in the year after the treatment. The numbers of *Harpalus* and *Feronia* were largely unaffected, but the abundance of the parasitic Staphylinid *Aleochara* was decreased.[11] Dieldrin applied to Illinois cropland at 2–3 lb/a to control Japanese beetle reduced the Carabid population to zero, and it had not fully recovered 3 yr later.[65] Where *Harpalus rufipes* and *Feronia melanaria* were actual pests by eating strawberry fruits, complete control was achieved by dieldrin at 5 lb/a

Table 3.6. **Effects of DDT treatments on Carabid populations in peach orchards: comparison of acre plots where DDT had been discontinued in 1950 with those where DDT was applied annually until 1955 (Herne, 1963).**

	1954 Trap census		1955 Trap census	
	Untreated	Treated	Untreated	Treated
Pterostichus melanarius	152	22	102	19
Harpalus pennsylvanicus	75	59	181	144
Agonoderus lecontei	25	27	34	28
Amara pennsylvanica	21	25	23	20
Anadaptus sanctaecrucis	13	26	17	23
Anisodactylus rusticus	7	16	5	25

or aldrin at 10 lb/a.[8] Heptachlor applied for fire ant control at 1.25 lb/a reduced the population of the valuable predator *Calosoma calidum* by 70%.[73]

DDT residues are accumulated by ground beetles (Table 3.7), and even by the seed-eating species *Harpalus pennsylvanicus*.[76] The Carabids in English fields began to show DDE as a metabolite when 12 hr had elapsed after the application;[14] DDD appeared in their bodies some 4 days later, but did not exceed the concentrations found for *o,p′*-DDT, the principal impurity in technical DDT (Table 3.8). In cornfields treated with aldrin, the dieldrin residues in the beetles came to exceed the dieldrin concentrations in the soil by at least four times.[57] When the organochlorine residues found in *Harpalus rufipes* are plotted against those in the treated soils they inhabited (Fig. 3.1), it is seen that in this species the concentration factor is only 0.5 times.[13]

Organophosphorus and Carbamate Insecticides. Parathion at 7 lb/a is highly destructive to the larvae of Carabids.[96] Diazinon and disulfoton at 8 lb/a, and phorate at 4 lb/a, reduced the beetle population in the soil by more than 50% (Table 3.4), but parathion gave the severest reduction.[29] Fenitrothion, on the other hand, resembled thionazin and fonofos in not reducing any Carabid species except *Feronia madida* (Table 3.9), a species which particularly suffered from aldrin treatments also.[38] In laboratory tests on six species of Carabids in treated soil, the following order of LC_{50} levels was found[66]: diazinon < thionazin < dieldrin < azinphosmethyl < chlorfenvinphos. When chlorfenvinphos (Birlane) was applied at 8 lb/a in the field,

Table 3.7. DDT residues found in Carabid beetles inhabiting cropland: ppm wet weight (Davis and French, 1969; El Sayed et al., 1967).

	DDT	DDE	Dosage Rate	Days after Application	Crop	Place
Feronia melanaria	1.36	0.53	2.5 lb/a	8	Peas	England
F. madida	4.3	0.84	"	"	"	"
Nebria brevicollis	0.21	66.0	1 lb/a	6	Strawberries	"
F. melanaria	0.16	0.29	"	56	Bruss. Sprouts	"
Harpalus rufipes	0.13	0.78	"	"	"	"
H. pennsylvanicus	0.10	5.25			Cotton	Louisiana
Calosoma alternous	4.34	2.93	(no data on treatments and collection times)			
Agonoderus lectonei	0.0	0.7	("	"	"	")

Table 3.8. **Residues of organochlorine insecticides found in Carabid beetles: ppm wet weight (Davis, 1968; Korschgen, 1970).**

	DDT	DDE	t-DDT	Aldrin	Dieldrin
English wheatfield treated with dieldrin at 2 oz/acre, and with DDT at 1 lb/a the previous year					
Harpalus rufipes	0	2.2	---	---	0.09
Agonum dorsale	0	0.06	---	---	0.06
Soil	0.3	0.1	---	---	0.1
Missouri cornfields treated for 16 years with aldrin at 1.5 lb/a/yr					
Harpalus pennsylvanicus	---	---	0.26	0.11	0.99
Poecilus chalcites	---	---	0.26	0.34	9.33
Soil	---	---	0.26	0.06	0.25

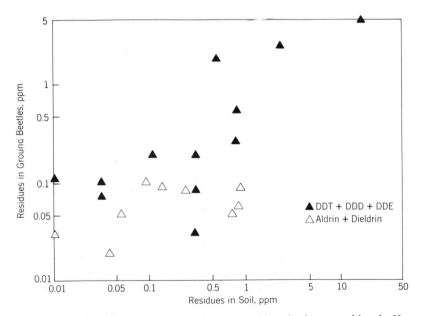

Fig. 3.1. Scatter diagram of organochlorine residues in the ground beetle *Harpalus rufipes* plotted against soil residues (after Davis, 1968).
Reprinted with permission from the Annals of Applied Biology 61:29–45.

73

Table 3.9.   Trap census of Carabids 1 yr after application of OP insecticides to soil at 1.5 lb/a (Griffiths, Raw, and Lofty, 1967).

	Control	Fenitrothion	Thionazin	Fonofos	Aldrin*
Feronia melanaria	188	209	189	235	188
Feronia madida	136	97	75	84	136
Bembidion sp.	40	22	20	22	40
Agonum dorsale	28	28	25	18	28
Harpalus rufipes	20	17	29	21	20

*2.25 lb/acre

the result was an increase in the trap census of Carabids[26]; this compound is 10 times less toxic to *Bembidion* than fensulfothion (Dasanit).[41] Malathion applied to forest clearings at 1.5 lb/a also stimulated an increase in the Carabid population.[92] Residues of OP compounds have been found in ground beetles, applications at 9 kg/ha leading to accumulations ranging from almost zero to 0.28 ppm with phorate, up to 0.55 ppm with diazinon, and up to 1.33 ppm with chlorfenvinphos.[24]

When carbaryl was applied to cleared forestland in the U.S.S.R. for tick control at 5 kg/ha, the population of *Carabus* and *Pterostichus* ground beetles was reduced by 90% for a full year.[92]

Other Soil Arthropods

Myriapoda.   Among the myriapods, the centipedes that inhabit the soil are useful as predators of pest insects. Although DDT applied at 10 lb/a reduced their numbers when well mixed into the soil,[25] surface applications at 25 lb/a applied to pastures for Japanese beetle control did not noticeably affect the centipede population,[32] nor did DDT sprayed at 2 lb/a onto deciduous forest.[45] Aldrin (Table 3.10) decisively reduced centipede numbers when applied at 4 lb/a to arable soil.[25] Parathion, diazinon, trichlorfon, and disulfoton at 8 lb/a decreased centipede populations to some extent; at 4 lb/a phorate and thionazin were strongly suppressive while chlorfenvinphos was harmless.[28] Malathion applied at 3 kg/ha from a dust to cleared woodland promoted an increase in the centipede population, whereas carbaryl at 5 kg/ha reduced *Lithobius* and *Geophilus* by nearly 50%, although they recovered their numbers a year later[92]; this carbamate insecticide strongly suppressed centipedes in arable soil.[25]

Symphyla. These are rangy centipedclike myriapods that feed on plant rootlets. *Scutigerella immaculata,* the so-called garden centipede of symphylid, is a pest of vegetable crops particularly in Washington State. Of the organochlorines, DDT and the cyclodiene derivatives are ineffective to control it at 10 lb/a,[48] although DDT at this dosage and aldrin at 4 lb/a have markedly suppressed symphylids in English arable fields.[20] Lindane at 4 lb/a is highly effective for controlling *Scutigerella.*[47] Of the OP insecticides, parathion at 5 lb/a is the most powerful control agent[49]; phorate and dichlofenthion (VC-13) are equally lethal but lack any residual effect. Carbophenothion, ethion, ronnel, and dioxathion are toxic to this symphylid, while diazinon, trichlorfon, disulfoton, phosphamidon, dimethoate, malathion, and azinphosmethyl form a series of decreasing activity.[47]

Diplopoda. The millipedes are the most abundant of the myriapods in damp soil, and some of them are pests of roots or tubers; they constitute an important element of the invertebrate biomass in soils.[18] Although their numbers were not noticeably reduced by DDT applied at 2 lb/a to deciduous forest[45] or even at 25 lb/a to pastures,[32] multiple annual applications at 0.7 lb/a against the cabbageworm progressively decreased the millipede population as DDT accumulated in the soil over a period of 3 yr.[15] Soil applications of DDT can give lasting control of the glasshouse millipede *Oxidus gracilis*[21]; so also can lindane, which at agricultural dosage rates caused 50% reductions in the diplopod population.[96] Lindane applied to the soil at 1 lb/a gave control of *Blaniulus guttulatus,* a pest of potato tubers in Washington State, although this millipede recovered its abundance in the following year.[60] Aldrin, like dieldrin, was toxic to *Oxidus,*[21] and reduced the population of Julid millipedes when applied to arable soil at 2 lb/a.[38]

The OP compound parathion is toxic to *Oxidus*[21] and to diplopods in arable soil treated at 8 kg/ha.[96] Diazinon is less toxic to *Oxidus* and mevin-

Table 3.10. **Effect of DDT and aldrin on Myriapoda in arable land: total numbers found in core samples (Edwards, Dennis, and Empson, 1967).**

	Untreated	DDT, 10 lb/a	Aldrin 4 lb/a
Diplopoda	139	35	19
Symphyla	54	18	15
Myriapoda	77	0	14

phos even less so,[24] while fenitrothion, thionazin, and fonofos at 1.5 lb/a are harmless to millipedes in the treated soil.[38] Malathion applied at 3 kg/ha to cleared forest slightly reduced the millipede population, while carbaryl at 5 kg/ha caused a 75% reduction, the numbers having not completely recovered 14 months later.[92]

Pauropoda.   These minute myriapods feed on fungi and plant debris; they are usually scarce but sometimes very abundant.[54] Reductions in their numbers have been observed after treatments with DDT at 10 lb/a[25] and chlordane at 2 lb/a,[64] and severe reductions have followed aldrin treatments at 4 lb/a.[25] Parathion, diazinon, and disulfoton applied at 8 lb/a, or phorate and chlorfenvinphos applied at 4 lb/a, virtually eliminated the soil population of pauropods.[28,29]

Arachnida.   The predaceous spiders (Araneida) and harvestmen (Phalangida) inhabit the surface litter rather than the soil itself. Phalangida are particularly sensitive to DDT, the species of *Leiobunum* inhabiting deciduous forest being almost eliminated at 2 lb/a; by contrast the spiders did not suffer until the DDT dosage was raised to 4 lb/a.[44] DDT applied several times a year at 0.7 lb/a to cabbage patches in England suppressed the spiders and *Phalangium opilio,* but the populations returned to about one-half their former level by recolonization from outside.[15] Applications of technical BHC to control the green rice leafhopper resulted in population reductions of the spider *Lycosa pseudoannulata* that preys on it, due to secondary poisoning because the leafhopper concentrates the gamma isomer it ingested from the sprayed rice plants.[55] Although aldrin had the effect of increasing the trap census of spiders in the soil,[38] heptachlor applied over wide areas for fire ant control resulted in a 75% reduction of Lycosid populations,[73] probably due to secondary poisoning. Applied to soil at 1.5 lb/a, fenitrothion, thionazin, fonofos, and trichloronat had no effect on spider populations, the first two compounds actually increasing the trap catches.[38] Carbaryl significantly reduced the spider population when applied to cleared forest at 5 kg/ha.[92]

C.   EFFECTS OF INSECTICIDES ON EARTHWORMS, SLUGS, AND NEMATODES

In temperate countries, earthworms pass through their gut and eject as surface castings about 10 tons of soil per acre each year.[54] Beneath a square meter of European grassland there may be as many as 2000 earthworms and 20,000 enchytraeids, as compared to 2000 dipterous and coleopterous larvae, 40,000 springtails, 120,000 mites and 120 million nematodes.[82]

Arsenical insecticides were toxic to earthworms at high dosages. Soil treated with lead arsenate at 650 lb/a, a level comparable to the accumulations in some orchards, was found to support a population barely half as great as untreated soil when examined 2 yr after the treatment.[70] Applications of calcium arsenate applied at 300 lb/a were just sufficient to control earthworms in sports turf.[31]

Organochlorine Insecticides

DDT applied to trefoil pasture at 5 lb/a did cause a one-third reduction in *Lumbricus terrestris*,[85] but even at 10 lb/a on arable land there was no obvious deleterious effect on earthworms,[25] and at 15 lb/a in pots little mortality was caused by DDT to *Eisenia foetida,* the English redworm or brandling (Table 3.11).[46] Application of DDT at 25 lb/a achieved a reduction in surface castings by *Lumbricus* and *Allolobophora* amounting to 85% for at least 5 months,[16] and soil treated at 38 lb/a supported an earthworm population slightly over half that in untreated soil 5 yr after the treatment.[70] Enchytraeid potworms were unaffected at DDT dosages up to 60 lb/a.[25]

Table 3.11. **Effect of insecticides on the survival and reproduction of *Eisenia* earthworms (Hopkins and Kirk, 1957).**

	LC_{50}	Earthworms exposed for 8 weeks		
	lb/a	Dosage lb/a	% Survival	No. of Progeny
DDT	n.d.	15	90	180
Lindane	1.8	5	96	90
Aldrin	4.3	7.5	83	16
Dieldrin	2.7	5	100	33
Endrin	n.d.	5	100	40
Heptachlor	4.1	7.5	93	0
Toxaphene	n.d.	30	76	0
Malathion	1.5	7.5	90	133
Azinphosmethyl	0.3	5	100	40
Control	---	---	100	103

Technical BHC (12% gamma isomer) had the effect of doubling the earthworm populations when applied at dosages between 4 and 25 kg/ha[39]; it did not reduce them at 60 kg/ha[63] and has been used as an insecticide in plant nurseries at 150–200 kg/ha without evident harm to earthworms.[71] Lindane at 5 lb/a did not reduce survival or progeny production,[46] and was evidently harmless to earthworm populations at 20 kg/ha.[71] Further tests with BHC and lindane are described by Bauer (1964) and listed by Edwards and Thompson (1973); the organochlorines in general are covered in the review by Davey (1963).

Aldrin did not affect populations of earthworms and potworms at 4 lb/a[25] and at one-half this dosage it increased the population (Table 3.14), particularly of *Allolobophora* spp.[38] At 7.5 lb/a aldrin almost completely inhibited the production of young by *Eisenia*[46] and a dose of 15 lb/a could be used for controlling earthworms in turf.[62] Dieldrin applied to cropland for Japanese beetle control at 2–3 lb/a had no effect on earthworm populations,[65] but at 15 lb/a it was an earthworm control agent.[62] Endrin resembled dieldrin in reducing progeny production without killing the adult *Eisenia* at 5 lb/a,[46] but was more deleterious in that an application of endrin to pasture at 1 lb/a reduced the *Lumbricus* populations by 60%.[85]

Chlordane is toxic to earthworms,[18] applications of 2 lb/a to control wireworms in sugarcane fields suppressing earthworms by 90%.[64] Chlordane at 10 lb/a reduced *Lumbricus* and *Allolobophora* castings to zero,[16] and at 18 lb/a it afforded almost complete control of earthworms in sports turf. Heptachlor, the principal insecticidal ingredient of chlordane, reduced earthworm populations by 25% when applied to cropland at 1.25 lb/a for fire ant control[73]; at 4 lb/a it suppressed Enchytraeid worms,[24] and at 7.5 lb/a it inhibited the production of young by *Eisenia*.[46]

Toxaphene resembles aldrin or dieldrin in its effect; in the higher dosages it reduces earthworms in turf at 8 lb/a[62] and completely inhibits *Eisenia* reproduction at 30 lb/a.[46] Isobenzan at 2 lb/a had no effect on earthworms in English soil,[17] but in New Zealand it completely eliminated the earthworm population of one soil for at least 3 yr.[53]

Organochlorine Residues in Earthworms

The earthworms that survive in soils treated with organochlorines accumulate considerable quantities of these insecticides and their metabolites in their tissues. When elm shade-trees at Urbana, Illinois were sprayed with DDT to control the elm barkbeetle vectors of Dutch elm disease, at dosage rates approximating 1 lb per tree, the soil under them contained about 10 ppm DDT and about 10 ppm DDE, and the earthworms accumulated the following average DDT + DDE residues (in ppm)[3]:

| *Lumbricus terrestris* | 33 + 24 | *Allolobophora caliginosa* | 39 + 14 |
| *Helodrilus zeteki* | 164 + 33 | *Octolasium lacteum* | 82 + 22 |

The largest amounts of the residues were in the crops and gizzards of the alimentary canal. Similar operations at East Lansing, Michigan, where it was proved that the earthworms were entirely unaffected by the DDT spray, resulted in average residue figures for DDT itself of about 30 ppm in soil, and about 65 ppm in *Lumbricus* and *Allolobophora*.[6] In a tract of elms at Madison, Wisconsin sprayed at 10–12 lb/a with DDT for 3 yr and methoxychlor for the next 3 yr,[50] the resulting residues in ppm were as follows:

	DDT	DDE	DDD	Methoxychlor
Soil	6.3	2.0	0.9	1.8
Lumbricus terrestris	11.3	4.8	2.5	6.0
Allolobophora caliginosa	28.0	6.3	4.3	7.0

Accumulations of residues by earthworms in cropland were more modest. *Lumbricus terrestris* inhabiting a peafield sprayed 24 days previously with DDT at 2.5 lb/a were found to have accumulated 20 ppm DDT and to have produced 1.5 ppm DDE, with 1.2 ppm DDD also being present in their bodies.[14] In a field in Oxfordshire, England, heavily treated with aldrin along with some DDT applications, the earthworms had accumulated up to 5 ppm dieldrin and 1 ppm DDE.[97] Body residues (Table 3.12) were highest in the small species of *Allolobophora* (especially *chlorotica*) that inhabit the upper soil layer; residue levels were lower in the deep-burrowing *Lumbricus*. When the earthworm concentrations (wet weight) were plotted against the soil concentrations (dry weight) of the organochlorine insecticides (Fig. 3.2), they were found to show a linear relationship on a log–log basis, the concentration factor being tenfold at the lowest concentration and falling to unity at the highest concentrations.

When 67 American soils were sampled for residues in the earthworms, slugs, or snails,[36] it was found that earthworms (*Allolobophora, Diplocardia, Helodrilus,* and *Lumbricus*) accumulated more than snails, and slugs (*Deroceras* and *Limax*) more than earthworms (Table 3.13). Since in this study the figures for both the earthworms and the soil were on the basis of dry weight, it was possible to calculate valid concentration factors for

Table 3.12.   Residues (ppm wet weight) found in earthworms taken from a calcareous loam field treated with aldrin and other organochlorines (Wheatley and Hardman, 1968).

	DDT	DDE	o,p'-DDT	Lindane	Aldrin	Dieldrin
<u>Lumbricus</u> <u>terrestris</u>	0.05	0.49	0.07	0.006	0.05	1.6
<u>Allolobophora</u> <u>longa</u>	0.28	0.38	0.19	0.006	0.28	2.2
<u>A</u>. <u>caliginosa</u>	0.52	0.65	0.35	0.011	0.52	3.8
<u>A</u>. <u>chlorotica</u>	0.98	1.00	0.72	0.013	0.98	4.6
<u>A</u>. <u>rosea</u>	0.64	0.70	0.30	0.017	0.64	3.9
<u>Octolasium</u> <u>roseum</u>	0.84	0.38	0.19	0.008	0.84	2.4
Soil (dry weight)	0.72	0.17	0.14	0.004	0.72	0.64

each insecticide, which turned out to be:

DDT + DDE + DDD (*t*-DDT)	9.0	Endrin	3.6
Aldrin + Dieldrin	9.2	Heptachlor + Epoxide	4.0

independent of the concentration. The earthworms continued to accumulate

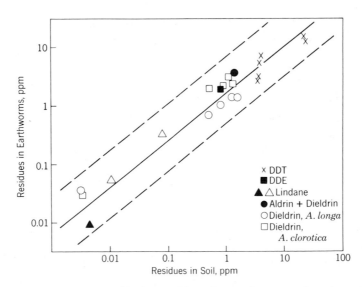

Fig. 3.2. Average organochlorine residues in earthworms plotted against soil residues: 95% confidence limits (after Wheatley and Hardman, 1968).

Table 3.13. **Residues of organochlorine insecticides (ppm dry weight) in earthworms, slugs, and snails inhabiting treated fields and orchards in the United States (Gish, 1968).**

	DDT	DDE	DDD	Aldrin	Dieldrin	Endrin
Cottonfield, Mississippi, clay loam, last treatment DDT at 14 lb/a						
Earthworms	7.4	1.4	0.8	---	0.1	0.1
Slugs	15.0	4.4	4.8	---	0.2	1.1
Soil	0.08	0.12	0.31	---	---	0.02
Cottonfield, Mississippi, silt, last treatment DDT at 3 lb/a						
Earthworms	20.8	6.0	4.9	---	0.04	0.04
Slugs	36.7	10.0	6.0	---	0.4	1.4
Snails	0.38	0.70	1.68	---	0.02	---
Soil	0.63	0.63	1.68	---	0.01	0.01
Cornfield, Missouri, silty clay, last treatments aldrin and DDT at 18 lb/a						
Earthworms	54.9	14.8	14.8	0.2	4.6	---
Soil	0.14	0.76	0.04	---	---	---
Orchard, Maryland, loam, treated with DDT, DDD and endrin at 5-10 lb/a/yr						
Earthworms	1.1	2.7	1.3	0.02	0.82	0.23
Slugs	11.9	4.2	2.6	---	2.8	11.6
Soil	2.50	0.89	n.d.	---	0.02	1.29
Orchard, Maryland, sandy loam, treated with DDT, DDD and endrin at 5-10 lb/a/yr						
Earthworms	1.1	17.6	18.7	---	---	11.0
Slugs	10.3	15.4	14.0	0.2	11.1	114.9
Snails	0.32	1.06	0.83	---	0.07	2.72
Soil	5.4	4.4	5.6	---	---	3.5

DDT as long as they were in the soil; they required 1 month in clean soil to void their body of DDT, and more than 2 months to eliminate the DDE.[30]

Organophosphorus Insecticides

The OP compounds now available in considerable variety as soil insecticides have little lasting effect on earthworm populations, although a few such as phorate and fensulfothion cause heavy initial kills. The numbers found in soil treated with various OPs at 4 lb/a were as follows,[23] expressed in percent of those in untreated controls:

Fenitrothion	95	Diazinon	86
Thionazin	93	Trichlorfon	74
Parathion	89	Disulfoton	70
Chlorfenvinphos	89	Phorate	19

Phosphorothioates. Parathion and diazinon had no effect at 8 lb/a on earthworms in England,[29] although parathion at 8 kg/ha was reported to be toxic to Lumbricids and Enchytraeids in Germany.[96] Application of fenitrothion at 1.5 lb/a actually increased earthworm populations (Table 3.14).[38] Chlorpyrifos (Dursban) did not induce obvious earthworm reductions even at 50 lb/a,[99] although a 15% reduction could be detected in pasture treated at 2 lb/a (Table 3.15). Fensulfothion (Dasanit) at 3 lb/a was very toxic to earthworms, reducing populations by 80–90%.[85]

Table 3.14.   **Effect of organophosphorus insecticides applied at 1.5 lb/a to loam soil: numbers of earthworms per 8 sq ft 7 months after application (Griffiths, Raw, and Lofty, 1967).**

	Control	Fenitrothion	Thionazin	Fonofos	Aldrin*
Lumbricus terrestris	55	63	98	82	85
Lumbricus castaneum	15	19	27	22	39
Allolobophora longa	120	124	155	144	131
Allolobophora caliginosa	114	153	133	117	196[†]
Allolobophora chlorotica	95	133	105	37[†]	138[†]
Octolasium cyaneum	9	7	3	15	11
Eisenia rosea	83	62	80	119[†]	113

* 2.25 lb/a   [†]significant decrease or increase

Table 3.15.   Effect of soil insecticides at normal dosage rates* on numbers and biomass of *Lumbricus terrestris* (Thompson, 1970).

	Dosage	Population per 40 ft^2		Biomass per 40 ft^2	
	lb/a	Numbers	% Reduction	Grams	% Reduction
Control	0	17.9	0	404	0
DDT	5	11.4	36	271	33
Endrin	1	7.9	56	242	67
Chlorpyrifos	2	14.1	21	344	15
Trichloronat	3	13.6	24	253	37
Fensulfothion	3	3.8	79	33	92
Carbaryl	2	7.2	59	128	68
Bux	1	8.6	52	132	40
Carbofuran	4	3.1	83	30	93

* Applied to cropped surface of a trefoil pasture on sandy loam, Ontario, Canada

Phosphorodithioates.   The orchard insecticide azinphosmethyl was harmless to *Eisenia* at 5 lb/a, but greatly impaired progeny production.[46] Malathion applied in cleared forestland at 3 kg/ha did reduce Lumbricids and Enchytraeids by 60%, but the populations more than recovered in the following year.[92] The exception among the phosphorodithioates is the systemic insecticide phorate (Thimet), which has almost eliminated earthworms when applied at normal agricultural rates of 4 lb/a[29]; the Enchytraeids, however, were increasing their populations in soil treated with 10 ppm phorate even as the Lumbricids were coming to the surface to die.[95]

New OP Compounds.   Applied to loam soil at 1.5 lb/a (Table 3.14), fonofos increased populations of *Eisenia* earthworms.[58] Thionazin had no effect at this dosage,[58] although it suppressed the earthworms at 10 ppm.[94] Menazon was harmless at 2 lb/a,[72] although it killed earthworms at 250 ppm in soil.[95] Leptophos had little effect, in contrast to the strong toxicity of phorate.[90] Slight suppression of earthworms resulted from applications of chlorfenvinphos at 4 lb/a,[28] disulfoton at 8 lb/a,[29] and dyfonate and tetrachlorvinphos (Gardona) at 8 kg/ha.[27] Applied to trefoil pasture at 3 lb/a, trichloronat reduced earthworm populations by 30%, fensulfothion (Dasanit) by 80%.[85] The new phosphorodithioate Counter was no less toxic than phorate, as the following percent reductions in earthworms after a 3 lb/a application[90] demonstrate:

	Numbers	Biomass
Leptophos	10	46
Phorate	92	92
Counter	93	96

In contrast to the organochlorines, there is little evidence that OP insecticides can be concentrated by earthworms in their tissues.[24] Earthworms of four different species exposed to chlorfenvinphos in the soil for 5 months contained 0.02 ppm of the metabolite 2,4-dichloroacetophenone, but not chlorfenvinphos itself or its two other principal metabolites.[28] By contrast, the survivors from an application of fensulfothion at 3 lb/a contained 22 ppm Dasanit sulfone, on the average.[41]

Carbamate Insecticides

Many of the carbamate insecticides are quite toxic to earthworms. Carbaryl at 2 or 4 lb/a reduced pasture populations by 60%,[23,85] and applied at 5 kg/ha on forest clearings it exerted a 66% reduction that lasted for at least 14 months.[92] Carbofuran at 2 lb/a raises dead worms to the surface of tobacco soils,[58] and at 4 lb/a causes a 95% reduction in pasture populations.[85] Although more toxic than fensulfothion, carbofuran causes less ChE inhibition; however it induces skin swellings and lesions in earthworms,[80] as carbaryl also does.[59] Aldicarb at 6 lb/a is harmless to earthworms but reduced Enchytraeids by 60%, while methomyl (Lannate) is almost harmless to both groups at 10 lb/a.[22] The following percent reductions in earthworms resulted 3 wk after pasture was treated with these carbamates at 3 lb/a[90]:

	Numbers	Biomass
Methomyl	28	14
Carbofuran	43	32

Bux causes a 50% reduction at 1 lb/a[85] and dead earthworms appear on the surface of soils treated at 4 lb/a.[58] The numbers of earthworms significantly reduced by carbofuran and Bux, as well as those reduced by fensulfothion and endrin, had recovered to their original level 1 yr after the treatment.[89]

Gastropod Molluscs

Organochlorines. Although DDT in a soil drench can kill slugs,[24] pasture treatments as high as 25 lb/a applied against the Japanese beetle did not affect the slug population.[32] Aldrin applied at 6 lb/a from a 1.2% dust could reduce damage by 75%.[81]

OP Compounds. Phorate applied as an insecticide to soil at 8 lb/a killed considerable numbers of slugs.[24] For the control of slugs, parathion is effective at 0.5% concentration in bran baits, while malathion or diazinon required an 8% concentration.[68]

Carbamates. In early experiments, isolan was found the most effective carbamate in baits to control the European brown snail *Helix aspersa* in California, followed by Pyrolan.[68] Carbaryl was discovered to be effective at 1.25% in baits against the grey garden slug *Deroceras reticulatum* in Colombia,[74] and mexacarbate (Zectran) applied as granular or spray formulation could control the grey ground slug *D. agreste* in ornamental plantings in Michigan at 1–2 lb/a.[40] The insecticide methiocarb (Mesurol) is now the carbamate of choice for use in baits against slugs and snails pestering garden plants and ornamental shrubs.[24]

Residues in Slugs. Slugs, like earthworms, are powerful concentrators of organochlorine residues, accumulating DDT and laying down more DDD than DDE. The following tissue concentrations (ppm) were found in an English orchard which had been under a DDT spray regime[13]:

	DDT	*o,p´*-DDT	DDE	DDD	DME
Earthworms (8 spp.)	4.6	0.3	1.5	4.5	0.2
Agriolimax reticulatus	12.8	0.9	1.2	3.8	0.3
Arion spp.	5.3	2.6	3.4	9.9	0

It can be seen that *Agriolimax* resembles earthworms in being able to dehydrochlorinate DDD (to DME) as well as DDT to DDE.

The slug *Milax budapestensis,* which ingests soil as well as plant tissue, accumulated *t*-DDT residues of 35 ppm in an orchard which had a soil concentration of 3.7 ppm.[86] All the available data[24] indicate that slugs concentrate *t*-DDT and aldrin–dieldrin soil residues by an average of seven to eight times (Fig. 3.3). The highest average residues of lindane found in slugs is only 0.25 ppm, whereas dieldrin has reached average concentrations of 18.3 ppm in slugs. In two orchards and two fields in the United States (Table 3.13), slugs concentrated organochlorines 2–10 times more than earthworms, and 20–100 times more than snails.[36] Slugs have a marked ability to concentrate OP compounds, tissue concentrations resulting from 8 lb/a soil applications reaching 162 ppm with diazinon and 280 ppm with chlorfenvinphos.[24]

Nematodes

Organochlorine Insecticides.   Applications of lindane or aldrin to a Wisconsin soil were found not to have changed the total numbers of nematodes when the plots were examined 3 yr later, even when the dosages were very high.[35] At the highest dosage of 1000 lb/a, DDT had the effect of increasing the population of the saprophagous nematode *Chiloplacus symmetricus* by eight times, while lindane at 100 lb/a resulted in this species eventually increasing by five times (Table 3.16). Whereas the immediate effect of DDT on total nematode numbers had been nil, that of the lindane application had been to reduce the nematode numbers by 85%.[35] Application of DDT at 1200 kg/ha to a Swiss soil also resulted in a threefold increase of nematodes 2 wk later, becoming sevenfold 1 month after treatment, while BHC (13%

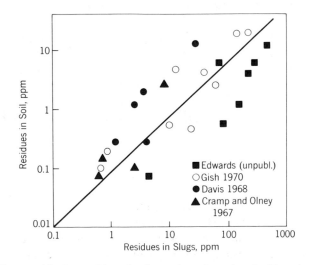

Fig. 3.3.   Organochlorine residues in slugs plotted against residues in the soil they inhabited (after Edwards and Thompson, 1973).

Table 3.16. Effect of organochlorine insecticides on soil population of nematodes: numbers per 500 grams soil (French, Lichtenstein, and Thorne 1959).

	Control,	DDT		Lindane		Aldrin
	Untreated	10 lb/a	1,000 lb/a	1 lb/a	100 lb/a	200 lb/a
Rhabditis spp.	98	14	9	164	49	30
Eucephalobus oxyuroides	61	11	184	59	17	134
Acrobeloides spp.	8	12	9	11	8	4
Chiloplacus symmetricus	16	14	124	38	84	17
Tylenchus spp.	31	21	0	32	7	21
Helicotylenchus spp.	20	60	10	6	24	n.d.
Boleodorus thylactus	37	84	0	54	3	17
Aphelenchus avenae	43	17	6	44	12	30
Dorylaimus spp.	68	74	5	91	152	76
Total (incl. other spp.)	492	352	454	539	399	428

gamma) at 400 kg/ha had no effect during this period.[83] In Russian soils, applications of technical BHC between 6 and 30 kg/ha increased the nematode population.[39] However, BHC (10% gamma) was sufficiently deleterious to the dagger nematode *Xiphinema* infesting the roots of apple and peach in West Virginia that it could be used for its control.[2]

OP and Carbamate Compounds. Parathion usually decreases nematode populations, and is sometimes used for this very purpose.[24] Phorate at 10 ppm in the soil significantly decreased the saprophagous as well as the parasitic nematodes.[95] When tested at 90–110 lb/a among possible control agents for the potato-cyst nematode *Heterodera rostochiensis*,[98] dicrotophos had no effect, but the reductions obtained with other OP compounds were as follows:

Strong: diazinon, chlorpyrifos, thionazin
Significant: chlorfenvinphos, phosphamidon, oxydemetonmethyl, schradan

Among the carbamate insecticides, methomyl was ineffective against this nematode, but aldicarb and isolan strongly reduced its numbers.[98] The carbamate carbofuran (Furadan) and the OP compound fensulfothion (Dasanit) have been employed as nematicides in soil.[91]

REFERENCES CITED

1. M. A. Abdellatif and H. T. Reynolds. 1967. Toxic effects of granulated disulfoton on soil arthropods. *J. Econ. Entomol.* **60**:281–283.

2. R. E. Adams. 1955. Injury to deciduous fruit trees by an ectoparasitic nematode and a promising control measure. *Phytopathol.* **45**:477–479.

3. R. J. Barker. 1958. Notes on some ecological effects of DDT sprayed on elms. *J. Wildl. Manage.* **22**:269–274.

4. G. W. Barrett. 1968. Effect of an acute insecticide stress on a semi-enclosed grassland ecosystem. *Ecology* **49**:1019–1035.

5. K. Bauer. 1964. Studien über Nebenwirkung von Pflanzenschutzmitteln auf die Boden-fauna. *Mitt. Biol. Bundesanst. Land. Forstwirtsch.* **112**:1–42.

6. E. A. Boykins. 1966. DDT residues in the food chains of birds. *Atl. Nat.* **21**:18–25.

7. C. H. Brett and W. C. Rhoades. 1946. Grasshopper control in alfalfa with hexachlorocyclohexane dust. *J. Econ. Entomol.* **39**:677–678.

8. J. B. Briggs and R. P. Tew. 1963. Control of ground beetles attacking strawberry fruits. *Bull. Entomol. Res.* **54**:83–92.

9. J. W. Butcher and R. M. Snider. 1975. The effect of DDT on the life history of *Folsomia candida* (Collembola: Isotomidae). *Pedobiol.* **15**:53–59.

10. J. W. Butcher, E. Kirknel, and M. Zabik. 1969. Conversion of DDT to DDE by *Folsomia candida.* Rev. Ecol. Biol. Soc. **6**(3):291–298.

11. T. H. Coaker. 1966. The effect of soil insecticides on the predators and parasites of the cabbage root fly. *Ann. Appl. Biol.* **57**:397–407.

12. S. P. Davey. 1963. Effects of chemicals on earthworms: A review of the literature. *Spec. Sci. Rep., U.S. Fish Wildl. Serv. 74.* 20 pp.

13. B. N. K. Davis. 1968. The soil macrofauna and organochlorine residues at twelve agri-cultural sites near Huntingdon. *Ann. Appl. Biol.* **61**:29–45.

14. B. N. K. Davis and M. C. French. 1969. The accumulation and loss of organochlorine insecticide residues by beetles, worms, and slugs in sprayed fields. *Soil Biol. Biochem.* **1**:45–55.

15. J. P. Dempster. 1967. A study of the effects of DDT applications against *Pieris rapae* on the crop fauna. *Proc. 4th Br. Insectic. Fungic. Conf.* **1**:19–25.

16. C. C. Doane. 1962. Effects of certain insecticides on earthworms. *J. Econ. Entomol.* **55**:416–418.

17. C. A. Edwards. 1965. Effects of pesticide residues on soil inverbebrates and plants. In *Ecology and the Industrial Society.* 5th Symp. Br. Ecol. Soc.

18. C. A. Edwards. 1965. Some side effects resulting from the use of persistent insecticides. *Ann. Appl. Biol.* **55**:329–331.

19. C. A. Edwards. 1969. Soil pollutants and soil animals. *Sci. Am.* **4**:92–99.

20. C. A. Edwards and E. B. Dennis. 1960. Some effects of aldrin and DDT on the soil fauna of arable land. *Nature* **188**:767.

21. C. A. Edwards and E. Gunn. 1961. Control of the glasshouse millipede. *Plant Pathol.* **10**:21–24.

22. C. A. Edwards and J. R. Lofty. 1971. Nematicides and the soil fauna. *Proc. 6th Br. Insectic. Fungic. Conf.* **1**:158–166.

23. C. A. Edwards and A. R. Thompson. 1969. Insecticides and the soil fauna. *Rep. Rothamsted Exp. Sta. 1968.* pp. 216–217.

24. C. A. Edwards and A. R. Thompson. 1973. Pesticides and the soil fauna. *Residue Rev.* **45**:1–79.

25. C. A. Edwards, E. B. Dennis, and D. W. Empson. 1967. Pesticides and the soil fauna: Effects of aldrin and DDT in an arable field. *Ann. Appl. Biol.* **60**:11–22.

26. C. A. Edwards, J. R. Lofty, and C. J. Stafford. 1970. Effects of pesticides on predatory beetles. *Rep. Rothamsted Exp. Sta. 1969.* p.246.

27. C. A. Edwards, J. R. Lofty, and C. J. Stafford. 1971. Pesticides and the soil fauna. *Rep. Rothamsted Exp. Sta. 1970.* p. 194.

28. C. A. Edwards, A. R. Thompson, and K. I. Beynon. 1968. Some effects of chlorfenvinphos on populations of soil animals. *Rev. Ecol. Biol. Soc.* **5**:199–224.

29. C. A. Edwards, A. R. Thompson, and J. R. Lofty. 1967. Changes in soil invertebrate populations caused by some OP insecticides. *Proc. 4th Br. Insectic. Fungic. Conf.* **1**:48–55.

30. C. A. Edwards, J. R. Lofty, A. E. Whiting, and K. A. Jeffs. 1971. Pesticides and earthworms. *Rep. Rothamsted Exp. Sta. 1970.* **1**:193–194.

31. J. R. Escritt. 1955. Calcium arsenate for earthworm control. *J. Sports Turf. Res. Inst.* **9**:28–34.

32. W. E. Fleming and I. M. Hawley. 1950. A large scale test with DDT to control the Japanese beetle. *J. Econ. Entomol.* **43**:586–590.

33. C. J. S. Fox. 1958. Some effects of insecticides on the wireworms and vegetation of grassland in Nova Scotia. *Proc. 10th Internat. Congr. Entomol.* **3**:297–300.

34. C. J. S. Fox. 1967. Effects of several chlorinated hydrocarbon insecticides on the springtails and mites of grassland soil. *J. Econ. Entomol.* **60**:77–79.

35. N. E. French, E. P. Lichtenstein, and G. Thorne. 1959. Effects of some chlorinated hydrocarbon insecticides on nematode populations in soils. *J. Econ. Entomol.* **52**:861–865.

36. C. D. Gish. 1970. Organochlorine insecticide residues in soils and soil invertebrates from agricultural lands. *Pestic. Monit. J.* **3**:241–252.

37. E. Gould and E. O. Hamstead. 1951. The toxicity of cumulative spray residues in soil. *J. Econ. Entomol.* **44**:713–717.

38. D. C. Griffiths, F. Raw, and J. R. Lofty. 1967. Effect on soil fauna of insecticides tested against wireworms in wheat. *Ann. Appl. Biol.* **60**:479–490.

39. T. G. Grigoreva. 1952. The action of hexachlorane introduced into the soil on soil fauna. *Rev. Appl Entomol. (A)* **41**:336. Also cited in Edwards and Thompson. 1973.

40. O. H. Hammer. 1962. Zectran pesticide: Some results from use of pests on ornamental plants. *Down to Earth* **17**(4):9–13.

41. C. R. Harris, A. R. Thompson, and C. M. Tu. 1972. Insecticides and the soil environment. *Proc. Entomol. Soc. Ontario* **102**:156–168.

42. R. C. Hartenstein. 1960. The effects of DDT and malathion upon forest soil arthropods. *J. Econ. Entomol.* **53**:357–362.

43. D. H. C. Herne. 1963. Carabids collected in a DDT-sprayed peach orchard in Ontario. *Can. Entomol.* **95**:357–362.

44. C. H. Hoffmann and E. P. Merkel. 1948. Fluctuations in insect populations associated with aerial applications of DDT to forests. *J. Econ. Entomol.* **41**:464–473.

45. C. H. Hoffmann, H. K. Townes, H. H. Swift, and R. I. Sailer. 1949. Field studies on the effect of airplane application of DDT. *Ecol. Monogr.* **19**:1–46.

46. A. R. Hopkins and V. M. Kirk. 1957. Effect of several insecticides on the English red worm. *J. Econ. Entomol.* **50**:699–700.

47. A. J. Howitt. 1959. Laboratory and greenhouse tests for evaluating compounds in the control of the garden symphylid. *J. Econ. Entomol.* **52**:672–677.

48. A. J. Howitt and R. M. Bullock. 1955. Control of the garden centipede. *J. Econ. Entomol.* **48**:246–250.

49. A. J. Howitt, J. S. Waterhouse, and R. M. Bullock. 1959. The utility of field tests for evaluating insecticides against the garden symphylid. *J. Econ. Entomol.* **52**:666–672.

50. L. B. Hunt and R. J. Sacho. 1969. Response of robins to DDT and methoxychlor. *J. Wildl. Manage.* **33**:336–345.

51. W. Karg. 1961. Uber die Wirkung von Hexachlorocyclohexan auf die Bodenbiozenose unter besonderer Berucksichtigung der Acarina. *Nachrichtenbl. Dtsch. Pflanzenschutzdst.* **15**:23–33.

52. W. Karg. 1963. Bodenbiologische Untersuchungen von Kohlfeldern nach Beregnungen mit HCH oder Trichlorphon. *Nachrichtenbl. Dtsch. Pflanzenschutz.* **17**:157–162.

53. J. M. Kelsey and E. Z. Arlidge. 1968. Effects of isobenzan on soil fauna and soil structure. *N.Z. J. Agric. Res.* **11**:245:260.

54. D. K. M. Kevan. 1962. *Soil Animals.* Witherby, London; Philosophical Library, New York. 237 pp.

55. K. Kiritani and S. Kawahara. 1973. Food-chain toxicity of granular formulations of insecticides to a predator of *Nephotettix cincticeps. Botyu Kagaku* **38**:69–75.

56. C. B. Knight and J. P. Chesson. 1966. Effects of DDT on the forest floor Collembola of a loblolly pine stand. *Rev. Ecol. Biol. Soc.* **3**(1):129–139.

57. L. J. Korschgen. 1970. Soil-foodchain-pesticide wildlife relationship in aldrin-treated fields. *J. Wildl. Manage.* **34**:186–199.

58. J. B. Kring. 1969. Mortality of the earthworm *Lumbricus terrestris* following soil applications of insecticides to a tobacco field. *J. Econ. Entomol.* **62**:963.

59. H. an der Lan and H. Aspock. 1962. Zur wirkung von Sevin auf Regenwurmer. *Anz. Schadlingsk.* **35**:180–182.

60. B. J. Landis and C. W. Getzendaner. 1959. Millipede injury to potatoes. *J. Econ. Entomol.* **52**:1021–1022.

61. E. R. Laygo and J. T. Schulz. 1963. Persistence of organophosphate insecticides and their effects on microfauna in soils. *Proc. N. Dak. Acad. Sci.* **17**:64–65.

62. D. C. Legg. 1968. Comparison of various worm-killing chemicals. *J. Sports Turf Res. Inst.* **44**:47–48.

63. J. J. Lipa. Effect on earthworm and diptera populations of BHC dust applied to soil. *Nature* **181**:863.

64. W. H. Long, H. L. Anderson, and A. L. Isa. 1967. Sugarcane growth responses to chlordane and microarthropods and effects of chlordane on soil microfauna. *J. Econ. Entomol.* **60**:623–629.

65. W. H. Luckmann and G. C. Decker. 1960. A five-year report of observations in the Japanese beetle control area of Sheldon, Illinois. *J. Econ. Entomol.* **53**:821–827.

66. D. J. Mowat and T. H. Coaker. 1967. The toxicity of some soil insecticides to carabid predators of the cabbage root fly. *Ann. Appl. Biol.* **59**:349–354.

67. P. G. Olivier and P. A. J. Ryke. 1969. The influence of citricultural practices on the composition of soil Acari and collembolan populations. *Pedobiologia* 9:277–281.

68. J. L. Pappas and G. E. Carman. 1955. Field screening tests with various materials against the European brown snail on citrus in California. *J. Econ. Entomol.* 48:698–700.

69. J. Phillipson. 1966. *Ecological Energetics.* St. Martin's Press. 57 pp.

70. J. B. Polivka. 1951. Effect of insecticides upon earthworm populations. *Ohio J. Sci.* 51:195–196.

71. A. A. Prisyazhnyuk. 1950. Use of 666 for the control of chafer grubs. *Agrobiologiya* 5:141–142.

72. F. Raw. 1965. Current work on side effects of soil-applied organophosphorus insecticides. *Ann. Appl. Biol.* 55:342–343.

73. W. C. Rhoades. 1962 and 1963. A synecological study of the effects of the imported fire ant control program: I and II. *Florida Entomol.* 45:161–173; 46:301–310.

74. R. F. Ruppel. 1959. Effectiveness of Sevin against the grey garden slug. *J. Econ. Entomol.* 52:360.

75. J. E. Satchell. 1955. Effects of BHC, DDT and parathion on soil fauna. *Soils and Fertilizers* 18:279–285.

76. E. I. el Sayed, J. B. Graves, and F. L. Bonner. 1967. Chlorinated hydrocarbon residues in selected insects and birds found in association with cotton fields. *J. Agric. Food Chem.* 15:1014–1017.

77. J. G. Sheals. 1957. Soil population studies. I. The effects of cultivation and treatment with insecticides. *Bull. Entomol. Res.* 47:803–822.

78. L. C. Stegeman. 1964. Effects of carbaryl upon forest soil mites and Collembola. *J. Econ. Entomol.* 57:803–808.

79. D. Steiner, F Wenzel, and D. Baumert. 1963. Zur Beeinflussung der Arthropoden Fauna nordwestdeutscher Kartoffelfelder durch die Anwendung synthetischer Kontaktinsectizide. *Mitt. Biol. Bundesanst. Land. Forstwirtsch.* 109:5–38.

80. J. Stenersen, A. Gilman, and A. Vardanis. 1973. Carbofuran; its toxicity for and metabolism by the earthworm (*Lumbricus terrestris*). *J. Agric. Food Chem.* 21:166–171.

81. J. W. Stephenson. 1959. Aldrin controlling slug and wireworm damage to potatoes. *Plant Pathol.* 8:53–54.

82. A. Stockli. 1946. Die biologische Komponente der Vererdung der Gare und der Nahrstoffpufferung. *Schweiz. Landwirts. Monatsh.* 24:3–19. (quoted in Thompson and Edwards 1974).

83. A. Stockli. 1952. Studien über Bodennematoden mit besonderer Berucksichtigung des Nematodengehaltes von Wald-, Grunland-, und ackerbaulich genutzten Boden. *Z. Pflanzenernahr.* 59:97–103.

84. A. Stringer, J. A. Pickard and C. H. Lyons. 1974. Accumulation of *p,p'*-DDT and related compounds in an apple orchard: I and II. *Pestic. Sci.* 5:587; 6:223–232.

85. A. R. Thompson. 1970. Effects of nine insecticides on the numbers and biomass of earthworms. *Bull. Environ. Contam. Toxicol.* 5:577–586.

86. A. R. Thompson. 1973. Pesticide residues in soil invertebrates. In *Environmental Pollution by Pesticides.* Ed. C. A. Edwards. Plenum. pp. 87–133.

87. A. R. Thompson and C. A. Edwards. 1974. Effects of pesticides on non-target invertebrates in freshwater and soil. In *Pesticides in Soil and Water.* Ed. W. D. Guenzi. *Soil Sci. Soc. Am.,* Madison, Wisconsin. pp. 341–386.

88. A. R. Thompson and F. L. Gore. 1972. Toxicity of twenty-nine insecticides to *Folsomia candida:* Laboratory studies. *J. Econ. Entomol.* **65**:1255–1260.

89. A. R. Thompson and W. W. Sans. 1974. Effects of soil insecticides in southwestern Ontario on non-target invertebrates: Earthworms in pasture. *Environ. Entomol.* **3**:305–308.

90. A. D. Tomlin and F. L. Gore. 1974. Effects of six insecticides and a fungicide on the numbers and biomass of earthworms in pasture. *Bull. Environ. Contam. Toxicol.* **12**:487–492.

91. D. C. Torgeson. 1972. Fungicides and nematicides: Their role now and in the future. *J. Environ. Qual.* **1**:14–17.

92. L. D. Voronova. 1968. Effect of some pesticides on the soil invertebrate fauna in the south taiga zone in the Perm region (U.S.S.R.). *Pedobiol.* **8**:507–525.

93. M. M. H. Wallace. 1954. The effect of DDT and BHC on the population of the lucerne flea and its control by predatory mites. *Austral. J. Agric. Res.* **5**:148–155.

94. M. J. Way and N. E. A. Scopes. 1965. Side effects of some soil-applied systemic insecticides. *Ann. Appl. Biol.* **55**:340–341.

95. M. J. Way and N. E. A. Scopes. 1968. Studies on the persistence and effects on soil fauna of some soil-applied systemic insecticides. *Ann. Appl. Biol.* **62**:199–214.

96. G. Weber. 1953. Die Makrofauna leichter und schwerer Ackerboden und ihre Beeinflussung durch Pflanzenschutzmitteln. *Z. Pflanzenernahr. Dung.* **61**:107–118.

97. G. A. Wheatley and J. A. Hardman. 1968. Organochlorine insecticide residues in earthworms from arable soils. *J. Sci. Food Agric.* **19**:219–225.

98. A. G. Whitehead and G. Storey. 1970. Control of potato-cyst nematode. *Rep. Rothamsted Exp. Sta. 1969.* p. 199.

99. W. K. Whitney. 1967. Laboratory tests with Dursban and other insecticides in soil. *J. Econ. Entomol.* **60**:68–74.

100. C. Wibo. 1973. Etude de l'action d'un insecticide organophosphore sur quelques populations de microarthropodes edaphiques. *Pedobiologia* **13**:150–163.

101. M. Witcamp and D. A. Crossley. 1966. The role of arthropods and microflora in breakdown of white oak litter. *Pedobiologia* **6**:293–303.

4

INSECTICIDES AND THE SOIL MICROFLORA

A. FATE OF PERSISTENT INSECTICIDES IN THE SOIL

Before the advent of the synthetic organic insecticides, the use of calcium arsenate on cotton, amounting to 50 lb/a-yr on some fields, has led to soil accumulations of 30 ppm arsenate, making them unfit for crops in general.[2] In some Colorado orchards, where arsenic-resistant populations of the codling moth had necessitated 12 sprays per season, accumulations of lead arsenate had reached 2000 ppm.[77] Residue determinations made during the twentieth century in representative North American orchards (Table 4.1) showed that soil contaminations of the order of 100 ppm arsenic were frequent, as compared to an average content of approximately 8 ppm As in untreated soils.[159] The added arsenate, being bound to the soil particles in insoluble form, did not harm the deep-rooted apple trees, but it was lethal to shallow-rooted groundcover crops, the effect being more pronounced in sandy and alkaline soils. The most susceptible crops, in which the yields are depressed by more than 25 or 50 ppm As in the soil, are onion, cucumber, pea, snap bean, lima bean, alfalfa, and other legumes.[159] Where the arsenate becomes reduced to the soluble arsenite, the phytotoxicity is increased; but

Table 4.1.   Total As content in orchard soils of North America, ppm (Walsh and Keeney, 1975).

	Untreated	Treated	Date Reported
Colorado	1.3-2.3	13-69	1908
Idaho	0-10	138-204	1971
Indiana	2-4	56-250	1971
Maryland	19-41	21-238	1940
New Jersey	10	92-270	1971
New York	3-12	90-625	1971
Nova Scotia	0-7.9	10-124	1962
Ontario	1.1-8.6	10-121	1968
Oregon	2.9-14.0	17-439	1945
	3-32	4-103	1971
Washington	4-13	48	1931
	6-13	106-830	1953
	8-80	106-2553	1971

Table 4.2. **Persistence of organochlorine insecticides in soil: applied at about 1 lb/a (Edwards, 1973).**

	Half-life	95% Disappearance
	years	years
Aldrin[*]	0.3	3
Isobenzan	0.4	4
Heptachlor	0.8	3.5
Chlordane	1.0	4
Lindane	1.2	6.5
Endrin	2.2	7
Dieldrin	2.5	8
DDT	2.8	10

* But much of the disappeared aldrin persists as dieldrin

arsenite is more liable, like the organic arsenical herbicides, to be converted to dimethylarsine.

Fate of the Organochlorines

The synthetic organochlorines proved to be effective and nonphytotoxic agents suitable for use as soil insecticides. The half-life in the soil for heptachlor, chlordane, and lindane was about 1 yr, and for dieldrin, endrin, and DDT in excess of 2 yr (Table 4.2). The short half-life for aldrin is due partly to its being volatilized and partly to its conversion to dieldrin, which is more persistent. Cultivation of the soil increases the first-year loss of aldrin, which is only 10% in noncultivated soil, to 53% in cultivated soil.[98] The greatest persistence is found in soils rich in organic matter, with high clay content, and with an acid pH. The disappearance of an insecticide applied to soil characteristically follows first-order kinetics, where the rate at any time after application is proportional to the amount remaining at that time; however, this is usually preceded by an initial burst of fast dissipation, when the insecticide is removed by translocation of some type from the treated site.[105]

On the average, soil applications of organochlorines at 2 lb/a annually will give residues levelling off at 4 lb/a with chlordane and lindane, at at 6 lb/a (3 ppm in the top 5 inches) with dieldrin and DDT.[35] Soils annually treated with aldrin or heptachlor at 5 lb/a came to reach an equilibrium after 5 yr where the concentration of these insecticides plus their epoxide metabolites each year was the same as that immediately after the first application, that is, the 3 ppm produced by 5 lb/a (Fig. 4.1), amounting to 20% of the total previously applied. With toxaphene, however, only 5–10% of the total accumulation of 12 weekly applications at 2 lb/a was found in the soil at the end of the season.[23] The fate of pesticides applied to crop soils is more fully discussed in Chapter 11. The subject has been recently covered in all its facets in the volume edited by Haque and Freed (1974).

Determinations of soil residues made mostly in the 1960s (Table 4.3) show that they were highest in orchard soils; among vegetable crops, they were highest for onion plots, and among field crops, highest in cottonfields. Heptachlor showed the lowest residues, even lower than toxaphene and lindane. The average level of DDT for all field-crop soils in the United States in 1972 was 0.15 ppm.[39]

The principal agent for reducing the concentration of insecticides added to the soil is the microflora. The subject matter of the following chapter has been largely covered in two recent reviews. Pfister (1972) has described the

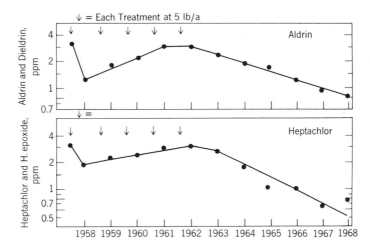

Fig. 4.1. Residues of aldrin (plus dieldrin) and of heptachlor (plus its epoxide) in soils after five annual applications, each at 5 lb/a (from Lichtenstein et al., 1970). Reprinted with permission from the Journal of Agricultural and Food Chemistry; copyright by the American Chemical Society.

Table 4.3. Soil residues in various types of crop lands, 1963–1972: range of average determinations in ppm (Edwards, 1973).

	t-DDT	Dieldrin	Heptachlor*
Orchards	6-123	0.1-3	0.02
Vegetables	1-24	.02-1.5	.02-.16
Field Crops	1-16	.02-2.3	.01-.08
Pasture	.06-2.6	.01-.20	.01-.02

* Including heptachlor epoxide

mutual interaction between the organochlorines and the soil microbiota, while Kaufman (1974) has reviewed the metabolic pathways of, and the microorganisms responsible for, the breakdown of pesticides in the soil.

Fate and Breakdown of DDT

DDT has an average half-life in soils of about 3 yr, and 5–10% of that applied still remains 10 yr after the application.[35] In a Maryland soil to which DDT had been thoroughly admixed and which was only occasionally in crop, as much as 40% of the insecticide was still present 17 yr after the application.[127] Some of it is lost by volatilization from the surface, about 0.5% of that applied being evaporated from a moist loam in 40 days at 30°C[55]; this loss is only one-fifth as much as with dieldrin.[69] There is some degradation to DDE by chemical catalysis on the surface of clay minerals[40] and in soils containing iron oxides[33,50] and to DDD by the reduced porphyrins which are present in the aqueous and organic fractions of soils.[31,121]

Bacterial Degradation. This is essentially an anaerobic process, producing DDD. When ^{14}C-DDT was incubated aerobically in a silt loam for 6 months there still remained 75% of the DDT intact while only 4% of the radioactivity had been converted to DDE and a trace to DDD; but when the treated soil was incubated anaerobically, more than 99% of the DDT had been degraded within 12 wk, with 41% of the radioactivity being found as DDD.[54] This conversion did not take place if the soil was sterilized.[86] When 25 species of bacteria associated with plants were incubated with DDT,[72] the following 23 species converted it to DDD, but only under anaerobic

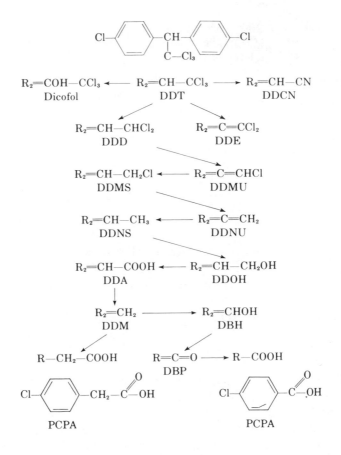

DDT	Dichlorodiphenyltrichloroethane	DDNS	Dichlorodiphenylethane
Dicofol	Dichlorodiphenyltrichloroethanol	DDOH	Dichlorodiphenylethanol
DDCN	Dichlorodiphenylacetonitrile	DDA	Dichlorodiphenylacetic acid
DDD	Dichlorodiphenyldichloroethane	DDM	Dichlorodiphenylmethane
DDE	Dichlorodiphenyldichloroethylene	DBH	Dichlorobenzhydrol
DDMU	Dichlorodiphenylchloroethylene	DBP	Dichlorobenzophenone
DDMS	Dichlorodiphenylchloroethane	PCPA	p-Chlorophenylacetic acid
DDNU	Dichlorodiphenylethylene	PCBA	p-Chlorobenzoic acid

This nomenclature is intended to give clues to the acronyms in most
cases; for the correct nomenclature, and for the references for the occurrence of each metabolite, see Table 1 in Kaufman (1974).

Fig. 4.2. Degradation of DDT in soils and in cultures of microorganisms (from
Matsumura, 1975).

conditions:

Pseudomonas, 6 spp.	*Aerobacter aerogenes*
Xanthomonas, 4 spp.	*Agrobacterium tumefaciens*
Erwinia, 4 spp.	*Clostridium pasterianum*
Bacillus, 3 spp.	*Corynebacterium michiganense*
Achromobacter sp.	*Kurthia zopfi*

Only *Clostridium sporogenes* and *Sarcina lutea* were incapable of this reductive dechlorination of DDT. Species of *Erwinia, Pseudomonas,* and *Xanthomonas* were the most active dechlorinators.[28] Systematic isolations from soils revealed 9 species of *Pseudomonas* and 3 of *Bacillus*, besides a *Micrococcus* and an *Arthrobacter*, that could degrade DDT to DDD under aerobic conditions; of these 14 isolates, 12 produced DDA also, while 9 yielded a compound resembling dicofol.[134] When 8 species of *Streptomyces* were incubated with DDT without anaerobic measures, 5 of them degraded it to DDD, and so did a species of the actinomycete *Nocardia*.[28]

Among the bacteria isolated from rat intestines, the coliform bacilli were the most active in producing DDD and a little DDE, followed by *Clostridium, Bacteroides, Pseudomonas, Bacillus, Lactobacillus,* and *Streptococcus,* in that order.[114] The gram-positive *Micrococcus* was inactive, while yeasts were more active than any of the bacteria.[24] *Escherichia coli* and *Aerobacter aerogenes* isolated from rat faeces also degraded DDT,[114] and this activity was shown by disrupted cells of *A. aerogenes*[161] and fractionated cell-membranes of *E. coli*.[45]

DDT Metabolites. Incubated anaerobically in silt loam for 4 wk, ^{14}C-DDT was 35% degraded to DDD, while 19% remained unchanged; only 0.25% was converted to DDE, but the following metabolites (Fig. 4.2) each contained about 0.6% of the radioactivity: dicofol, DDA, DBP, and PCBA.[53] *Aerobacter aerogenes* incubated with DDT in broth was found to produce the metabolites DDMU, DDMS, DDNU, and DDOH in addition[162] and to degrade DDA through DDM and DBH to DBP.[163] *Proteus vulgaris* from the gut flora of the mouse produces DDD and then further metabolizes it to DDMU, DDMS, and DDNS.[12] A species of *Hydrogenomonas* isolated from sewage can degrade DDT, but only under anaerobic conditions,[4] producing the above-mentioned metabolites[135]; it can then produce PCPA from DDM,[43] and this may be degraded by another bacterial species such as *Arthrobacter*.[135] The intermediate metabolite DDNS was discovered to be produced from DDT, and from DDD, by half of the bacterial isolates obtained from Lake Michigan sediments.[113]

Fungal Degradation. Isolates of the common soil fungi were inactive where the actinomycetes were active.[28] But 12 out of 18 isolates of *Trichoderma viride* could metabolize DDT under aerobic conditions, 8 producing DDD plus dicofol (Fig. 4.2), 1 producing DDD plus DDE, and 3 producing DDD only; an isolate from an Ohio orchard, degraded 90% of the DDT in 3 days.[109] The fungus *Fusarium oxysporum* could produce DDD from either DDT or DDE, the metabolic pathway then passing through DDMU, DDOH, and DDA to DBH.[36] A species of hyaline Moniliaceous fungus was able to break down DDM and PCPA to CO_2 and chloride when it was in company with *Hydrogenomonas,* utilizing the breakdown products of this bacterium as nutrient.[42] *Mucor alternans* was capable of taking DDT to water-soluble metabolites which were apparently chlorinefree.[78] This fungus species degraded DDT in the laboratory so rapidly that applications of its spores were made to soil with the objective of decontaminating the DDT residues, but in the soil ecosystem this *Mucor* apparently lost its degradative capacity.[6] On the other hand, bacterial cultures of *Aerobacter aerogenes* added to soils roughly doubled the rate of loss of DDT, being more effective in clay than in loams.[83] Yeasts are also very active in degrading DDT, *Saccharomyces cerevisiae* dechlorinating more than half of it in 50 hr, while leaving DDE intact.[79]

In soil, it is difficult to detect in any quantity the metabolites beyond DDD and DDE, possibly because the more polar compounds cannot be recovered from soil samples.[131] Whereas DDD is further metabolized, DDE is evidently an end product; *Aerobacter,* for example, cannot degrade it under either aerobic or anaerobic conditions.[162] In sewage and water–sediment mixtures, DBP can be detected in addition to DDD.[135] A by-product of DDT in sewage is the toxic nitrile DDCN,[3] which has also been discovered to be produced in lake sediments in Sweden.[71]

The structural isomer *o,p'*-DDT, present as an impurity of technical grade DDT in amounts approximating 15%, is degraded by *Aerobacter* to *o,p'*-DDD.[115] Methoxychlor, an analog of DDT in which the *para* chlorines are replaced by methoxy groups, is not reductively dechlorinated by *A. aerogenes* to the analog of DDD, but a certain amount is dechlorinated to the methoxy analog of DDE.[115] Chlorobenzilate, an ethyl ester of 2-hydroxy DDA that is used as an acaricide, is hydrolyzed by the yeast *Rhodotorula* to *p,p'*-dichlorobenzilic acid and then decarboxylated to DBP.[123]

Fate and Breakdown of Other Organochlorines

The cyclodiene insecticides, particularly aldrin and heptachlor, have been extensively applied to control soil insects, while endrin was a staple cotton insecticide until the boll weevil became resistant to it. Endrin and dieldrin are almost as persistent as DDT (Table 4.2), chlordane and its active constituent heptachlor have a half-life not exceeding 1 yr, while aldrin is the least persistent because of its appreciable volatility and its conversion to dieldrin.

Aldrin. A year after the application of ^{14}C-aldrin to a cornfield at 1 lb/a, about 25% of the radioactivity remained in the soil and it was nearly all dieldrin; this dieldrin persisted almost unchanged for at least 4 yr, when 0.1 ppm still remained in treated soil.[90] In cornfields 6 yr after annual applications of aldrin at 5 lb/a had been terminated, the soil residues stood at 0.9 ppm dieldrin, 0.015 ppm photodieldrin, and 0.005 ppm aldrin, a decline from the initial level of 3 ppm aldrin (Fig. 4.1); 6 yr after similar applications of heptachlor, the residues were 0.1 ppm heptachlor and 0.5 ppm heptachlor epoxide.[100] In silt loam 9 yr after applications of aldrin or of heptachlor at 5 lb/a had been tilled in, the following residues (in ppb) were still present despite the annual raising of crops:

Aldrin	5	Heptachlor	9
Dieldrin	98	Heptachlor epoxide	169

These levels were sufficient to immobilize the wireworms that the original applications were designed to control.[165] In sandy loam 14 yr after application of aldrin at 100 ppm, the dieldrin concentration was still 40 ppm despite occasional cropping; heptachlor at 100 ppm resulted in a final concentration of 16 ppm heptachlor plus its epoxide at the end of the 14-yr period.[127]

When aldrin is applied at 0.5 lb/a in the spring, there is already twice as much dieldrin as aldrin in the treated soil by the autumn, and five times as much dieldrin as aldrin by the following year; with heptachlor applied at that rate, about one-sixth of it has been converted to heptachlor epoxide (Fig. 4.3) by the end of the growing season.[46] This epoxidation, to a product which is no less insecticidal, is caused by microorganisms (Table 4.4); the actinomycetes *Nocardia* and *Streptomyces,* and the fungi in the genera *Trichoderma, Fusarium,* and *Penicillium,* are the most effective to convert aldrin to dieldrin.[155] The epoxidation of heptachlor could be accomplished by 26 of 45 soil bacteria and actinomycetes tested, and by 35 of the 47 fungal isolates.[120] The maximum epoxidation achieved in 6 wk was 70% aldrin to dieldrin by a *Fusarium* isolate, and 18% heptachlor to its epoxide by a *Rhizopus* isolate.

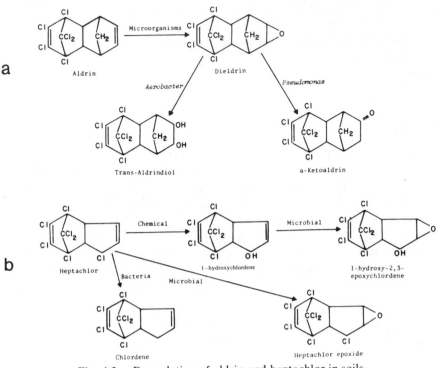

Fig. 4.3.   Degradation of aldrin and heptachlor in soils.

Dieldrin.   This stable compound is not degraded in soils sterilized of microorganisms,[96] although it is decomposed to a slight extent on the clay minerals employed in wettable powders.[44] Of 577 isolates of soil organisms, only 10 were found capable of degrading dieldrin, namely six *Pseudomonas,* two *Bacillus,* and two *Trichoderma* isolates. *Trans*-aldrin diol (6,7-*trans*-dihydroxydihydroaldrin) was the principal metabolite (Fig. 4.3); but only 1–6% of the dieldrin was converted to the diol and six other water-soluble compounds by these microorganisms.[108] A culture of *Aerobacter aerogenes* could convert 12% of the dieldrin to *trans*-aldrindiol in 3 days.[164] An isolate of *Trichoderma viride* from an Ohio orchard produced *trans*-aldrindiol among five metabolites, and converted 41% of the added dieldrin into solvent-soluble and 2% into water-soluble products in 1 month.[109] On the other hand, a *Pseudomonas* isolated from a soil sample taken from the cyclodiene-manufacturing plant at Denver, Colorado produced a different aldrindiol, besides one aldehyde and two keto derivatives; these ketoaldrins were found along with dieldrin in the soil sample itself, and the more abundant of the two was characterized by internal cyclization.[110]

Table 4.4. Proportion of isolates of microorganisms found to degrade cyclodiene insecticides (references in brackets).

	Ald-Dld (155)	Dld-CO$_2$ (70)	Ald-Diol (134)	End-Keto (134)	Hept-Epox (120)
Arthrobacter	2/5	19/35	0/1	1/1	x/5
Bacillus	6/12	1/1	1/3	3/3	x/12
Pseudomonas		10/24	5/9	9/9	
Mycobacterium		24/47			
Corynebacterium		22/44			0/1
Mycococcus		3/5			
Micrococcus		1/1	1/1	1/1	
Nocardia	13/16	8/15			x/16
Trichoderma	15/15	1/1*	2/2	2/2	x/15

* Ref. (16)

	Ald-Dld (155)	Hept-Epox (120)
Micronospora	1/1	1/1
Streptomyces	6/7	x/7
Actinomyces	3/3	x/3
Aspergillus		0/2
Fusarium	8/16	x/16
Mucor	0/1	0/1
Penicillium	6/11	x/11
Rhizopus	1/2	x/2

An isolate of *Trichoderma koningi* from cranberry mold could convert 3% of the dieldrin into CO_2 in 16 days, and if the culture was enriched the conversion was 50% in that period.[16] A German soil treated with 10 ppm dieldrin converted 2% of it to CO_2 in 7 wk; of 175 isolates of bacteria and actinomycetes (Table 4.4), 91 could degrade dieldrin to CO_2 and at least seven water-soluble metabolites, a *Micrococcus* for example converting 0.11% and a *Nocardia* 0.05%, into CO_2 within 5 wk.[70]

It is clear that the breakdown of dieldrin in the soil is very slow, the chlorinated-ring moiety being particularly stable; moreover, none of the products mentioned (except CO_2) have been shown to be less toxic.[106] Photodieldrin, a product of ultraviolet irradiation that is more toxic than dieldrin to houseflies and mice,[140] has been found in soil along with hydrophilic metabolites.[100] Of 500 isolates from dieldrin-contaminated soils that were tested, 42 produced photodieldrin.[112] This product is degraded by microorganisms much faster than dieldrin,[106] and several water-soluble metabolites have been produced from it by *Penicillium notatum* and *Aspergillus flavus*.[91]

Heptachlor. In soil, heptachlor is liable to chemical hydroxylation (hydrolysis) to 1-hydroxychlordene (Fig. 4.3), which often reaches much greater concentrations than heptachlor epoxide, even accounting for 60% of the heptachlor applied.[26] When heptachlor was incubated in a salt solution for 6 wk, it was all converted to hydroxychlordene; when microorganisms were added to it, the metabolite 1-hydroxy-2,3-epoxychlordene was produced from the hydroxychlordene, as well as heptachlor epoxide from the heptachlor. Chlordene was among the metabolites produced by actinomycetes and bacteria, but not by fungi. The maximum concentration of each metabolite produced from 1 ppm heptachlor by any microorganism was 0.36 ppm chlordene produced by a *Bacillus* and 0.25 ppm hydroxyepoxychlordene produced by a *Fusarium*. Complete conversion of hydroxychlordene to its epoxy derivative could be effected by *Micromonospora*.[120] Hydroxyepoxychlordene appeared at deeper levels in a heptachlor-treated Oregon soil 4 yr after the application, presumably originating from the hydroxychlordene already produced by chemical hydroxylation.[27] Both the above breakdown products are noninsecticidal, and thus the degradation is probably a general detoxication.

Endrin. Of 150 soil isolates, 25 of them could degrade endrin to keto-endrin and six other metabolites, the extent of the conversion ranging from 5% with a *Bacillus* to 46% with a *Pseudomonas*.[111] Soil microorganisms that could degrade dieldrin, including species of *Arthrobacter, Micrococcus,* and *Trichoderma* (Table 4.4), could also convert endrin to ketoendrin.[134] Considerable conversion of endrin to the aldehyde and ketone can occur even in air-dry soils.[10] Telodrin (isobenzan), a highly toxic relative of heptachlor epoxide but with the oxygen in the pentane ring, could be degraded to a water-soluble metabolite by *Aspergillus flavus, A. niger,* and *Penicillium notatum.*[92]

Mirex. This polycyclic organochlorine, employed in bait granules to control the imported fire ant, is not broken down by any soil organisms, nor does it affect the population levels of soil microorganisms.[76] But in sewage sludge under anaerobic conditions in the dark, mirex is about 80% degraded into an unknown metabolite.[7]

Gamma-HCH. Lindane, a technical product containing more than 99% of the gamma isomer of BHC, is the least persistent organochlorine insecticide in soil. The rate of loss by volatilization is at least 17 times greater for lindane than for DDT,[69] about 20% of that applied to soil being evaporated in 40 days.[55] After 3 yr in a silt loam, 10–14% of the lindane applied is still present. By comparison, after 14 yr in a sandy loam about 10% of the HCH isomers in an application of technical BHC still remained in the soil.[95]

In flooded paddy soils in the Philippines, 45–100% of the lindane applied was broken down after 1 month, while in unflooded soil the degradation was only 2% in that period.[171] With technical BHC, the loss of HCH-isomers was 99% after 70 days, but only 30% if the soil had been sterilized.[104] In hydrosoils under anaerobic conditions, gamma-HCH isomerizes to the alpha and delta isomers, respectively 4 and 50 times less insecticidal,[130] but this did not occur in the Philippine rice soils.[171]

In an Ontario soil, lindane was dehydrochlorinated to gamma-PCCH (2,3,4,5,6-pentachlorocyclohexene), which is 1000 times less insecticidal than gamma-HCH; this conversion, amounting to 9% in 2 months, could be effected by *Bacillus cereus* isolated from this soil.[172] In a flooded clay loam, gamma-HCH was degraded to gamma-BTC (3,4,5,6-tetrachlorocyclohexene); the conversion, amounting to 20% in 2 wk, was inhibited by the antibiotic sodium azide.[153] A *Clostridium* isolated from a paddy soil dechlorinated gamma-HCH to pentachlorocyclohexene, and broke down gamma-PCCH faster than gamma-HCH. In submerged tropical soils, microorganisms were active in producing CO_2 from lindane[104]; but 20 soils that could degrade dieldrin, endrin, and DDT were unable to break down gamma-HCH.

B. FATE OF NONPERSISTENT INSECTICIDES IN SOIL

Duration of Activity of OP Compounds

The organophosphorus insecticides as a class have much less persistence in soil than the organochlorines, and thus it may be expressed in terms of weeks rather than years (Fig. 4.4). They may be applied to soil directly to control rootworms, root maggots, and cutworms (e.g., diazinon, phorate), or they may reach the soil incidental to their application to field crops or orchards (e.g., parathion, azinphosmethyl). Their fate in insects and plants has been covered in the review by Menzie (1972). It is very unusual for an OP insecticide to persist from one year to the next, as shown by the virtual disappearance of parathion from the soil long before the next application is made (Fig. 4.5). However, spray treatments at the high rate of 30 lb/a applied for four consecutive years resulted in parathion still being found in traces in the soil 16 yr after they had ceased.[151]

Diazinon is an OP insecticide that persists long enough to give control of soil insects for an entire growing season (see Fig. 4.7). Its half-life in a loam soil was 20–40 days,[25] and it was 85% lost after 20 wk in a silt loam.[47] It has all gone in 180 days when applied at 3 lb/a, and in 70 days when applied at 0.3 lb/a.[58] It may be generalized that its "persistence," the time it takes for more than 75% to be lost,[83] is 12 wk for diazinon, as compared to 1 wk for parathion and malathion.

Azinphosmethyl, which reaches the soil as a result of the spraying of orchard foliage, resembles diazinon in its persistence (Fig. 4.5), being 90% lost after 5 months in a silt loam.[141] Its half-life ranges from 1 month in

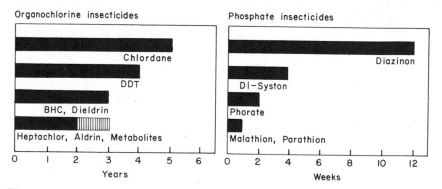

Fig. 4.4. Persistence of insecticides in soils: time taken for a compound applied at the normal dosage to decrease by 75% (from Kearney et al., 1969).

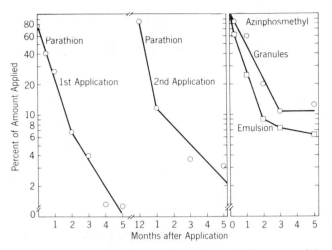

Fig. 4.5. Residues of parathion and of azinphosmethyl recovered from soil after application at 5 lb/a (from Lichtenstein and Schulz, 1964; Schulz et al., 1970). Reprinted from the Journal of Economic Entomology by permission of the copyright owners, the Entomological Society of America.

Florida muck-sand to 3 months in Louisiana clay.[5] After an initial latent period of 3–10 days, its disappearance proceeds according to first-order kinetics, declining to 4% of the starting level after 8 wk in a silt loam at 25°C; in sterilized soil 20% remains, and in dry soil 63% is still present, at the end of the 8-wk period.[170] The decided effect that the edaphic type exerts on the persistence of an OP insecticide is exemplified by chlorfenvinphos (Birlane), whose half-life can range from 2 wk to more than 23 wk depending on the soil.[15] Thus valid comparisons of insecticides for their persistence can be drawn only where they were made in the same soil. The longest series of OP compounds to be tested uniformly (Table 4.5) assessed the amount of insecticide remaining by bioassay with an insect, either the eye gnat *Hippelates*[126] or the field cricket *Acheta*.[61-64] In the latter series (Table 4.6), where the tests are based on the speed of descent from the (minimal) LC_{100} level to lower percent mortalities, diazinon has become classed along with parathion and chlorpyrifos as slightly residual insecticides. It contrasts with dimethoate, which in sandy loam showed a half-life of 2 days.[17]

There are a few cases where different insecticides were compared under the same conditions by means of chemical determination. When applied at 1–25 lb/a, the period in days for which the insecticide remained detectable in clay soil was 2 for malathion, 9 for diazinon, and more than 23 days for

Table 4.5. Persistence of organophosphorus insecticides in Coachella fine sand as bioassayed by *Hippelates collusor* larvae (Mulla et al., 1961; Mulla, 1964).

< 2 months	> 2 months	> 6 months
Parathion	Fenitrothion	Diazinon
Fenthion	Zinophos	Coumaphos
Ronnel	Phorate	Trichloronat
Azinphosmethyl	Phosmet	Carbophenothion
Disulfoton		Ethion
Dimethoate		Crotoxyphos
		Chlorfenvinphos

demeton and phorate.[93] Applied to silt loam, methyl parathion was 97% lost in 1 month, while parathion took 3 months to lose this much.[97] After 20 wk in silt loam, 90% of the thionazin (Zinophos) applied had been lost as compared to a loss of 85% with diazinon.[47] The half-life of the systemic aphicide menazon in sandy soil was 23 days, as compared to 57 and 68 days, respectively, for thionazin and phorate.[160] Four months after applications had been made to a soil cover crop of rye only 3% of a chlorpyrifos application could be found, whereas 15% of the leptophos and 20% of the temephos (Biothion) applied were still remaining.[65]

Chlorpyrifos and fonofos have half-lives in peaty loam of 18 and 22 wk, respectively, as compared to 5 wk for diazinon; phorate has a similar residual longevity, only that it has all been converted in a few days to the insecticidal sulfoxide and sulfone.[152] Fensulfothion (Dasanit) is slower in being converted to the sulfone, which is insecticidal and takes a year to disappear.[67] Disulfoton is less persistent,[64] although it is converted to the sulfoxide which is fairly stable in soil.[66]

Chemical Hydrolysis

Among the least persistent insecticides are those that are hydrolyzed chemically even before soil microorganisms get to work on them. Thus the alkali-labile phosphorodithioate malathion is degraded within 24 hr by 50% in clay or silty clay loam, and by as much as 90% in loamy sand.[89] It is more usual for complete breakdown of malathion to take about 10 days, of which as much as 90% in loam and 75% in clay is due to microorganisms.[157] The phosphate dicrotophos (Bidrin) is less ephemeral, 15% remaining after 8 days in sandy loam, while 10% of the phosphorodithioate Imidan (phosmet) remains after 11 days in a loam soil.[117] Of the more persistent OP soil insecticides, only 1% of the thionazin (Zinophos) applied remains after 1 month in an alkaline clay soil or 3 months in an acid sandy soil,[132] while a trace of the phorate applied to silt loam remains after 5 months.[94] On the average, depending on the soil, 20% of the chlorfenvinphos (Birlane) applied is found remaining after 4 months,[14] and 40% of the carbophenothion (Trithion) remains after 6 months in loamy sand.[116] Diazinon, a fairly persistent insecticide that depends on microorganisms for much of its decomposition, is hydrolyzed chemically to a certain extent in acid soils, in contrast to malathion and monocrotophos, which are alkali hydrolyzed.[8,68]

The positive correlation between chemical hydrolytic breakdown and the adsorptive organic-matter content of the soil is demonstrated by the decomposition rates found for diazinon in three Wisconsin soils, as follows[88]:

	Poygan	Kewaunee	Ella
Percent Organic Content	10.0	3.8	1.6
Percent Breakdown	11	7	6

Acceleration of breakdown in soils of greater adsorptive capacity was also shown for malathion.[89] For the systemic insecticide disulfoton (Di-Syston),

Table 4.6. Persistence of organophosphorus insecticides in sandy loam, as bioassayed by *Acheta pennsylvanica* nymphs (Harris, 1969, 1970, 1973; Harris and Hitchon, 1970).

Slightly residual	Moderately residual	Highly residual
Parathion	Fensulfothion	Mephosfolan
Diazinon	Dimethoate	Phosfolan
Chlorpyrifos	Ethoprop	Pirimiphos-ethyl
Bromophos	Trichloronat	
Phoxim	Pirimiphos-methyl	
Phenthoate		
Phorate		
Disulfoton		

although its adsorption was proportional to the organic matter content of the soil, breakdown was being accomplished by microorganisms in aqueous solution.[51]

Relative Importance of Biological Degradation

Autoclaving the soil, which *inter alia* destroys the microorganisms, reduces the percentage breakdown of OP insecticides in soil, as demonstrated by the following loss figures:

	Unautoclaved	Autoclaved
Dicrotophos, 8 days in sandy loam[59]	85	40
Phosmet, 11 days in loam[117]	90	80
Carbophenothion, 200 days in loamy sand[116]	60	20

Autoclaving reduces the loss rate of diazinon in soil by about two-thirds,[142] while steam treatment of the soil extends its half-life by 60–90%[25] Sterilization of the soil by irradiation reduced the loss rate of crotoxyphos (Ciodrin), but only because it decreased the amount absorbed.[87]

Destruction of the microorganisms in the soil by exposing it to gamma-radiation did not decrease the power of a sandy loam to degrade OP insecticides nearly to the extent that autoclaving did (Table 4.7). Getzin and Rosefield (1968), therefore, made extracts from irradiated and autoclaved soil in 0.2 *N* NaOH, and found that addition of these two extracts, and an extract from the untreated soil, resulted in the following percentage breakdowns of malathion incubated for 20 hr in *tris* buffer: untreated 88, irradiated 75, autoclaved 25. This indicated to them the existence of a heat-labile factor promoting malathion breakdown. It is possible that this factor may be related to the exoenzymes known to exist free in the soil.[147] Water from ricefields that had been repeatedly treated with diazinon was found by Sethunathan and Pathak (1971, 1972) to be capable of mineralizing diazinon within 4 days; the factor was inactivated by streptomycin, and strains of *Flavobacterium* and *Arthrobacter* were isolated from the water.

The role of soil microorganisms in breaking down OP insecticides was first investigated by Ahmed and Casida (1958), who found that phorate (Thimet) incubated for 8 days in cultures of two species of bacteria, an alga and a yeast, showed the following percentage breakdowns:

Thiobacillus	75	*Chlorella*	22
Pseudomonas	58	*Torulopsis*	78

Table 4.7. Effect of microorganisms and their thermolabile products on the degradation rate in clay loam (Getzin and Rosefield, 1968).

	Incubation Period	Autoclaved soil	Irradiated soil	Non-sterile soil
Malathion	1 day	7	90	97
Ciodrin	1 day	4	34	87
Dichlorvos	1 day	17	88	99
Mevinphos	1 day	1	38	95
Methyl parathion	1 wk	20	26	95
Methidathion	1 wk	17	20	50
Parathion	2 wk	17	16	35
Dimethoate	2 wk	18	20	77
Zinophos	4 wk	17	24	71
Dursban	4 wk	33	38	62

The alga *Chlorella pyrenoidosa* could also degrade parathion, ronnel, schradan, and dimefox, while the yeast *Torulopsis utilis* could also break down methyl parathion, ronnel, chlorthion, and dicapthon. Since that time many species of bacteria and fungi, and one actinomycete (*Streptomyces*), have been added to the list of isolates from soil capable of degrading OP insecticides (Table 4.8). One of the most active is the common fungus *Trichoderma viride*,[109] which breaks down the OP compounds faster than carbamates and much faster than the organochlorine insecticides (Table 4.9).

Metabolites of OP Compounds

The possibility that the metabolites produced by breakdown of the OP insecticides in soil might be a source of poisonous pollution was raised by the finding that phorate gave rise to soil residues that were more insecticidal to *Drosophila* than the original material applied.[32] There was particular concern about the possible production of the strongly ChE-inhibiting oxon derivatives of thiophosphate insecticides. Therefore, the metabolic products found in treated soil or produced by microorganisms isolated from soil have been investigated in some detail.

Table 4.8.   Soil bacteria, actinomycetes, fungi, yeasts, and algae found capable of degrading organophosphorus insecticides (references in brackets).

Pseudomonas sp.	Methyl parathion (129)
Pseudomonas fluorescens	Phorate (1)
Bacillus subtilis	Methyl parathion, Fenitrothion, EPN (122)
Thiobacillus thiooxidans	Phorate (1)
Flavobacterium sp.	Diazinon (144)
Arthrobacter sp.	Diazinon (143), Malathion (158)
Rhizobium spp.	Parathion (119), Malathion (125)
Streptomyces sp.	Diazinon (142)
Trichoderma lignorum	Bromophos (150)
Trichoderma viride	Parathion, Diazinon, Dichlorvos (109), Malathion (107)
Aspergillus niger	Malathion (124), Trichlorfon (173)
Penicillium notatum	Malathion (124), Trichlorfon (173)
Alternaria tenuis	Bromophos (150)
Rhizoctonia solani	Malathion (124)
Fusarium sp.	Trichlorfon (173)
Torulopsis utilis	Methyl parathion, Ronnel, Chlorthion, Dicapthon, Phorate (1)
Chlorella pyrenoidosa	Parathion (174), Ronnel, Phorate, Schradan, Dimefox (1)

Phosphorothioates.   The breakdown products found in soil treated with parathion consist principally of aminoparathion (Fig. 4.6), produced mainly by the reducing capacity of yeasts, and *p*-nitrophenol, produced by chemical hydrolysis and the hydrolytic capability of soil bacteria; these products by themselves are destroyed in moist soils in 2 and 16 days, respectively.[97] In some soils small amounts of *p*-nitrophenol persist where aminoparathion is no longer found[168]; but *Corynebacterium simplex* is able to convert it completely to nitrite in 1 wk.[56] Paraoxon is either not found[97] or only in extremely small concentrations[168]; added to soil, it is all hydrolyzed within 12 hr.[97] Cultures of *Rhizobium* converted 85% of ^{14}C-parathion to aminoparathion in 24 hr, but 10% of the radioactivity appeared as the hydrolysis product DEPTA (diethylphosphorothioic acid).[119] Cultures of *Chlorella* converted 53% of the parathion in the culture medium to aminoparathion in 7 days; although 3 unidentified metabolites were also produced, *p*-nitrophenol and *p*-aminophenol appeared in only traces.[174]

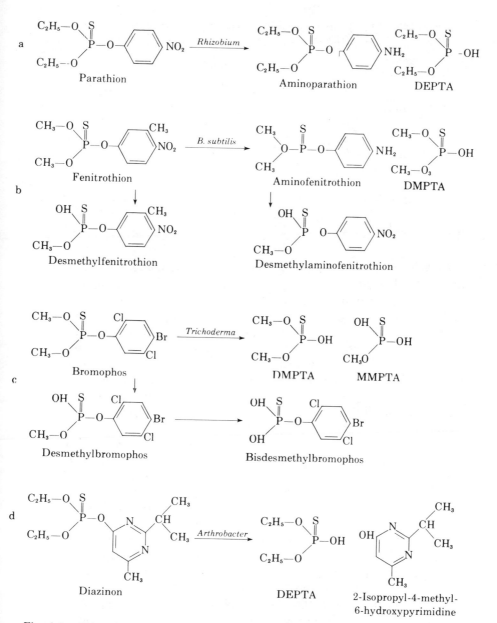

Fig. 4.6. Degradation of phosphorothioate insecticides by soil microorganisms.

Table 4.9. **Degradation of insecticides by the soil fungus**
Trichoderma viride (Matsumura and Boush, 1968).

	Days Exposure	% Apolar Metabolites	% Polar Metabolites
Diazinon	1	0.1	24.4
Parathion	1	3.3	22.2
Dichlorvos	1	0.5	95.4
Carbaryl	1	9.2	2.1
DDT	3	75.5	13.0
Dieldrin	30	40.8	1.9

Cultures of *Bacillus subtilis* converted 65% of radioactive fenitrothion to its amino derivative in 2 days; DMPTA (dimethylphosphorothioic acid) also appeared, as well as the desmethyl derivatives of fenitrothion and of the amino derivative (Fig. 4.6). Methyl parathion was similarly converted to analogous metabolites, but faster, and no oxon derivatives were found.[122] The fungi *Trichoderma* and *Alternaria* dealkylated bromophos into its desmethyl and bidesmethyl derivatives, and hydrolyzed it to dimethyl and monomethyl phosphorothioic acids.[150] It may be generalized that microorganisms hydrolyze rather than oxidize the OP insecticides, while fungi can in addition dealkylate them, and yeasts can reduce nitro groups where they are present.

Diazinon sprayed into silt loam was yielding the hydrolysis product 2-isopropyl-4-methyl-6-hydroxypyrimidine 3 wk later, but no diazoxon. Most of the hydrolysis products were concurrently converted to CO_2 (Fig. 4.7), but if the soil was fumigated to kill the microorganisms, then more hydrolysis products accumulated and less CO_2 was produced.[47] Hydrolysis to DEPTA (diethylthiophosphoric acid) and the substituted pyridine mentioned previously (Fig. 4.6) was the major mechanism in diazinon degradation in Wisconsin soils.[88] In submerged paddy soils in the Philippines, the pyrimidine hydrolysis product came to exceed the diazinon by 12 days after the application[145]; a *Streptomyces* was isolated, but there was little conversion of the pyrimidine ring to CO_2.[142] An *Arthrobacter* having been isolated that metabolizes diazinon in presence of glucose,[143] it was found that *Streptomyces* when in company with *Arthrobacter* can break the ring to CO_2, while *Arthrobacter* alone can convert only the DEPTA moiety into CO_2.[57]

When ^{14}C-thionazin was applied to silt loam, much of the radioactivity ended up as CO_2 (Fig. 4.7); the hydrolysis product 2-pyrazinol could not be identified,[47] but this was probably because it is rapidly leached and the remainder is strongly absorbed.[85] When fensulfothion, a phosphorothioate with a sulfoxide group, was applied to soil it was converted to the more stable Dasanit sulfone; in 4 wk there was more sulfone than sulfoxide, and 52 wk the residues of both fensulfothion and its oxidative metabolite had disappeared.[67] The conversion of the sulfur atom was similar to what occurred to phorate, a phosphorodithioate (see Fig. 4.9 below).

Phosphorodithioates. The effect of soil microorganisms on malathion, which is largely degraded by chemical hydrolysis, was investigated by Matsumura and Boush (1966) in three soils and two organisms isolated from two of them (Table 4.10). Both the *Pseudomonas* bacterium and the *Trichoderma viride* fungus isolated were most active in deethylating the carboxylic acid side chain of the malathion, but dealkylation of the O-methyl substituents was also important, especially with *Trichoderma*. When *Rhizobium trifolii* or *leguminosarum* were incubated with ^{14}C-malathion for 1 wk, about 25% of the activity had become malathion monoacid, 20% malathion diacid, 5% dimethyl phosphorodithioate, 10% dimethyl phosphorothioate (DMPTA), and 5% dimethyl phosphate (Fig. 4.8), while 30% had become completely mineralized to inorganic phosphate.[125] A species of *Arthrobacter* was isolated from Mississippi soil that broke down malathion to the first four metabolites mentioned.[158] When *Penicillium, Rhizoctonia,*

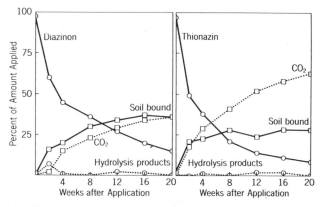

Fig. 4.7. Distribution of radioactivity resulting from the degradation of ^{14}C-diazinon and ^{14}C-thionazin in silt loam (after Getzin, 1967).
Reprinted from the Journal of Economic Entomology by permission of the copyright owner, the Entomological Society of America.

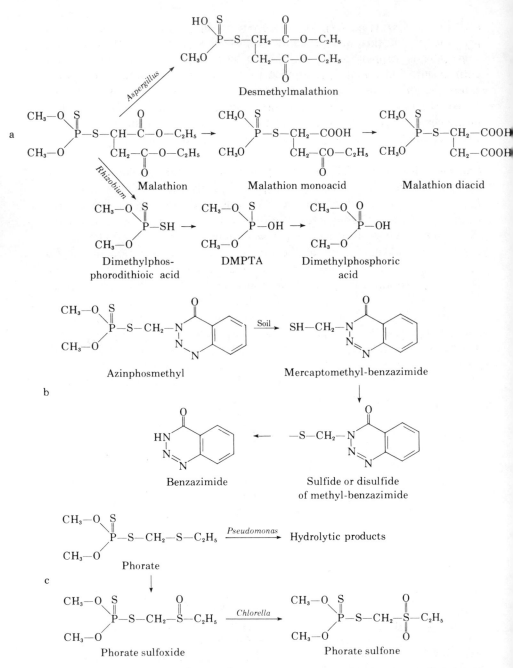

Fig. 4.8. Degradation of phosphorodithioate insecticides in soil and by soil microorganisms.

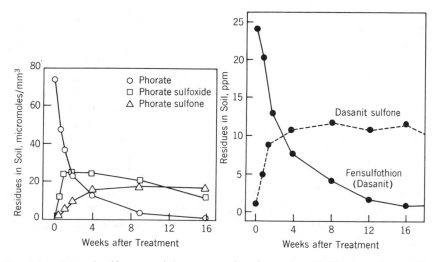

Fig. 4.9. Fate of sulfur-containing organophosphorus insecticides in soil: oxidation of phorate and of fensulfothion (from Getzin and Shanks, 1970; Harris et al., 1972).

and notably *Aspergillus* fungi were incubated with malathion, the metabolites were similar except for the lack of dimethyl phosphorodithioate and the presence of the dealkylated product desmethyl phosphorodithioate.[124] When the phosphorodithioate azinphosmethyl was applied at 5 lb/a to silt loam, the 13% of the radioactivity remaining 1 yr later consisted not so much of azinphosmethyl as of its metabolites (Fig. 4.8), among which the mercaptomethyl and methyl derivatives, and the sulfide and disulfide, were detected.[100]

Table 4.10. Degradation of malathion by soils and soil isolates of *Pseudomonas* sp. and *Trichoderma viride*: percent converted after 24 hr (Matsumura and Boush, 1966).

	Tope orchard, Ohio		Snyder orchard, Ohio		Shell plant
	Soil	*Pseudomonas*	Soil	*Trichoderma*	Soil
Desmethyl malathion	19.3	11.3	8.9	27.6	0.8
Carboxylesterase products	50.7	80.4	41.0	43.3	25.2
Diethyl malate	0.4	3.1	0.3	11.4	0.0
Other hydrolysis products	15.8	3.2	22.8	1.2	1.9
Malathion remaining	13.8	2.0	27.0	16.5	71.9

Phorate, an aliphatic phosphorodithioate, was the first OP insecticide to be studied for microorganismal breakdown, and it was found that *Chlorella* and *Torulopsis* oxidized the S in the side chain to the sulfoxide and the sulfone (Fig. 4.8), while *Thiobacillus* and *Pseudomonas* hydrolyzed the side chain from the phosphorodithioic acid moiety.[1] Phorate sulfoxide came to exceed phorate only 1 hr after the insecticide had been applied to a silt loam soil.[94] By 8 wk after the application, the sulfoxide and the sulfone each exceeded the phorate remaining by four times (Fig. 4.9), and after 16 wk only the sulfoxide and sulfone could be found, although when sulfoxide was added to the soil some original phorate appeared.[49] The oxon derivative of phorate appearing in the soil treated by Dewey and Parker (1965) could have been an impurity in the technical formulation applied, since the only oxon that microorganisms produced was that of the phorate sulfoxide.[1] Phorate oxon is slightly more insecticidal, phorate sulfoxide slightly less insecticidal, than phorate.[32]

Phosphonates. The aliphatic phosphonate trichlorfon (Dipterex), when exposed to cultures of *Aspergillus, Penicillium,* or *Fusarium* (Fig. 4.10), was dealkylated to monomethyl trichlorohydroxyethyl phosphonate, and to a metabolite which was probably the trichlorohydroxyethylphosphonic acid.[173] Dyfonate, a phosphonodithioate, evidently is stripped of its O-ethyl substituent which is volatilized, while the S-phenyl moiety is removed by leaching.[99]

Phosphates. Whereas phosphate insecticides are typically ephemeral in soil due to chemical hydrolysis, the aryl phosphate chlorfenvinphos (Birlane) is quite persistent and has been studied in four different soils for the fate of its phenylethane group.[14] After 4 months, when 1–5% of the applied Birlane remained, about 0.1% of the radioactivity was found in each of three breakdown products, namely the monodesethyl derivative, the phenylethanol, and the acetophenone, with traces of the phenyl ethandiol and the phenacyl chloride (Fig. 4.10). In soils rich in organic matter, the first metabolite phenacyl chloride accumulates to the 0.1 ppm level.[15]

Nitrogenous Insecticides

Carbamates. The N-methylcarbamate carbaryl (Sevin) has been found to have a half-life of some 7 days in soils.[74] Depending on the edaphic type, between 2 and 38% of its carbon has been converted to CO_2 in 30 days.[82]

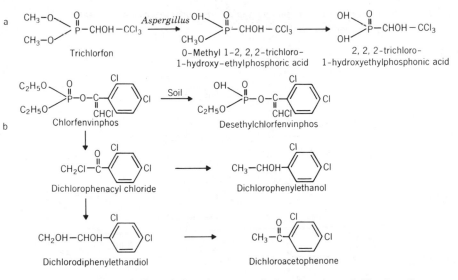

Fig. 4.10. Degradation of phosphonate and phosphate insecticides in soil.

Carbaryl is hydrolyzed to 1-naphthol in soil waters,[101] and in soils themselves[82] probably mostly by chemical hydrolysis, but it is also degraded by a variety of soil microorganisms (Table 4.11). Hydrolysis to 1-naphthol was accelerated by the fungus *Fusarium solani*, and by a gram-negative coccus and a gram-positive rodlike *Bacterium*.[18] Carbaryl is degraded by the fungus *Gliocladium roseum* to its N-hydroxymethyl carbamate (hydroxymethylcarbaryl, Fig. 4.11), and to two derivatives (4- and 5-hydroxycarbaryl) in which the naphthalene ring is hydroxylated.[102] Eight other species of fungi proved capable of hydroxylating carbaryl in these three positions on the molecule.[19] *Aspergillus terreus* produced 1-naphthylcarbamate (desmethylcarbaryl) in addition, and it could hydrolyze both this compound and the hydroxymethylcarbaryl to 1-naphthol.[103] The naphthol can be quickly metabolized to CO_2 by *Fusarium solani*,[20] by a gram-negative coccus,[18] and by a species of *Pseudomonas* which produces coumarin and three other metabolites in the process of converting 1-naphthol to CO_2.[82]

Mexacarbate (Zectran) is degraded by soil microorganisms by two routes. In five isolates, including the bacterium HF_3, it was demethylated to the aminophenyl derivative (Fig. 4.11). In nine isolates it was decarbamylated to dimethylaminoxylenol; both the bacterium *Pseudomonas* and the fungus *Trichoderma viride* belonged to this group.[13] Propoxur,

Table 4.11.   Soil microorganisms found capable of degrading carbamate and formamidine insecticides.

BACTERIA		FUNGI (cont'd)	
Aerobacter aerogenes	Chlordimeform[a]	F. moniliforme	Chlordimeform[a]
Pseudomonas sp.	Mexacarbate[b]	Gliocladium roseum	Carbaryl[g]
Serratia marcescens	Chlordimeform[a]	G. catenulatum	Aldicarb[e]
ACTINOMYCETES		Mucor racemosus	Carbaryl[d]
Streptomyces griseus	Chlordimeform[a]	Penicillium multicolor	Aldicarb[e]
FUNGI		Penicillium sp.	Carbaryl[d]
Aspergillus terreus	Carbaryl[c]	Rhizoctonia sp.	Aldicarb[e]
A. flavus	Carbaryl[c]	Rhizopus nigricans	Chlordimeform[a]
A. niger	Carbaryl[d]	Rhizopus sp.	Carbaryl[d]
Cunninghamella elegans	Aldicarb[e]	Trichoderma viride	Carbaryl[d], Mexacarbate[b]
Fusarium solani	Carbaryl[f]	T. harzianum	Aldicarb[e]
F. roseum	Carbaryl[d]		

[a](73)   [b](13)   [c](103)   [d](19)   [e](75)   [f](18)   [g](102)

however, was not degraded by isolates of *Trichoderma, Pseudomonas,* and *Bacillus* which could degrade endrin and dieldrin.[134]

Aldicarb has a half-life in soil of 1–2 wk, being oxidized to aldicarb sulfoxide (Fig. 4.11); the relative percentages of either compound to be found in the soil after aldicarb treatments at 20 ppm were as follows:

	Clay		Silty Clay Loam		Fine Sand	
	1 wk	12 wk	1 wk	12 wk	1 wk	12 wk
Aldicarb	57	0.7	53	0.4	65	3.6
Aldicarb Sulfoxide	32	9.5	42	25	28	50

Also present were traces of aldicarb sulfone, and of the oxime, oxime sulfoxide, nitrile sulfoxide, and nitrile sulfone of aldicarb.[30] Pure cultures of *Gliocladium, Penicillium, Rhizoctonia, Cunninghamella,* and *Trichoderma* fungi (Table 4.11) further degraded aldicarb to water-soluble metabolites of

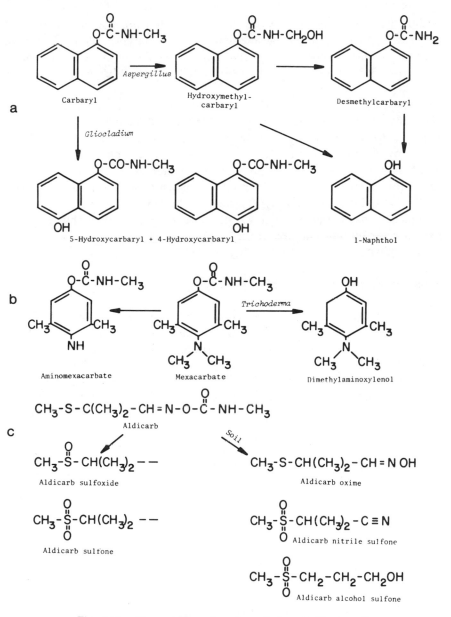

Fig. 4.11. Degradation of carbamate insecticides in soil.

which the most important were methyl-(methylsulfonyl)-propionamide
(aldicarb amide sulfone) and methyl-(methylsulfonyl)-propanol (aldicarb
alcohol sulfone); *Gliocladium catenulatum* was the most active of them,
converting 35% of the aldicarb to water-soluble metabolites in 3 wk.[75]

Formamidines. The insecticide and acaricide known as chlordimeform
(Galecron) was degraded, by removal of the dimethylamine group, to
formylchlorotoluidine by the bacteria *Aerobacter* and *Serratia,* the
actinomycete *Streptomyces,* and the fungi *Fusarium* and *Rhizopus* (Fig.
4.12); *Streptomyces* and *Serratia* in particular degraded it further to
chlorotoluidine.[73] Formetanate, simultaneously a formamidine and a carbamate insecticide, was similarly degraded by removal of the dimethylamine
group, and in addition by removal of the carbamate group; thus after 16
days in river-bottom soil 58% of the formetanate had been converted to for-
maminophenol, with 19% further degraded to aminophenol.[9]

Effect of Insecticides on Soil Microorganisms

Chlorinated Hydrocarbons. Applied at 10 lb/a to silty loam,[21] DDT did
not materially reduce the numbers of soil bacteria, actinomycetes, or fungi
(Table 4.12). Even at 400 lb/a in silt loam plots, DDT caused no reduction
in the numbers of bacteria, fungi, or protozoa.[149] At levels of the order of
100 ppm, as could be given by a 200-lb/a application, DDT did not affect
nitrate production,[38] although there was a stimulation of ammonification

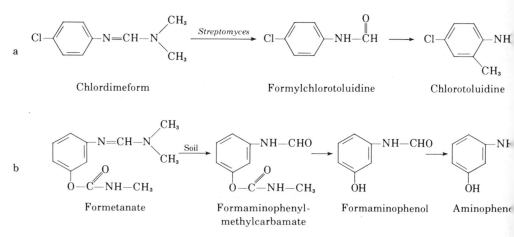

Fig. 4.12.   Degradation of formamidine insecticides in soil.

Table 4.12. **Effect of insecticides on numbers of soil microorganisms: silty clay loam at 10 lb/a (Bollen et al., 1954).**

	Population per gm dry soil			Percentage of mold spp. among fungi		
	Bacteria	Actinomycetes	Fungi	Penicillia	Mucors	Aspergillus
	millions	millions	thousands			
DDT	3.0	3.2	22.3	71	9	4
BHC	3.5	3.7	23.2	74	12	4
Aldrin	2.1	2.6	11.0	82	4	2
Dieldrin	3.6	3.7	42.2	83	3	1
Chlordane	5.0	3.7	40.2	85	3	5
Toxaphene	4.5	3.6	37.0	82	3	1
EPN	4.0	3.1	26.7	82	9	1
Parathion	4.3	5.5	20.8	82	10	1
Untreated	3.5	2.9	26.0	77	8	5

and in total numbers of microorganisms; there was no reduction in nodulation of legumes except in the common bean.[37,166]

At such unrealistically high levels in soil, technical DDT could have deleterious effects on certain sensitive crop plants, such as cowpeas, tobacco, and some varieties of rye; the p,p'-DDT that constitutes about 15% of technical DDT is about four times as phytotoxic as p,p'-DDT. When 10 species of bacteria and 10 species of actinomycetes were exposed to 10 ppm DDT in nutrient agar, 5 of the actinomycetes but only 2 of the bacteria were inhibited. DDD at the same concentration inhibited 8 of the actinomycetes and 4 of the bacteria. At 1 ppm, DDT inhibited *Corynebacterium fascians* and DDD inhibited *Streptomyces aureofaciens*.[86]

Technical BHC, a mixture of four HCH isomers, had no effect on the total numbers of bacteria and fungi when applied to silt loam at 400 lb/a; but it did suppress urea-hydrolyzing bacteria at 200 ppm,[52] and reduce nitrate production at 100 ppm,[38] while an application at 100 lb/a reduced the numbers of nitrate-forming bacteria in the soil.[149] Reductions in nitrifiers, and fungi, were not evident at dosages lower than 500 lb/a.[38] The ammonifying bacteria, and the total microorganismal count, remained high even in soil containing 1000 ppm[34] or 1500 ppm[167] of technical BHC. By contrast, only 12.5 ppm of the BHC product in soil was enough to inhibit nodulation of sweet clover by bacteria,[38] although at 1000 ppm in agar medium it did not affect *Rhizobium* or *Azotobacter* while it inhibited *Bacillus subtilis*.[34] A modest application of 1 lb/a of technical BHC was enough to control damping-off to *Rhizoctonia* in pine seedlings. Many crop plants are adversely affected if grown in soil containing 200 lb/a or more of this mixture of BHC isomers.[37]

Lindane, a gamma-HCH product containing less than 1% of the other BHC isomers, is less bacteriostatic. Lindane failed to inhibit the growth of 35 species of saprophytic bacteria at levels where technical BHC did;[52] inhibition of sweet clover nodulation was reached at 100 ppm lindane, and with BHC concentrations amounting to 25 ppm lindane.[37] Applied to Philippine ricefields at dosages from 2 to 20 kg/ha, lindane stimulated bacterial nitrogen fixation and increased the population of phosphate-dissolving bacteria.[137] Levels around 100 ppm decreased nitrification in some soils[22,38] but not in orders,[52] while ammonification was stimulated.[22] Even at 1000 ppm in clay soil, gamma-HCH increased the total numbers of bacteria, while reducing the numbers of mold fungi and *Streptomyces* actinomycetes.[22]

Aldrin had no effect on the overall bacterial and fungal population, nor on nitrogen fixation, when applied to soil at 100 lb/a[133] or at 100 ppm.[38] At 1000 ppm in Oregon soils, it had only minor and irregular effects on nitrification, and occasional stimulatory effects on ammonifiers and actinomycetes.[41] Nevertheless, aldrin inhibited sweet clover nodulation at 25 ppm.[38] Treatment of an Oregon silty clay loam at 10 lb/a reduced the numbers of mold fungi by more than 50% (Table 4.12).[21] Thus aldrin has had the side effect of reducing infection by barley root-rot from *Helminthosporium sativum*,[138] wheat damping-off from *F. culmorum*,[11] clubroot of cabbage from *Plasmodiophora brassicae*,[29] and take-all disease of wheat caused by *Ophiobolus graminis*,[148] as well as early blight and Fusarium wilt of tomato.[139] Dieldrin, applied at 10 lb/a[21] or at the regular rate of 1 lb/a for three successive years[37] had no deleterious effect on soil microorganisms or nitrate production. At 100 ppm in fine sand, dieldrin increased the numbers of fungi.[38]

Technical chlordane, a mixture of cyclodiene isomers and analogs, had no deleterious effects on soil bacteria and fungi, nor on nitrification or ammonification, at concentrations ranging up to 220 ppm.[21,37,38,133] Applied at 300 lb/a, where it did not suppress nitrification, chlordane caused a severe setback to Sudan grass and a tomato crop,[146] and even at 25 lb/a it depressed the growth of most vegetable crops.[37] Heptachlor, the most insecticidal compound in chlordane and marketed at technical-grade purity, had no overall effect on microorganismal numbers at 100 ppm, but it decreased nitrate production and inhibited sweet clover nodulation.[38] At the lower dosage of 100 lb/a, there was no effect on nitrification.[146] Fields treated with three successive annual applications at 300 lb/a had normal nitrification, crop yield, and fungal populations.[38] Although it increased the incidence of barley root-rot,[138] heptachlor was the most effective of the cyclodiene insecticides in controlling the fungal take-all disease of wheat due to *Plasmodiophora brassicae*.[148]

Toxaphene, a mixture of chlorinated terpenes, had no adverse effect on soil microorganisms at 100 ppm,[38] although at 500 lb/a it caused a slight reduction in nitrite-forming bacteria.[149] Indeed at levels of 100 and 500 lb/a it considerably increased the numbers of bacteria and fungi, all indications being that the toxaphene was being used as an energy source.[149] Despite an appreciable susceptibility to breakdown by microorganisms in soil, toxaphene had depressed the germination of vegetable crops at 100 ppm, and slightly reduced their yields at 20 ppm.[37]

Organophosphorus Insecticides. Parathion applied at 10 lb/a,[21] or at 2 lb/a for three successive years,[37] had no effect on the numbers of soil fungi or bacteria (Table 4.12), nor on nitrification. Applications up to 100 lb/a had no evident effect on nitrifying and denitrifying bacteria beyond a slight but temporary reduction at the highest application level.[97] Soils treated with 200 ppm parathion developed higher populations of yeasts and of bacteria (Fig. 4.13) including the nitrogen-fixers and nitrifiers,[129] Only where there had been spills at the loading sites of sprayers, leading to soil contaminations approximating 50,000 ppm, did parathion cause decisive and long-lasting reductions in the microorganismal flora.[168] However, *Rhizobium japonicum* and *meliloti* were suppressed by parathion at 10^{-6} *M*, even although these nitrogen-fixing bacteria could break it down.[119] Methyl parathion increased the proliferation of soil bacteria in the genus *Pseudomonas*.[128]

Among systemic insecticides, demeton applied at 1.7 lb/a for three successive years caused no reductions in fungi or in nitrate production,[37] while schradan at 500 ppm increased the soil fungi and bacteria, and especially *Azotobacter*.[156] Applied at 10 lb/a, EPN was without appreciable effect on soil microorganisms.[21] Diazinon applied at 3 lb/a elicited a high population of a coccoidal rod-shaped bacterium that broke it down; 6 months later the soil showed along with a normal spectrum of bacterial and fungal numbers a 350-fold increase in *Streptomyces* actinomycetes.[58] Mixed in soil at 10 ppm, diazinon caused a temporary reduction in bacteria and fungal molds lasting for 1 wk only; a similar effect was obtained with thionazin, while with chlorpyrifos and trichloronat the reduction lasted for 2 wk. All these insecticides caused a slight decrease in nitrification, which was not further

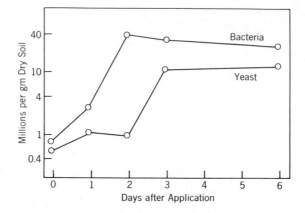

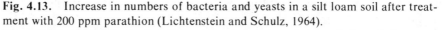

Fig. 4.13. Increase in numbers of bacteria and yeasts in a silt loam soil after treatment with 200 ppm parathion (Lichtenstein and Schulz, 1964).
Reprinted from the Journal of Economic Entomology by permission of the copyright owners, the Entomological Society of America.

aggravated when the concentration was raised to 100 ppm, and all of them increased ammonium production. At the 100 ppm level, the respiration of the microorganismal populations was increased by all these OP insecticides, suggesting that they were utilized as an energy source.[154]

Organophosphorus insecticides present in the soil have not had deleterious effects on crop plants. Lettuce seed shown in a clay loam treated with parathion at 100 lb/a germinated completely though slightly more slowly, and the resulting plants accumulated a higher P content than normal.[80] A potato crop which had received an aggregate of either 20 lb/a of parathion or of 5 lb/a tepp in 9 sprays applied during the season showed at 25% increase in yield.[169]

REFERENCES CITED

1. M. K. Ahmed and J. E. Casida. 1958. Metabolism of some organophosphorus insecticides by microorganisms. *J. Econ. Entomol.* **51**:59–63.

2. W. B. Albert and W. R. Paden. 1931. Calcium arsenate and unproductiveness in certain soils. *Science* **73**:622.

3. E. S. Albone, G. Eglinton, N. C. Evans, and M. M. Rhead. 1972. Formation of *p,p′*-DDCN from *p,p′*-DDT in anaerobic sewage sludge. *Nature* **240**:420–421.

4. M. Alexander. 1972. Microbial degradation of pesticides. In *Environmental Toxicology of Pesticides.* Ed. F. Matsumura, G. M. Boush, and T. Misato. Academic. pp. 365–383.

5. C. A. Anderson et al. 1974. Guthion (azinphosmethyl: Organophosphorus insecticide. *Residue Rev.* **51**:123–180.

6. J. P. E. Anderson, E. P. Lichtenstein, and W. F. Whittingham. 1970. Effect of *Mucor alternans* on the persistence of DDT in culture and in soil. *J. Econ. Entomol.* **63**:1595–1599.

7. P. S. L. Andrade and W. B. Wheeler. 1974. Biodegradation of mirex by sewage sludge microorganisms. *Bull. Environ. Contam. Toxicol.* **11**:415–416.

8. D. E. Armstrong and J. G. Konrad. 1974. Nonbiological degradation of pesticides. In *Pesticides in Soil and Water.* Ed. W. D. Guenzi. Soil Sci. Soc. Am., Madison, Wisconsin. pp. 123–131.

9. S. K. Arurkar and C. O. Knowles. 1970. Decomposition of formetanate acaricide in soil. *Bull. Environ. Contam. Toxicol.* **5**:324–328.

10. R. I. Asai, W. E. Westlake, and F. A. Gunther. 1969. Endrin decomposition on air-dried soils. *Bull. Environ. Contam. Toxicol.* **4**:278–284.

11. A. M. Baghdadi. 1970. Colonization by *Fusarium culmorum* of seedling roots of wheat in relation to chemical treatments. *Trans. Br. Mycol. Soc.* **54**:473–477.

12. P. S. Barker and F. O. Morrison. 1965. The metabolism of TDE by *Proteus vulgaris. Can. J. Zool.* **43**:652–654.

13. H. J. Benezet and F. Matsumura. 1974. Factors influencing the metabolism of mexacarbate by microorganisms. *J. Agric. Food Chem.* **22**:427–430.

14. K. I. Beynon and A. N. Wright. 1967. Breakdown of carbon-14-chlorfenvinphos in soils and crops grown in the soil. *J. Sci. Food Agric.* **18**:143–150.

15. K. I. Beynon, L. Davies, and K. Elgar. 1966. Analysis of crops and soils for residues of chlorfenvinphos. *J. Sci. Food Agric.* **17**:167–174.

16. N. W. Bixby, G. M. Boush, and F. Matsumura. 1971. Degradation of dieldrin to carbon dioxide by a soil fungus. *Bull. Environ. Contam. Toxicol.* **6**:491–494.

17. W. R. Bohn. 1964. The disappearance of dimethoate from soil. *J. Econ. Entomol.* **57**:798–799.

18. J. M. Bollag and S. Y. Liu. 1971. Degradation of Sevin by soil microorganisms. *Soil Biol. Biochem.* **3**:337–345.

19. J. M. Bollag and S. Y. Liu. 1972. Hydroxylations of carbaryl by soil fungi. *Nature* **236**:177–178.

20. J. M. Bollag and S. Y. Liu. 1972. Fungal degradation of 1-naphthol. *Can. J. Microbiol.* **18**:1113.

21. W. B. Bollen, H. E. Morrison, and H. H. Crowell. 1954. Effect of field treatments of insecticides on numbers of bacteria, *Streptomyces,* and molds in the soil. *J. Econ. Entomol.* **47**:302–306.

22. W. B. Bollen, H. E. Morrison, and H. H. Crowell. 1954. Effect of field and laboratory treatments with BHC and DDT on nitrogen transformations and soil respiration. *J. Econ. Entomol.* **47**:307–312.

23. J. R. Bradley, T. J. Sheets, and M. D. Jackson. 1972. DDT and toxaphene movements in surface water from cotton plots. *J. Environ. Quality* **1**:102–105.

24. R. C. Braunberg and V. Beck. 1968. Interaction of DDT and gastrointestinal flora of the rat. *J. Agric. Food Chem.* **16**:451–453.

25. F. Bro-Rasmussen, E. Noddegaard and K. Voldum-Clausen. 1968. Degradation of diazinon in soil. *J. Sci. Food Agric.* **19**:278–282.

26. F. L. Carter and C. A. Stringer. 1970. Residues and degradation products of technical heptachlor in various soil types. *J. Econ. Entomol.* **63**:625–628.

27. F. L. Carter, C. A. Stringer, and D. Heintzelman. 1971. 1-Hydroxy-2,3-epoxychlordene in Oregon soil previously treated with technical heptachlor. *Bull. Environ. Contam. Toxicol.* **6**:249–254.

28. C. I. Chacko, J. L. Lockwood, and M. J. Zabik. 1966. Chlorinated hydrocarbon pesticides: degradation by microbes. *Science* **154**:893–895.

29. A. G. Channon and W. G. Keyworth. 1960. Field trials on the effect of aldrin on club-root of summer cabbage. *Ann. Appl. Biol.* **48**:1–7.

30. J. R. Coppedge, D. A. Lindquist, D. L. Bull, and H. W. Dorough. 1967. Fate of Temik in cotton plants and soil. *J. Agric. Food Chem.* **15**:902–910.

31. D. G. Crosby. 1969. The non-metabolic decomposition of pesticides. *Ann. N. Y. Acad. Sci.* **160**:82–90.

32. J. E. Dewey and B. L. Parker. 1965. Increase in toxicity to *Drosophila melanogaster* of phorate-treated soils. *J. Econ. Entomol.* **58**:491–497.

33. W. G. Downs, E. Bordas, and L. Navarro. 1951. Duration of action of residual DDT deposits on adobe surfaces. *Science* **114**:259–262.

34. J. Duda. 1958. The effect of hexachlorocyclohexane and chlordane on soil microflora. *Acta Microbiol. Polonica.* **7**:237–244.

35. C. A. Edwards. 1973. *Persistent Pesticides in the Environment.* Chemical Rubber Co. Press, Cleveland. 2nd edition. 170 pp.

36. R. Engst and M. Kujawa. 1967. Enzymatische Abbau des DDT durch Schimmelpilze: Reaktionsverlauf des enzymatischen DDT-Abbaues. *Nahrung* **11**:751–760.

37. C. F. Eno. 1958. Insecticides and the soil. *J. Agric. Food Chem.* **6**:348–351.

38. C. F. Eno and P. H. Everett. 1968. Effects of soil applications of 10 chlorinated hydrocarbon insecticides on soil microorganisms. *Proc. Soil Sci. Soc. Am.* **22**:235–238.

39. Environmental Protection Agency. 1972. *Consolidated DDT Hearing; Hearing Examiner's Findings, Conclusions, and Orders.* 40 CRS 164.32, Washington, D.C., 25 April.

40. E. E. Fleck and H. L. Haller. 1945. Compatibility of DDT with insecticides, fungicides, and fertilizers. *Industr. Eng. Chem.* **37**:403–405.

41. D. W. Fletcher and W. B. Bollen. 1954. The effects of aldrin on soil microorganisms and some of their activities related to fertility. *Appl. Microbiol.* **2**:349–354.

42. D. C. Focht. 1972. Microbial degradation of DDT metabolites to carbon dioxide, water, and chloride. *Bull. Environ. Contam. Toxicol.* **7**:52–56.

43. D. C. Focht and M. Alexander. 1970. DDT metabolites and analogs: Ring fission by *Hydrogenomonas. Science* **170**:91–92.

44. F. M. Fowkes et al. 1960. Clay-catalyzed decomposition of insecticides. *J. Agric. Food Chem.* **8**:203–209.

45. A. L. French and R. Hoopingarner. 1970. Dechlorination of DDT by membranes isolated from *Escherichia coli. J. Econ. Entomol.* **63**:756–759.

46. N. Gannon and J. H. Bigger. 1958. The conversion of aldrin and heptachlor to their epoxides in soil. *J. Econ. Entomol.* **51**:1–2.

47. L. W. Getzin. 1967. Metabolism of diazinon and zinophos in soils. *J. Econ. Entomol.* **60**:505–508.

48. L. W. Getzin and I. Rosefield. 1968. Organophosphate insecticide degradation by heat-labile substances in soil. *J. Agric. Food Chem.* **16**:598–601.

49. L. W. Getzin and C. H. Shanks. 1970. Persistence, degradation, and bioactivity of phorate and its oxidative analogues in soil. *J. Econ. Entomol.* **63**:52–58.

50. B. L. Glass. 1972. Relation between the degradation of DDT and the iron redox system in soils. *J. Agric. Food Chem.* **20**:324–327.

51. I. J. Graham-Bryce. 1967. Adsorption of disulfoton by soil. *J. Sci. Food Agric.* **18**:72–77.

52. P. H. H. Gray. 1954. Effects of benzenehexachloride on soil microorganisms. I and II. *Can. J. Bot.* **32**:1–9, 10–15.

53. W. D. Guenzi and W. E. Beard. 1967. Anaerobic biodegradation of DDT to DDD in soil. *Science* **156**:1116–1117.

54. W. D. Guenzi and W. E. Beard. 1968. Anaerobic conversion of DDT to DDD and aerobic stability of DDT in soil. *Soil Sci. Soc. Am., Proc.* **32**:522–524.

55. W. D. Guenzi and W. E. Beard. 1970. Volatilization of lindane and DDT from soil. *Soil Sci. Soc. Am., Proc.* **34**:443–447.

56. K. Gundersen and H. L. Jensen. 1956. A soil bacterium decomposing organic nitro-compounds. *Acta Agric. Scand.* **6**:100 114.

57. H. B. Gunner and B. M. Zuckerman. 1968. Degradation of diazinon by synergistic microbial action. *Nature* **217**:1183–1184.

58. H. B. Gunner, B. M. Zuckerman, R. W. Walker, C. W. Miller, K. H. Deubert, and R. E. Langley. 1966. The distribution and persistence of diazinon applied to plant and soil and its influence on rhizosphere and soil microflora. *Plant and Soil* **25**:249–264.

59. W. E. Hall and Y. P. Sun. 1965. Mechanism of detoxication and synergism of Bidrin insecticide in house flies and soil. *J. Econ. Entomol.* **58**:845–849.

60. R. Haque and V. H. Freed (Eds.). 1974. *Environmental Dynamics of Pesticides. Environmental Science Research*, Vol. 6., Plenum. 387 pp.

61. C. R. Harris. 1969. Laboratory studies on the persistence of biological activity of some insecticides in soils. *J. Econ. Entomol.* **62**:1437–1441.

62. C. R. Harris. 1970. Laboratory evaluation of candidate materials as potential soil insecticides III. *J. Econ. Entomol.* **63**:782–787.

63. C. R. Harris. 1973. Laboratory evaluation of candidate materials as potential soil insecticides. IV. *J. Econ. Entomol.* **66**:216–221.

64. C. R. Harris and J. L. Hitchon. 1970. Laboratory evaluation of candidate materials as potential soil insecticides. *J. Econ. Entomol.* **63**:2–7.

65. C. R. Harris and J. R. W. Miles. 1975. Pesticide residues in the Great Lakes region of Canada. *Residue Rev.* **57**:27–29.

66. C. R. Harris, H. J. Svec, and W. W. Sans. 1973. Laboratory and microplot field studies on effectiveness and persistence of some experimental insecticides used for control of the darksided cutworm. *J. Econ. Entomol.* **66**:199–203.

67. C. R. Harris, A. R. Thompson, and C. M. Tu. 1972. Insecticides and the soil environment. *Proc. Entomol. Soc. Ont.* **102**:156–168.

68. C. S. Helling, P. C. Kearney, and M. Alexander. 1970. Behavior of pesticides in soils. *Adv. Agron.* **23**:147–240.

69. K. Igue, W. J. Farmer, W. F. Spencer, and J. P. Martin. 1969. Volatility of organochlorine pesticides from soil. *Agron. Abstr. 61st Ann. Meet.* pp. 77–78.

70. G. Jagnow and K. Haider. 1972. Evolution of $^{14}CO_2$ from soil incubated with dieldrin-^{14}C. *Soil Biol. Biochem.* **4**:43–49.

71. S. Jensen, R. Gothe, and M. O. Kindstedt. 1972. DDN, a new DDT derivative formed in anaerobic digested sewage sludge and lake sediment. *Nature* **240**:421–422.

72. B. T. Johnson, R. N. Goodman, and H. S. Goldberg. 1967. Conversion of DDT to DDD by pathogenic and saprophytic bacteria associated with plants. *Science* **157**:560–561.

73. B. T. Johnson and C. O. Knowles. 1970. Microbial degradation of the acaricide N-(4-chloro-o-tolyl)-N, N-dimethylformamidine. *Bull. Environ. Contam. Toxicol.* **5**:158–163.

74. D. P. Johnson and H. A. Stansbury. 1965. Adaptation of carbaryl residue method to various crops. *J. Agric. Food Chem.* **13**:235–238.

75. A. S. Jones. 1976. Metabolism of aldicarb by five soil fungi. *J. Agric. Food Chem.* **24**:115–117.

76. A. S. Jones and C. S. Hodges. 1974. Persistence of mirex and its effects on soil microorganisms. *J. Agric. Food Chem.* **22**:435–439.

77. J. S. Jones and M. B. Hatch. 1937. The significance of inorganic spray residue accumulations in orchard soils. *Soil Sci.* **44**:37–61.

78. F. W. Juengst and M. Alexander. 1976. Conversion of DDT to water-soluble products by microorganisms. *J. Agric. Food Chem.* **24**:111–115.

79. B. J. Kallman and A. K. Andrews. 1963. Reductive dehydrochlorination of DDT to DDD by yeast. *Science* **141**:1050–1051.

80. R. Kasting and J. C. Woodward. 1951. Persistence and toxicity of parathion when added to the soil. *Sci. Agric.* **31**:133–138.

81. D. D. Kaufman. 1974. Degradation of pesticides by soil microorganisms. In *Pesticides in Soil and Water.* Ed. W. D. Guenzi. Soil Sci. Soc. Am., Madison, Wisconsin. pp. 133–202.

82. H. Kazano, P. C. Kearney, and D. D. Kaufman. 1972. Metabolism of methylcarbamate insecticides in soils. *J. Agric. Food Chem.* **20**:975–979.

83. P. C. Kearney, E. A. Woolson, J. R. Plimmer, and A. R. Isensee. 1969. Decontamination of pesticides in soils. *Residue Rev.* **29**:137–149.

84. P. C. Kearney, R. G. Nash, and A. R. Isensee. 1969. Persistence of pesticide residues in soils. In *Chemical Fallout.* Ed. M. W. Miller and G. G. Berg. Charles C. Thomas. pp. 54–67.

85. U. Kiigemagi and L. C. Terriere. 1963. The spectrophotometric determination of Zinophos and its oxygen analog in soil and plant tissues. *J. Agric. Food Chem.* **11**:293–297.

86. W. H. Ko and J. L. Lockwood. 1968. Conversion of DDT to DDD in soil and the effect of these compounds on soil microorganisms. *Can. J. Microbiol.* **14**:1069–1073.

87. J. G. Konrad and G. Chesters. 1969. Degradation in soils of Ciodrin, an organophosphate insecticide. *J. Agric. Food Chem.* **17**:226–230.

88. J. G. Konrad, D. E. Armstrong, and G. Chesters. 1967. Soil degradation of diazinon, a phosphorothioate insecticide. *Agron. J.* **59**:591–594.

89. J. G. Konrad, G. Chesters, and D. E. Armstrong. 1969. Soil degradation of malathion, a phosphorodithioate insecticide. *Soil Sci. Soc. Am., Proc.* **33**:259–262.

90. L. J. Korschgen. 1971. Disappearance and persistence of aldrin after five annual applications. *J. Wildl. Manage.* **35**:494–500.

91. F. Korte and P. E. Porter. 1970. Evaluation: Biotransformation by microorganisms. *J. Assoc. Office Anal. Chem.* **53**:494–500.

92. F. Korte and M. Stiasni. 1964. Unwandlung von Telodrin durch Mikroorganismen und Moskito-larven. *Ann. Chemie.* **673**:146–152.

93. E. R. Laygo and J. T. Schulz. 1963. Persistence of organophosphate insecticides and their effect on microfauna in soils. *Proc. N. Dak. Acad. Sci.* **17**:64–65.

94. E. P. Lichtenstein. 1966. Persistence and degradation of pesticides in the environment. In *Scientific Aspects of Pest Control*. Nat. Acad. Sci. Publ. 1402. pp. 221–229.

95. E. P. Lichtenstein and K. R. Schulz. 1959. Breakdown of lindane and aldrin in soils. *J. Econ. Entomol.* **52**:118–124.

96. E. P. Lichtenstein and K. R. Schulz. 1960. Epoxidation of aldrin and heptachlor in soils as influenced by autoclaving, moisture, and soil types. *J. Econ. Entomol.* **53**:192–197.

97. E. P. Lichtenstein and K. R. Schulz. 1964. The effects of moisture and microorganisms on the persistence and metabolism of some organophosphorus insecticides in soil. *J. Econ. Entomol.* **57**:618–627.

98. E. P. Lichtenstein, G. R. Myrdal, and K. R. Schulz. 1964. Effect of formulation and mode of application of aldrin on the loss of aldrin and its epoxide from soils. *J. Econ. Entomol.* **57**:133–136.

99. E. P. Lichtenstein, K. R. Schulz, and T. W. Fuhremann. 1972. Management and fate of dyfonate in soils under leaching and non-leaching conditions. *J. Agric. Food Chem.* **20**:831–838.

100. E. P. Lichtenstein, K. R. Schulz, T. W. Fuhremann, and T. T. Liang. 1970. Degradation of aldrin and heptachlor in field soils during a ten-year period. *J. Agric. Food Chem.* **18**:100–106.

101. E. P. Lichtenstein, K. R. Schulz, R. F. Skrentny, and Y. Tsukano. 1966. Toxicity and fate of insecticide residues in water. *Arch. Environ. Health* **12**:199–212.

102. S. Y. Liu and J. M. Bollag. 1971. Metabolism of carbaryl by a soil fungus. *J. Agric. Food Chem.* **19**:487–490.

103. S. Y. Liu and J. M. Bollag. 1971. Carbaryl decomposition to 1-napthyl carbamate by *Aspergillus terreus*. *Pestic. Biochem. Physiol.* **1**:366–372.

104. I. C. MacRae, K. Raghu, and T. F. Castro. 1967. Persistence and biodegradation of four common isomers of benzene hexachloride in submerged soils. *J. Agric. Food Chem.* **15**:911–914.

105. F. Matsumura. 1972. Metabolism of insecticides in microorganisms and insects. *Environ. Quality and Safety* **1**:96–106.

106. F. Matsumura. 1975. *Toxicology of Insecticides*. Plenum. 503 pp.

107. F. Matsumura and G. M. Boush. 1966. Malathion degradation by *Trichoderma viride* and a *Pseudomonas* species. *Science* **153**:1278–1280.

108. F. Matsumura and G. M. Boush. 1967. Dieldrin: Degradation by soil microorganisms. *Science* **156**:959–961.

109. F. Matsumura and G. M. Boush. 1968. Degradation of insecticides by a soil fungus, *Trichoderma viride. J. Econ. Entomol.* **61**:610–612.

110. F. Matsumura, G. M. Boush, and A. Tai. 1968. Breakdown of dieldrin in the soil by a microorganism. *Nature.* **219**:965–967.

111. F. Matsumura, V. G. Khanvilkar, K. C. Patil, and G. M. Boush. 1971. Metabolism of endrin by certain soil microorganisms. *J. Agric. Food Chem.* **19**:27–31.

112. F. Matsumura, K. C. Patil, and G. M. Boush. 1970. Formation of "photodieldrin" by microorganisms. *Science* **170**:1206–1207.

113. F. Matsumura, K. C. Patil and G. M. Boush. 1971. DDT metabolized by microorganisms from Lake Michigan. *Nature* **230**:325–326.

114. J. L. Mendel and M. S. Walton. 1966. Conversion of *p,p´*-DDT to *p,p´*-DDD by intestinal flora of the rat. *Science* **151**:1527–1528.

115. J. L. Mendel, A. K. Klein, J. T. Chen, and M. S. Walton. 1967. Metabolism of DDT and some other chlorinated organic compounds by *Aerobacter aerogenes. J. Assoc. Offic. Anal. Chemists* **50**:897–903.

116. J. J. Menn, G. G. Patchett, and G. H. Batchelder. 1960. The persistence of Trithion, an organophosphorus insecticide. *J. Econ. Entomol.* **53**:1080–1082.

117. J. J. Menn, J. B. McBain, B. J. Adelson, and G. G. Patchett. 1965. Degradation of Imidan in soil. *J. Econ. Entomol.* **58**:875–878.

118. C. M. Menzie. 1972. Fate of pesticides in the environment. *Annu. Rev. Entomol.* **17**:199–222.

119. D. L. Mick and P. A. Dahm. 1970. Metabolism of parathion by two species of *Rhizobium. J. Econ. Entomol.* **63**:1155–1159.

120. J. R. W. Miles, C. M. Tu, and C. R. Harris. 1969. Metabolism of heptachlor and its degradation products by soil microorganisms. *J. Econ. Entomol.* **62**:1334–1338.

121. R. P. Miskus, D. P. Blair, and J. E. Casida. 1965. Conversion of DDT to DDD by bovine rumen fluid, lake water, and reduced porphyrins. *J. Agric. Food Chem.* **13**:481–483.

122. J. Miyamoto, K. Kitagawa, and Y. Sato. 1966. Metabolism of organophosphorus insecticides by *Bacillus subtilis. Jap. J. Exp. Med.* **36**:211–225.

123. S. Miyazaki, G. M. Boush, and F. Matsumura. 1969. Metabolism of chlorobenzilate and chloropropylate by *Rhodotorula gracilis. Appl. Microbiol.* **18**:972–976.

124. I. Y. Mostafa, M. R. E. Bahig, I. M. I. Fakhr, and Y. Adam. 1972. Malathion breakdown by soil fungi. *Z. Naturforsch.* **27b,** 1115–1116.

125. I. Y. Mostafa, I. M. I. Fakhr, M. R. E. Bahig, and Y. A. El-Zawahry. 1972. Degradation of malathion by *Rhizobium* spp. *Arch. Mikrobiol.* **85**:221–224.

126. M. S. Mulla, G. P. Georghiou, and H. W. Cramer. 1961. Residual activity of organophosphorus insecticides in soil as tested against the eye gnat. *J. Econ. Entomol.* **54**:865–870.

127. R. G. Nash and E. A. Woolson. 1967. Persistence of chlorinated hydrocarbon insecticides in soils. *Science* **157**:924–927.

128. K. Naumann. 1959. Einfluss von Pflanzenschutzmittel auf die Bodenmikroflora. *Mitt. Biol. BundesAnst. Berlin* **97**:109–117.

129. K. Naumann. 1967. Uber den Parathionabbau durch Bodenbakterien. *Phytopathologische Z.* **60**:343–357.

130. L. W. Newland, G. Chesters, and G. B. Lee. 1969. Degradation of BHC in simulated lake impoundments as affected by aeration. *J. Water Pollut. Contr. Fed.* **41**:R174–188.

131. D. E. Ott and F. A. Gunther. 1965. DDD as a decomposition product of DDT. *Residue Rev.* **10**:70–84.

132. B. F. Pain and R. F. Skrentny. 1969. Persistence and effectiveness of thionazin against potato aphids in three soils in southern England. *J. Sci. Food Agric.* **20**:485–488.

133. A. N. Pathak, H. Shankar, and K. S. Aswathi. 1961. Effect of some pesticides on available nutrients and soil microflora. *J. Indian. Soc. Soil Sci.* **9**:197–200.

134. K. C. Patil, F. Matsumura and G. M. Boush. 1970. Degradation of endrin, aldrin, and DDT by soil microorganisms. *App. Microbiol.* **19**:879–881.

135. F. K. Pfaender and M. Alexander. 1972. Extensive microbial degradation of DDT *in vitro* and DDT metabolism by natural communities. *J. Agric. Food Chem.* **20**:842–846.

136. R. M. Pfister. 1972. Interactions of halogenated pesticides and microorganisms: A review. *Critic Rev. Microbiol.* **2**:1–33.

137. K. Raghu and I. C. MacRae. 1967. The effect of the gamma isomer of benzene hexachloride upon nitrogen mineralization and fixation and selected bacteria. *Can. J. Microbiol.* **13**:621–627.

138. L. T. Richardson. 1957. Effect of insecticides and herbicides applied to soil on the development of plant diseases. I. The seedling disease of barley caused by *Helminthosporium sativum*. *Can. J. Plant Sci.* **37**:196–204.

139. L. T. Richardson. 1959. Ibid. II. Early blight and Fusarium wilt of tomato. *Can. J. Plant Sci.* **39**:30–38.

140. J. D. Rosen and D. J. Sutherland. 1967. The nature and toxicity of the photoconversion products of aldrin. *Bull. Environ. Contam. Toxicol.* **2**:1–9.

141. K. R. Schulz, E. P. Lichtenstein, T. T. Liang, and T. W. Fuhremann. 1970. Persistence and degradation of azinphosmethyl in soils. *J. Econ. Entomol.* **63**:432–438.

142. N. Sethunathan and I. C. Macrae. 1969. Persistence and biodegradation of diazinon in submerged soils. *J. Agric. Food Chem.* **17**:221–225.

143. N. Sethunathan and M. D. Pathak. 1971. Development of a diazinon-degrading bacterium in paddy water after repeated application of diazinon. *Can. J. Microbiol.* **17**:699–702.

144. N. Sethunathan and M. D. Pathak. 1972. Increased biological hydrolysis of diazinon after repeated application in rice fields. *J. Agric. Food Chem.* **20**:586–589.

145. N. Sethunathan and T. Yoshida. 1969. Fate of diazinon in submerged soil: Accumulation of hydrolysis product. *J. Agric. Food Chem.* **17**:1192–1195.

146. W. M. Shaw and B. Robinson. 1960. Pesticide effects in soil on nitrification and plant growth. *Soil Sci.* **90**:320–323.

147. J. J. Skujins. 1967. Enzymes in soil. In *Soil Biochemistry*. Ed. A. D. McLaren and G. H. Peterson. Dekker. pp. 371–414.

148. D. B. Slope and F. T. Last. 1963. Effects of some chlorinated hydrocarbons on the development of take-all of wheat. *Plant Pathol.* **12**:37–39.

149. N. R. Smith and M. E. Wenzel. 1947. Soil microorganisms are affected by some of the new insecticides. *Soil Sci. Soc. Am., Proc.* **12**:227–233.

150. J. Stenersen. 1969. Degradation of P^{32}-bromophos by microorganisms and seedlings. *Bull. Environ. Contam. Toxicol.* **4**:104–112.

151. D. K. R. Stewart, C. Chisholm, and M. T. H. Ragab. 1971. Long term persistence of parathion in soil. *Nature* **229**:47.

152. D. L. Suett. 1971. Persistence and degradation of chlorfenvinphos, diazinon, fonofos, and phorate in soils. *Pestic. Sci.* **2**:105–111.

153. U. Tsukano and A. Kobayashi. 1972. Formation of gamma-BTC in flooded ricefield soils treated with gamma-BHC. *Agric. Biol. Chem.* **36**:166–167.

154. C. M. Tu. 1970. Effect of four organophosphorus insecticides on microbial activities in soils. *Appl. Microbiol.* **19**:479–484.

155. C. M. Tu, J. R. Miles, and C. R. Harris. 1968. Soil microbial degradation of aldrin. *Life Sci.* **7**:311–323.

156. O. Verona and G. Picci. 1952. Intorno all' azione esercitata dagli insetticidi systemici sulla microflora del terreno. *Agric. Ital.* **52**:61–70.

157. W. W. Walker and B. J. Stojanovic. 1973. Microbial versus chemical degradation of malathion in soil. *J. Environ. Quality* **2**:229–232.

158. W. W. Walker and B. J. Stojanovic. 1974. Malathion degradation by an *Arthrobacter* species. *J. Environ. Quality* **3**:4–13.

159. L. M. Walsh and D. R. Keeney. 1975. Behavior and phytotoxicity of inorganic arsenical in soils. In *Arsenical Pesticides.* Ed. E. A. Woolson. Am. Chem. Soc. Symposium Series 7. pp. 35–52.

160. M. J. Way and N. E. A. Scopes. 1968. Studies on the persistence and effects on soil fauna of some soil-applied systemic insecticides. *Ann. Appl. Biol.* **62**:199–214.

161. G. Wedemeyer. 1966. Dechlorination of DDT by *Aerobacter aerogenes*. *Science* **152**:647.

162. G. Wedemeyer. 1967. Dechlorination of DDT by *Aerobacter aerogenes*. I. Metabolic products. *Appl. Microbiol.* **15**:569–574.

163. G. Wedemeyer. 1967. Biodegradation of DDT; intermediates in dichlorodiphenyl-acetic acid metabolism by *Aerobacter aerogenes*. *Appl. Microbial.* **15**:1494–1495.

164. G. Wedemeyer. 1968. Partial hydrolysis of dieldrin by *Aerobacter aerogenes*. *Appl. Microbiol.* **16**:661–662.

165. A. T. S. Wilkinson, D. G. Finlayson, and H. V. Morley. 1964. Toxic residues in soil 9 years after treatment with aldrin and heptachlor. *Science* **143**:681–682.

166. J. K. Wilson and R. S. Choudhri. 1946. Effects of DDT on certain microbiological processes in the soil. *J. Econ. Entomol.* **39**:537–538.

167. J. K. Wilson and R. S. Choudhri. 1948. The effect of benzene hexachloride on soil organisms. *J. Agric. Res.* **77**:25–32.

168. H. R. Wolfe, D. C. Staiff, J. F. Armstrong, and S. W. Comer. 1973. Persistence of parathion in soil. *Bull. Environ. Contam. Toxicol.* **10**:1–9.

169. D. O. Wolfenbarger. 1948. Nutritional values of phosphatic insecticides. *J. Econ. Entomol.* **41**:818–819.

170. B. Yaron, B. Heuer and Y. Birk. 1974. Kinetics of azinphosmethyl losses in the soil environment. *J. Agric. Food Chem.* **22**:439–441.

171. T. Yoshida and T. F. Castro. 1970. Degradation of gamma-BHC in rice soils. *Soil Sci. Soc. Am. Proc.* **34**:440–442.

172. W. N. Yule, M. Chiba, and H. V. Morley. 1967. Fate of insecticide residues: Decomposition of lindane in soil. *J. Agric. Food Chem.* **15**:1000–1004.

173. S. M. A. D. Zayed, I. Y. Mostafa, and A. Hassan. 1965. Metabolism of organophosphorus insecticides. VII. Transformation of ^{32}P-labeled Dipterex through microorganisms. *Archiv. Microbiol.* **51**:188–191.

174. B. M. Zuckerman, K. Deubert, M. Mackiewicz, and H. Gunner. 1970. Studies on the biodegradation of parathion. *Plant and Soil* **33**:273–281.

5

INSECTICIDES AND AQUATIC INVERTEBRATE BIOTA

A. AREA TREATMENTS OVER FORESTS AND OTHER TERRAIN

Effects on Stream Arthropods

DDT. With the introduction of DDT at 1 lb/a to control lepidopteran defoliators of forests, the insect fauna of the streams so valuable as food for game fish suffered heavily from the spray, although in shaded stretches only 25% of that applied from the air reached the water. Reductions in invertebrate numbers were often 70–90% in hardwood forest sprayed against the gypsy moth,[55] while the reduction in biomass of stream insects was 87–92% in coniferous forest sprayed to control the black-headed budworm in British Columbia.[25] The invertebrate food found in trout stomachs in a sprayed area in Maine came to lack the usual blackfly and caddisfly larvae, and to be deficient in mayfly and stonefly nymphs; instead, more of the midge and other dipterous larvae were eaten, and in some cases terrestrial insects and snails.[133]

When an unprotected stream was sprayed directly at 1 lb/a, the nymphs of all species of mayflies were exterminated, and larvae of every species of caddisfly suffered to some extent (Table 5.1). Nymphs of a few species of stoneflies were unaffected, as were nymphs of Odonata[54] and larvae of the dobson fly *Corydalis*.[56] The fish-fly *Nigronia* subsequently increased its numbers, probably because of the removal of the competing predaceous species of stoneflies.[52] In forest treatments in northwestern Ontario, Sialid larvae, dragonfly nymphs, and some Copepods and Cladocerans resisted the killing effect of the DDT spray.[117] However, most of the microcrustacea suffered heavily from DDT, although they usually recovered their numbers in the same season.[57] Crayfish are especially sensitive to DDT, being killed even when dosages are restricted to 0.5 lb/a.[23]

It was noted that mayfly populations would completely recover their numbers by the year following the 1 lb/a spraying, as was the experience after spruce-budworm spraying in Maine[43] and the gypsy moth spraying in

Table 5.1.   Effect on aquatic insects of DDT sprayed directly at 1 lb/a on a warm-water bass stream in West Virginia (Hoffmann and Merkel, 1948).

Ephemeroptera		Plecoptera	
Heptagenia maculipennis	D	Acroneuria internata	C
Stenonema fuscum	D	Neoperla clymene	A
S. rubrum	D	Neophasganophora capitata	B
Stenonema spp.	D	Trichoptera	
Baetis spp.	D	Chimarra obscura	B
Iron rubidus	D	C. socia	D
Megaloptera		Cheumatopsyche spp.	D
Corydalis cornutus	A	Hydropsyche spp.	C
Nigronia serricornis	B	Macronemum zebratum	C

A, unaffected;   B, reduced;   C, greatly reduced;   D, exterminated

Pennsylvania[54]; only the Baetines, which browse on stream vegetation, lagged in this respect. However the caddisflies had only partially recovered their numbers in the postspray year in Pennsylvania, with *Chimarra* still remaining eliminated,[54] while in Maine a few streams stayed deficient in their Trichopteran and Plecopteran fauna.[43] In New Brunswick, where DDT was airsprayed at 0.5 lb/a over large areas infested by the spruce budworm, the stream fauna that had been initially cut back to Chironomid midges almost exclusively,[66] had recovered to only 45% of its prespray level by the following year, the caddisflies still remaining scarce in the year after that.[67] After applications of DDT at 1 lb/a in Montana, even the mayflies took 3 yr to recover their numbers completely, and stoneflies of the genus *Leptocera* never did return to the streams in the treated area.[47]

Organophosphorus Insecticides.   Among the compounds used as alternatives to DDT in forest spraying, malathion has a 48-hr LC_{50} to *Baetis* mayfly nymphs of 6 ppb, as compared to 12 ppb for DDT (and higher values for cyclodiene insecticides).[20] To stoneflies, the LC_{50} for malathion as well as for DDT is of the order of 10 ppb for short-term exposure (Tables 5.2, 5.3) and 1 ppb for long-term exposure (Table 5.4). An aerial spray of malathion at 2 lb/a over broadleaved forest in Ohio greatly decreased the stream insect fauna, but the numbers recovered soon after the treatment,

Table 5.2. 96-hr LC_{50} figures in ppb (static test) to nymphs of the stonefly *Pteronarcys californica* (Sanders and Cope, 1968).

Endrin	0.25	Trichloronat	0.1	Diazinon	25
Dieldrin	0.5	Dichlorvos	0.1	Trichlorfon	35
Strobane	0.5	Azinphosmethyl	1.5	Dimethoate	43
Heptachlor	1.1	Ethion	2.8	Phosphamidon	150
Aldrin	1.3	Fenitrothion	4.0	Dicrotophos	430
Methoxychlor	1.4	Fenthion	4.5	Carbaryl	4.8
Toxaphene	2.3	Disulfoton	5.0	Methiocarb	5.4
Endosulfan	2.3	Monocrotophos	5.0	Mexacarbate	10
Lindane	4.5	Parathion	5.4	Propoxur	13
DDT	7.0	Naled	8.0	Allethrin	1.0
Chlordane	15	Chlorpyrifos	10	Pyrethrum	2.1
DDD	380	Malathion	10	Rotenone	380
		Temephos	10		

Table 5.3. LC_{50} figures in ppb of insecticides to nymphs of two species of stoneflies (Sanders and Cope, 1968).

	Pteronarcella badia			Claassenia sabulosa		
	24 h.	48 h.	96 h.	24 h.	48 h.	96 h.
Endrin	2.8	1.7	0.54	3.2	0.84	0.76
Dieldrin	3.0	1.5	0.5	4.5	2.3	0.58
Heptachlor	6.0	4.0	0.9	9.0	6.4	2.8
Toxaphene	9.2	5.6	3.0	6.0	3.2	1.3
DDT	12	9.0	1.9	16	6.4	3.5
Chlorpyrifos	4.2	1.8	0.38	8.2	1.8	0.57
Parathion	8.0	5.6	4.2	8.8	3.5	1.5
Malathion	10	6.0	1.1	13	6.0	2.8
Trichlorfon	50	22	11	110	70	22
Carbaryl	5.0	3.6	1.7	12	6.8	5.6

and the crayfish had been unaffected.[107] Trichlorfon, with 24-hr LC_{50} figures of 910 ppb for *Hexagenia* mayflies and 17 ppb for *Hydropsyche* caddisflies,[15] and ranging from 50 to 110 ppb for stoneflies (Table 5.3), evidently had little effect on stream fauna when applied at 1 lb/a, judging from the drift-net captures after the spray. It is more suitable than malathion for application against these species where the abundance of the emerging adults are a nuisance to lakeside or riverside towns, since it is 100 times less toxic to fish such as bluegills.[15]

Phosphamidon, applied at 1 lb/a to control hemlock looper in British Columbia, killed all the caddisfly larvae in the creeks.[119] However, it is one-tenth as toxic as DDT to the crayfish *Procambarus clarki*.[96] Fenitrothion, used effectively as a spruce-budworm control agent at 4 oz/a, killed large amounts of stream insects, especially Tendipedid larvae, and cut the Oligochaete population in half.[40] Methyl parathion is very toxic to *Procambarus*, having a 48-hr LC_{50} of only 40 ppb; by contrast, dicrotophos and naled are harmless to crayfish.[96]

Carbamate Compounds. Carbaryl, with LC_{50} values of the order of 5 ppb for stoneflies, kills 50–95% of the stream insect biomass when applied at 1.25 lb/a from a suspension in fuel oil to deciduous forest for gypsy moth control; although recovery does not commence until at least 1 month after the application, this carbamate is not considered to affect the fish production in the streams.[5] Carbaryl, with a 48-hr LC_{50} of 3 ppm for *Procambarus clarki*, is virtually safe for crayfish.[96]

Effects on Molluscs and Annelids

Molluscs in general survived forest spraying with DDT, even at dosages of 5 lb/a, along with the oligochaetes and *Planaria* in the pools.[56] But concentrations reaching 1 ppm, such as would arise from an application of 0.8 lb/a to water 3 inches deep, can kill *Physa* snails.[33] DDT is not accumulated in snails as much as in slugs; *Cepaea hortensis*, for example, excreted up to one-half the DDT it ingested.[31] However, Roman snails (*Helix pomatia*) fed on lettuce lightly treated with DDT laid down thinner shells and opercula.[19] Toxaphene applied as a piscicide to a small Michigan lake at 0.1 ppm did no apparent harm to the snails, Unionid and Sphaeriid clams, or Tubificid worms[49]; but sprayed at 2 lb/a in oil solution onto a North Dakota marsh, it caused the aquatic snail population to progressively decline to zero during the ensuing 10 days, and recolonization did not commence until 1 month after the treatment.[45] In Mississippi, the 72-hr LC_{50} levels for toxaphene and endrin to a snail and a clam species were somewhat higher in a heavily sprayed area (Belzoni) than at an unsprayed locality (State College), as follows (in ppm)[97]:

	Toxaphene		Endrin	
	State Coll.	Belzoni	State Coll.	Belzoni
Physa gyrina	0.35	0.48	0.45	0.6
Eupera singleyi	0.4	0.6	0.06	0.32

Diazinon at 0.32 ppm did not kill the freshwater mussel *Elliptio complanatus*, which accumulated residues of 0.6 ppm in its tissues.[84]

Molluscicides. The delta isomer of benzene hexachloride, present in the technical BHC preparations used in the 1950s, is a powerful molluscicide. But for control of the aquatic snails that are the alternate hosts of *Schistosoma* parasites of man, mainly in the genera *Bulinus, Biomphalaria,* and *Oncomelania,* compounds have been discovered and developed for their particular activity. Thus the former cheap control chemicals such as copper sulfate and the sodium salt of PCP are now largely replaced by niclosamine (Bayluscide), a nitrated derivative of benzamide, and Frescon which is N-trityl morpholine. While niclosamine is as toxic to fish as to snails,[92] Frescon is nontoxic to those species of *Tilapia* bream that eat the snails and the vegetation that harbors them.[121] One advantage of niclosamine is that it has no effect on freshwater crabs or dragonfly nymphs, while the cladocerans and copepods recover rapidly.[46] Cuprous oxide and cuprous chloride also kill the schistosome-bearing snails without killing fish.[29]

Annelids. The maximum concentrations in which the oligochaete *Tubifex tubifex* could survive for 72 hr without mortality were 6 ppm for toxaphene, endrin, dieldrin, and methyl parathion, 4 ppm for lindane, 3 ppm for aldrin and DDT, 2 ppm for parathion and chlorpyrifos, 1.5 ppm for chlordane, ethion, and carbaryl, 1 ppm for azinphosmethyl and monocrotophos, and only 0.5 ppm for the acaricides dicofol and DMC.[97]

B. TREATMENT OF WATER BODIES FOR DISEASE-VECTOR CONTROL

Effects of Blackfly Larvicides

The application of DDT to control the larvae of *Simulium* blackflies in running water, the dosage being such as to give approximately 0.1 ppm in that

Table 5.4.   LC$_{50}$ figures in ppb for long-term and short-term exposure of stoneflies to insecticides (Jensen and Gaufin 1964, 1966).

	Acroneuria pacifica		Pteronarcys californica	
	96-hr. Static	30-day Cont. Flow	96-hr. Static	30-day Cont. Flow
Endrin	0.32	0.035	2.4	1.2
Dieldrin	24	0.20	39	2.0
Aldrin	143	22	180	2.5
DDT	320	72	1800	265
Fenthion	5.1	0.64	26	3.6
Malathion	7.0	0.80	50	8.8
Disulfoton	8.2	1.4	28	1.9
Azinphosmethyl	8.5	0.24	22	1.3
Trichlorfon	16.5*	8.7	69	9.8

* continuous-flow test

amount of water that passes a given point in 15–30 min, resulted in a 90–100% kill of mayflies, stoneflies, and especially of caddisflies, in a Labrador brook[48]; This dosage had caused a 77–86% reduction of mayflies and stoneflies, decreasing to a 7% reduction 17 miles downstream, in the South Saskatchewan River.[1] Applications of 0.4 ppm/30 min, repeated every week for 10 wk to control *Simulium damnosum* in the Victoria Nile, almost completely eliminated the populations of Ephemeroptera and Trichoptera, but most of the affected species had become reestablished within 40 days of the last treatment.[22] An application of 0.1 ppm/30 min to control *S. neavei* in a stream draining Mount Elgon, Uganda resulted in an increase of the *Simulium* larvae and of *Baetis* mayflies consequent on the destruction of the predaceous *Neoperla* stonefly nymphs and *Hydropsyche* and *Cheumatopsyche* caddisfly larvae.[65] Annual treatments repeated for 10 yr in the Adirondack Mountains resulted in a stream fauna in which the abundance of the Plecoptera was no less, the Trichoptera insignificantly less, and the Ephemeroptera one-half, of the original.[69] Applications of DDT at 0.1 ppm in the Hudson Bay area also killed small crustacea such as Cladocera, Copepoda, Anostraca, and Conchostraca.[53]

Methoxychlor, sprayed from the air as a blackfly larvicide, had about the same initial effect on the stream fauna as DDT, while not posing the same residue problems.[6] Fenthion, which does not induce residues in fish, did not kill mayflies, stoneflies, and caddisflies at 0.15 ppm, although it did at

0.5 ppm; temephos (Abate) was completely harmless to these aquatic insects when applied to creeks at 0.1 ppm, a dosage that is completely effective against blackfly larvae.[125] Temephos is the blackfly larvicide selected for use against *Simulium damnosum* in the seven-nation onchocerciasis control program in West Africa, the only OP insecticide approaching it in effectiveness and in safety to other aquatic fauna being chlorpyrifos-methyl.

Effects of Mosquito Larvicides

DDT and Methoxychlor. While the arsenical Paris green formerly employed to control anopheline larvae had caused no serious destruction of the aquatic organisms that constituted fish food,[3] the introduction of DDT sprayed from aircraft at the usual rate of 0.25 lb/a was highly destructive to the predaceous Dytiscid larvae and adults, although it did not affect the predaceous Chaoborine larvae and dragonfly nymphs.[127] Applied directly to mosquito breeding areas in Tennessee at 0.1 lb/a every week, DDT reduced the numbers of Ephemeroptera, Coleoptera, Hemiptera, and Tendipedids in the water, but caused an increase in oligochaete, nematode, and copepod populations.[126] Applied weekly in a very fine aircraft spray at 0.1 lb/a to give deposits of the order of 0.01 lb/a, DDT had no effect on aquatic insects; however, it eliminated the surface Hemiptera such as *Gerris, Trepobates,* and *Hydrometra,* while the Amphipoda were the only crustacea to be killed.[50] To cladoceran crustaceans, DDT is the most toxic of the organochlorine insecticides (Tables 5.5, 5.6).

Methoxychlor, a biodegradable substitute for DDT for controlling aquatic insects, is more toxic than DDT to *Gammarus* amphipods (Table 5.7). When introduced into ponds at 0.04 ppm, it eliminated Heptageniid mayflies but scarcely affected the Baetids; it also reduced Coenagrionid damselfly populations by about 50%. Chironomid larvae increased their numbers by about 10 times in the first 2 months after the methoxychlor treatment, while *Physa* snails started to increase 1 month after the spray and reached extremely high levels; presumably both these forms had been released from control by their predators.[75]

Other Organochlorines. The cyclodiene insecticides and lindane are all considerably more toxic than DDT to the insectan fish-food species (Tables 5.1, 5.2). Heptachlor applied at 0.25 lb/a in North Dakota caused heavy kills of crayfish.[41] One year after the late-fall treatment of an Illinois farming area with aldrin at 2 lb/a for Japanese-beetle control, the Ephemeroptera in the streams were severely reduced, although the Trichoptera and Tendipedids recovered from an initial decimation to reach higher levels than ever before. In the second year after the treatment, the

Table 5.5.　　LC$_{50}$ (EC$_{50}$ immobilization) figures in ppb for insecticides to *Daphnia magna* and *D. carinata* (Sanders and Cope 1966, and FWPCA, 1968).

	D. magna	D. carinata		D. magna	D. carinata
	48-50 h.	64 h.		48-50 h.	64 h.
DDT	1.4	2.2	Ethion	0.01	---
Methoxychlor	3.6	---	Trichlorfon	0.12*	0.25[†]
DDD	4.6*	---	Azinphosmethyl	0.2	---
Aldrin	29	4.0[†]	Parathion	0.8	0.5
Heptachlor	58	20	Malathion	0.9	0.2
Dieldrin	330	250[†]	Phosphamidon	4.0	---
Aramite	345	---	Diazinon	4.3	0.8[†]
Endrin	352	50	Methyl parathion	4.8	---
Lindane	1,100	---	Dimethoate	2,500	---
Dicofol	39,000	---			

* 24-h. figure　　　　　† 32-h. figure

Trichoptera and Tendipedids reverted to normal levels, while the Ephemeroptera still remained at one-sixth the normal population density. Elmid beetles, very sensitive to pollution, did not suffer at all from the aldrin.[91]

The minute ostracod crustaceans, which inhabit the mud–water interface where insecticide residues accumulate, are quite susceptible to cyclodiene insecticides, the 24-hr LC$_{50}$ levels for *Chlamydotheca arcuata* being only 1.2 ppb for aldrin and 2.4 ppm for dieldrin; exposed to 0.01 ppb dieldrin in the water for 2 months, these ostracods accumulated 260 ppb dieldrin in their bodies.[74] Organochlorine residues in adults of mayflies and diving beetles in Louisiana farming areas were found in 1964 to range from 0.1 to 2.6 ppm DDE, with about 0.4 ppm endrin, 0.2 ppm dieldrin, and traces of toxaphene.[118]

Toxaphene employed at 0.1 ppm for the elimination of trash fish in lakes is lethal to aquatic arthropods such as *Chaoborus,* but bottom-feeders such as copepods escape the effect while the chironomid larvae and *Physa* snails multiply exceedingly, not to return to normal until a new fish population (e.g., some species of trout) is introduced into the lake.[51] Where the dead

Table 5.6. 48-hr LC_{50} (EC_{50} immobilization) figures in ppb for insecticides to the Cladocerans *Daphnia pulex* and *Simocephalus serrulatus* at 60°F (Sanders and Cope, 1966).

	Daphnia	Simocephalus
DDT	0.4	2.5
Methoxychlor	0.8	5
DDD	3.2	4.5
Toxaphene	15	19
Endrin	20	26
Aldrin	28	23
Chlordane	29	20
Heptachlor	42	47
Aramite	160	180
Endosulfan*	240	---
Dieldrin	250	240
Lindane	460	520
Chlorobenzilate	870	550

	Daphnia	Simocephalus
Dichlorvos	0.07	0.26
Mevinphos	0.16	0.43
Trichlorfcn	0.18	0.70
Naled	0.35	1.1
Parathion	0.60	0.37
Fenthion	0.80	0.9
Diazinon	0.90	1.8
Malathion	1.8	3.5
Azinphosmethyl	3.2	4.2
Phosphamidon	8.8	12.0

	Daphnia	Simocephalus
Carbaryl	6.4	7.6
Mexacarbate	10	13
Allethrin	21	56
Pyrethrins	25	42
Rotenone	100	190
Cryolite	5,000	10,000
Lime-sulfur	10,000	11,000

* Federal Water Pollution Control Administration 1968

Table 5.7. LC$_{50}$ figures in ppb (static test) for insecticides to the Amphipod crustacean *Gammarus lacustris* (Sanders, 1969).

Insecticide				Insecticide			
Methoxychlor	2.9	1.3	0.8	Coumaphos	0.32	0.14	0.07
DDD	5.6	1.8	0.64	Azinphosmethyl	0.56	0.25	0.15
DDT	4.7	2.1	1.0	Chlorpyrifos	0.76	0.40	0.11
Endrin	6.4	4.7	3.0	Dichlorvos	2.0	1.0	0.5
Endosulfan	9.2	6.4	5.8	Malathion	3.8	1.8	1.0
Toxaphene	180	70	26	Ethion	5.6	3.2	1.8
Chlordane	160	80	26	Phosphamidon	8.4	3.8	2.8
Lindane	120	88	48	Parathion	12	6	3.5
Aramite	350	100	60	Fenthion	15	11	8.4
Heptachlor	150	100	29	Phorate	24	14	9.0
Tetradifon	370	140	110	Carbophenothion	45	28	5.2
Dieldrin	1,400	1,000	460	Tepp	74	52	39
Aldrin	45,000	12,000	9,800	Trichlorforn	92	60	40
				Disulfoton	110	70	52
Carbaryl	40	22	16	Naled	240	160	110
Propoxur	66	50	34	Mevinphos	650	310	130
Mexacarbate	86	76	46	Dimethoate	900	400	200
				Diazinon	800	500	200
Pyrethrum	28	18	12	Dioxathion	830	690	270
Allethrin	38	20	11	Temephos	960	640	82
Rotenone	6,000	3,500	2,600	Dicrotophos	2,200	790	540

fish are left in the lake, the resulting fertilization produces a rapid increase of zooplankton.[49] Toxaphene applied at 100 ppb in a Colorado reservoir progressively eliminated the entomostracan microcrustacea over the ensuing 2-month postapplication period, while the rotifers were moderately reduced; protozoa were also progressively eliminated, and took longer to recover their numbers than the Entomostraca.[58]

Contamination of a stream in Britain with BHC from excess sheep dip had the effect of eliminating all *Gammarus* amphipods, *Leuctra* stoneflies, *Baetis* mayflies, and *Hydropsyche* caddisflies, leaving the field to three species of *Chironomus*.[64] Small crustacea are quite susceptible to lindane poisoning, the LC_{50} for *Daphnia magna* being 0.48 ppm and for *Gammarus fasciatus* 0.04 ppm; with a 2-month continuous-flow exposure, 20 ppb slightly but significantly reduces the survival of *Daphnia,* while 9 ppb reduces pairing and offspring survival in *Gammarus*.[80]

Among the PCB pollutants, Aroclor 1254 showed an LC_{50} of 31 ppb to *Daphnia magna* in a 3-wk static test, and only 1.8 ppb in a 2-wk continuous-flow test.[101] Taking reproductive inhibition into account, the 14-day static LC_{50} values were 10 ppb for Aroclor 1254 and 0.3 ppb for *p,p'*-DDT, the effects of the two when combined being slightly less than additive.[81]

Organochlorine Resistance.   In Mississippi, populations of *Cyclops* copepods taken from a heavily sprayed area were about four times as tolerant to cyclodiene insecticides as those from an unsprayed area; of the organochlorines DDT was the most toxic, lindane the least.[97] A population of the shrimp *Palaemonetes kadakiensis* in one of the sprayed areas showed elevenfold resistance to toxaphene, besides a sixfold resistance to carbaryl and a fourfold resistance to methyl parathion; the resistance to other cyclodienes and lindane was high, while the tolerance to other OP compounds and to DDT was low.[98] Nymphs of the mayfly *Heptagenia hebe* in the DDT-sprayed forest areas of New Brunswick were 12–40 times more resistant to DDT, and converted 15 times more DDT to DDE, than those from unsprayed areas.[44] DDE is essentially noninsecticidal, but the egg masses from female *Chironomus tentans* midges reared in 30 ppb DDE showed a reduced viability in terms of the number of F_1 adults eventually produced.[28] When exposed to *p,p'*-DDT at 50–80 ppt for 3 days, *Hexagenia bilineata* nymphs converted 85% of the DDT they absorbed into DDE, while *Daphnia magna* and *Chironomus* larvae converted 30%.[72]

Organophosphorus Compounds.   The first OP insecticides to be substituted for DDT as larvicides were malathion at 0.5 lb/a and parathion (later

fenthion and naled) at 0.1 lb/a. At laboratory exposures corresponding to
these dosages malathion caused 100% mortality among the predaceous
Hydrophilid larvae whereas parathion and fenthion caused only 20%
mortality[76]; fenthion, however, was lethal to nymphs of the large dragonfly
Libellula.[134] Applications of fenthion at 0.05 ppm directed against the vector mosquito *Culex tarsalis* resulted in these larvae recovering to twice their
original numbers 1 month after treatment, probably because of the elimination of three predators, namely the beetles *Rhantus* and *Tropisternus* and
the dragonfly *Tarnetrum corruptum* (Fanara, 1971[60]). Naled is quite toxic
to stoneflies, the 24-hr LC_{50} to *Hydroperla crosbyi* being 11 ppb by flowing-
water test, and sublethal exposure to three-quarters of the LC_{50} resulted in it
becoming less tolerant to oxygen deficiency.[82]

Malathion and parathion are no less toxic than DDT to cladoceran crustacea in the genera *Daphnia* and *Simocephalus* (Tables 5.5, 5.6). Applied
at 0.2 lb/a, fenthion eliminated the Cladocera, which took more than 2
months to recover their numbers.[106] Copepods, however, were not killed
even at 1 lb/a applied in granules to a New Jersey salt marsh, *Microcyclops
bicolor* not showing any mortality from fenthion at 0.5 ppm. The ostracod
Cypridopsis vidua was also resistant, surviving fenthion dosages nearly 3000
times higher than the cladoceran *Ceriodaphnia quadrangula*.[111]

Chlorpyrifos.   When Dursban was introduced for use in California at 0.01
lb/a it was destructive to nymphs of the Baetid mayfly *Siphlonurus* and the
Dytiscid diving beetle *Laccophilus*[135]; it also considerably reduced the numbers of the Corixid water boatman *Corisella,* although they recovered in 2
wk.[62] At 0.025 lb/a, there was some reduction among predators such as
Coenagrionid damselfly nymphs, the backswimmers *Buenoa* and
Notonecta, and the Dytiscid *Thermonectes* as well as *Laccophilus, Corisella*
and *Siphlonurus,* but the numbers recovered 1–2 months later.[60] In Tennessee, chlorpyrifos was harmless at 0.005 lb/a but harmful at 0.01 lb/a to
the insectan predators of mosquito larvae.[87]

Among the microcrustacea inhabiting New Jersey salt marshes, two
species of copepods and one ostracod (*Cypronotus*) showed the following
approximate 24-hr LC_{50} figures (ppm)[112]:

	Malathion	Fenthion	Naled	Chlorpyrifos	Temephos
Diaptomus sp.	<0.01	0.1	0.007	0.03	0.002
Cyclops spartinus	>5.0	>5.0	1.0	0.1	>5.0
Cypronotus incongruens	0.08	1.5	0.15	0.01	2.5

In California, populations of the cladoceran *Moina micrura* and the copepod *Cyclops vernalis* were decimated by chlorpyrifos at 0.1 lb/a, but recovered in 2 wk; after repeat applications at higher doses these species did not recover and were replaced by the rapid multiplication of the copepod *Diaptomus pallidus* and the predaceous rotifer *Asplanchna brightwelli*.[62] The planktonic herbivorous rotifers, principally *Brachionus, Hexarthra,* and *Filinia,* increased by about five times in the treated ponds due to the reduction in *Cyclops* which preys on them and *Moina* which competes with them. These herbivorous rotifers in turn provide prey for the large rotifer *Asplanchna,* which becomes 35 times more abundant in chlorpyrifos-treated ponds. Where *Asplancha* predation has cut back the herbivorous rotifers, this coupled with the reduction in the herbivore *Moina* allows the development of blooms of blue-green algae or of the diatom *Synedra*.[60]

Temephos. The larvicide sold as Abate was harmless at 0.01 lb/a to Odonata, Dytiscids, Hydrophilids, and Gerrids.[87] Applied at 0.03 lb/a to pools in Wisconsin, temephos was toxic to young Libellulid nymphs, the caddisfly *Limnephilus,* and to cladocerans, although nontoxic to copepods and ostracods.[108] Treatment of a 1.3-acre lake to control *Notonecta* backswimmers, and thus protect human bathers, with temephos at 0.4 lb/a killed nearly all the aquatic insects.[37] Temephos is 50 times safer than fenthion for rotifers, and at least 100 times safer for crustaceans, as the following 24-hr LC_{50} figures in ppm show[131].

	Fenthion	Temephos
Grass shrimp, *Palaemonetes paludosus*	0.011	2.0
Amphipod, *Hyalella azteca*	0.016	2.2
Rotifers	1.0	50

It is one of the least toxic of the OP compounds to the amphipod *Gammarus lacustris* (Table 5.7). When added in granules to stocked aquaria, temephos at 0.05 and 0.1 lb/a killed the cladocerans *Ceriodaphnia* and *Simocephalus,* but caused little or no mortality among the copepods *Cyclops, Eucyclops,* and *Ectocyclops,* ostracods, Hydrophilids, Chaoborines, or *Physa* snails.[30]

The mosquitofish *Gambusia affinis,* often used in larval control and frequently in combination with temephos and other OP insecticides, is of itself so effective a predator of the aquatic insects, microcrustacea, and rotifers that their presence may contribute to the growth of phytoplankton, but not of the benthic *Spirogyra*.[63] A species of silversides from Mississippi,

Menidia audens, can devour phytoplankton as well as zooplankton, and was successfully introduced in 1967 into Clear Lake, California to control the Clear Lake midge, *Chaoborus astictopus* without aggravating the algal blooms.[18]

New Alternative Larvicides. The juvenile-hormone mimic methoprene applied at 0.05 lb/a controlled mosquito larvae without affecting populations of *Cyclops* or of *Daphnia,*[85] while the various crustacea and aquatic insects were at least 1000 times as tolerant as the target species (Table 5.8). The chitin-synthesis inhibitor diflubenzuron (Dimilin) applied at 0.025 lb/a markedly reduced the cladocerans *Daphnia* and *Moina* by the fourth day after treatment, but they recovered their numbers in 3 wk; the copepods *Cyclops* and *Diaptomus* showed a slight transitory reduction while the seed shrimps and mayfly nymphs were unaffected.[86] It will be noted (Table 5.8)

Table 5.8. **Toxicity of insect development inhibitors to nontarget organisms: 24-hr LC$_{50}$ figures in ppm (Miura and Takahashi, 1973, 1974).**

		Diflubenzuron *	Methoprene
Mosquito larvae[†]	Aedes nigromaculis	0.0005	0.000008
Clam shrimp	Eulimnadia sp.	0.00015	1.0
Tadpole shrimp	Triops longicaudatus	0.0008	5.0
Cladoceran	Daphnia magna	0.0015	0.9
Mayfly nymphs	Callibaetis sp.	0.05	---
Dragonfly nymphs	Orthemis sp.	0.05	20
Midge larvae	Chironomus stigmaterus	0.01	0.01
Backswimmers	Notonecta unifasciata	0.01	1.2
Protozoan	Paramoecium sp.	---	1.25
Amphipod	Hyalella azteca	---	1.25
Water-boatmen	Corisella decolor	---	1.65
Beetle larvae	Hydrophilus triangularis	0.1	50
Copepod	Cyclops sp.	0.2	4.6
Seed shrimp	Cypricercus sp.	0.5	1.5

*approximate figures, exposure periods

† larger organism

that diflubenzuron is much more toxic to microcrustacea than methoprene, and that there is no safety factor for the clam shrimp *Eulimnadia* and the tadpole shrimp *Triops* as compared to the target mosquito larvae.

The synthetic pyrethroids resmethrin and dimethrin, and the pyrethrins or allethrin synergized with piperonyl butoxide, are especially harmful to mayfly nymphs when applied at 0.1 lb/a, although harmless to dragonfly naiads.[95] The high-boiling petroleum–hydrocarbon fractions (e.g., Flit MLO) now used to combat insecticide-resistant mosquito larvae have no effect on mayfly or dragonfly nymphs, nor on Dytiscid or Hydrophilid larvae, at the recommended dosage of 2 gal/a.[94] The long-chain aliphatic amines (e.g., Alamine 11, Armeen 15) also employed have no effect on these nontarget insects, even when applied at 2 lb/a in the aforementioned type of oil.[93]

C. CONTAMINATION OF ESTUARINE AND MARINE WATERS

Effects on Crustacea

Salt Marsh Fauna. In the salt-water environment, DDT applied to New Jersey salt marshes at 0.25 lb/a caused 20% mortality of the blue crab *Callinectes sapidus,* and even greater mortality among the smaller specimens of the marsh fiddler crabs *Uca pugnax* and the bait shrimp, *Palaemonetes pugio.* The amphibious crustacean *Orchestes gryllus,* an amphipod known as the marsh flea, and the isopod *Philosia* known as the salt-marsh sowbug, were almost eliminated. While there was some mortality of the oligochaete clamworm *Nereis,* there was no mortality of the mussel *Volsella demissa* and the salt-marsh snail *Melampus lineatus.*[122] Applied at the mosquito-control dosage of 0.2 lb/a to salt-marsh ditches in Florida, DDT drastically reduced the population of *Palaemonetes* shrimps, and caused some mortality of the oligochaete worm *Laonereis culveri.*[24] In Delaware the larger inhabitants of streambanks in salt marshes, namely the marsh crab *Sesarma reticulatum* and the red-jointed fiddler crab *Uca minax,* did not suffer as much as the marsh fiddler.[42] Residues of DDT in the plant detritus of salt marshes, amounting to about 10 ppm on Long Island, so affected the fiddler crabs feeding on this material that they lost their equilibrium when making their normal escape movements.[104] Aerial spraying operations against the gypsy moth on Long Island have caused considerable mortality of crabs in the tidal marshes adjoining the treated woodlands.[21] While juvenile blue crabs can grow normally in continuous-flow water containing 0.25 ppb DDT, at double this concentration they were all killed within 3 wk.[78]

Dieldrin at 1 ppb caused some mortality among zoea larvae of *Leptodius floridanus* but not of *Panopeius herbstii,* while neither species could complete their larval stages in 10 ppb dieldrin.[36] Strobane, a mixture of chlorinated terpenes rather similar to toxaphene, had about the same effect on crabs in a Delaware salt marsh when applied at a slightly higher dosage than that of DDT (Table 5.9). BHC applied at a dosage of 0.1 lb gamma isomer per acre was no less destructive.[42]

Marine Fauna. To three species of marine decapod crustacea, DDT was the most toxic of the organochlorines (Table 5.10), while methyl parathion was the most toxic of the organophosphorus insecticides tested.[34] However, there was no damage to the lobster fishery of the Gulf of St. Lawrence from the extensive application of DDT against spruce budworm in New Brunswick, because the runoff was minimized by the heavy duff cover over the forest soil. DDT is very toxic to the white shrimp *Penaeus setiferus* and the brown shrimp *P. aztecus,* while lindane is extremely toxic, showing a 48-hr LC_{50} of only 0.4 ppb for the pink shrimp *P. duorarum.*[13] These three species form the basis of the shrimp fishery of the southeastern United States bordering the Gulf of Mexico, and the juvenile forms inhabit estuaries liable to be contaminated by organochlorines. The fact that residues of these insecticides are seldom found in these shrimp may be due to the poisoned juveniles having been removed by death.[11] Moreover, *P. aztecus* and *P. duorarum* living in water contaminated by 0.1 ppb *p,p′*-DDT, within the range of actual pollution levels, showed a 25% reduction in the concentra-

Table 5.9. Percentage mortalities of caged crabs exposed to antimosquito sprays in Delaware salt marshes (George, Darsie, and Springer, 1957).

	BHC	DDT	Strobane	Untreated
	0.1 lb/a gamma	0.2 lb/a tech.	0.3 lb/a	Control
Marsh Fiddler	80	75	68	16
Red-jointed Fiddler	35	36	20	44
Marsh Crab	50	50	46	33
Blue Crab, in ponds	10	17	27	25
, in streams	52	10	20	10

Table 5.10. **96-hr LC$_{50}$ figures in ppb (static test) for organochlorine and organophosphorus insecticides to decapod crustaceans (Eisler, 1969).**

		Sand Shrimp	Grass Shrimp	Hermit Crab
OC1	p,p'-DDT	0.6	2.0	3
	Endrin	1.7	1.8	5
	Methoxychlor	4	12	1
	Lindane	5	10	3
	Dieldrin	7	50	5
	Aldrin	8	9	11
	Heptachlor	8	440	32
OP	Methyl parathion	2	3	5
	Dichlorvos	4	15	28
	Monocrotophos	11	69	10
	Malathion	33	82	50
	Dioxathion	38	285	65

tions of Na and K ions in the hepatopancreas.[102] Nevertheless, this shrimp fishery increased from a $100 million industry in 1967 to a $1 billion industry in 1973.[100]

The polychlorinated biphenyl Aroclor 1254 is about one-tenth as toxic as DDT to marine shrimps, the 15-day LC$_{50}$ for juvenile *Penaeus duorarum* being about 1 ppb,[103] and to *Gammarus oceanicus* about 10 ppb.[136] Leakage of this PCB from a heat-exchange unit into an estuary did not kill the pink shrimp, but it was concentrated at least 5000 times in the shrimps' bodies, especially in the hepatopancreas.[103]

The 48-hr LC$_{50}$ figures for the toxicity of organochlorines and OP compounds to the brown shrimp, *Crangon crangon,* of British waters were as follows (in ppb)[109].

BHC	1–3.3	Azinphosmethyl	0.3–1
Aroclor 1254	3–10	Dimethoate	0.3–1
DDT	3.3–10	Parathion	3.3–10
Endosulfan	10	Malathion	330–1000
Dieldrin	10–33	Morphothion	1000–3300

The use of parathion in ricefields in the Chikugo drainage basin of Japan in 1952 was followed by a very poor harvest of Euphausid krill.[68] A comparison of organochlorines and OP compounds for their effect on the Korean shrimp *Palaemon macrodactylus* is also afforded by the following 96-hr LC_{50} figures (ppb) obtained:

DDT	0.86	Chlorpyrifos	0.25
Endrin	4.7	Fenthion	5.3
Toxaphene	20.3	Malathion	82

If the static test employed had been replaced by a continuous or intermittent flow test, these figures could have been lower by a factor of 10 (Earnest, 1971[99]).

Carbaryl has been employed to control the pests and predators of the oyster, particularly starfish (Echinodermata), oyster drills (Mollusca), and the ghost shrimp and mud shrimp among crustacea.[77] The 24-hr EC_{50} levels for the two predators as compared with two desirable crustacea[124] were as follows (in ppm):

Mud shrimp, *Upogebia pugetensis*	0.13
Ghost shrimp, *Callianassa californiensis*	0.47
Shore crab, *Hemigrapsus oregonensis*	0.27
Dungeness crab, *Cancer magister*	0.60

as compared to 48-hr EC_{50} levels of 0.027 ppm carbaryl for the brown shrimp and 0.013 ppm for the white shrimp.

Effects on Molluscs

Oysters. Organochlorine insecticides can depress the growth rate of oysters at very low concentrations. Exposure to 1 ppm DDT for 1 wk was enough to reduce the growth rate of the eastern oyster *Crassostrea virginica* by 95%,[129] and the LC_{50} for its larvae is only 35 ppb DDT.[26] Although the oyster grows at a normal rate of 10 ppb, its shell deposition is inhibited; indeed, only 0.1 ppb reduces the shell deposition by 80%, although this effect is reversible in uncontaminated water.[10]

Aldrin, dieldrin, and endrin also inhibit shell deposition at this low concentration, while heptachlor and chlordane can inhibit at 0.01 ppb.[8] Toxaphene at 0.1 ppb reduces shell deposition by one-third, at 1 ppb reduces the growth rate, and at 100 ppb kills the oysters in 4 wk. Like aldrin, toxaphene inhibits the development of clam eggs at 1 ppb.[129]

Oysters in contaminated sea water can concentrate DDT by as much as 70,000 times, 0.1 ppb in the water giving rise to 7 ppm in the oyster.[12] However, tissue accumulations of 150 ppb acquired from higher water concentrations were almost entirely excreted after 3 months in clean water.[10] Methoxychlor can also be accumulated, a 10-day exposure to 50 ppb resulting in a 5800-fold concentration in the oyster.[137]

Shell deposition in oysters is seldom reduced by organophosphorus or carbamate insecticides even at 1 ppm,[14] although carbaryl can inhibit the development of clam eggs at 1 ppb.[129] Carbaryl also can kill marine molluscs, the 48-hr LC_{50} being 2.25 ppm for the Pacific oyster *C. gigas* and the bay mussel *Mytilus edulis*.[124] For the larvae and the eggs, the LC_{50} is 3 ppm for both stages of the eastern oyster *C. virginica*[26]; as compared to carbaryl, aldrin and dieldrin are more toxic, lindane and malathion less toxic (Table 5.11).

Clams. The hard clam *Mercenaria mercenaria* appeared to be fairly resistant to the organochlorines tested, this species and the gastropod *Nassa obsoleta* being more resistant than fish to organochlorines by several orders of magnitude.[35] Like the oyster, the hard clam in its immature stages is quite susceptible to stable OP compounds such as azinphosmethyl,[26] while being resistant to lindane in the water (Table 5.11). The cockle clam *Clinocardium nuttalli* is evidently more susceptible to carbaryl than the hard clam, since all its larvae are killed by a 7-day exposure to 0.8 ppm, and 0.4 ppm reduces the larval growth rate; it is interesting that for this mollusc the metabolite 1-naphthol is slightly more toxic than the original carbamate.[7]

D. EFFECTS ON PHYTOPLANKTON ORGANISMS

Inhibition of Growth and Respiration

In screening the available pesticides for their possible use to abate algae, it was found that neither DDT nor lindane were toxic at 2 ppm concentration to cultures of *Microcystis, Cylindrospermum, Scenedesmus, Gomphionema, Nitzschia,* or *Chlorella,* while toxaphene showed some toxicity to five of these species, particularly to *Microcystis,* but not to *Chlorella*.[105] In screening insecticides for their side effects on a mixed culture of marine phytoplankton, it was found that all the organochlorines were inhibitory at 1 ppm for 4 hr, lindane, endrin, and mirex being the least so. The OP compounds were not inhibitory, with the exception of ethion, naled and disulfoton which caused 50–70% decreases. Among the carbamates, methiocarb was

Table 5.11. LC$_{50}$ figures in ppm for insecticides on immature stages of oysters and clams (Davis and Hidu, 1969).

	American Oyster (Crassostrea virginica)		Hard Clam (Mercenaria mercenaria)	
	Eggs, 48-hr.	Larvae, 14-day	Eggs, 48-hr.	Larvae, 12-day
Coumaphos	0.11	>1	0.12	5.2
Azinphosmethyl	0.62	---	0.86	0.86
Trichlorfon	---	1.0		
Toxaphene			1.12	<0.25
Dieldrin	0.64	>10		
Dicapthon			3.3	5.7
Endrin	0.79	>10		
Carbaryl	3.0	3.0	3.8	>2.5
Disulfoton	5.9	3.7	5.3	1.3
Aldrin			>10	0.41
Malathion	9.1	2.7		
Lindane	9.1	---	>10	>10
Tepp	>10	>10		

inhibitory, propoxur and aminocarb not, while carbaryl was intermediate in effect.[9] When five species of marine phytoplankton were compared (Table 5.12), the flagellate *Monochrysis* (*lutheri*) was the most susceptible and the nonmotile unicellular *Protococcus* was the most resistant.[128]

Among freshwater algae, *Anabaena, Oedogonium* and *Scenedesmus* were resistant to DDT, aldrin, dieldrin, and endrin at 15 ppm, but *Microcystis* (*aeruginosa*) was susceptible to these cyclodiene compounds at 5 ppm.[130] Taken from waste stabilization ponds, *Scenedesmus, Chlorella, Ankistrodesmus,* and *Euglena* were resistant to 100 ppm DDT, but not to malathion or carbaryl.[16] Carbaryl at 1 ppm, however, considerably increased the growth rate and the biomass attained by *Scenedesmus quadricaudata* after 1 wk of exposure; toxaphene and diazinon at this concentration had no effect, while DDT and dieldrin at 0.1 ppm decreased the growth slightly but significantly.[123]

By contrast, the marine diatoms *Skeletonema costatum* and *Cyclotella nana* were sensitive to 50 ppb DDT, which inhibited their utilization of ^{14}C in photosynthesis by 30-80%.[83,138] In a third marine diatom, *Thalassiosira pseudonana,* as well as in *Skeletonema,* a dose-response relationship could be demonstrated, the inhibition decreasing from 70% at 50 ppb down to zero at 1 ppb.[39] No inhibitory effect at 50 ppb was shown on the marine alga *Coccolithus huxleyi,*[39,83,138] nor on the estuarine flagellate *Dunaliella tertiolecta*[83]; the latter species, along with *Scenedesmus quadricaudata,* was unaffected even by 1 ppm DDT, DDE, or DDD.[79] Although DDT and DDE have been demonstrated to inhibit the electron transport between the two light reactions of photosynthesis in isolated chloroplast preparations,[4] it is probable that DDT inhibits phytoplankton by depressing the rate of cell division, since between the species mentioned growth-rate effects and ^{14}C-utilization effects ran in parallel,[83] and the CO_2 uptake per cell was not reduced by the DDT.[39]

The PCB Aroclor 1254 caused partial inhibition of ^{14}C uptake in *Dunaliella* and *Scenedesmus* at 1 ppm,[79] and in *Thalassiosira* at 50 ppb.[39] When compared for their susceptibility to a mixture of PCBs, the ranking was *Skeletonema* > *Thalassiosira* > *Chlamydomonas* > *Euglena* > *Dunaliella.*[89] For a phytoplankton community (mainly diatoms) taken from the Gulf coast of Florida, Aroclor 1254 had an inhibitory effect down to 1 ppb, making it more toxic than *p,p´*-DDT which did not inhibit below 10 ppb.[88] A leakage of this PCB from the heat exchange unit of an industrial plant on the shore of Escambia Bay on this coast gave water concentrations of 275 ppb at the outfall, decreasing to less than 1 ppb in the bay itself.[32]

Endrin and dieldrin were more inhibitory than DDT for the growth of the diatom *Cyclotella.*[83] For the marine blue-green alga *Anacystis nidulans,* concentrations of 500 ppb were required for growth inhibition with dieldrin

Table 5.12. Effect of insecticides applied to growing colonies of five species of marine phytoplankton: density of growth as compared to untreated colonies (Ukeles, 1962).

	Conc'n ppm	Dunaliella	Phaeodactylum	Monochrysis	Protococcus	Chlorella
Untreated	0.0	1.00	1.00	1.00	1.00	1.00
DDT	0.6	0.74	0.91	0.28	0.50	1.00
Lindane	9.0	0.60	0.00	0.00	1.00	0.33
Toxaphene	0.07	0.53	0.00	0.00	0.80	0.00
Trichlorfon	100	0.42	0.39	0.00	0.54	0.68
Tepp	300	0.49	0.25	0.38	0.17	0.27
Carbaryl	1	0.65	0.00	0.00	0.74	0.80

and photodieldrin, endrin and ketoendrin, while aldrin and photoaldrin were only slightly inhibitory at 1 ppm.[2] When two organophosphorus compounds and a carbamate insecticide were compared with DDT for their effect on four phytoplankton species taken from the Atlantic, the following percent reductions in O_2 output were obtained at 100 ppb[27]:

	Skeletonema	*Cyclotella*	*Dunaliella*	*Phaeodactylum*
Fenthion	51	48	27	29
Temephos	23	13	23	28
Propoxur	23	13	32	28
DDT	32	33	32	10

Since the effect of DDT (and presumably other organochlorines) is on cell multiplication rather than photosynthesis itself, it is unlikely that their presence in the oceans would reduce O_2 production, since the more tolerant species would make up for decreases in the very susceptible species.[39,138] It could be demonstrated in the laboratory that the addition of 100 ppb DDT or 25 ppb PCBs prevented the fast-growing but susceptible diatom *Thalassiosira* from overpowering the slow-growing but comparatively resistant flagellate *Dunaliella*.[90]

Induction of Phytoplankton Blooms

It has been the general experience that DDT treatments of forests for defoliator control or of streams for *Simulium* control, or rotenone treatments of lakes for trash-fish elimination, are followed by upsurges (blooms) of phytoplankton as a result of the decimation of aquatic arthropods that graze upon them.[60] Repeated aerial treatments of a reservoir at Wilmington, Delaware with DDT as a public health measure was followed by the multiplication of the diatom *Synedra* to 100 times the normal level as a result of the disappearance of *Cyclops* and *Daphnia*.[120] Phytoplankton activity in a New Jersey salt marsh treated with DDT at 1 lb/a increased nearly four times because the larvae of the mosquito *Aedes sollicitans* were eliminated; with fenthion at 1 lb/a, which far from reducing *Cyclops* slightly increased its abundance, the phytoplankton production was only 40% more than the normal.[113]

An experimental application of chlorpyrifos at 0.25 lb/a so reduced *Cyclops, Moina,* and other microcrustaceans in a California pond that a large phytoplankton bloom resulted; unfortunately the most important species was the blue-green alga *Anabaena*,[61] which has been frequently responsible for intoxicating and killing livestock and wildlife in farm ponds.

Heavy blooms of *Anabaena* also followed three successive applications of methyl parathion at 3 ppb to Clear Lake, California, consequent upon the reduction in the numbers of *Daphnia, Cyclops,* and *Diaptomus.*[17] Ponds at Minneapolis treated with DDT at 1 lb/a in granules for mosquito control provided an example where ostracod, copepod, and cladoceran numbers were normal 2 wk after the application; here the only phytoplankton form which responded was *Volvox,* which increased to seven times its normal level.[73]

Benthic algae may also increase to the point where they constitute a pollution carpet on the bottom of ponds, streams, or flooded fields. A stream in the Isle of Man contaminated by BHC discharged from sheep dips developed an overgrowth of *Cladophora,* accompanied by some *Spirogyra,* because of the destruction of the herbivorous arthropods such as mayfly nymphs and *Gammarus* amphipods.[64] Application of lindane at 5 kg/ha to Philippine ricefields resulted in a bloom of blue-green algae due to the elimination of a species of ostracod.[110] While trichlorfon at 0.2 ppm killed all Japanese microcrustacea exposed except certain copepods (and rotifers), it had no observable effect on phytoplankton at 10 ppm.[132] Fenthion applied to California ponds at 0.05 ppm caused so much mortality in the mayfly *Callibaetis pacificus* that there were considerable increases in the filamentous algae *Zygnema* and *Mougeotia* on which the naiads feed.[60]

REFERENCES CITED

1. A. P. Arnason, A. W. A. Brown, F. J. H. Fredeen, W. W. Hopewell, and J. G. Rempel. 1949. Experiments in the control of *Simulium arcticum* by means of DDT in the Saskatchewan River. *Sci. Agric.* **29:**527–537.

2. J. C. Batterton, G. M. Boush, and F. Matsumura. 1971. Growth response of blue-green algae to aldrin, dieldrin, endrin, and their metabolites. *Bull. Environ. Contam. Toxicol.* **6:**589–594.

3. E. L. Bishop. 1940. Cooperative investigations of the relation between mosquito control and wildlife conservation. *Science* **92:**201–202.

4. G. W. Bowes and R. W. Gee. 1971. Inhibition of photosynthetic electron transport in the chloroplast. *Bioenergetics* **2:**47–60.

5. G. E. Burdick, H. J. Dean, and E. J. Harris. 1960. Effect of Sevin upon the aquatic environment. *N.Y. Fish Game J.* **7:**14–25.

6. G. E. Burdick, H. J. Dean, E. J. Harris, J. Skea, C. Frisa, and C. Sweeney. 1968. Methoxychlor as a blackfly larvicide: Persistence of its residues in fish and its effect on stream arthropods. *N.Y. Fish Game J.* **15:**121–142.

7. J. A. Butler, R. E. Millemann, and N. E. Stewart. 1968. Effect of the insecticide Sevin on survival and growth of the cockle clam. *J. Fish Res. Bd. Can.* **25:**1621–1635.

8. P. A. Butler. 1962. In *Effects of Pesticides on Fish and Wildlife, 1960 Investigations.* U.S. Fish Wildl. Serv., Bur. Sport Fish. Wildl. Circ. **143:**20–24, 42–44.

9. P. A. Butler. 1963. Commercial fisheries investigations. In *Pesticide–Wildlife Studies, 1961 and 1962*. U.S. Dept. Int., Fish Wildl. Serv. Circ. **167**:11:25.

10. P. A. Butler. 1966. Pesticides in the marine environment. *J. Appl. Ecol.* **3**(Suppl.):253–259.

11. P. A. Butler. 1968. Pesticides in the estuary. *Proc. Marsh and Estuary Management Symposium*, L.S.U., Baton Rouge, July 19–20, 1967. pp. 120–124.

12. P. A. Butler. 1969. The significance of DDT residues in estuarine fauna. In *Chemical Fallout*. Ed. M. W. Miller and G. G. Berg. Charles C Thomas. pp. 205–220.

13. P. A. Butler and P. F. Springer. 1963. Pesticides, a new factor in coastal environments. *Wildl. Nat. Res. Conf. Trans.* **28**:378–390.

14. P. A. Butler, A. J. Wilson and A. J. Rick. 1962. Effect of pesticides on oysters. *Proc. Natl. Shellfish Assoc.* **51**:23–32.

15. C. A. Carlson. 1966. Effects of three organophosphorus insecticides on immature *Hexagenia* and *Hydropsyche* of the Upper Mississippi River. *Trans. Am. Fish Soc.* **95**:1–5.

16. A. E. Christie. 1969. Effects of insecticides on algae. *Water and Sewage Works* **116**:172–176.

17. S. F. Cook and J. D. Conners. 1963. The short-term side effects of the insecticidal treatment of Clear Lake, California in 1962. *Ann. Entomol. Soc. Am.* **56**:819–824.

18. S. F. Cook and R. L. Moore. 1970. Mississippi silversides, *Menidia audens* (Atherinidae) established in California. *Trans. Am. Fish Soc.* **99**:70–73.

19. A. S. Cooke and E. Pollard. 1973. Shell and operculum formation by immature Roman snails when treated with p,p'-DDT. *Pestic. Biochem. Physiol.* **3**:230–236.

20. O. B. Cope. 1966. Contamination of the freshwater system by pesticides. *J. Appl. Ecol.* **3**(Suppl.):33–44.

21. O. B. Cope and P. F. Springer. 1958. Mass control of insects: The effects on fish and wildlife. *Bull. Entomol. Soc. Am.* **4**:52–56.

22. P. S. Corbet. 1958. Some effects of DDT on the fauna of the Victoria Nile. *Rev. Zool. Bot. Afr.* **57**:73–95.

23. C. Cottam and E. Higgins. 1946. DDT: Its effect on fish and wildlife. *J. Econ. Entomol.* **39**:44–52.

24. R. A. Croker and A. J. Wilson. 1965. Kinetics and effects of DDT in a tidal marsh ditch. *J. Am. Fish. Soc.* **94**:152–159.

25. R. A. Crouter and E. H. Vernon. 1959. Effects of black-headed budworm control on salmon and trout in British Columbia. *Can. Fish. Cult.* No. **24**:23–40.

26. H. C. Davis and H. Hidu. 1969. Effects of pesticides on embryonic development of clams and oysters and on survival and growth of the larvae. *Fishery Bull. (U.S. Fish Wildl. Serv.)* **67**:293–404.

27. S. B. Derby and E. Ruber. 1970. Depression of oxygen evolution in algal cultures by organophosphorus insecticides. *Bull. Environ. Contam. toxicol.* **5**:553–558.

28. S. K. Derr and M. J. Zabik. 1972. The uptake and distribution of DDE by *Chironomus tentans*. *Trans. Am. Fish. Soc.* **101**:323–329.

29. R. Deschiens, H. Floch, and Y. Le Coroller. 1965. Les molluscicides cuivreux dans la prophylaxie des bilharzioses. *Bull. World Health Org.* **33**:73–88.

30. V. Didia, R. LaSalle, and K. Liem. 1975. The effects of Abate mosquito larvicide on selected non-target aquatic organisms collected from forested temporary pools. *Mosq. News* **35**:227–228.

31. D. L. Dindal and K. H. Wurzinger. 1971. Accumulation and excretion of DDT by the terrestrial snail *Cepaea hortensis. Bull. Environ. Contam. Toxicol.* **6**:362–371.

32. T. W. Duke, J. I. Lowe, and A. J. Wilson. 1970. A polychlorinated biphenyl in the water, sediment, and biota of Escambia Bay, Florida. *Bull. Environ. Contam. Toxicol.* **5**:171–180.

33. P. M. Eide, C. C. Deonier, and R. W. Burrell. 1945. The toxicity of DDT to certain forms of aquatic life. *J. Econ. Entomol.* **38**:492–493.

34. R. Eisler. 1969. Acute toxicities of insecticides to marine decapod crustaceans. *Crustaceana* **16**:302–310.

35. R. Eisler. 1970. *Acute Toxicities of Organochlorine and Organophosphorus Insecticides to Estuarine Fishes.* U.S. Dept. Interior, Bur. Sport Fish Wildl. Tech. Paper 46. 12 pp.

36. C. E. Epifanio. 1971. Effects of dieldrin in seawater on the development of two species of crab larvae. *Marine Biol.* **11**:356–362.

37. J. H. Fales, P. J. Spangler, O. F. Bodenstein, G. D. Mills, and C. G. Durbin. 1968. Laboratory and field evaluation of Abate against a backswimmer, *Notonecta undulata. Mosq. News* **28**:77–81.

38. FWPCA. 1968. *Water Quality Criteria.* Report of the National Tech. Adm. Comm. to Secretary of the Interior, Federal Water Pollution Control Administration. USDI. 234 pp.

39. N. S. Fisher. 1975. Chlorinated hydrocarbon pollutants and photosynthesis of marine plankton: A reassessment. *Science* **189**:463–464.

40. J. F. Flannagan. 1973. Field and laboratory studies of the effect of exposure to fenitrothion of freshwater aquatic invertebrates. *Manitoba Entomol.* **7**:15–25.

41. J. L. George. 1959. Effect on fish and wildlife of chemical treatments of large areas. *J. Forestry* **57**:250–254.

42. J. L. George, R. F. Darsie, and P. F. Springer. 1957. Effects on wildlife of aerial applications of Strobane, DDT, and BHC to tidal marshes in Delaware. *J. Wildl. Manage.* **21**:42–53.

43. J. R. Gorham. 1961. *Aquatic Insects and DDT Forest Spraying in Maine.* Maine Forest Service Bull. 19. 49 pp.

44. C. D. Grant and A. W. A. Brown. 1967. Development of DDT resistance in certain mayflies in New Brunswick. *Can Entomol.* **99**:1040–1050.

45. W. R. Hanson. 1952. Effects of some herbicides and insecticides on biota of North Dakota marshes. *J. Wildl. Manage.* **16**:299–308.

46. A. D. Harrison. 1966. The effects of Bayluscid on gastropod snails and other aquatic fauna in Rhodesia. *Hydrobiologia* **28**:371–384.

47. E. Hastings, W. H. Kittams and J. H. Pepper. 1961. Repopulation by aquatic insects in streams sprayed with DDT. *Ann. Entomol. Soc. Am.* **54**:436–437.

48. C. T. Hatfield. 1969. Effects of DDT larviciding on aquatic fauna of Bobby's Brook, Labrador. *Can Fish. Cult.* **40**:61–72.

49. J. E. Hemphill. 1954. Toxaphene as a fish toxin. *Progress. Fish.-Cult.* **16**:41–42.

50. A. D. Hess and G. G. Keener. 1947. Effects of airplane-distributed DDT thermal aerosols on fish and fish food organisms. *J. Wild. Manage.* **11**:1–10.

51. W. L. Hilsenhoff. 1965. The effect of toxaphene on the benthos in a thermally stratified lake. *Trans. Am. Fish. Soc.* **94**:210–213.

52. S. W. Hitchcock. 1965. *Field and Laboratory Studies of DDT and Aquatic Insects.* Conn. Agric. Exp. Sta. Bull. 668. 25 pp.

53. B. Hocking, C. R. Twinn, and W. C. McDuffie. 1949. A preliminary evaluation of some insecticides against immature stages of blackflies. *Sci. Agric.* **29**:69–80.

54. C. H. Hoffmann and A. T. Drooz. 1953. Effects of a C-47 airplane application of DDT on fish food organisms in two Pennsylvania watersheds. *Am. Midl. Nat.* **50**:172–188.

55. C. H. Hoffmann and J. P. Linduska. 1949. Some considerations of the biological effects of DDT. *Sci. Monthly* **69**:104–114.

56. C. H. Hoffmann and E. P. Merkel. 1948. Fluctuations in insect populations associated with aerial applications of DDT to forests. *J. Econ. Entomol.* **41**:464–473.

57. C. H. Hoffmann, H. K. Townes, R. I. Sailer, and H. H. Swift. 1946. *Field Studies on the Effect of DDT on Aquatic Insects.* U.S. Dept. Agric. Bur. Entomol. Plant Quar. E-702. 20 pp.

58. D. A. Hoffman and J. R. Olive. 1961. The effects of rotenone and toxaphene upon plankton of two Colorado reservoirs. *Limnol. Oceanogr.* **6**:219–222.

59. F. F. Hooper and A. R. Grzenda. 1957. The use of toxaphene as a fish poison. *Trans. Am. Fish Soc.* **85**:180–190.

60. S. H. Hurlbert. 1975. Secondary effects of pesticides on aquatic ecosystems. *Residue Rev.* **57**:81–148.

61. S. H. Hurlbert, M. S. Mulla, and H. R. Willson. 1972. Effects of an organophosphorus insecticide on the phytoplankton, zooplankton, and insect populations of fresh-water ponds. *Ecol. Monogr.* **42**:269–299.

62. S. H. Hurlbert, M. S. Mulla, J. O. Keith, W. E. Westlake, and M. E. Duesch. 1970. Biological effects and persistance of Dursban in freshwater ponds. *J. Econ. Entomol.* **63**:43–52.

63. S. H. Hurlbert, J. Zedler and D. Fairbanks. 1972. Ecosystem alteration by mosquitofish (*Gambusia affinis*) predation. *Science* **175**:639–641.

64. H. B. N. Hynes. 1961. The effect of sheep-dip containing the insecticide BHC on the fauna of a small stream. *Ann. Trop. Med. Parasit.* **55**:192–196.

65. H. B. N. Hynes and T. R. Williams. 1962. The effect of DDT on the fauna of a central African stream. *Ann. Trop. Med. Parasit.* **56**:78–91.

66. F. P. Ide. 1957. Effects of forest spraying with DDT on aquatic insects of salmon streams. *Trans. Am. Fish Soc.* **86**:208–219.

67. F. P. Ide. 1967. Effects of forest spraying with DDT on aquatic insects of salmon streams in New Brunswick. *J. Fish. Res. Bd. Can.* **24**:769–805.

68. H. Ishikura. 1972. Impact of pesticide use on the Japanese environment. In *Environmental Toxicology of Pesticides.* Ed. F. Matsumura, G. M. Boush, and T. Misato. Academic. pp. 1–32.

69. H. Jamnback and H. S. Eabry. 1962. Effects of DDT, as used in black fly larval control, on stream arthropods. *J. Econ. Entomol.* **55**:636–639.

70. L. D. Jensen and A. R. Gaufin. 1964. Long-term effects of organic insecticides on two species of stonefly naiads. *Trans. Am. Fish. Soc.* **93**:357–363.

71. L. D. Jensen and A. R. Gaufin. 1966. Acute and long-term effects of organic insecticides on two species of stonefly naiads. *J. Water Pollut. Contr. Fed.* **38**:1273–1286.

72. B. T. Johnson, C. R. Saunders, H. O. Sanders, and R. S. Campbell. 1971. Biological

magnification and degradation of DDT and aldrin by freshwater invertebrates. *J. Fish. Res. Bd. Can.* **28**:705–709.

73. B. R. Jones and J. B. Moyle. 1963. Populations of plankton animals and residual chlorinated hydrocarbons in soils of six Minnesota ponds treated for control of mosquito larvae. *Trans. Am. Fish. Soc.* **92**:211–215.

74. J. A. Kawatski and J. C. Schmulbach. 1971. Accumulation of insecticides in freshwater ostracods. *Trans. Am. Fish. Soc.* **100**:565–569.

75. H. D. Kennedy, L. F. Eller, and D. F. Walsh. 1970. *Chronic effects of Methoxychlor on Bluegills and Aquatic Invertebrates.* U.S. Dept. Int., Bur. Sport Fish. Wildl., Tech. Paper 53. 18 pp.

76. L. L. Lewallen. 1962. Toxicity of certain insecticides to Hydrophilid larvae. *Mosq. News* **22**:112–113.

77. V. L. Loosanoff. 1960. Recent advances in the control of shellfish predators and competitors. *Proc. Gulf Caribbean Fisheries Inst.* **13**:113–128.

78. J. I. Lowe. 1965. Chronic exposure of blue crabs *Callinectes sapidus* to sublethal concentrations of DDT. *Ecology* **46**:899–900.

79. E. J. Luard. 1973. Sensitivity of *Dunaliella* and *Scenedesmus* (Chlorophyceae) to chlorinated hydrocarbons. *Phycologia* **12**:29–33.

80. K. J. Macek, K. S. Buxton, S. K. Derr, J. W. Dean, and S. Sauter. 1976. *Chronic Toxicity of Lindane to Selected Aquatic Invertebrates and Fishes.* Environmental Protection Agency, Washington. Processed Publ. EPA-600/3-76-046. 50 pp.

81. A. W. Maki and H. E. Johnson. 1975. Effects of PCB (Aroclor 1254) and *p,p'*-DDT on production and survival of *Daphnia magna. Bull. Environ. Contam. Toxicol.* **13**:412–416.

82. A. W. Maki, K. W. Stewart, and J. K. G. Silvey. 1973. The effects of Dibrom on respiratory activity of a stonefly, the hellgrammite, and the golden shiner. *Trans. Am. Fish Soc.* **102**:806–815.

83. D. W. Menzel, J. Anderson, and A. Randtke. 1970. Marine phytoplankton vary in their response to chlorinated hydrocarbons. *Science* **167**:1724–1726.

84. C. W. Miller, B. M. Zuckerman, and A. J. Charig. 1966. Water translocation of diazinon and parathion off a model cranberry bog and subsequent occurrence in fish and mussels. *Trans. Am. Fish. Soc.* **95**:345–349.

85. T. Miura and R. M. Takahashi. 1973. Insect development inhibitors. 3. Effects on nontarget aquatic organisms. *J. Econ. Entomol.* **66**:917–922.

86. T. Miura and R. M. Takahashi. 1974. Insect developmental inhibitors: Effects of candidate mosquito control compounds on nontarget aquatic organisms. *Environ. Entomol.* **4**:631–636.

87. J. B. Moore and S. G. Breeland. 1967. Field evaluation of two mosquito larvicides against *Anopheles quadrimaculatus* and associated *Culex* species. *Mosq. News* **27**:105–111.

88. S. A. Moore and R. C. Harriss. 1972. Effects of polychlorinated biphenyl on marine phytoplankton communities. *Nature* **240**:356–357.

89. J. L. Mosser, N. S. Fisher, T. C. Cheng, and C. F. Wurster. 1972. Polychlorinated biphenyls: Toxicity to certain phytoplankters. *Science* **175**:191–192.

90. J. L. Mosser, N. S. Fisher, and C. F. Wurster. 1972. Polychlorinated biphenyls and DDT alter species composition in mixed cultures of algae. *Science* **176**:533–535.

91. W. C. Moye and W. H. Luckmann. 1964. Fluctuations in populations of certain aquatic insects following applications of aldrin granules. *J. Econ. Entomol.* **57**:318–322.

92. R. C. Muirhead-Thomson. 1971. *Pesticides and Freshwater Fauna.* Academic. pp. 63–67.

93. M. S. Mulla and H. A. Darwazeh. 1971. Influence of aliphatic amines–petroleum oil formulations on aquatic non-target insects. *Proc. Calif. Mosq. Control Assoc.* **39**:126–131.

94. M. S. Mulla, J. R. Arias, and H. A. Darwazeh. 1971. Petroleum oil formulations against mosquitoes and their effects on some non-target insects. *Proc. Calif. Mosq. Control Assoc.* **39**:131–135.

95. M. S. Mulla, R. D. Sjogren, and J. R. Arias. 1972. Mosquito larvicides: Efficacy under field conditions and effects on non-target organisms. *Proc. Calif. Mosq. Control Assoc.* **40**:139–145.

96. R. J. Muncy and A. D. Oliver. 1963. Toxicity of ten insecticides to the red crawfish, *Procambarus clarki. Trans. Am. Fish. Soc.* **92**:428–431.

97. S. M. Naqvi and D. E. Ferguson. 1968. Pesticide tolerances of selected freshwater invertebrates. *J. Miss. Acad. Sci.* **14**:121–127.

98. S. M. Naqvi and D. E. Ferguson. 1970. Levels of insecticide resistance in fresh-water shrimp, *Palaemonetes kadakiensis. Trans. Am. Fish. Soc.* **99**:696–699.

99. National Academy of Sciences. 1973. *Water Quality Criteria, 1972.* Report EPA-R3-73-033. Sup't. of Documents Stock No 5501-00520. pp. 420–518.

100. National Academy of Sciences. 1976. *Pest Control: An Assessment of Present and Alternative Technologies.* Vol. 5, p. 101–109.

101. A. V. Nebeker and F. A. Puglisi. 1974. Effect of PCB's on survival and reproduction of *Daphnia, Gammarus* and *Tanytarsus. Trans. Am. Fish. Soc.* **103**:722–728.·

102. D. R. Nimmo and R. R. Blackman. 1972. Effects of DDT and cations in the hepatopancreas of penaeid shrimp. *Trans. Am. Fish. Soc.* **101**:547–549.

103. D. R. Nimmo, R. R. Blackman, A. J. Wilson, and J. Forester. 1971. Toxicity and distribution of Aroclor 1254 in the pink shrimp. *Marine Biol.* **11**:191–197.

104. W. E. Odum, G. M. Woodwell, and C. F. Wurster. 1969. DDT residues absorbed from organic detritus by fiddler crabs. *Science* **64**:576–577.

105. C. Palmer and T. E. Maloney. 1955. Preliminary screening for potential algicides. *Ohio J. Sci.* **55**:1–8.

106. R. S. Patterson and D. L. Von Windeguth. 1964. The effects of Baytex on some aquatic organisms. *Mosq. News* **24**:46–49.

107. T. J. Peterle and R. H. Giles. 1964. *New Tracer Techniques for Evaluating the Effects of an Insecticide on the Ecology of a Forest Fauna.* Ohio State Univ. Res. Found'n. Rep. 1207 (to USAEC). 435 pp.

108. C. H. Porter and W. L. Gojmerac. 1969. Field observations with Abate and bromophos: Their effect on mosquitos and aquatic arthropods in a Wisconsin park. *Mosq. News* **29**:617–620.

109. J. E. Portmann and K. W. Wilson. 1971. *The Toxicity of 140 Substances to the Brown Shrimp and Other Marine Animals.* Min. Agric. Fish. Food, U.K., Shellfish Information Leaflet No. 22. 11 pp.

110. K. Raghu and I. C. MacRae. 1967. The effect of the gamma isomer of benzene hexachloride upon algae in submerged rice soils. *Can. J. Microbiol.* **13**:173–180.

111.   E. Ruber. 1963. The effects of certain mosquito larvicides on microcrustacean populations. *Proc. 50th Ann. Mtg. N.J. Mosq. Exterm. Assoc.* pp. 256–263.

112.   E. Ruber and J. Baskar. 1968. Sensitivities of selected microcrustacea to eight mosquito toxicants. *Proc. N.J. Mosq. Exterm. Assoc.* **55:**99–103.

113.   E. Ruber and F. Ferrigno. 1964. Some effects of DDT, Baytex, and endrin on salt marsh productivities, copepods, and *Aedes* mosquito larvae. *Proc. N.J. Mosq. Exterm. Assoc.* **51:**84–93.

114.   H. O. Sanders. 1969. *Toxicity of Pesticides to the Crustacean* Gammarus lacustris. Tech. paper 25, Bur. Sport Fish. Wildl., U.S. Dept. of Interior. 18 pp.

115.   H. O. Sanders and O. B. Cope. 1966. Toxicities of several pesticides to two species of cladocerans. *Trans. Am. Fish. Soc.* **95:**165–169.

116.   H. O. Sanders and O. B. Cope. 1968. The relative toxicities of several pesticides to naiads of three species of stoneflies. *Limnol. Oceanogr.* **13:**112–117.

117.   J. Savage. 1949. *Aquatic Insects: Mortality Due to DDT and Subsequent Reestablishment.* Dept. Lands Forests Ontario, Biological Bull. No. 2. pp. 39–47.

118.   E. I. el Sayed, J. B. Graves, and F. L Bonner. 1967. Chlorinated hydrocarbon residues in selected insects and birds found in association with cotton-fields. *J. Agric. Food Chem.* **15:**1014–1017.

119.   W. J. Schouwenburg and K. J. Jackson. 1966. A field assessment of the effects of spraying a small coastal coho salmon stream with phosphamidon. *Can. Fish-Cult.* **37:**35–43.

120.   M. S. Shane. 1948. Effect of DDT spray on reservoir biological balance. *J. Am. Water Works Assoc.* **40:**333–336.

121.   C. J. Shiff, N. O. Crossland, and D. R. Miller. 1967. The susceptibilities of various species of fish to the molluscicide N-tritymorpholine. *Bull. Wld. Health Org.* **36:**500–507.

122.   P. F. Springer and J. R. Webster. 1951. Biological effects of DDT applications on tidal salt marshes. *Mosq. News* **11:**67–74.

123.   L. Stadnyk, R. S. Campbell and B. T. Johnson. 1971. Pesticide effect on growth and ^{14}C assimilation in a freshwater alga. *Bull. Environ. Contam. Toxicol.* **6:**1–8.

124.   N. E. Stewart, R. E. Millemann, and W. B. Breese. 1967. Acute toxicity of the insecticide Sevin and its hydrolytic product 1-naphthol to some marine organisms. *J. Am. Fish. Soc.* **96:**25–30.

125.   Y. H. Swabey, C. F. Schenke, and G. L. Parker. 1967. Evaluation of two organophosphorus compounds as blackfly larvicides. *Mosq. News* **27:**149–155.

126.   C. M. Tarzwell. 1947. The effects on surface organisms of the routine land application of DDT larvicides for mosquito control. *U.S. Public Health Reports* **62:**525–554.

127.   C. R. Twinn, A. W. A. Brown, and H. Hurtig. 1950. Area control of mosquitoes by aircraft in sub-arctic Canada. *Proc. N.J. Mosq. Exterm. Assoc. 1950.* pp. 113–140.

128.   R. Ukeles. 1962. Growth of pure cultures of marine phytoplankton in the presence of toxicants. *Appl. Microbiol.* **10:**532–537.

129.   USDI. 1960. *Effects of pesticides on Fish and Wildlife.* Fish. Wildl. Serv., Bur. Sport Fish. Wildl. Circ. 143. 52 pp.

130.   B. D. Vance and W. Drummond. 1969. Biological concentration of pesticides by algae. *J. Am. Water Works Assoc.* **61:**360–362.

131.   D. L. Von Windeguth and R. S. Patterson. 1966. The effect of two organophosphorus insecticides on segments of the aquatic biota. *Mosq. News* **26:**377–380.

132. G. W. Ware and C. C. Roan. 1970. Interaction of pesticides with aquatic microorganisms and plankton. *Residue Rev.* **33**:15–45.

133. K. L. Warner and O. C. Fenderson. 1962. Effect of DDT spraying for forest insects on Maine trout streams. *J. Wildl. Manage.* **26**:86–93.

134. S. L. Warnick, R. F. Gaufin, and A. R. Gaufin. 1966. Concentration and effects of pesticides on aquatic environments. *J. Am. Water Works Assoc.* **58**:601–608.

135. R. K. Washino, W. Ahmed, J. D. Linn, and K. G. Whitesell. 1972. Effects of low volume Dursban sprays upon aquatic non-target organisms. *Mosq. News* **32**:531–537.

136. D. J. Wildish. 1970. The toxicity of polychlorinated biphenyls (PCB) in sea water to *Gammarus oceanicus. Bull. Environ. Contam. Toxicol.* **5**:202–204.

137. A. J. Wilson. 1965. *Chemical Assays.* Ann. Rep't. Gulfbreeze Lab., U.S. Bur. Comm. Fish. Circ. 247. pp. 6–7.

138. C. F. Wurster. 1968. DDT reduces photosynthesis by marine phytoplankton. *Science* **159**:1474–1475.

6

INSECTICIDES AND FISH

A. EFFECTS OF ORGANOCHLORINE INSECTICIDES ON FISH

Introduction: Susceptibility Tests

Fish in standing water are unable to escape from an insecticide once it has been added to water, and have to submit to its physiological insults until it has been removed by adsorption and sedimentation or other mechanisms. Even when offered the opportunity of proceeding to water less contaminated with DDT, *Gambusia* and *Cyprinodon* minnows do not choose to do so.[70,73] Thus the toxic effects increase, and the median lethal concentration (TL_m or LC_{50}) decreases, with the duration of exposure.

In assessing the toxicities of organochlorines, the 96-hr exposure period is preferred to one of 48 hr or 24 hr. With salmonids, the 96-hr LC_{50} usually turns out to be about 25% lower than the 24 hr figure.[90] With *Fundulus heteroclitus,* a top-minnow known as the mummichog, the 10-day LC_{50} is usually somewhat more than one-half the 96 hr figure, although one-fifth with heptachlor and one-twelfth with lindane.[50] For these longer exposure periods, a continuous-flow method of test is essential. For the shorter periods, it is true that for DDT on the fathead minnow the 48-hr LC_{50} by static test was much lower than by continuous flow test with DDT, and that there was no difference between the two figures for endrin.[100] With OP compounds also, the 24-hr LC_{50} for naled to the golden shiner obtained by flowing-water test was 6.1 ppm as compared to 6.5 for the static test.[108] But it is usual to find that continuous-flow or intermittent-flow tests give lower LC_{50} figures than static ones, as shown by the following 96-hr values (ppb) obtained on estuarine shiner perch and dwarf perch[48]:

	Cymatogaster aggregata		*Micrometrus minimus*	
	Static	Intermittent-flow	Static	Intermittent-flow
Endrin	0.8	0.12	0.6	0.13
Dieldrin	3.7	1.50	4.6	2.44
Aldrin	7.4	2.26	5.0	2.03
DDT	7.6	0.45	18.0	0.26

A half-century ago, when calcium arsenate was employed for forest spraying, no effects were noted among fish in the streams. Indeed when Paris green was applied at 1 lb/a to standing water to control the larvae of anopheline mosquitoes, the fish were unaffected apart from a slight decline in their arthropod food.[16] With the introduction of DDT, a fish poison was added to the aquatic environment, its short term LC_{50} to salmonids being about 5 ppb; the centrarchid sunfish were intermediate in susceptibility, these family differences holding true for insecticides in general (Table 6.1).

DDT, DDD, and Methoxychlor

Mosquito Larviciding.   When applied every week at 0.1 lb/a in Tennessee, DDT killed bluegills and other *Lepomis* sunfish, but not the cyprinodont *Gambusia affinis* nor the eel *Anguilla bostoniensis*.[148] When pools were treated, bluegills, darters (*Catonotus*), and sculpins (*Cottus*) would be killed where blacknose dace, bluntnose minnows, and various species of shiners would survive. Brook trout (*Salvelinus fontinalis*), brown trout (*Salmo trutta*), and rainbow trout (*S. gairdnerii*) succumbed to 14 ppb in the water unless some mud was added to adsorb it.[147] Goldfish and fathead minnows are fairly tolerant to DDT, and so are the ictalurids, although in fish farming the channel catfish (*I. punctatus*) has suffered from DDT because of its habit of bottom-feeding at the sediment level.[38]

Single applications of DDT at 0.2 lb/a could control mosquito larvae without killing fish, although dead fish were found when shallow salt marshes were treated in New Jersey.[66] When a ditch in a tidal salt marsh in Florida was experimentally treated at this dosage, a careful search yielded 642 dead striped mullet and 3830 dead sheepshead minnows found along its 0.4-mile length, and among caged fish exposed to the treatment the mortality was complete in tidewater silversides, longnosed killifish, and young striped mullet.[41] Although the fish population in the ditch had returned to normal 3 months later, it is probable that the DDT in the corpses had found its way into fish-eating birds,[37] while the residues in the survivors were of the order of 10–40 ppm.[41] In treated standing water, the onset of autumn may bring on a recrudescence of mortality, especially among channel catfish, because the cessation of feeding results in mobilization of DDT from the fish's fat reserves,[38] and DDT has a negative temperature coefficient in its toxic action.[36]

Control of Other Aquatic Insects.   DDT has been applied to water to kill the larvae of *Simulium* blackflies in streams and rivers, for which it was highly effective at a dosage of 0.1 ppm/30 min (i.e., that body of water that takes 30 min to pass a given point) which did not kill fish. The repeated applications at higher dosages necessary to control *S. damnosum,* the vector complex of onchocerciasis in Africa, have caused fish kills in Guinea,[136] and a single application of DDT in oil to the Blue Nile for control of pest midges killed many fish because of the formation of a continuous oil film over a considerable length of the river.[23] Since blackfly larval control involved the uptake of DDT residues into the fish in the treated streams,

Table 6.1. Comparative susceptibility of fish families to insecticides: 96-hr LC_{50} (TL_m) figures by static test (Macek and McAllister, 1970).

	Organochlorines, ppb			Methyl Parathion	Organophosphorus cpds. ⟵ ppm ⟶ Carbamates		Azinphos-methyl	Carbaryl	Mexacarbate
	DDT	Lindane	Toxaphene		Fenthion	Malathion			
Salmonidae									
Rainbow Trout	7	27	11	2.8	0.93	0.17	0.014	4.3	10.2
Brown Trout	2	2	3	4.7	1.3	0.20	0.004	2.0	8.1
Coho Salmon	4	41	8	5.3	1.3	0.10	0.017	0.8	1.7
Percidae									
Yellow Perch	9	68	12	3.1	1.6	0.26	0.013	0.7	2.5
Centrarchidae									
Red-ear Sunfish	5	83	13	5.2	1.9	0.17	0.052	11.2	16.7
Bluegill Sunfish	8	68	18	5.7	1.4	0.10	0.022	6.8	11.2
Largemouth Bass	2	32	2	5.2	1.5	0.28	0.005	6.4	14.7
Ictaluridae									
Channel Catfish	16	44	13	5.7	1.7	9.0	3.3	15.8	11.4
Black Bullhead	5	64	5	6.6	1.6	12.9	3.5	20.0	16.7
Cyprinidae									
Goldfish	21	131	14	9.0	3.4	10.7	4.3	13.2	19.1
Fathead Minnow	19	87	14	8.9	2.4	8.6	0.24	14.6	17.0
Carp	10	90	4	7.1	1.2	6.6	0.70	5.3	13.4

DDT was replaced by methoxychlor which had been found to leave no residues in bluegills and bullheads.[26] Although the continuous-flow 96-hr LC_{50} levels of methoxychlor for fathead minnows and yellow perch were as low as 7.5 and 20 ppb, respectively, methoxychlor in natural waters has a half-life of only about 2 wk.[112] Applied to ponds at 0.04 ppm, methoxychlor did not kill bluegills but induced the formation in their blood vessels and capillaries of microproteinoid precipitation products which appeared within 3 days of the application, to disappear 2 months later; methoxychlor residues in the fish, amounting to 15–20 ppm in the second half of the first week, also had disappeared in 2 months.[92] For midge larval control, DDD was substituted on the Blue Nile,[20] and was employed for the four applications at 0.4 ppm/15 min made to the St. Lawrence to protect the International Exposition of 1967, the residues resulting in the fish being only 0.2 ppm additive to the 0.15 ppm already present in the fish before the river was treated.[62]

Forest Spraying. The application of DDT at 1 lb/a to broadleaved forest against the gypsy moth in the northeastern United States resulted in only sporadic fish kills; these were restricted to open pools, streams, or marshes since the leaf canopy normally allows only about one-quarter of the dosage to reach the forest floor.[39,162] In pine–oak–palmetto forest on the Atlantic coast, even 3 lb/a did not kill fish in the pools, since only 0.14 lb/a reached the forest floor.[67] But applied over coniferous forest in British Columbia to control the black-headed budworm this dosage killed some steelhead trout and totally eliminated the young-of-the-year of coho salmon,[42] and in a recent operation against the fir tussock moth in Idaho deaths of sculpin fry resulted from off-target drift.[56] Considerable mortality was observed in Montana during spruce budworm spraying at this dosage among cutthroat trout, an especially susceptible species (Table 6.2); in Idaho and Wyoming, where there were no fish mortalities in some streams and up to 80% mortality in others, the bottom-feeders such as suckers and bullheads suffered more than trout.[1] Moreover, the stream-water draining the 2-million-acre territory sprayed annually since 1952 had so denuded the Yellowstone River of aquatic invertebrates that extensive mortalities of whitefish, brown trout, and rainbow trout developed in the autumn of young-of-the-year brook trout, and of suckers, sculpins, sticklebacks, and minnows in 1958.[35] Aerial application of DDT at 1 lb/a over Maine forests directly resulted in some mortality among such species, although in the following year the surviving brook trout put on more weight than usual and there was unusually high survival of the young-of-the-year due to the removal of intraspecies competition and predation.[68,151]

Table 6.2. Toxicity of insecticides to salmonid fishes and the three-spine stickleback: 96-hr LC_{50} (TL_m) figures by static test in ppb (Katz, 1961; Post and Schroeder, 1971).

	Coho Salmon	Chinook Salmon	Brook Trout	Rainbow Trout	Cutthroat Trout	Threespine Stickleback
Endrin	0.5	1	0.4	0.4	0.1	0.4
Toxaphene	9	2	--	8	--	9
Dieldrin	11	6	--	10	--	15
DDT	18	11	7	2	0.8	18
Aldrin	46	8	--	18	--	40
Lindane	50	40	--	38	--	44
Chlordane	56	57	--	44	--	90
Heptachlor	59	17	--	19	--	112
Methoxychlor	66	28	--	63	--	86
Azinphosmethyl	4	4	--	3	--	12
Malathion	265	23	130	122	150	94
Coumaphos	15,000	--	--	1,500	--	1,862
Carbaryl	1,300	--	1,070	1,470	1,500	3,990

In New Brunswick, where there had been mortalities of brook trout and lake trout, the dosage employed against the spruce budworm was reduced to 0.5 lb/a for fear of harming the Atlantic salmon; yet the loss among parr (salmon before they go to sea) was 40%, and among young-of-the year was 98%.[54] A dosage of 0.25 lb/a, which occasionally failed to achieve budworm control and made a second spray necessary, killed no parr but still involved 6% loss among fingerlings and 78% among the salmon fry.[93]

Cyclodiene Insecticides

Aldrin and related compounds, as well as toxaphene and lindane, are no less toxic than DDT to trout (Table 6.3) and to other species of fish (Table 6.4). By 1950, when their use had been added to that of DDT, runoff from regular agricultural operations was causing simultaneous fish kills in 14 tributaries of the Tennessee River. In the crop season of 1960, 73 out of 185 fish kills reported from 31 states were attributed to pesticides.[87] Among the 193 cases of fish-kill reported to the U.S. Government in 1964, which involved a total of 18.3 million fish, 1.5 million were killed by agricultural

Table 6.3.   Toxicity of insecticides to rainbow trout; 48-hr LC$_{50}$ figures in ppb (from compilation of Pimentel, 1971).

Endosulfan	1	Azinphosmethyl	4	Carbaryl	1,350
Endrin	1	Chlorpyrifos	20	Mexacarbate	8,000
Toxaphene	3	Parathion	47	DNOC	210
Aldrin	3	Diazinon	170	Pyrethrins	54
DDT	3	Malathion	196	Dimethrin	7,000
Methoxychlor	7	Naled[1]	240	Silvex	650
Heptachlor	9	Fenthion[2]	930	2,4-D ester	1,100
TDE (DDD)	9	Temephos[3]	1,500	2,4,5-T acid	1,300
Chlordane	10	Methyl parathion	2,750	Diuron	4,300
Dieldrin	13	Trichlorfon	3,200	Simazine	5,000
Lindane	18	Monocrotophos	7,000	Atrazine	12,600
Chlorobenzilate	710	Phosphamidon	8,000	Na arsenite	36,500
Dicofol	100,000	Dimethoate[1]	19,000	2,4-D amine[1]	250,000

[1] 24-hr.   [2] 96-hr.   [3] Brook trout

Table 6.4. Comparative toxicities of organochlorine and organophosphorus insecticides to freshwater fish: 96-hr LC_{50} (TL_m) by static test (Henderson et al., 1959; Pickering et al. 1962).

Organochlorines: parts per billion (ppb)			
Fathead Minnow, *Pimephales promelas*	Bluegill Sunfish, *Lepomis macrochirus*	Goldfish, *Carassius auratus*	Guppy, *Poecilia reticulata*
Endrin	1.0	0.6	1.9	1.5
Toxaphene	7.5	3.5	5.6	20
Dieldrin	16	7.9	37	22
DDT (tech.)	32	16	27	43
Aldrin	33	13	28	33
Chlordane	52	22	82	190
Lindane	62	77	152	138
Methoxychlor	64	62	56	120
Heptachlor	94	19	230	107
BHC (16% γ)	2300	790	2300	2170

Organophosphorus compounds: parts per million (ppm)				
	Minnows	Bluegills	Goldfish	Guppies
Azinphosmethyl	0.09	0.005	1.3	0.11
EPN	0.25	0.10	0.45	0.03
Parathion	1.3	0.10	2.7	0.05
Tepp	1.9	1.1	21	1.8
Chlorthion	3.0	0.7	2.3	1.2
Demeton	3.2	0.10	11	0.61
Disulfoton	3.7	0.06	6.5	0.25
Methyl parathion	8.0	1.9	9.6	7.8
Dioxathion	10	0.03	32	0.21
Malathion	23	0.09	0.45	0.84
Coumaphos	> 18	0.18	> 18	0.56
Schradan	100	110	610	20
Trichlorfon	140	3.8	99	7.1

operations, as compared to 4.1 million by municipal wastes and 12.7 million by industrial wastes. By 1970 pesticides were deemed to be responsible for about 2% of the fish kills (D. L. Stalling, U.S. Department of the Interior, 1972).

The cyclodiene insecticides were very effective mosquito larvicides at a field dosage of 0.1–0.2 lb/a. Endrin was too toxic to fish, but the others caused only partial kills of *Gambusia* mosquitofish in treated ponds,[124] the following percentage mortalities being obtained from their exposure for 24 hr:

lb/acre	Chlordane	Heptachlor	Toxaphene	Aldrin	Dieldrin
0.1	18	25	6	6	18
0.5	26	28	100	90	100

Treatment of ponds with heptachlor to a concentration of 0.05 ppm eventually killed 90% of the bluegills in them.[9] Application of toxaphene every week at 0.1 lb/a for control of anopheline larvae in Alabama had killed all the fish (mainly sunfish) by the second or third application.[148]

Aldrin and Dieldrin. The application of aldrin in granules at 2 lb/a to eliminate an infestation of the Japanese beetle over 35 sq mi of farmland around Milford, Illinois caused extensive mortalities of fish in the creek running through the area; although they were mainly shiners and darters, 14 other fish species showed mortalities, but not the carp and catfish. The fish population, however, had returned to normal 7 months later.[122] The application of heptachlor or dieldrin, which are less toxic than aldrin, at 2 lb/a to control the imported fire ant in the southeastern United States caused frequent fish kills in the creeks.[11,21] The use of dieldrin to control rice leafminer in California killed an aggregate of 60,000 fish in the flooded ricefields in 1 yr.[141] Dieldrin applied at 1 lb/a to 3 square miles of a Florida tidal salt-marsh to control larvae of the sandfly *Culicoides* killed 25 tons of fish of 30 species, including young tarpon; the fiddler crabs feeding on their corpses died of secondary poisoning, leaving it to the aquatic snails to clean up the carnage.[74]

Endrin and Endosulfan. These are the most toxic insecticides for fish, their acute LC_{50} levels being of the order of parts per billion (Table 6.3). Endrin was widely used in cotton-growing areas in the southeastern United States against the boll weevil before this insect becomes resistant to it. The mortality among the fish in the drainage ditches and creeks resulted in the development of endrin-resistant strains in several species (see below). Endrin was considered to be the major factor in the extensive fish kills in the lower Mississippi River in the autumn and winter of 1960 (3.5 million dead) and of 1963 (5–10 million dead). Extracts of Mississippi mud taken in 1963 were found to be lethal to healthy fish, as were extracts of the fish from the river.[40] Dying channel catfish from the Mississippi contained about 0.5 ppm endrin in their blood, twice as much as the level found to separate survivors from deaths when catfish were exposed to endrin in the laboratory.[119,121] Endosulfan in its turn was involved in the die-off of 40 million fish in the Rhine in 1967, attributed to the emptying of a drum into the river near Bonn, Germany.[15] The kills occurring at that time in the lower Rhine in Holland could not, however, be definitely connected with endosulfan, nor did the Mississippi fish kills show good correlation with endrin concentrations in time and place. However, there were no fish kills in the Mississippi in 1964 and 1965, after a manufacturing plant at Memphis, Tennessee suspected as the source of the contamination had reduced the concentration of endrin emitting from the city sewers (at the same time the use of endrin against the boll weevil had been sharply curtailed).

Toxaphene. This chlorinated camphene product has been widely employed as a piscicide to remove trash fish or undesirable fish predators; an example is its use in Alaska streams to remove fry-devouring sculpins while the pink salmon (*Oncorhynchus*) are still in the ocean.[123] While applications at 0.05 ppm or greater may be employed to remove all trash fish,[89] an application of 0.005 ppm toxaphene to a Michigan lake exercised a selectivity in that it killed the smaller fish and left the bigger bluegills, yellow perch, and largemouth bass unharmed.[64] Toxaphene is more toxic to fish than rotenone, the piscicide formerly used in the form of derris or cube powder; for bluegills the 96-hr LC_{50} for toxaphene is only 3.5 ppb, as compared to the 48-hr LC_{50} of 22 ppb recorded for rotenone.[135] The use of toxaphene as an agricultural insecticide between 1957 and 1961 in the Klamath Basin of northern California led to deaths of fish-eating birds in the Lower Klamath and Tule Lake wetland refuges, notably white pelicans, American egrets, herons, and grebes; although DDT had been used more extensively and showed residues in the birds much higher than toxaphene, yet the experimental feeding of contaminated food showed that toxaphene could kill pelicans where DDT did not, and indicated that it was the effect of toxaphene on the fish that was the main cause of the bird kills.[91]

Fry of fathead minnows and brook trout exposed to toxaphene as they grew up developed weaknesses in the backbone (due to collagen deficiency) at concentration rates down to about 50 ppt, and there was a slight decrease in growth rate. Toxaphene residues of this order of magnitude have been found in United States waters in the national pesticide monitoring program.[110,111]

Lindane and BHC. Gamma-BHC (gamma-HCH) is one of the less toxic organochlorines to fish, but its LC_{50} of about 100 ppb is quite low (Table 6.4), and even lower concentrations cause the viviparous *Gambusia* to abort their young.[17] Lindane is scarcely cumulative, its half-life in fish being less than 2 days.[105] Long-term exposure to lindane in continuous-flow tanks revealed slight reductions in the survival rate of concentrations down to 24 ppb in the case of the fathead minnow, 17 ppb in brook trout fry, and 12 ppb in adult bluegills.[106] BHC and lindane were among the cotton insecticides formulated at a plant in Austin, Texas which so contaminated the Colorado River of that state that fish, including 80-lb catfish, were killed for 140 miles downstream.[32]

Cyclodiene Resistance. The use of endrin and toxaphene for controlling the boll weevil had led to the development of resistance in the fish inhabiting the drainage ditches and bayous in the lower Mississippi valley. A population of the mosquitofish *Gambusia affinis* at Sidon, Mississippi was found in 1963 to have developed a fourfold tolerance to DDT[149] and DDD, and as much as 50-fold resistance to toxaphene and the cyclodiene insecticides.[18] Populations of the bluegill sunfish *Lepomis macrochirus,* the green sunfish *L. cyanellus* and the golden shiner *Notemigonus chrysoleucas* at Twin Bayou also showed about 50-fold cyclodiene resistance while remaining susceptible to DDT.[59] The yellow bullhead *Ictalurus natalis* near Belzoni was found to have developed a 60-fold resistance to toxaphene and endrin,[57] while the endrin resistance in green sunfish and golden shiners was between 50- and 100-fold (Table 6.5). By 1972, populations of *Gambusia affinis* on the lower Brazos River of Texas, where cottonfields had been heavily treated with methyl parathion as well as toxaphene, showed twelvefold resistance to DDT as compared to tenfold and fivefold resistance to toxaphene and aldrin, respectively.[47] The cyclodiene-resistant *Gambusia* were twice as active as the normal in producing water-soluble metabolites from aldrin, and took up 30% less dieldrin into a bound form in the brain.[159] Even lower uptakes and retentions in the resistant strain as compared to the normal were shown with endrin in all tissues and with aldrin in myelin preparations.[163]

Although development of cyclodiene resistance protected certain fish as a species, they became a hazard to predators. Single resistant mosquitofish that had survived exposure to endrin for 7 days contained enough of the

Table 6.5. Cross-resistance ratios of insecticide-resistant fish at Belzoni, Mississippi (Minchew and Ferguson, 1969).

	Lepomis cyanellus Green sunfish	*Notemigonus crysoleucas* Golden shiners	*Gambusia affinis* Mosquito fish
Endrin	52	103	499
Aldrin	53	59	71
Lindane	39	21	42
Chlordane	38	6	94
Dieldrin	37	36	54
Toxaphene	37	40	389
Heptachlor	30	5	358
Strobane	19	37	559
Parathion	1.3	1.5	5

toxicant to kill 100-gram largemouth bass and 400-gram water-snakes when force-fed to them.[140]

Sublethal Effects on Organochlorines

In his review on the impact of pesticides on fishes, Johnson (1968) was particularly concerned with the sublethal effects of the organochlorine insecticides, which are not only toxic but also persistent. Such effects should be studied in order to determine those water pollution levels which are acceptable for long-term exposure of the fish fauna.

Behavioral Changes. In sublethal concentrations of DDT, preexposure to 20 ppb for 24 hr is enough to make brackish-water *Gambusia* prefer a level of salinity twice as high as the normal preferendum.[71] Fish sublethally poisoned have been observed to prefer warmer water, and fingerling salmon exposed to 5 ppb for only 24 hr violently avoid entering cooler water where the temperature is less than 5°C.[131] That DDT causes the lateral-line nerve to become hypersensitive was indicated by the results of exposing brook trout to 0.1 ppm for 24 hr.[5] However, rainbow trout can be poisoned by DDT in the water without the pattern of action–potential discharges in the lateral line being affected.[10] Involvement of the central nervous system itself was demonstrated by the fact that exposure of brook trout to 20 ppb for 24 hr rendered them less capable, some incapable, of acquiring a conditioned reflex (a propellerlike movement of the tail normally inducible by electroshock to the gular region) to the switching-on of an electric light.[7] A conditioned response to avoid the illumination level they normally prefer which could be induced by electric shock in normal trout could not be induced in trout previously exposed to 0.02 ppm DDT.[6] However, since it could be induced in DDT-exposed brook trout and salmon parr under certain experimental conditions, it was considered that the effect originally observed was not on learning but on the ability to perform the conditioned response.[83] Nevertheless, exposure of goldfish to 10 ppb for 4 days permanently abolished a locomotory pattern of turning behavior which involved a memory process.[46]

Physiological Changes.   Another effect of DDT is to impair osmoregulation, which in marine fishes is achieved by the absorption of water and ions through the intestinal wall, followed by excretion of the ions through the gills. The addition of 50 ppm to an intestinal preparation from the eel *Anguilla rostrata* inhibited its absorption of water by about 50%.[85] It also inhibited the Na-K-ATPase activity of the intestinal mucosa, responsible for the active transport, by 60% in the eel and by 38% in preparations from two species of flounders.[84] Rainbow trout fed a diet containing the equivalent of 17 ppm DDT for 2 wk had only half the normal Na-K-ATPase activity in their gills and kidneys, so that when placed in sea water the effect of this enzyme reduction was a considerable rise in serum Na^+.[31,98] Nevertheless, young brook trout raised on a DDT-contaminated diet put on 15% more weight than normal.[104]

Endrin at 100 ppm in the diet induced complete osmoregulatory failure in the goldfish, while the serum Na^+ level was increased by diets containing between 1 and 33 ppm endrin.[69] Only 0.5 ppb endrin in seawater was sufficient to increase the levels of Na^+, K^+, and Ca^{2+} in the serum at the expense of those in the liver.[52] Like DDT, the PCBs Aroclor 1242 and 1254 in low concentration inhibited the Na-K-ATPase activity in the gills and kidneys of fathead minnows.[94]

An instance of histopathological change caused by DDT was found in coho salmon fed 25 and 100 ppm p,p'-DDT in their diet for long periods; the nose became ulcerated and the distal convoluted tubule of the kidney degenerated.[22]

Uptake and Metabolism of DDT.   When fingerling Atlantic salmon are exposed to 1 ppm DDT, they absorb enough in 5 min to give 1.6 ppm DDT in their liver and spleen.[138] Brown trout exposed to 2 ppb DDT concentrated it about 500 times in the gill tissues and about 3000 times in the muscle.[79]

The gills of such 2-lb fish pass about 700 liters of water a day.[81] The transfer of the organochlorine through the gills from water to the vascular system was proved with dieldrin in isolated gill preparations perfused internally with plasma; since it did not occur when the perfusate was Ringer's saline, it was considered that it is the blood lipoprotein that carries the organochlorine from the gills to the tissues.[63]

After 4 hr of exposure to 1 ppm DDT, fingerling Atlantic salmon have already converted two-thirds of the uptake into noninsecticidal metabolites.[138] After 3 wk on a diet containing 90 ppm DDT, goldfish contain 6.8 ppm DDE as compared to 2.3 ppm DDT, as well as 0.9 ppm DDD.[164] In a pond treated with DDT at 0.02 ppm, the rainbow trout 16 months later contained 1.3 ppm DDE and 1.4 ppm DDD as compared to 0.13 ppm DDT, while the black bullhead contained only DDE and DDD.[19] Cutthroat trout intermittently exposed to DDT for 20 months accumulated the following concentrations (ppm) in the brain:

Exposure	DDT	DDE	DDD
1 ppm in water for 30 min, every 4 wk	6.2	6.3	1.0
1 mg/kg in food for 1 day, every week	28.0	7.0	1.0

It was considered that the origin of the DDD might be bacterial.[3]

The toxic effect of DDT and other organochlorine insecticides is on the nervous system, resulting in uncoordinated movements, sluggishness alternating with hyperexcitability, and difficulty in respiration.[80] The action of DDT on fish, as on insects, has a negative temperature coefficient, being more toxic to rainbow trout and bluegill sunfish at 13°C than at 19 or 23°C.[36] The first conspicuous sign of DDT poisoning in brook trout and Atlantic salmon is that the fish come to the surface, sometimes on their side, and show occasional convulsive darting movements; this was observed in the Miramichi River after the application of DDT to the New Brunswick forest,[93] and in Mill River, Prince Edward Island, as a result of the spillage of endrin used in agriculture.[142]

Continuous exposure of cutthroat trout to very low levels of DDT in water resulted in the residues of *t*-DDT (DDT + DDE + DDD) coming to a plateau proportional to the level in the water. Although even the highest DDT concentration did not decrease the number of ova produced by the female trout, nor was the embryonic development affected, the young fry which hatched often died.[3,4] In brook trout, the addition of DDT at sublethal levels in the diet of yearlings did result in their eventually laying fewer eggs, along with mortality of the hatched fry.[104]

Sac-Fry Mortality. This phenomenon was first observed in the lake trout of Lake George, New York State, which was receiving the runoff from the application of about 5 tons of DDT annually to its drainage basin for gypsy moth control. The eggs contained from 3 to 355 ppm DDT derived from the DDT residues in the parent females, and those fry that contained more than 3 ppm died at the time of final absorption of the yolk sac.[27] Mortalities of sac-fry of brook, rainbow, and cutthroat trout ranging from 30 to 90% were observed in the hatchery at Jasper, Alberta, where the feed supplied for the brood stock was sufficiently contaminated with DDT to give more than 0.4 ppm in the eggs.[43] In New Zealand the rainbow trout taken from a contaminated lake yielded eggs containing sufficient DDT to cause 68% fry mortality.[82] In Lake Taupo in that country a 5-ppm contamination of rainbow trout eggs caused 45% mortality of the fry.[44] Coho salmon taken as they ascended the rivers on the eastern shore of Lake Michigan in the autumn of 1968 yielded eggs which, having DDT contents ranging from 1.1 to 2.8 ppm, showed fry mortalities ranging from 15 to 75%, respectively, the death occurring just after the absorption of the last of the yolk which was rich both in fat and DDT.[88]

An actual decline in fish population was noted in the land-locked salmon of Sebago Lake, Maine, which over the years had received 59 DDT applications to control mosquitoes; the whole-fish residues of 21 ppm *t*-DDT (18 ppm of which was DDE) would be enough to contaminate the egg yolk to partially lethal levels; cessation of spraying brought the residues down by 50% in 1 yr.[8] The fathead minnow, which is fairly tolerant of DDT, could be reared in 2 ppb DDT for a 9-month period; it concentrated the insecticide more than 100,000 times and experienced a mortality rate of fry that was twice the normal.[86]

Investigations conducted at the Duluth laboratory of the U.S. Environmental Protection Agency indicate that the PCBs, frequently present as water contaminants along with DDT, do not cause reductions in egg-hatch, nor mortality in sac-fry.

Sublethal Effects of the Cyclodiene Group. As exemplified by the effect of heptachlor on bluegills (*Lepomis macrochirus*), the cyclodiene compounds caused the fish to show nervousness and hypersensitivity and then become sluggish before the onset of more serious symptoms, which included degenerative lesions in the liver. There was no reduction in spawning among those which survived in ponds treated with heptachlor at 0.05 ppm.[9] Guppies (*Poecilia reticulata*) exposed to 0.5 ppb endrin developed normal gonads, but the resulting hypersensitivity and increased activity of the fish prevented them from reproducing normally.[117]

Long-term exposure to only 1.7 ppb dieldrin caused pumpkinseed sunfish

(*L. gibbosus*) to cruise at lower speeds despite an increased oxygen consumption.[30] The black darter (*Etheostoma nigrum*) continuously exposed to 2.3 ppb dieldrin first showed increased blood glucose and a decreased O_2 consumption; these symptoms returned to normal in 2 wk, but the fish became less able to withstand heat stress.[145] Although dieldrin is the least toxic of the cyclodiene insecticides to rainbow trout, exposure to only 0.12 ppb is enough to reduce their growth rate.[33] Green sunfish (*L. cyanellus*) exposed to 6 ppb dieldrin reached their LT_{50} in 5-6 days, those that died having accumulated more than 8 ppm in the liver and brain.[78] The sailfin molly (*Poecilia latipinna*), which could withstand 7.5 ppb dieldrin in salt water, accumulated 1 ppm in its gills in the first hour of exposure, and 10 ppm in the brain after 6 days.[97] In bluntnose minnows exposed to sublethal concentrations of endrin, however, the insecticide was not found in the gills but was abundant in the digestive tract.[117]

Toxaphene at a concentration of 1.8 ppm, about one-eighth the LC_{50} level for goldfish, induced in this fish an aversion response to light which eventually disappeared on habituation.[152] A significant thickening of the gill lamellae was observed in spot (*Leiostomus xanthurus*) after long exposure to 0.1 ppb in seawater.[102]

When goldfish or bluegills were held in flowing clean water after they had taken up a sublethal dose of an organochlorine from a 30 ppb concentration, it was found that whereas lindane was totally eliminated in 2-4 days, and virtually all the dieldrin after 1 month, there still remained more than 50% of the *t*-DDT unexcreted by that time (Fig. 6.1). If the fish were held in static water, there was little net loss of these organochlorines and their metabolites to the water. When 60 bluegills were confined in 100 liters of water containing 30 ppb, they took up in 5 hr some 60% of the DDT (or dieldrin) present.[65] In heptachlor-treated ponds inhabited by bluegills, heptachlor residues reached their peak in the fish 3 days after the treatment and no heptachlor or heptachlor epoxide remained in the bluegills after 2 months.[9] Toxaphene has a half-life in fish approaching 6 months, but that for methoxychlor is only 1 wk; the OP compounds are less persistent in fish, the half-lives for parathion, diazinon, azinphosmethyl, and chlorpyrifos being less than 1 wk, and for malathion less than 1 day.[105]

B.   EFFECTS OF NONPERSISTENT INSECTICIDES

Susceptibilities to Organophosphorus Compounds

Whereas the LC_{50} figures for organochlorines are commonly expressed in parts per billion, OP compounds are so much less toxic to fish that they are

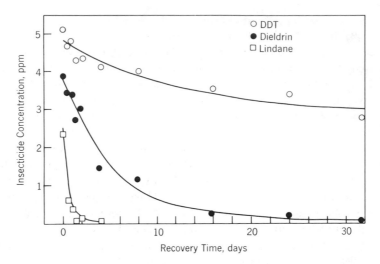

Fig. 6.1. Whole-body concentrations of DDT, dieldrin, and lindane in goldfish after transfer from treated to clean continuous-flow water (from Gakstatter and Weiss, 1967).

expressed in parts per million (Tables 6.1, 6.4). Azinphosmethyl is the most toxic, followed by parathion, carbophenothion, and diazinon; chlorfenvinphos, formothion, ethion, and bromophos are also notably toxic to fish (Table 6.6). The salmonids are the most susceptible to OP compounds, while the ictalurids and cyprinids are the most resistant (Table 6.1). The centrarchids occupy an intermediate position, but even then the bluegill sunfish is 250 times more susceptible to malathion than the fathead minnow, and 250 times more susceptible to azinphosmethyl than goldfish. It is to such phosphorodithioates that the difference between species in susceptibility is greatest, while it is less to the phosphorothioates and especially to those with a benzene ring in the leaving group.[107]

In fish farming, channel catfish *Ictalurus punctatus* in Louisiana may be freed of its competitors, including bass, sunfish, shiners, gar, shad, and even carp, by treating the rearing ponds with 1 ppm azinphosmethyl which, however, leaves residues up to 5 ppm in the fish.[113] The cultivable carp *Labeo rohita* in eastern India may be similarly favored by applying dichlorvos, for which its 7-day LC_{50} is 22 ppm while its competitors have LC_{50}'s between 2 and 18 ppm.[95] The susceptibilities of the various salmons and trouts are very similar, except that the larger fish which are two to four times heavier have an LC_{50} higher by 60–70% than the smaller fish[137]; all of the salmonids are about half as susceptible as the threespine stickleback (Table 6.2).

Forest Spraying

From the data on rainbow trout (Table 6.3) it can be seen why attention was paid to substituting trichlorfon or phosphamidon for DDT in spraying forests containing trout streams, since they are 1000 times safer to these fish than such organochlorines. The aerial spraying of an Ohio deciduous forest with malathion at 2 lb/a did not affect the fish in the streams.[132a] Phosphamidon applied at 1 lb/a to coniferous forest in British Columbia did not kill coho salmon fry in the creeks.[143] Airsprays of either phosphamidon at this dosage, or of monocrotophos at 0.25 lb/a to control the forest tent caterpillar in Louisiana, had no effect on green sunfish (*Lepomis cyanellus*) or other sunfish species, nor on the flagfin shiner *Notropis signipinnis*.[132] Trichlorfon applied at 1 lb/a for gypsy moth control in New York State had no effect on fish; the brook trout could take up about 1 ppm, a body residue that was virtually all lost in 4 days.[61]

Fenitrothion applied to a Colorado forest at 0.4 lb/a had no effect on rainbow trout, although it visibly affected minnows without killing them.[134] Rainbow trout exposed to a 0.25-lb/a spray in a Manitoba forest took up about 0.5 ppm fenitrothion in their bodies, which they lost 4 days later without having experienced any ChE reduction.[101] Brook trout and Atlantic salmon remained unaffected by a fenitrothion spray at this dosage in a Newfoundland forest,[76] although kills of salmon with fenitrothion spray have been reported from Anticosti Island.

Among the carbamate insecticides, a suspension of carbaryl (Sevin) in fuel oil caused no mortality of fish in the streams when applied at 1.25 lb/a

Table 6.6. **Toxicity of organophosphorus insecticides to the harlequin fish *Rasbora heteromorpha*: 48-hr LC_{50} figures by continuous-flow test, in ppm (Alabaster, 1969).**

Azinphosmethyl	(20%)*	0.38	Dichlorvos	(83%)	7.8
Chlorfenvinphos	(92%)	0.27	Demeton-methyl	(50%)	13
Formothion	(25%)	1.2	Malathion	(50%)	13
Ethion		0.52	Mevinphos	(99.5%)	11.5
Bromophos		0.62	Dimethoate	(32%)	27
Phosalone		2.4	Menazon	(70%)	220
Stirofos		4.3	Dicrotophos	(100%)	>1,000

* Percent active ingredient in the formulation tested; the LC_{50} figures apply to the formulation

for gypsy moth control in New York State.[24] Carbaryl is nontoxic to fingerling brown trout at water concentrations below 1.5 ppm, which such an application could produce only in open water less than 4 inches deep.[25] Mexacarbate (Zectran), resembles carbaryl in its toxicity to fish (Table 6.1); employed for forest spraying against spruce budworm in the western United States at 1 lb/a, it has not caused fish kills.

Pyrethrins applied to a Colorado forest at 0.2 lb/a, from a mineral oil solution in which they were stabilized with antioxidant, did not affect rainbow trout or other fish in the streams; this lack of effect in such piscicidal compounds was probably due to their failure to partition into the water before they were degraded or otherwise dissipated.[134]

The chitin-synthesis inhibitor diflubenzuron (Dimilin), which is highly suitable for aerial spraying against forest defoliators such as the gypsy moth, spruce budworm, and Douglas-fir tussock moth, has a very low toxicity to fish, the 96-hr LC_{50} figures being 660 ppm for bluegills and 240 ppm for rainbow trout.

Mosquito Larviciding

It is for control of mosquito larvae that OP compounds are applied directly to water, the usual dosages being 0.25 lb/a for malathion, 0.1 lb/a for parathion, methyl parathion, naled, fenitrothion, and fenthion, and 0.05 lb/a for chlorpyrifos (Dursban) and temephos (Abate). Their LC_{50} figures for pool or pond fish are in terms of ppm, whereas their LC_{50}'s for the mosquito larvae are of the order of ppb (Table 6.7). Nevertheless at 1 ppm, a common operational dosage, naled or malathion causes about 25% mortality among the sac-fry of rainbow trout.[99] The mosquitofish *Gambusia affinis*, which has been widely distributed in North American and southern Europe for biological control of mosquito larvae, is one of the less susceptible species. Field trials at dosages not less, and frequently many times more, than the recommended rates gave the following percentage mortalities for *Gambusia* exposed for 24 hr immediately after the treatment:[126]

Diazinon (0.3 lb/a)	100	Fenitrothion (1.6 lb/a)	44
Carbophenothion (0.2 lb/a)	98	Naled (2 lb/a)	20
Parathion (0.1 lb/a)	18	Fenthion (0.8 lb/a)	0
Methyl parathion (0.8 lb/a)	10	Ronnel (0.8 lb/a)	0

Of the other candidate OP compounds (Table 6.7), all except malathion have safety factors of at least 300 times for fish as compared to mosquito larvae; larvicidal applications of malathion have occasionally killed bluegills and even *Gambusia*.[125] Methyl parathion at 0.4 lb/a killed no *Gam-*

Table 6.7. **Comparative toxicities of organophosphorus insecticides to** ***Culex***
mosquito larvae and to their fish predators: 24-hr LC$_{50}$ figures by static
test (Rongsriyam et al., 1968; Shim and Self, 1972; Ferguson et al.,
1966; Von Windeguth and Patterson, 1966).

	Culex	*Aplocheilus*	*Poecilia*	*Gambusia*	*Lepomis*
	ppm	ppm	ppm	ppm	ppm
Malathion	50	2.8	0.05		
Naled	9.4*	3.0	---		
Fenitrothion	7.4	4.5	2.2		
Fenthion	6.0	2.6	3.3	2.0	1.8
Methyl parathion	0.54*	2.9	---		
Chlorpyrifos	0.8	0.31	0.22	0.2[†]	0.02[†]
Abate	0.5	---	>200	>200	>200

* <u>Culex</u> <u>tritaeniorhynchus</u>; all other values for <u>C</u>. <u>quinquefasciatus</u>

[†] 36-hr LC$_{50}$ figure

busia while parathion at 0.1 lb/a caused 18% mortality; fenitrothion is even
safer, at 0.8 lb/a killing none where methyl parathion killed 10% of the
Gambusia in the pool.[126] Fenthion was an improvement on parathion at 1
lb/a, the *Gambusia* kills being 6% as compared to 30%.[125] Applied at 0.15
lb/a in Ohio, fenthion killed no fish although it reduced their brain ChE
level.[53] Applied at 0.4 lb/a to fish ponds in Tanzania, neither fenthion nor
methyl parathion killed any of the three species of *Tilapia* bream being cultivated in them.[154]

Chlorpyrifos is more toxic than fenthion, causing 85% mortality at 1 ppm
where fenthion kills no *Gambusia*, but is usually used at one-half that
dosage, at which it kills none.[45] Applied to control ricefield mosquitoes at
0.0125 lb/a every 6 wk, chlorpyrifos killed no *Gambusia* but some 35% of
the bluegills exposed.[153] On salt marshes, a single application of chlorpyrifos
at 0.025 lb/a was harmless, but one of 0.05 lb/a caused a moderate and
transitory decrease in the numbers of *Fundulus* and *Cyprinodon* top-
minnows.[103] Temephos (Abate) is completely safe for fish, being at least 100
times less toxic to *Gambusia* and other species than fenthion[150]; and no
mosquitofish are killed even at 5 ppm, which is 1000 times more than the
dosage of temephos required to eliminate mosquito larvae.[45]

A population of *Gambusia affinis* sampled in 1964 near Sidon,

Mississippi, where cottonfields had been heavily sprayed with methyl parathion for several years, proved to be at least five times as resistant as the normal to methyl parathion, but showed no resistance to parathion, malathion, or other OP compounds.[58] Another population of *Gambusia* at Belzoni, and green sunfish and golden shiners at Indianola, proved to be about three times as tolerant to chlorpyrifos as normal strains at State College, Mississippi.[60] Five years later, the green sunfish and golden shiners at Belzoni were highly resistant to methyl parathion but still susceptible to parathion.[115] Ten years later, the cyclodiene-resistant population of *Gambusia* at Belzoni showed the following 48-hr LC_{50} figures (ppm)[34]:

	Parathion	Methyl Parathion
Susceptible (Starkville)	0.35	13.5
Resistant (Belzoni)	0.39	17.5

The juvenile hormone mimic methoprene (Altosid) at the operational dosage of 0.1 lb/a had no effect on *Gambusia,* for which its LC_{50} is 80 ppm; the chitin synthesis inhibitor diflubenzuron (Dimilin) had no effect on mosquitofish at the operational dosage of 0.0025 lb/a, nor at 1 ppm for 48 hr.[116]

Cholinesterase Inhibition

Fish under the influence of OP compounds in water respond to stimuli by rapid swimming followed by tetanic immobility; malathion-poisoned bluegills extend their pectoral fins forward to the limit before dying,[49] while trichlorfon-poisoned rainbow trout larvae no longer avoid the light.[109] Bluegills showing toxic symptoms are found to have 30–50% of their brain cholinesterase inhibited.[49] In the menhaden (*Brevoortia tyrannus*) of the Ashley River near Charleston, South Carolina, which were being poisoned by wastes from a plant manufacturing organophosphorus insecticides, the brain ChE had been inhibited by some 47% in those which died and by some 16% in those still alive.[160] In largemouth bass exposed to the point that one-half of them have succumbed (i.e., the LT_{50}), the brain ChE has been reduced to about 20% with malathion and only 5% with azinphosmethyl (Table 6.8). When goldfish were exposed for 15 days to various OP compounds at 1 ppb, parathion, malathion, and azinphosmethyl caused significant reductions in brain ChE, while bluegills suffered reductions from several other OP compounds as well.[157] EPN was found to act synergistically with malathion in golden shiners, as in insects,[155] but no synergism was found between EPN and azinphosmethyl.[157]

Table 6.8. Effect of organophosphorus insecticides on the cholinesterase activity of fish brains (Weiss, 1959*; Weiss, 1961[+]; Weiss and Gakstatter, 1964‡).

	Percent ChE Activity remaining after 24 h. in 100 ppb* (goldfish)	at LT$_{50}$ in 500 ppb‡ (largemouth bass)		Response or no-response by ChE reduction after 15 days in 1 ppb‡		
		Live	Dead	Goldfish	Bluegill	Shiner
Azinphosmethyl	49	5	4	R	R	--
EPN	53	42	60	NR	R	NR
Parathion	38	24	24	R	R	--
Malathion	67	24	1?	R	R	R
Diazinon	38	18	28	NR	NR	NR
Demeton	75	--	--	NR	NR	NR
Dioxathion	--	--	--	NR	R	--
Disulfoton	--	24	31	NR	--	--
Fenthion (20 ppb)	--	--	--	NR	R	--
Methyl parathion	--	--	--	NR	?R	--
Coumaphos	--	--	--	NR	?R	--
Dichlorvos	--	24	19	NR	--	--

R: ChE < 60% of normal level   ?R: 60-80% of normal

189

Bluegill sunfish, which are 250 times more sensitive than goldfish to azinphosmethyl and than fathead minnows to malathion,[107] accumulate 15 times more malaoxon in their liver than black bullheads, because of a greater production rather than a lower breakdown of this anticholinesterase metabolite.[128] The injection of malaoxon at 2.5 mg/kg causes 48% inhibition of the brain ChE in both species, but the bluegills all die while none of the bullheads die; the same species difference was found with the 90% inhibition induced *in vivo* by azinphosmethyl.[129] It was in brown bullheads that no evidence could be found of the formation of the anticholinesterase paraoxon on exposure to parathion,[118] only parathion being found in the blood, at a concentration more than 50 times greater than that in the water.

With shorter exposures to OP compounds insufficient to kill, the brain usually recovers its original cholinesterase level; this usually occurs within a month of the transfer of the fish to uncontaminated water, except with parathion (Fig. 6.2). With azinphosmethyl also, the recovery of the brain ChE in the pumpkinseed sunfish was so slow that another exposure given 2 wk later had a cumulative effect.[14]

Sublethal Symptoms

Golden shiners that had been exposed to a sublethal concentration of parathion for 2 days showed a 20% reduction in their blood haematocrit and a 50–60% reduction in lymphocytes, heterophils, and total leucocytes.[28] Exposed to naled for 24 hr at a concentration only three-quarters of the LC_{50}, they became less tolerant to oxygen deficiency.[108] Atlantic salmon parr exposed for only 24 hr to the 96-hr LC_{50} concentrations of temephos and fenitrothion became less active; with temephos they were slower to learn a conditioned response to light, while fenitrothion completely inhibited such learning.[75] By contrast, DDT made these fish hypersensitive and quicker to learn, while methoxychlor had no effect.

In a flooded borrow-pit sprayed with parathion at 0.1 lb/a the mosquitofish *Gambusia affinis* had accumulated residues of 15 ppm parathion 1 day later, but these decreased to 3 ppm in the next 2 days.[127] Carp exposed to 5 ppm malathion for 4 days accumulated 28 ppm in their flesh but only 2.5 ppm in the brain; after 4 days in clean water the flesh residues had decreased to less than 1 ppm. When malathion is employed for mosquito control, the residues in the fish do not exceed 1 ppm and some disappear,[12] but when azinphosmethyl is applied as a selective piscicide at 1 ppm, the channel catfish that it is designed to preserve show residues above the permissible tolerance limits.[113] Diazinon applied to cranberry bogs at 0.3 ppm accumulated in the top-minnow *Fundulus heteroclitus* to 10 times its

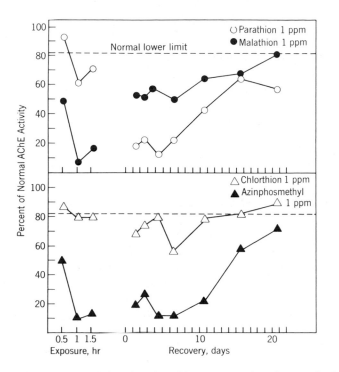

Fig. 6.2. Cholinesterase inhibition by OP compounds: changes in brain ChE activity in largemouth bass during exposure and recovery periods (from Weiss, 1959).

© (1959) by WPCF, reprinted with permission.

concentration in water, but 50% of the body residues were lost in less than a week.[114]

Although the 96-hr LC_{50} of malathion for bluegills is 0.11 ppm, continuous exposure to dosages above 20 ppb are deleterious to spawning and 10 ppb causes about one-third of the offspring to develop spinal deformations; thus the MATC (maximum acceptable toxic concentration) for malathion is considered to lie between 7.4 and 3.5 ppb.[49] For fathead minnows the 96-hr LC_{50} by static test is as high as 9.0 ppm; but in continuous-flow exposures for a 10-month period, normal F_1 production was not obtained unless the concentration was below 2 ppb.[120] The malathion metabolites dimethylphosphoro-dithioic and -thioic acids are, respectively, one-half and one-quarter as toxic as malathion for fathead minnows; diethyl fumarate, however, is twice as toxic as malathion,[13] but it has been demonstrated as a product of chemical but not of biological hydrolysis, and malathion is not hydrolyzed in waters unless they are notably acid.[158] In an earlier study, diethyl succinate, which is a likely metabolite *in vivo*, was about as toxic as malathion itself.[161]

Unlike DDT, malathion in sublethal concentrations does not induce in *Gambusia* a preference for water with higher salinities.[71] Whereas *Gambusia* avoids water containing malathion when offered the choice, the sheepshead minnow *Cyprinodon* does not; both species, however, avoid water containing chlorpyrifos.[70,71]

C.   EFFECTS OF INSECTICIDES ON SALT-WATER FISH

To marine and estuarine fish (Tables 6.9, 6.10), endrin is again the most toxic organochlorine, followed by DDT. The Atlantic silverside was the most susceptible species tested, the northern puffer was the most resistant.[51] Winter flounder exposed for some time to 2 ppb DDT laid eggs that frequently showed failure of gastrulation in the embryos, or severe vertebral deformities in the larval fry which hatched; these deformities were not seen after exposures to dieldrin, even at 2 ppm.[146]

In speckled sea trout, DDT residues in the parent females inhibited spawning in the Laguna Madre estuary, Texas, where by 1969 the runoff from agricultural lands had led to an accumulation of about 8 ppm *t*-DDT in the ovaries.[29] The population had been decreasing from the normal 30 down to 0.2 per acre in 1969; it rose to a level above normal in 1971,[55] after the suspension of DDT.

When Aroclor 1254, one of the most toxic of the PCBs, was tested for its toxicity to the estuarine pinfish (*Lagodon rhomboides*) and spot (*Leiostomus xanthurus*), it was found that exposure to 5 ppb for 1 month killed about 50% of the juveniles, during which period the pinfish accumulated 109 ppb of this polychlorinated biphenyl in their bodies.[72] With fathead minnows (*Pimephales promelas*), the 96-hr LC_{50} of Aroclor 1254 for the newly hatched fry was 7.7 ppb, and juvenile fatheads were all killed by exposure for 1 month to Aroclor 1248 at 5 ppb.[130]

Methyl parathion is characterized by a very low toxicity to marine fish, while malathion shows an LC_{50} of less than 1 ppm to nearly all species (Table 6.7). To the striped bass, chlorpyrifos was especially toxic, and methyl parathion notably nontoxic.[96] Sculpins are slow to produce the anticholinesterase oxons from malathion, parathion, or azinphosmethyl, and especially fast to break down the anticholinesterase activity in the liver, as compared to winter flounders.[128]

Table 6.9. Toxicity of organochlorine and organophosphorus insecticides for estuarine fish*: 96-hr LC$_{50}$ figures (ppb) by static test (Eisler, 1970).

	Silverside	Bluehead	Mullet	Killifish	Eel	Mummichog	Puffer
Endrin	0.05	0.1	0.3	0.3	0.6	1.0	3.1
p,p'-DDT	0.4	7	1.5	1	4	5	89
Heptachlor	3	0.8	194	32	10	50	188
Dieldrin	5	6	23	4	0.9	10	34
Lindane	9	14	66	28	56	60	35
Aldrin	13	12	100	17	5	8	36
Methoxychlor	33	13	63	30	12	46	150
Dioxathion	6	35	39	15	6	20	75
Malathion	125	27	550	250	82	240	3,250
Monocrotophos	320	74	300	75	65	300	800
Dichlorvos	1250	1,440	225	2,300	1,800	2,680	2,250
Methyl parathion	5700	12,300	5,200	13,800	16,900	58,000	75,800

*Menidia menidia, Atlantic silverside; Thalassoma bifasciatum, bluehead; Mugil cephalus, striped mullet; Fundulus majalis, striped killifish; Anguilla rostrata, American eel; Fundulus heteroclitus, mummichog; Sphaeroides maculatus, northern puffer

Table 6.10.   Toxicity of insecticides for striped bass (*Morone saxatilis*): 96-hr LC_{50} figures (ppb) by intermediate-flow test (Korn and Earnest, 1974).

Endrin	0.094	Chlorpyrifos	0.58
Endosulfan	0.10	Malathion	14.0
DDT	0.53	Parathion	17.8
Heptachlor	3.0	EPN	60
Methoxychlor	3.3	Fenthion	453
Toxaphene	4.4	Naled	500
Lindane	7.3	Methyl parathion	790
Chlordane	11.8	Temephos	1000
Dieldrin	19.7	Carbaryl	1000

REFERENCES CITED

1. L. Adams, M. G. Hanavan, N. W. Hosley, and D. W. Johnston. 1949. The effects on fish, birds, and mammals of DDT used in the control of forest insects in Idaho and Wyoming. *J. Wildl. Manage.* 13: 245–254.

2. J. S. Alabaster. 1969. Survival of fish in 164 herbicides, insecticides, fungicides, wetting agents, and miscellaneous substances. *Internat. Pest Control* 11 (2): 29–35.

3. D. B. Allison, B. J. Kallman, O. B. Cope, and C. C. Van Valin. 1963. Insecticides: Effects on cutthroat trout of repeated exposures to DDT. *Science* 142:958–961.

4. D. B. Allison, B. J. Kallman, O. B. Cope, and C. C. Van Valin. 1964. *Some Chronic Effects of DDT on Cutthroat Trout.* U.S. Bur. Sports Fish. Wildl., Res. Rep. 64. 30 pp.

5. J. M. Anderson. 1968. Effect of sublethal DDT on the lateral line of brook trout. *J. Fish. Res. Bd. Can.* 25:2677–2682.

6. J. M. Anderson and M. R. Peterson. 1969. DDT: Sublethal effects on brook trout nervous system. *Science* 164:440–441.

7. J. M. Anderson and H. B. Prins. 1970. Effects of sublethal DDT on a simple reflex in brook trout. *J. Fish. Res. Bd. Can.* 27:331–334.

8. R. B. Anderson and W. H. Everhart. 1966. Concentrations of DDT in landlocked salmon in Sebago Lake, Maine. *Trans. Am. Fish. Soc.* 95:160–164.

9. A. K. Andrews, C. C. Van Valin, and B. F. Stebbings. 1966. Some effects of heptachlor on bluegills (*Lepomis macrochirus*). *Trans. Am. Fish. Soc.* 95:297–309.

10. T. G. Bahr and R. C. Ball. 1971. Action of DDT on evoked and spontaneous activity from the rainbow trout lateral line nerve. *Comp. Biochem. Physiol.* 38A:279–284.

11. M. F. Baker. 1958. Observations of effects of an application of heptachlor or dieldrin on wildlife. *Proc. Symposium on Fire Ant Eradication Program,* 12th Ann. Conf. SE Assoc. Game and Fish Commissioners, Columbia, S.C. pp. 18–20.

12. M. E. Bender. 1969. Uptake and retention of malathion by the carp. *Progr. Fish-Cult.* **31**:155–159.

13. M. E. Bender. 1969. The toxicity of the hydrolysis and breakdown products of malathion to the fathead minnow. *Water Res.* **3**:571–582.

14. G. M. Benke and S. D. Murphy. 1974. Anticholinesterase action of methyl parathion, parathion and azinphosmethyl on mice and fish. *Bull. Env. Contam. Toxicol.* **12**:117–122.

15. D. Binder. 1969. *New York Times,* 26 June.

16. E. L. Bishop. 1940. Cooperative investigations of the relation between mosquito control and wildlife conservation. *Science* **92**:201–202.

17. C. E. Boyd. 1964. Insecticides cause mosquito fish to abort. *Progr. Fish-Cult.* **26**:138.

18. C. E. Boyd and D. E. Ferguson. 1964. Susceptibility and resistance of mosquito fish to several insecticides. *J. Econ. Entomol.* **57**:430–431.

19. W. R. Bridges, B. J. Kallman, and A. K. Andrews. 1963. Persistence of DDT and its metabolites in a farm pond. *Trans. Am. Fish. Soc.* **92**:421–427.

20. A. W. A. Brown, D. J. McKinley, H. W. Bedford, and M. Qutubuddin. 1961. Insecticidal operations against chironomid midges along the Blue Nile. *Bull. Entomol. Res.* **51**:789–801.

21. W. L. Brown. 1961. Mass insect control programs. *Psyche* **68**:75–109.

22. D. R. Buhler, M. E. Rasmussen, and W. E. Shanks. 1969. Chronic oral DDT toxicity in juvenile Coho and chinook salmon. *Toxicol. Appl. Pharmacol.* **14**:535–555.

23. E. H. W. J. Burden. 1956. A case of DDT poisoning in fish. *Nature* **178**:546–547.

24. G. E. Burdick, H. J. Dean, and E. J. Harris. 1960. Effect of Sevin upon the aquatic environment. *N.Y. Fish Game J.* **7**:14–25.

25. G. E. Burdick, H. J. Dean, E. J. Harris, J. Skea, and D. Colby. 1965. Toxicity of Sevin (carbaryl) to fingerling trout. *N.Y. Fish Game J.* **12**:127–146.

26. G. E. Burdick, H. J. Dean, E. J. Harris, J. Skea, C. Frison, and C. Sweeney. 1968. Methoxychlor as a blackfly larvicide: Persistence of its residues in fish and its effect on stream arthropods. *N.Y. Fish Game J.* **15**:121–142.

27. G. E. Burdick, E. J. Harris, J. H. Dean, T. M. Walker, J. Skea, and D. Colby. 1964. The accumulation of DDT in lake trout and the effect on reproduction. *Trans. Am. Fish. Soc.* **93**:127–136.

28. G. W. Butler, D. E. Ferguson, and C. R. Sadler. 1969. Effects of sublethal parathion exposure on the blood of golden shiners. *J. Miss. Acad. Sci.* **15**:33–36.

29. P. A. Butler. 1969. The significance of DDT residues in estuarine fauna. In *Chemical Fallout.* Ed. M. W. Miller and G. G. Berg. Charles C. Thomas. pp. 205–220.

30. J. Cairns and A. Scheier. 1964. The effect upon pumpkinseed sunfish of chronic exposure to lethal and sublethal concentrations of dieldrin. *Not. Nat. Acad. Sci. Philadelphia.* No. 370, pp. 1–10.

31. R. D. Campbell, T. P. Leadem and D. W. Johnson. 1974. The *in vivo* effect of *p,p* - DDT on Na-K-activated ATPase activity in rainbow trout. *Bull. Environ. Contam. Toxicol.* **11**:425–428.

32. R. Carson. 1962. Silent Spring. Houghton-Mifflin Co. 360 pp.

33. G. G. Chadwick and D. L. Shumway. 1969. Effects of dieldrin on the growth and development of steelhead trout. In *The Biological Impact of Pesticides in the Environment.* Oregon State Univ., Environ. Health Series No. 1. pp. 90–96.

34. J. E. Chambers and J. D. Yarbrough. 1974. Parathion and methyl parathion toxicity to insecticide-resistant and susceptible mosquito fish. *Bull. Environ. Contam. Toxicol.* **11**:312–320.

35. O. B. Cope. 1961. Effects of DDT spraying for spruce budworm on fish in the Yellowstone river system. *Trans. Am. Fish. Soc.* **90**:239–251.

36. O. B. Cope. 1965. Agricultural chemicals and freshwater ecological systems. In *Research in Pesticides*. Ed. C. O. Chichester. Academic. pp. 115–128.

37. O. B. Cope. 1966. Contamination of the freshwater ecosystem by pesticides. *J. Appl. Ecol.* **3**(Suppl.):33–44.

38. O. B. Cope. 1971. Interactions between pesticides and wildlife. *Annu. Rev. Entomol.* **16**:325–364.

39. O. B. Cope and P. F. Springer. 1958. Mass control of insects: The effects on fish and wildlife. *Bull. Entomol. Soc. Am.* **4**:52–56.

40. C. Cottam. 1965. The ecologists' role in problems of pesticide pollution. *BioScience* **15**:457–463.

41. R. A. Croker and A. J. Wilson. 1965. Kinetics and effects of DDT on a tidal marsh ditch. *J. Am. Fish. Soc.* **94**:152–159.

42. R. A. Crouter and E. H. Vernon. 1959. Effects of black-headed budworm control on salmon and trout in British Columbia. *Can. Fish-Cult.* **1959**(24):23–40.

43. J. P. Cuerrier, J. A. Keith and E. Stone. 1967. Problems with DDT in fish culture operations. *Naturaliste Can.* **94**:315–320.

44. J. C. Dacre and D. Scott. Possible DDT mortality in young rainbow trout. *N.Z. J. Mar. Freshwater Res.* **5**:58–65.

45. H. A. Darwazeh and M. S. Mulla. 1974. Biological activity of organophosphorus compounds and synthetic pyrethroids against immature mosquitoes. *Mosq. News* **34**:151–155.

46. F. B. Davy, H. Kleerekoper, and P. Gensler. 1972. Effects of sublethal exposure to DDT on the locomotor behavior of the goldfish. *J. Fish. Res. Bd. Can.* **29**:1333–1336.

47. L. J. Dzuik and F. W. Plapp. 1973. Insecticide resistance in mosquitofish from Texas. *Bull. Environ. Contam. Toxicol.* **9**:15–19.

48. R. D. Earnest and P. E. Benville. 1972. Acute toxicity of four organochlorine insecticides to two species of surf perch. *Calif. Fish Game* **58**:127–132.

49. J. G. Eaton. 1970. Chronic malathion toxicity to the bluegill (*Lepomis macrochirus*). *Water Res.* **4**:673–684.

50. R. Eisler. 1970. *Factors Affecting Pesticide-Induced Toxicity in an Estuarine Fish*. U.S. Dept. Interior, Bur. Sport Fish Wildl., Tech. Paper 45. 20 pp.

51. R. Eisler. 1970. *Acute Toxicities of Organochlorine and Organophosphorus Insecticides to Estuarine Fishes*. U.S. Dept. Interior, Bur. Sport Fish. Wildl., Tech. Paper 46. 12 pp.

52. R. Eisler and P. H. Edmonds. 1966. Effects of endrin on blood and tissue chemistry of a marine fish. *Trans. Am. Fish. Soc.* **95**:153–159.

53. J. B. Elder and C. Henderson. 1969. *Field Appraisal of ULV Baytex Mosquito Control Applications on Fish and Wildlife*. U.S. Bur. Sports Fish. Wildl. and Ohio Dept. Nat. Res., Special Rep. 74 pp.

54. F. P. Elson. 1967. Effects on wild young salmon of spraying DDT over New Brunswick forests. *J. Fish. Res. Bd. Can.* **24**:731–767.

55. Environmental Protection Agency. 1971. *Public Hearing on DDT.* Transcript of evidence. Washington, D.C.

56. Environmental Protection Agency. 1975. *DDT: A Review of Scientific and Economic Aspects of the Decision to Ban Its Use as a Pesticide.* Washington, D. C. Document EPA-540/1-75-022. 300 pp.

57. D. E. Ferguson and C. R. Bingham. 1966. Endrin resistance in the yellow bullhead, *Ictalurus natalis. Trans. Am. Fish. Soc.* **95**:325–326.

58. D. E. Ferguson and C. E. Boyd. 1964. Apparent resistance to methyl parathion in mosquitofish, *Gambusia affinis. Copeia* **1964**:706.

59. D. E. Ferguson, D. D. Culley, W. D. Cotton, and R. P. Dodds. 1964. Resistance to chlorinated hydrocarbon insecticides in three species of freshwater fish. *BioScience* **14**:43–44.

60. D. E. Ferguson, D. T. Gardner, and A. L. Lindley. 1966. Toxicity of Dursban to three species of fish. *Mosq. News* **26**:80–82.

61. T. Finger and R. Werner. 1970. *Environmental Impact and Efficacy of Dylox Used for Gypsy Moth Suppression in New York State.* State Univ. N.Y., Appl. For. Res. Inst. (Unpublished.)

62. F. J. H. Fredeen and J. R. Duffy. 1970. Insecticide residues in some components of the St. Lawrence river ecosystem following insecticide applications of DDD. *Pestic. Monit. J.* **3**:219–226.

63. P. O. Fromm and R. C. Hunter. 1969. Uptake of dieldrin by isolated perfused gills of rainbow trout. *J. Fish. Res. Bd. Can.* **26**:1939–1942.

64. K. G. Fukano and F. F. Hooper. 1958. Toxaphene as a selective fish poison. *Progr. Fish-Cult.* **20**:189–190.

65. J. H. Gakstatter and C. M. Weiss. 1967. The elimination of DDT-C^{14}, dieldrin-C^{14}, and lindane-C^{14} from fish following sublethal exposure in aquaria. *Trans. Am. Fish. Soc.* **96**:301–307.

66. J. L. George, R. F. Darsie, and P. F. Springer. 1957. Effects of wildlife of aerial applications of Strohane, DDT and BHC to tidal marshes in Delaware. *J. Wildl. Manage.* **21**:42–53.

67. P. Goodrum, W. P. Baldwin, and J. W. Aldrich. 1949. Effect of DDT on animal life of Bull's Island, South Carolina. *J. Wildl. Manage.* **13**:1–10.

68. J. R. Gorham. 1961. *Aquatic Insects and DDT Forest Spraying in Maine.* Maine Forest Service Bull. 19. 49 pp.

69. B. F. Grant and P. M. Nehrle. 1970. Chronic endrin poisoning in goldfish, *Carassius auratus. J. Fish. Res. Bd. Can.* **27**:2225–2232.

70. D. J. Hansen. 1969. Avoidance of pesticides by untrained sheepshead minnows. *Trans. Am. Fish. Soc.* **98**:426–429.

71. D. J. Hansen. 1972. DDT and malathion: Effect on salinity selection by mosquitofish. *Trans. Am. Fish. Soc.* **101**:346–350.

72. D. J. Hansen, P. R. Parish, J. I. Lowe, A. J. Wilson, and P. D. Wilson. 1971. Chronic toxicity of Aroclor 1254 in two estuarine fishes. *Bull. Environ. Contam. Toxicol.* **6**:113–119.

73. D. J. Hansen, E. Matthews, S. L. Nall, and D. P. Dumas. 1972. Avoidance of pesticides by untrained mosquitofish, *Gambusia affinis. Bull. Environ. Contam. Toxicol.* **8**:46–51.

74. R. W. Harrington and W. L. Bidlingmayer. 1958. Effect of dieldrin on fishes and invertebrates of a salt marsh. *J. Wildl. Manage.* **22**:76–82.

75. C. T. Hatfield and P. H. Johansen. 1972. Effects of four insecticides on the ability of Atlantic salmon parr to learn and retain a single conditioned response. *J. Fish. Res. Bd. Can.* **29**:315–321.

76. C. T. Hatfield and L. G. Riche. 1970. Effects of aerial Sumithion spraying on juvenile Atlantic salmon and brook trout in Newfoundland. *Bull. Environ. Contam. Toxicol.* **5**:440–442.

77. C. Henderson, Q. H. Pickering, and C. M. Tarzwell. 1959. Relative toxicity of ten chlorinated hydrocarbon insecticides to four species of fish. *Trans. Am. Fish. Soc.* **88**:23–32.

78. R. L. Hogan and E. W. Roelofs. 1971. Concentration of dieldrin in the blood and brain of the green sunfish at death. *J. Fish. Res. Bd. Can.* **28**:610–612.

79. A. V. Holden. 1962. A study of the absorption of C^{14}-labelled DDT from water by fish. *Ann. Appl. Biol.* **50**:467–477.

80. A. V. Holden. 1965. Contamination of freshwater by persistent insecticides and their effects on fish. *Ann. Appl. Biol.* **55**:332–335.

81. A. V. Holden. 1966. Organochlorine residues in salmonid fish *J. Appl. Ecol.* **3**(Suppl.):45–53.

82. C. L. Hopkins, S. R. B. Solly, and A. R. Ritchie. 1969. DDT in trout and its possible effect on reproductive potential. *N.Z. J. Marine Freshwater Res.* **3**:220–229.

83. D. A. Jackson, J. M. Anderson, and D. R. Gardner. 1970. Further investigations of the effect of DDT on learning in fish. *Can. J. Zool.* **48**:577–580.

84. R. Janicki and W. B. Kinter. 1970. DDT inhibition of intestinal salt and water absorption in teleosts. *Am. Zool.* **10**:540–541.

85. R. Janicki and W. B. Kinter. 1971. DDT: Disrupted osomoregulatory events in the intestines of the eel adapted to sea-water. *Science* **173**:1146–1148.

86. A. W. Jarvinen, M. J. Hoffmann, and T. W. Thorslund. 1974. *Significance to Fathead Minnows of Food and Water Exposure to DDT*. Environ. Prot. Agency, Duluth Laboratory Project No. 16AAK, Task A 010, Program Element 1BA021.

87. D. W. Johnson. 1968. Pesticides and fishes: A review of selected literature. *Trans. Am. Fish. Soc.* **97**:398–424.

88. H. E. Johnson and C. Pecor. 1969. Coho salmon mortality and DDT in Lake Michigan. *N. Am. Wildl. Nat. Res. Conf. Trans.* **34**:159–166.

89. B. J. Kallman, O. B. Cope, and R. J. Navarre. 1962. Distribution and detoxication of toxaphene in Clayton Lake, New Mexico. *Trans. Am. Fish. Soc.* **91**:14–22.

90. M. Katz. 1961. Acute toxicity of some organic insecticides to three species of salmonids and to the three-spine stickleback. *Trans. Am. Fish. Soc.* **40**:264–268.

91. J. O. Keith. 1966. Insecticide contaminants in wetland habitats and their effects on fish-eating birds. *J. Appl. Ecol.* **3**(Suppl.):71–85.

92. H. D. Kennedy, L. F. Eller, and D. F. Walsh. 1970. *Chronic Effects of Methoxychlor on Bluegills and Aquatic Invertebrates*. U.S. Dept. Interior, Bur. Sport Fish. Wildl. Tech. Paper 53. 18 pp.

93. C. J. Kerswill and H. E. Edwards. 1967. Fish losses after forest spraying with insecticides in New Brunswick, 1952–62. *J. Fish. Res. Bd. Can.* **24**:709–729.

94. R. B. Koch, P. Desaiah, H. H. Yap, and L. K. Cutkomp. 1972. Polychlorinated biphenyls; effect of long-term exposure on ATPase activity in fish, *Pimephales promelas. Bull. Environ. Contam. Toxicol.* **7**:87–92.

95. S. K. Konar. 1969. Laboratory studies on two organophosphorus insecticides, DDVP and phosphamidon, as selective toxicants. *Trans. Am. Fish. Soc.* **98**:430–437.

96. S. Korn and R. D. Earnest. 1974. Acute toxicity of twenty insecticides to striped bass, *Morone saxatilis. Calif. Fish Game* **60**:128–131.

97. C. E. Lane and R. J. Livingston. 1970. Some acute and chronic effects of dieldrin on the sailfin molly. *Trans. Am. Fish. Soc.* **99**:489–495.

98. T. P. Leadem, R. D. Campbell, and D. W. Johnson. 1974. Osmoregulatory responses to DDT and varying salinities in *Salmo gairdneri.* I. Gill Na-K-ATPase. *Comp. Biochem. Physiol.* **49A**:197–205.

99. L. L. Lewallen and W. H. Wilder. 1962. Toxicity of certain organophosphorus and carbamate insecticides to rainbow trout. *Mosq. News* **22**:369–372.

100. J. Lincer, J. M. Solon, and J. H. Nair. 1970. DDT and endrin fish toxicity under static versus dynamic bioassay conditions. *Trans. Am. Fish. Soc.* **99**:13–19.

101. W. L. Lockhart, D. A. Metner, and N. Grift. 1973. Biochemical and residue studies on rainbow trout following field and laboratory exposures to fenitrothion. *Manitoba Entomol.* **7**:26–36.

102. J. I. Lowe. 1964. Chronic exposure of spot (*Leiostomus xanthurus*) to sublethal concentrations of toxaphene in sea water. *Trans. Am. Fish. Soc.* **93**:396–399.

103. P. D. Ludwig, H. J. Dishburger, J. C. McNeill, W. O. Miller, and J. R. Rice. 1968. Biological effects and persistence of Dursban insecticide in a salt-marsh habitat. *J. Econ. Entomol.* **61**:626–633.

104. K. J. Macek. 1968. Reproduction in the brook trout (*Salvelinus fontinalis*) fed sublethal concentrations of DDT. *J. Fish. Res. Bd. Can.* **25**:1787–1796.

105. K. J. Macek. 1970. Biological magnification of pesticide residues in food chains. In *Biological Impact of Pesticides in the Environment.* Ed. J. W. Gillett. Oregon State Univ., Environ. Hlth. Series No. 1 pp. 17–21.

106. K. J. Macek, K. S. Buxton, S. K. Derr, J. W. Dean, and S. Sauter. 1976. *Chronic Toxicity of Lindane to Selected Aquatic Invertebrates and Fishes.* Environ. Protection Agency, Washington. Ecological Research Series, EPA 600/3-76-046.

107. K. J. Macek and W. A. McAllister. 1970. Insecticide susceptibility of some common fish family representatives. *Trans. Am. Fish. Soc.* **99**:20–27.

108. A. W. Maki, K. W. Stewart, and J. G. K. Silvey. 1973. The effects of Dibrom on respiratory activity of a stonefly, the hellgrammite, and the golden shiner. *Trans. Am. Fish. Soc.* **102**:806–815.

109. P. Matton and O. N. Lettam. 1969. Effect of the organophosphate Dylox on rainbow trout larvae. *J. Fish. Res. Bd. Can.* **26**:2193–2220.

110. P. M. Mehrle and F. L. Mayer. 1975. Toxaphene effects on growth and bone composition of fathead minnows. *J. Fish. Res. Bd. Can.* **32**:593–598.

111. P. M. Mehrle and F. L. Mayer. 1975. Toxaphene effects on growth and development of brook trout. *J. Fish. Res. Bd. Can.* **32**:609–613.

112. J. W. Merna, M. E. Bender, and J. R. Novy. 1972. The effects of methoxychlor on fishes. I. Acute toxicity and breakdown studies. *Trans. Am. Fish. Soc.* **101**:298–301.

113. F. P. Meyer. 1965. The experimental use of Guthion as a selective fish eradicator. *Trans. Am. Fish. Soc.* **94**:203–209.

114. C. W. Miller, B. M. Zuckerman, and A. J. Charig. 1966. Water translocation of diazinon-C^{14} and parathion-S^{35} off a model cranberry bog and subsequent occurrence in fish and mussels. *Trans. Am. Fish. Soc.* **95**:345–349.

115. C. D. Minchew and D. E. Ferguson. 1969. Toxicity of six insecticides to resistant and susceptible green sunfish and golden shiners in static bioassays. *J. Miss. Acad. Sci.* **15**:29–32.

116. T. Miura and R. M. Takahashi. 1974. Insect development inhibitors: Effects of candidate mosquito control compounds on non-target aquatic organisms. *Environ. Entomol.* **4**:631–636.

117. D. I. Mount. 1962. *Chronic Effects of Endrin on Bluntnose Minnows.* U.S. Fish Wildl. Serv. Res. Rept. 58. 38 pp.

118. D. I. Mount and H. W. Boyle. 1969. Parathion: Use of blood concentration to diagnose mortality of fish. *Environ. Sci. Technol.* **3**:1183–1185.

119. D. I. Mount and G. J. Putnicki. 1966. Summary report of the 1963 Mississippi fish kills. *Trans. N. Am. Wildl. Conf.* **31**:177–184.

120. D. I. Mount and C. E. Stephan. 1967. A method for establishing acceptable toxicant limits of fish: malathion and the BE ester of 2,4-D. *Trans. Am. Fish. Soc.* **96**:185–195.

121. D. I. Mount, L. W. Vigor, and M. L. Schafer. 1966. Endrin: Use of concentration in blood to diagnose acute toxicity to fish. *Science* **152**:1388–1390.

122. W. C. Moye and W. H. Luckmann. 1964. Fluctuations in populations of certain aquatic insects following application of aldrin granules to Sugar Creek, Iroquois county, Illinois. *J. Econ. Entomol.* **57**:318–322.

123. R. C. Muirhead-Thomson. 1971. *Pesticides and Freshwater Fauna.* Academic. p. 73.

124. M. S. Mulla. 1963. Toxicity of organochlorine insecticides to the mosquito fish and the bullfrog. *Mosq. News* **23**:299–303.

125. M. S. Mulla and L. W. Isaak. 1961. Field studies on the toxicity of insecticides to the mosquito fish *Gambusia affinis. J. Econ. Entomol.* **54**:1237–1242.

126. M. S. Mulla, L. W. Isaak, and H. Axelrod. 1963. Field studies on the effects of insecticides on some aquatic wildlife species. *J. Econ. Entomol.* **56**:184–188.

127. M. S. Mulla, J. O. Keith, and F. A. Gunther. 1966. Persistence and biological effects of parathion residues in waterfowl habitats. *J. Econ. Entomol.* **59**:1085–1090.

128. S. D. Murphy. 1966. Liver metabolism and toxicity of thiophosphate insecticides in mammalian, avian, and piscine species. *Proc. Soc. Exp. Biol. Med.* **123**:392–398.

129. S. D. Murphy, R. R. Laywerys, and K. L. Cheever. 1968. Comparative anticholinesterase action of organophosphorus insecticides in vertebrates. *Toxicol. Appl. Pharmacol.* **12**:22–35.

130. A. V. Nebeker, F. A. Puglisi, and D. L. DeFoe. 1974. Effect of polychlorinated biphenyl compounds on survival and reproduction of the fathead minnow and flagfish. *Trans. Am. Fish. Soc.* **103**:562–568.

131. D. M. Ogilvie and J. M. Anderson. 1965. Effect of DDT on temperature selection by young Atlantic salmon. *J. Fish. Res. Bd. Can.* **22**:503–512.

132. A. D. Oliver. 1964. Control studies of the forest tent caterpillar, *Malacosoma disstria,* in Louisiana. *J. Econ. Entomol.* **57**:157–160.

132a. T. J. Peterle and R. H. Giles. 1964. *New Tracer Techniques for Evaluating the Effects of an Insecticide on the Ecology of a Forest Fauna.* Ohio State Univ. Res. Found. Rep. 1207 (to USAEC). 435 pp.

133. Q. H. Pickering, C. Henderson, and E. A. Lemke. 1962. The toxicity of organic phosphorus insecticides to different species of warmwater fishes. *Trans. Am. Fish. Soc.* **91**:175–184.

134. R. E. Pillmore. 1973. Toxicity of pyrethrum to fish and wildlife. In *Pyrethrum, the Natural Insecticide.* Ed. J. E. Casida. Academic. pp. 143–165.

135. D. Pimentel. 1971. *Ecological Effects of Pesticides on Non-target Species.* Exec. Off. President, Off. Sci. Technol. Sup't. Documents, Washington. 220 pp.

136. A. Post and R. Garms. 1966. Die Empfindlichkeit einiger tropisch Susswasserfische gegenuber DDT und Baytex. *Z. Angew. Zool.* **53**:487–494.

137. G. Post and T. R. Schroeder. 1971. The toxicity of four insecticides to four salmonid species. *Bull. Environ. Contam. Toxicol.* **6**:144–155.

138. F. H. Premdas and J. M. Anderson. 1963. The uptake and detoxification of C^{14}-labelled DDT in Atlantic salmon, *Salmo salar. J. Fish. Res. Bd. Can.* **20**:827–836.

139. Y. Rongsriyam, S. Prownebon, and S. Hirakoso. 1968. Effects of insecticides on the feeding activity of the guppy, a mosquito-eating fish. *Bull. Wld. Hlth. Org.* **39**:977–980.

140. P. Rosato and D. E. Ferguson. 1968. The toxicity of endrin-resistant mosquitofish to eleven species of vertebrates. *BioScience* **18**:783–784.

141. R. L. Rudd and R. E. Genelly. 1956. *Pesticides: Their Use and Toxicity in Relation to Wildlife.* Calif. Fish and Game Bull. 7. 309 pp.

142. J. W. Saunders. 1969. Mass mortalities and behaviour of brook trout and Atlantic salmon in a stream polluted by agricultural pesticides. *J. Fish. Res. Bd. Can.* **26**:695–699.

143. W. J. Schouwenburg and K. J. Jackson. 1966. A field assessment of the effects of spraying a small coastal coho salmon stream with phosphamidon. *Can. Fish-Cult.* **37**:35–43.

144. J. C. Shim and L. S. Self. 1972. *Toxicity of Agricultural Chemicals to Larvivorous Fish in Korean Rice Fields.* Unpublished document WHO/VBC/72.342, Geneva. 7 pp.

145. E. K. Silbergeld. 1973. Dieldrin: Effects of chronic sublethal exposure on adaptation to thermal stress in freshwater fish. *Environ. Sci. Tech.* **7**:846–849.

146. R. M. Smith and C. F. Cole. 1973. Effects of egg concentrations of DDT and dieldrin on development in winter flounder. *J. Fish. Res. Bd. Can.* **30**:1894–1898.

147. E. W. Surber. 1948. Chemical control agents and their effects on fish. *Progr. Fish-Cult.* July 1948. pp. 125–131.

148. C. M. Tarzwell. 1950. Effects on fishes of the routine manual and airplane application of DDT and other mosquito larvicides. *U.S. Public Health REsp.* **65**:231–255.

149. S. B. Vinson, C. E. Boyd, and D. E. Ferguson. 1963. Resistance to DDT in the mosquito fish, *Gambusia affinis. Science* **139**:217–218.

150. D. L. Von Windeguth and R. S. Patterson. 1966. The effect of two organophosphorus insecticides on segments of the aquatic biota. *Mosq. News* **26**:377–380.

151. K. Warner and O. C. Fenderson. 1962. Effects of DDT spraying for forest insects on Maine trout streams. *J. Wildl. Manage.* **26**:86–93.

152. R. E. Warner, K. K. Petersen, and L. Borgman. 1966. Behavorial pathology in fish: A quantitative study of sublethal pesticides to toxication. *J. Appl. Ecol.* **3**(Suppl.):223–251.

153. R. K. Washino, W. Ahmed, J. D. Linn, and K. G. Whitesell. 1972. Ricefield mosquito control studies. IV. Effects upon aquatic non-target organisms. *Mosq. News* **32**:531–537.

154. G. Webbe. 1961. Field trials of phosphoric acid esters as larvicides and their toxicity to fish. *Ann. Trop. Med. Parasit.* **55**:187–191.

155. C. M. Weiss. 1959. Response of fish to sub-lethal exposures of organic phosphorus insecticides. *Sewage Indust. Wastes (J. Water Pollut. Contr. Fed.)* **31**:580–593.

156. C. M. Weiss. 1961. Physiological effect of organic phosphorus insecticides on several species of fish. *Trans. Am. Fish. Soc.* **90**:143–152.

157. C. M. Weiss and J. H. Gakstatter. 1964. Detection of pesticides in water by biochemical assay. *J. Water Pollut. Contr. Fed.* **36**:240–253.

158. C. M. Weiss and J. H. Gakstatter. 1965. The decay of anticholinesterase activity of organic phosphorus insecticides on storage in water of different pH. *Proc. 2nd Internat. Water. Poll. Res. Conf. Tokyo.* pp. 83–95.

159. M. R. Wells, J. L. Ludke, and J. D. Yarbrough. 1973. Epoxidation and fate of ^{14}C-aldrin in insecticide-resistant and susceptible populations of mosquitofish. *J. Agric. Food Chem.* **21**:428–429.

160. A. K. Williams and C. R. Sova. 1966. Acetylcholinesterase levels in brains of fishes from polluted waters. *Bull. Environ. Contam. Toxicol.* **1**:198–204.

161. B. R. Wilson. 1966. Fate of pesticides in the environment: A progress report. *Trans. N.Y. Acad. Sci.* **28**:694–705.

162. A. C. Worrell. 1960. Pests, pesticides, and people. *Am. Forests.* **66**(7):39–81.

163. J. D. Yarbrough. 1974. Insecticide resistance in invertebrates. In *Survival in Toxic Environments.* Ed. M. A. Q. Khan and J. P. Bederka. Academic. pp. 373–397.

164. R. G. Young, L. St. John, and D. J. Lisk. 1971. Degradation of DDT by goldfish. *Bull. Environ. Contam. Toxicol.* **6**:351–354.

7

INSECTICIDES AND TERRESTRIAL VERTEBRATES

A. EFFECT OF INSECTICIDES ON MAMMALIAN WILDLIFE

When the practice of treating forests from aircraft was commenced in the early 1920s, the application of calcium arsenate dusts (15% As_2O_3 content) at 50 kg/ha led to no observable wildlife fatalities in Europe, although some rabbits were killed in Canada.[77] There was one instance where calcium arsenate dust containing 40% As_2O_3 was used at this dosage at Haste, Germany in 1926, resulting in extensive mortality of hares, rabbits, and especially roe deer,[22] but when this grade was applied at 20–30 kg/ha there was no mortality of quadrupeds.[48] Sodium arsenite as dusted from the air for locust control in Africa was more fatal than the arsenates to big-game animals. Pastured in an orchard sprayed with lead arsenate at three times the normal concentration, sheep developed symptoms of poisoning, while a calf died.[60] However, in orchards sprayed with lead arsenate for 20–30 yr where the soil had accumulated 1300–6200 ppm of lead, the average lead contents in meadow voles were 5 ppm in the liver and 230 ppm in the bones; white-footed mice, which burrow less, had lower residues, while pine voles which burrow more contained more lead.[25]

Area Spraying with DDT

Forest Treatments.   When DDT was sprayed from the air on hardwood forest in Maryland, dosage rates of 2 lb/a had no effect on populations of white-footed deermice and short-tailed shrews,[72] nor did such sprayings repeated annually for five successive years.[73] Applications of DDT at 4 lb/a to a Texas prairie for tick control had no effect on raccoons, rabbits, armadillos or skunks.[34] Treatment of spruce–fir forest in Ontario at 6 lb/a did no harm to any species of mammal either directly or through their feeding on poisoned insects or contaminated fruits in the area[54]; surely, however, it would be expected that raccoons would be deprived of the crayfish that are their favorite food. But antimosquito larvicidal applications of DDT at 0.015 lb/a repeated every week in South Carolina induced no differences from the normal in populations of cottontail rabbits, cotton rats, or even raccoons.[27] Even at 7.5 lb/a, DDT applications in Montana caused no significant decrease in populations of deermice, voles, chipmunks, pocket gophers, deer, or bear, although a few chipmunks were showing mild symptoms of poisoning.[1] The dose had to be raised to the equivalent of 100 lb/a for direct spraying to induce symptoms or mortality in mice, shrews, or chipmunks.[54]

Feeding tests showed that neither *Microtus* nor cottontails developed symptoms until 2000 ppm DDT was added to the diet, while *Peromyscus* did not develop symptoms even at 10,000 ppm.[17] In New Jersey, DDT applications against the Dutch elm disease at dosages up to 3 lb/tree caused some mortality among gray squirrels and the red bat *Lasiurus borealis.*[9] DDT dust (25%) had been successfully employed to control house mice, a species in which the natural death rate is significantly increased by 200 ppm DDT in the diet.[13] Bats are considerably more susceptible to DDT than mice, the oral LD_{50} for the big brown bat *Eptesicus fuscus* being 30 mg/kg as compared to 400 mg/kg for mice,[52] and residual sprays of DDT wettable powder are now employed in the United States to eliminate bat colonies carrying rabies infections.

White-footed mice in a forest sprayed with DDT at 2 lb/a, where about 12% of that applied reached the forest floor, came to accumulate about 22 ppm *t*-DDT whole-body residues within 7–10 days of the spraying.[39] In Maine forests treated at 1 lb/a for spruce budworm control, the accumulations in *Peromyscus* and the vole *Clethrionomys* were not as great, and were greatly exceeded by those in the shrews *Microsorex* and *Blarina,* viz.:

	Mice and Voles	Shrews
Year of Treatment	1.06	15.6
3–4 yr After	0.08	2.5
6–7 Yr After	0.05	1.8
After 3 Sprays in 7 yr	0.17	4.8

The figures also show that these small mammals pick up residues remaining in the forest floor biota years after the application of the DDT.[23] A similar difference between the herbivorous and the carnivorous animals was shown in the larger mammals, where in the year of treatment the *t*-DDT residues were 0.08 ppm in snowshoe hare but 8.5 ppm in mink; 9–10 yr later they were, respectively, 0.02 and 1.6 ppm in the two species.[70] The average DDT residues in the brains of small mammals in areas of New Brunswick treated annually at 0.5 lb/a against the spruce budworm were 39, 24, and 27 ppb for shrews, voles, and mice, respectively.[26]

Ranch mink fed on a diet one-quarter of which consisted of fish taken from the Miramichi River draining the sprayed areas and containing 1.5–13.2 ppm DDT residues, accumulated 1.0–1.7 ppm in the liver and 2.3–14.2 ppm in the adipose fat, and they produced 8% fewer kits than normal due to increased deaths of embryos *in utero.*[35] Ranch mink whose diet included Lake Michigan fish containing DDT residues of the order of 10 ppm suffered from heavy kit mortality, ranging up to 80% although the mothers were not affected.[4] It is true that the *o,p´*-DDT, constituting more than one-quarter of the material in technical DDT, is as uterotropic as oestrone, causing a 100% increase in uterine weight when injected into mink at 10 mg/kg.[24] However, when a mixture of *p,p´*-DDT, *o,p´*-DDT, DDE, and DDD was added experimentally to their diet to give a total residue in excess of 20 ppm, it did not induce any kit mortality in the Michigan ranch mink.[4] This mortality was found to result from the contamination of the fish in the diet with PCBs,[66] present in Lake Michigan to about the same extent as the DDT residue. A concentration of only 0.64 ppm Aroclor 1254 in the beef diet of ranch mink was enough to inhibit the reproduction of all but one of the 12 inseminated females fed on it, and she produced kits which all died on the first day *post partum.*[65]

Agricultural Treatments. When an abandoned field in Ohio was sprayed with DDT at 1 lb/a, the population of the short-tailed shrew *Blarina* during the ensuing 2 yr showed the following levels in organs and tissues: brain 4 ppm, liver and muscle 10 ppm, adipose 135 ppm.[32] In prairie situations, a survey of whole-body *t*-DDT residues during 2 yr in the 1960s showed the following average figures in North Dakota[74]:

	1966	1967
Short-Tailed Shrew, *Blarina brevicauda*	5.2	25.2
Ground Squirrels (gophers), *Citellus* sp.	0.08	0.69
Meadow Voles, *Microtus pennsylvanicus*	0.20	0.16
Jumping Mice, *Zapus hudsonius*	0.08	0.18
Pocket Gophers, *Geomys bussarius*	Nil	Nil

Residues in pocket mice (*Perognathus*) near cottonfields in Texas reached 17 ppm *t*-DDT in August, about equally divided between DDT, DDE and DDD[20]; residues in cotton rats and ground squirrels were similar.[3] Among mammals in the lower Mississippi valley, DDT residues were lowest in cottontail rabbits and muskrats, notably higher in harvest mice (*Reithrodontomys*) and house mice, and highest in opossums.[74]

In agricultural areas in central Iowa, wild mink taken in the winter of 1970–1971 were found to have the following average residues (ppm) in their adipose tissue:

DDE	DDD	DDT	*o,p*′-DDT	Dieldrin
0.98	0.26	0.30	0.14	0.34

No DDT was found in the brain, and the DDE and DDD levels were below 0.08 and 0.05 ppm, respectively.[33]

In a thriving colony of the free-tailed bat (*Tadarida brasiliensis*) in Bracken cave, Texas, the whole-body residues in parturient females were 27 ppm DDE, 1.3 ppm DDT, and 5.9 ppm dieldrin.[16] But the liver residues of 6–29 ppm DDT plus 8–54 ppm DDE found in overwintered pipistrelle bats at Ramsey in the agricultural area of East Anglia, U.K. in 1969 were just under the lethal limit.[44] In a colony of the big brown bat *Eptesicus fuscus* near Laurel, Maryland, where 10% of the young were born dead and whole-body DDE content of the mothers and their litters were, respectively, 2.6 and 0.4 ppm, the PCB residues (identified as Aroclor 1260 in this case) were, respectively, 2.0 and 1.2 ppm; since the only pregnant female that failed to reproduce at all contained the highest residue of this PCB, it was apparently the PCB transferred through the placenta that caused the stillbirths.[15]

Big-Game Animals. Elk and mule deer taken in the vicinity of operations against the spruce budworm with DDT at 1 lb/a in the Rocky Mountain states during the early 1960s showed maximum residues of 20–40 ppm *t*-DDT in the adipose fat; a year after the spraying the mule deer had lost all their residues, except for one animal which carried small amounts of the metabolites DDE, DDD, and DDMU.[63] Mule deer inhabiting the Salmon National Forest, Idaho showed the following average DDT residues (ppm) in their adipose tissue[8]:

	DDE	DDD	DDT	
1964	0.52	1.62	17.22	3 months after spray
1969	0.03	0.11	0.03	5 yr after spray

Deer from land bordering the agricultural lowlands between Tucson and Nogales, Arizona when analyzed in 1970 showed even lower average residues (ppb) in their body fat,[50] as follows:

	DDE	DDD	DDT	*o,p′*-DDT	Dieldrin
Mule Deer	16	8	47	5	4
Whitetail Deer	5	8	39	3	3

However, the flesh of whitetail deer shot in Alabama soybean areas in 1968 and 1969 contained an average of 3.0 ppm DDT and 5.5 ppm *t*-DDT.[14]

Black bear, a general feeder, sampled in Idaho in the fall of 1972, contained only 0.3–1.6 ppb DDT and 0.03–2.1 ppb DDE in the adipose tissue, 1000 times less than the Hg content.[7] This is a long way from the brain residue of 50 ppm DDT which, judging from the laboratory rat, is the threshold level of danger to the life of mammals.[21] On the other hand, a population of European hares in the forest-steppe region of the Ukraine, reported to contain DDT residues as high as 21 ppm in the brain and 23 ppm in the gonads, was characterized by a low proportion of males, reduced litter size, and diminished population growth.[2]

Cyclodiene Insecticides

Some members of this group of organochlorines, which include several that have proved useful in rodent control, have caused considerable mortality among nontarget wild mammals. Heptachlor applied at 2 lb/a for fire ant control, usually as granules, killed many raccoons and rabbits,[67] and opossums were conspicuous among the 53 corpses of 12 quadruped species found on four treated farms in Louisiana.[71] Dieldrin applied at 3 lb/a to eradicate the Japanese beetle in Illinois virtually eliminated ground squirrels, cottontails, and muskrats, caused heavy mortality in fox squirrels, woodchucks, voles, and shrews, and some mortality of moles and opossums; white-footed mice were in the first to recover their numbers.[69] No poisoning of small mammals occurred with treatments of either chlordane or toxaphene at 1.5 lb/a for grasshopper control.[12]

The use of heptachlor in seed dressings against the wheat bulb fly resulted in the death of some 1300 foxes in England during the winter of 1959–1960; they were found wandering aimlessly about and dying in convulsions.[75] Analysis of the corpses revealed 10–90 ppm heptachlor epoxide (plus 3–13 ppm dieldrin) in their livers, and feeding experiments established that three to six wood pigeons taken from the treated areas in East Anglia were sufficient to kill a fox in 1–2 wk.[10] During the 1960s, many badgers were found in emaciated condition and undergoing tremors, and occasionally drowned in streams; that this was due to dieldrin was indicated by the fact that their livers contained 17–46 ppm dieldrin, which would have been acquired mainly from dead wood pigeons that had eaten treated seed.[43] Feeding studies with dogs indicated that the fatal level for dieldrin in the brain was approximately 5 ppm,[38] much the same as in birds. The regular annual spraying of cornfields with aldrin at 1.5 lb/a for 15 successive years resulted in the white-footed mice accumulating slightly less than 1 ppm dieldrin, and a trace of aldrin, in their bodies.[47]

Among the cyclodiene derivatives, endrin has been used to control rodents in orchards, 1.3 lb/a being sufficient to eliminate most wild mice in Washington State.[80] In Virginia, where dosages of 2.5 lb/a applied in November had been effective in controlling the pine mouse (*Pitymys*) probably by dermal contact,[41] certain populations (e.g., at Berryville) had become 12 times as endrin resistant as those from untreated orchards.[79] Dieldrin at 3 lb/a was found to be an effective barrier spray to protect cornfields against gophers (*Citellus* spp.) in Illinois.[53] On laboratory mice, either endrin or dieldrin at 5 ppm in the diet significantly reduced the size of the litters.[36] Dieldrin is considerably less toxic than endrin to large herbivores, although not so safe as toxaphene, as the following acute oral LD_{50} figures (mg/kg) show[78]:

	Mule Deer	Domestic Goats
Endrin	—	25–50
Dieldrin	75–150	100–200
Toxaphene	139–240	>160

White-tailed deer on a diet containing 25 ppm dieldrin grew more slowly than normal, and the death rate was higher than normal in their own fawns and other fawns suckling on them.[64] Heptachlor in the diet of laboratory rats resulted in smaller litters and a higher death rate among their young.[55]

Organophosphorus and Carbamate Insecticides

It is surprising that there are no records of wild mammal kills with parathion or other of the more toxic OP insecticides (Table 7.1). Pocket gophers in areas near cottonfields in Texas came to contain whole-body residues of approximately 5 ppm parathion and 5 ppm methyl parathion.[20] White-footed mice in New Jersey woodland adjacent to vegetable crops dusted from the air with parathion at 0.4 lb/a, and thus receiving about 0.03 lb/a by drift, were not measurably affected in population density or production of young.[42] In an Ohio woodland sprayed with malathion at 2 lb/a, there were 40–45% population reductions in white-footed mice (*Peromyscus*) and striped chipmunks (*Tamias*), due not to direct lethality but to diminished productivity and survival; there were no reductions in short-tailed shrews (*Blarina*) or large animals such as raccoons.[62] In red-clover fields in Indiana, the application of dimethoate at 0.5 lb/a was without effect on the population of the prairie deermouse (*Peromyscus*), but it did reduce the omnivorous house mouse (*Mus*) to one-fifth of its former numbers, presumably owing to the sudden loss of the insect component of its food supply; at the same time it increased the population of the herbivorous prairie vole (*Microtus*) by four times.[6]

Of the two carbamate insecticides which have been employed in forest spraying, mexacarbate is quite toxic to deer, in contrast to carbaryl (Table

Table 7.1. Acute oral LD_{50} (in mg/kg) of OP compounds and carbamates to large quadrupeds when given in capsules (Tucker and Crabtree, 1970).

OP Compound	Mule Deer	Domestic Goats	Carbamate	Mule Deer	Domestic Goats
Demeton	--	13	Aminocarb	11	--
Parathion	33	42	Methomyl	16	--
Monocrotophos	38	35	Mexacarbate	25	22
Naled	200	--	Landrin	75	--
Dimethoate	>200	--	Propoxur	225	>800
Chlorpyrifos	--	>500	Carbaryl	300	--
Fenitrothion	727	--			

7.1). Application of carbaryl at 2 lb/a to grassland had no effect on populations of *Mus* or *Microtus*, but did reduce the numbers of the cotton rat (*Sigmodon*), a species whose reproductive rate was unusually susceptible to reduction by this carbamate.[5] Applied at 1.25 lb/a from suspension in fuel oil to a deciduous forest in New York State, carbaryl had no effect on the abundance, condition and reproduction of the small mammal fauna.[18]

B. EFFECT OF INSECTICIDES ON AMPHIBIANS AND REPTILES

Organochlorine Insecticides

Amphibians. Frogs, toads, and their tadpoles are considerably less susceptible to insecticide poisoning than fish. Whereas DDT at 6 lb/a caused 50% mortality of frogs and toads in coniferous forest,[51] at 3 lb/a it caused no mortality in salt marshes.[76] Applications of DDT at 1 lb/a were safe for frogs and their tadpoles except in pools where the water was less than 3 inches deep.[49] However, tadpoles in the hyperactive or ataxic phases of sublethal poisoning with DDT are subject to preferential predation by newts.[19] A case of secondary poisoning has been described where forest tent caterpillars killed at 1 lb/a were lethal to populations of the wood frog (*Rana sylvatica*) around pools in the sprayed area.[28]

Tests of the tadpoles of the western chorus frog and Fowler's toad (Table 7.2) showed methoxychlor and DDD to be more toxic than DDT to these anurans, the reverse of the situation in fish. Although methoxychlor was more toxic than heptachlor and aldrin for 96-hr exposure of the toad tadpoles,[68] all of the cyclodiene insecticides were more toxic than methoxychlor for 30-day exposure of adult leopard frogs (*R. pipiens*) fed periodically on uncontaminated food.[46] For 36-hr exposure to filter paper impregnated with the insecticide, adult Fowler's toads and the cricket frog *Acris crepitans* were susceptible to organochlorines in the following descending order: endrin > dieldrin > aldrin > toxaphene > DDT.[30] The effect of the cyclodienes on the frogs was to induce hyperresponsiveness to stimuli, followed by convulsions.[46] When heptachlor was applied at 2 lb/a in granules to Mississippi farmland for fire ant control, anuran kills were rare, only one dead bullfrog and two distressed leopard frogs being found; Fowler's toads successfully metamorphosed in pools where bluegill sunfish had been killed, and this species along with narrow-mouth toads, southern cricket frogs, crayfish frogs (*R. areolata*), and bullfrogs (*R. catesbeiana*) remained abundant.[29] Applied as mosquito larvicides at 0.1 lb/a, endrin, dieldrin, aldrin, and endosulfan caused mortality of bullfrog tadpoles, while

Table 7.2. LC_{50} (TL_{50}) values in ppm for technical grade insecticides to tadpoles (Sanders, 1970).

	Western Chorus Frog *Pseudacris triseriata*		Fowler's Toad *Bufo woodhousii fowleri*	
	24 hr	96 hr	24 hr	96 hr
Organochlorine cpds.				
Dieldrin	0.23	0.10	1.1	0.15
Endrin	0.29	0.18	0.57	0.12
Methoxychlor	0.44	0.33	0.76	---
DDD	0.61	0.40	0.70	0.14
DDT (tech.)	1.4	0.80	2.4	1.0
Toxaphene	1.7	0.50	0.60	0.14
Heptachlor	---	---	0.85	0.44
Aldrin	---	---	2.0	0.15
Lindane	4.0	2.7	14	44
BHC (tech.)	---	---	13	3.2
Organophosphorus cpds.				
Carbophenothion	0.10	0.028	---	---
Azinphosmethyl	---	---	0.68	0.13
Malathion	0.56	0.20	1.9	0.42
Parathion	1.6	1.0	---	---
Naled	2.2	1.7	---	---

chlordane, heptachlor, and toxaphene did not.[56] Lindane and technical BHC (Table 7.2) are the least toxic organochlorine materials for frogs.[46,56]

Developed Resistance. In cottonfields heavily treated with organochlorine insecticides in the early 1960s, it was observed that some cricket frogs, *A. crepitans* and *A. gryllus,* were surviving in field pools where there was a heavy mortality of green tree-frogs (*Hyla cinerea*), diamond-backed water snakes (*Natrix rhombifera*), and red-eared turtles (*Pseudemys scripta*). On being tested, both species of cricket frogs taken from treated areas proved to be more DDT resistant than those from untreated localities, showing a fivefold higher LC_{50} upon exposure to residues.[11]

Reptiles. Direct treatment of ponds at DDT concentrations of 2 ppm or more has killed water snakes and turtles, as well as salamanders,[12] and in forest sprayed with DDT at 6 lb/a a garter snake died from tertiary poisoning by eating a frog that had been feeding on poisoned insects.[51] When broadleaved forest was sprayed at 1 lb/a, water snakes were killed in small numbers,[40] whereas box turtle populations remained apparently unaffected in deciduous forest sprayed with DDT at 2 lb/a for 5 consecutive years.[73] In pine–oak palmetto forest sprayed with DDT at 3 lb/a, occasional specimens of the black snake, the common water snake, and the chameleon were found dead in the area; green tree-frogs suffered most, and some groups of leopard frogs were killed.[37] In the Brazos River flood-plain of Texas, where cottonfields had been heavily treated with DDT as well as toxaphene and methyl parathion, the average residues (ppm) in the fat bodies of aquatic snakes in 1971 were as follows:

	DDE	DDD	DDT	Dieldrin
Cottonmouth	569	0.1	6.6	6.1
Ribbon Snake	537	2.3	17.6	8.9
Blotched Water Snake	346	0.4	19.6	2.4
Common Water Snake	590	3.3	20.4	4.1

The *t*-DDT residues in the brain did not exceed 1.5 ppm, while fat-body residues in terrestrial snakes (e.g., copperheads, king snakes) were much lower than in aquatic snakes.[31] In cornfields treated with aldrin at 1.5 lb/a annually for the preceding 15 yr, the garter snakes contained about 12 ppm dieldrin, but no aldrin.[47]

Organophosphorus and Carbamate Insecticides

Among the organophosphorus insecticides, carbophenothion and azinphosmethyl show appreciable toxicity to tadpoles (Table 7.2). The spraying of swampy woodland in Louisiana did not kill tadpoles or adults of the leopard frogs and bullfrogs when bicrotophos at 0.25 lb/a or phosphosphamidon at 1.0 lb/a was applied to control the forest tent caterpillar.[60] Tests of the effect of mosquito larvicides at 0.4 lb/a on the tadpoles of the bullfrog and western toad (*Bufo boreas*) showed that only carbophenothion caused 100% mortality, while azinphosmethyl, parathion, naled, methyl parathion, fenthion, fenitrothion, ronnel, and trichloronat caused no mortality.[58] Even at 1.0 lb/a, a dosage that kills most of the mosquitofish, parathion was harmless to juvenile bullfrogs.[59] The highly effective larvicide temephos (Abate) was harmless to bullfrog tadpoles at 0.4 lb/a.[57] Abate and chlorpyrifos at sublethal concentrations slightly reduce the thermal tolerance of juvenile western toads, while fenthion lowers the harmful threshold from 36°C down to 34°C.[45]

The spraying of a deciduous forest in Ohio with malathion at 2 lb/a had no effect on the amphibians and reptiles inhabiting it.[62] Neither phosphamidon at 1 lb/a nor dicrotophos at 0.25 lb/a had any effect on *Natrix* water snakes, cottonmouths, or copperheads in Louisiana woodland.[61] In deciduous forest in New York State sprayed with carbaryl at 1.25 lb/a, the frogs, toads, snakes, and salamanders remained unaffected.[18]

REFERENCES CITED

1. L. Adams, M. G. Hanavan, N. W. Hosley, and D. W. Johnston. 1949. The effects on fish, birds, and mammals of DDT used in the control of forest insects in Idaho and Wyoming. *J. Wildl. Manage.* **13**:245–254.

2. L. V. Aleeva, B. A. Galaka, A. P. Fedorenko, and L. S. Shevchenko. 1972. On the effect of agricultural chemicals on reproduction of hares (*Lepus europaeus*). *Vestn. Zool.* **6**:58–61.

3. H. G. Applegate. 1970. Insecticides in the Big Bend National Park. *Pestic. Monit. J.* **4**:2–7.

4. R. J. Aulerich, R. K. Ringer, H. L. Seagran, and W. G. Youatt. 1971. Effects of feeding coho salmon and other Great Lakes fish on mink reproduction. *Can. J. Zool.* **49**:611–616.

5. G. W. Barrett. 1968. The effects of an acute insecticide stress on a semi-enclosed grassland ecosystem. *Ecology* **49**:1019–1035.

6. G. W. Barrett and R. M. Darnell. 1967. Effects of dimethoate on small mammal populations. *Am. Midl. Nat.* **77**:164–175.

7. W. W. Benson, J. Gabica, and J. Beecham. 1974. Pesticide and mercury levels in bear. *Bull. Environ. Contam. Toxicol.* **11**:1–4.

8. W. W. Benson and P. Smith. 1972. Pesticide levels in deer. *Bull. Environ. Contam. Toxicol.* **8**:1–9.

9. A. H. Benton. 1951. Effects on wildlife of DDT used for control of Dutch elm disease. *J. Wildl. Manage.* **15**:20–27.

10. D. K. Blackmore. 1963. The toxicity of some chlorinated hydrocarbon insecticides to British wild foxes. *J. Comp. Path. Therap.* **73**:391–409.

11. C. E. Boyd, S. B. Vinson and D. E. Ferguson. 1963. Possible DDT-resistance in two species of frogs. *Copeia* **1963**(2):426–429.

12. A. W. A. Brown. 1951. Insecticides and the balance of animal populations. In *Insect Control by Chemicals*. Wiley. pp. 720–780.

13. M. S. Cannon and L. C. Holcomb. 1968. The effect of DDT on reproduction in mice. *Ohio J. Sci.* **68**:19–24.

14. K. Causey, S. C. McIntyre, and R. W. Richburg. 1972. Organochlorine insecticide residues in quail, rabbits and deer from selected Alabama soybean fields. *J. Agric. Food Chem.* **20**:1205–1209.

15. D. R. Clark and T. G. Lamont. 1976. Organochlorine residues and reproduction of the big brown bat. *J. Wildl. Manage.* **40**:249–254.

16. D. R. Clark, C. O. Martin, and D. M. Swinford. 1975. Organochlorine residues in the free-tailed bat at Bracken cave, Texas. *J. Mammal.* **56**:429–443.

17. D. R. Coburn and R. Treichler. 1946. Experiments on toxicity of DDT to wildlife. *J. Wildl. Manage.* **10**:208–216.

18. P. F. Connor. 1960. A study of small mammals, birds, and other wildlife in an area sprayed with Sevin. *N.Y. Fish Game J.* **7**:26–32.

19. A. S. Cooke. 1971. Selective predation by newts on frog tadpoles treated with DDT. *Nature* **229**:275–276.

20. D. D. Culley and H. G. Applegate. 1967. Insecticide concentrations in wildlife at Presidio, Texas. *Pestic. Monit. J.* **1**(2):21–28.

21. W. E. Dale, T. B. Gaines, and W. J. Hayes. 1962. Storage and excretion of DDT in starved rats. *Toxicol. Appl. Pharmacol.* **4**:89–106.

22. P. W. Danckwortt and E. Pfau. 1926. Massenvergiftungen von Tieren durch Arsenbestaubung vom Flugzeug. *Z. Angew. Chem.* **39**:1486–1487.

23. J. B. Dimond and J. A. Sherburne. 1969. Persistence of DDT in wild populations of small mammals. *Nature* **221**:486–487.

24. R. T. Duby, H. F. Travis, and C. E. Terrill. 1971. Uterotropic activity of DDT in rats and mink and its influence on reproduction in the rat. *Toxicol. Appl. Pharmacol.* **18**:348–355.

25. D. C. Elfring, W. M. Haschek, R. A. Stehn, C. A. Bache, W. H. Gutenmann, and D. J. Lisk. 1976. Lead in plants and animals inhabiting old orchard soils. *N.Y. Food Life Sci.* **9**(3):14–15.

26. Environmental Protection Agency. 1972. *DDT: A Review of Scientific and Economic Aspects of the Decision to Ban Its Use as a Pesticide.* Washington, D.C. Document EPA-540/1-75-022. 300 pp.

27. A. B. Erickson. 1947. Effects of DDT mosquito larviciding on wildlife, Part II. *U.S. Public Health Rep.* **62**:1254–1262.

28. B. A. Fashingbauer. 1957. The effects of aerial spraying with DDT on wood frogs. *Flicker* **29**:160.

29. D. E. Ferguson. 1963. Notes concerning the effects of heptachlor on certain poikilotherms. *Copeia* **1963**(4):441–443.

30. D. E. Ferguson and C. C. Gilbert. 1967. Tolerances of three species of anuran amphibians to five chlorinated hydrocarbon insecticides. *J. Miss. Acad. Sci.* **13**:135–138.

31. R. R. Fleet, D. R. Clark, and F. W. Plapp. 1972. Residues of DDT and dieldrin in snakes from two Texas agro-ecosystems. *BioScience* **22**:664:665.

32. D. J. Forsyth and T. J. Peterle. 1973. Accumulation of chlorine-36 ring-labeled DDT residues in various tissues of two species of shrew. *Arch. Environ. Contam. Toxicol.* **1**:1–17.

33. J. C. Franson, P. A. Dahm, and L. D. Wing. 1974. Chlorinated hydrocarbon insecticides in adipose, liver, and brain samples from Iowa mink. *Bull. Environ. Contam. Toxicol.* **11**:379–385.

34. J. L. George and W. H. Stickel. 1949. Wildlife effects of DDT used for tick control on a Texas prairie. *Am. Midl. Nat.* **42**:228–237.

35. F. F. Gilbert. 1969. Physiological effects of natural DDT residues and metabolites on ranch mink. *J. Wildl. Manage.* **33**:933–943.

36. E. E. Good and G. W. Ware. 1969. Effects of insecticides on reproduction in the laboratory mouse. IV. Endrin and dieldrin. *Toxicol. Appl. Pharmacol.* **14**:201–203.

37. P. Goodrum, W. P. Baldwin, and J. W. Aldrich. 1949. Effect of DDT on animal life of Bull's Island, South Carolina. *J. Wildl. Manage.* **13**:1–10.

38. D. L. Harrison, P. E. G. Maskell, and D. F. L. Money. 1963. Dieldrin poisoning of dogs. 2. Experimental studies. *New Zealand Vet. J.* **11**:23–31.

39. D. W. Hayne, 1970. DDT body burden of forest *Peromyscus* after spraying. Final Rep. to Bur. Sport Fish & Wildlife, USDI. Contract No. 14-16-0008-653. 65 pp. Cited by Pimentel (1971).

40. C. H. Hoffmann and J. P. Linduska. 1949. Some considerations on the biological effects of DDT. *Sci. Mon.* **69**:104–114.

41. F. Horsfall. 1956. Rodenticidal effect on pine mice of endrin used as a ground spray. *Science* **123**:61.

42. W. B. Jackson. 1952. Populations of the wood mouse (*Peromyscus leucopus*) subjected to the applications of DDT and parathion. *Ecol. Monogr.* **22**:259–281.

43. D. J. Jefferies. 1969. Causes of badger mortality in eastern counties of England. *J. Zool., London* **157**:429–436.

44. D. J. Jefferies. 1972. Organochlorine insecticide residues in British bats and their significance. *J. Zool., London* **166**:245–263.

45. C. R. Johnson and J. E. Prine. 1976. The effects of sublethal concentrations of organophosphorus insecticides on temperature tolerance in juvenile western toads. *Comp. Biochem. Physiol.* **53A**:147–149.

46. H. M. Kaplan and J. G. Overpeck. 1964. Toxicity of halogenated hydrocarbon insecticides for the frog *Rana pipiens*. *Herpetologica* **20**:163–169.

47. L. J. Korschgen, 1970. Soil-food-chain-pesticide wildlife relationship in aldrin-treated fields. *J. Wildl. Manage.* **34**:186–199.

48. H. Krieg. 1925. Die Bekampfung forstlicher Schadlinge durch Abwurf von Calziumarseniat vom Flugzeug. *Anz. Schadlingsk.* **1**:97–98.

49. J. B. Lackey and M. L. Steinle. 1945. Effects of DDT upon some aquatic organisms other than insect larvae. *U.S. Public Health Rep. Suppl.* **186**:80–89.

50. J. A. Laubscher, G. R. Butt, and C. C. Roan. 1971. Chlorinated insecticide residues in wildlife and soil as a function of distance from application. *Pestic. Monit. J.* **5**:251–258.

51. E. B. S. Logier. 1949. *Effect of DDT on Amphibians and Reptiles, 1944*. Dept. Lands and Forests, Ontario, Biol. Bull. No 2. pp. 49–56.

52. M. M. Luckens and W. H. Davis. 1964. Bats: sensitivity to DDT. *Science* **146**:948.

53. W. H. Luckmann. 1957. Control of ground squirrels with ground sprays of dieldrin. *Agron. J.* **49**:107.

54. R. E. Mackie. 1949. *The Effect of DDT on Mammals*. Dept. Lands and Forests Ontario, Biol. Bull. No. 2. pp. 63–70.

55. M. Mestitzova. 1966. On reproduction studies and the development of cataracts after long-term feeding of the insecticide heptachlor. *Experientia* **23**:42–43.

56. M. S. Mulla. 1963. Toxicity of organochlorine insecticides to the mosquito fish and bull-frog. *Mosq. News* **23**:299–303.

57. M. S. Mulla. 1966. Toxicity of some new organic insecticides to mosquito fish and some other aquatic organisms. *Mosq. News* **26**:87–91.

58. M. S. Mulla, L. W. Isaak, and H. Axelrod. 1963. Field studies on the effects of insecticides on some aquatic wildlife species. *J. Econ. Entomol.* **56**:184–188.

59. M. S. Mulla, J. O. Keith, and F. A. Gunther. 1966. Persistence and biological effects of parathion residues in waterfowl habitats. *J. Econ. Entomol.* **59**:1085–1090.

60. W. C. O'Kane, C. H. Hadley, and W. A. Osgood. 1917. *Arsenical Residues after Spraying.* New Hampshire Agric. Exp. Sta. Bull. 183. 62 pp.

61. A. D. Oliver. 1964. Control studies on the forest tent caterpillars in Louisiana. *J. Econ. Entomol.* **57**:157–160.

62. T. J. Peterle and R. H. Giles. 1964. *New Tracer Techniques for Evaluating the Effects of an Insecticide on the Ecology of a Forest Fauna.* Ohio State Univ. Res. Found. Rep. 1207 (to U.S. Atomic Energy Comm.). 455 pp.

63. R. E. Pillmore and R. B. Finley. 1963. Residues in game animals resulting from forest and range insecticide applications. *Trans. 28th N. Am. Wildl. Nat. Resour. Conf.* pp. 409–421.

64. D. Pimentel. 1971. *Ecological Effects of Pesticides on Nontarget Species.* Exec. Office President, Off. Sci. Technol. Supt. Doc., Washington. 220 pp.

65. N. S. Platonow and L. H. Karstad. 1973. Dietary effects of polychlorinated biphenyls on mink. *Can. J. Comp. Med.* **37**:391–401.

66. R. K. Ringer, R. J. Aulerich, and M. Zabik. 1972. Effect of dietary polychlorinated biphenyls on growth and reproduction of mink. Preprint of paper presented at *164th Ann. Meet. Am. Chem. Soc.* **12**(2):149–154.

67. R. L. Rudd. 1964. *Pesticides and the Living Landscape.* Univ. Wisconsin Press. pp. 34–36.

68. H. O. Sanders. 1970. Pesticide toxicities to tadpoles of the western chorus frog and Fowler's toad. *Copeia* **1970**(2):246–251.

69. T. G. Scott, Y. L. Willis, and J. A. Ellis. 1959. Some effects of a field application of dieldrin on wildlife. *J. Wildl. Manage.* **23**:409–427.

70. J. A. Sherburne and J. B. Dimond. 1969. DDT persistence in hares and mink. *J. Wildl. Manage.* **33**:944–948.

71. R. D. Smith and L. L. Glasgow. 1963. Effects of heptachlor on wildlife in Louisiana. *Ann. Conf. South. Assoc. Game Fish. Comm.* **17**:140–154.

72. L. F. Stickel. 1946. Field studies of a *Peromyscus* population in an area treated with DDT. *J. Wildl. Manage.* **10**:216–218.

73. L. F. Stickel. 1951. Wood mouse and box turtle populations in an area treated annually with DDT for 5 years. *J. Wildl. Manage.* **15**:161–164.

74. L. F. Stickel. 1973. Pesticide residues in birds and mammals. In *Environmental Pollution by Pesticides.* Ed. C. A. Edwards. Plenum Press. pp. 254–312.

75. J. C. Taylor and D. K. Blackmore. 1961. A short note on the heavy mortality of foxes during the winter 1959–60. *Vet. Rec.* **73**:232–233.

76. R. E. Tiller and E. N. Cory. 1947. Effects of DDT on some tidewater aquatic animals. *J. Econ. Entomol.* **40**:431–433.

77. I. Tragardh. 1935. The economic possibilities of aircraft dusting against forest insects. *Bull. Entomol. Res.* **26:**487–495.

78. R. K. Tucker and D. G. Crabtree. 1970. *Handbook of Toxicity of Pesticides to Wildlife.* U.S. Dept. Int., Fish Wildl. Serv., Resour. Publ. No 84. 131 pp.

79. R. E. Webb and F. Horsfall. 1967. Endrin resistance in the pine mouse. *Science* **156:**1762.

80. H. R. Wolfe. 1957. *Orchard Mouse Control with Endrin Sprays.* State Coll. Washington, Ext. Circ. 282. 8 pp.

8

INSECTICIDES AND BIRDS

A. DIRECT EFFECTS OF DDT-TYPE ORGANOCHLORINES

Introductory: Before DDT

A half-century ago, when forest insects were controlled by aerial dusting with calcium arsenate, no mortality of birds was found in Austria with a dust containing 15% As_2O_3 which was applied at 50 kg/ha,[130] and no reduction in birdsong in Germany when a calcium arsenate of 40% As_2O_3 content was applied at 20–30 kg/ha.[84] However the application of the 40% grade at 50 kg/ha to an oak forest at Haste in Minden district resulted in kills of woodlarks and whitethroats.[33]

The use of calcium arsenate or Paris green to control the introduced Colorado potato beetle in France killed no pheasant or partridge, which are seed eaters, but there were a few instances (only eight confirmed) of lethal secondary poisoning among domestic turkeys, guineafowl, and hens that had fed on the poisoned beetle larvae.[22] Orchards sprayed with lead arsenate at 25 lb/a were safe pasture for chickens, which could withstand over 800 mg/day of this insecticide in their diet.[129] Sodium arsenite, however, when used as grasshopper baits in Manitoba killed the Franklin's gulls that fed on the poisoned grasshoppers.[13]

The use of DNOC in England around 1950 as an insecticide, and also a herbicide, at 1–6 lb/a resulted in kills of songbirds and pheasants foraging on the treated crop.[29,105]

DDT Applied to Arboreal Habitats

Forest Spraying. The acute toxicity of DDT to birds is one of the lowest among the insecticides (see Table 8.6) and its chronic toxicity is quite low (Table 8.1). Nevertheless the application of DDT to 200 acres of broad-leaved forest at the 25-lb/a dosage necessary to eliminate the Japanese beetle resulted in 42 dead or dying robins and catbirds being found, and the abandonment of 14 nests.[28] With DDT at 8 lb/a applied against pine bark beetle in Wyoming there were bird mortalities and population suppression,[2] although this dosage applied to larch forest in Switzerland had no noticeable effect on bird life.[115] Treatment of deciduous forest at 5 lb/a caused deaths of vireos, warblers, and tanagers and temporarily reduced the number of singing males by 98% in Pennsylvania,[64] and in Maryland where population reductions of yellowthroats, wrens, and warblers amounting to 80% were associated with their eating the poisoned insects.[110] At 3 lb/a the only mortality was in nestling birds, and no nesting territories were abandoned.[120]

Table 8.1. Chronic toxicities of insecticides to bobwhite quail: LC_{50} values, ppm in diet* (Heath et al., 1972).

Endrin	14	Phosphamidon	24
Aldrin	37	Fenthion	30
Dieldrin	39	Methyl parathion	90
Heptachlor	92	Temephos	92
Chlordane	330	Fenitrothion	160
DDT	610	Parathion	190
Endosulfan	800	Diazinon	245
Toxaphene	830	Phorate	370
Lindane	880	Azinphosmethyl	490
TDE	2200	Phosmet	500
Mirex	2500	Demeton	600
Dicofol	3000	Trichlorfon	720
Tetradifon	>5000	Naled	2100
Methoxychlor	>5000	Malathion	3500
Paris green	480	Propoxur	210
		Carbaryl	>5000

* 5 days on treated diet <u>ad libitum</u>, plus 3 days on clean diet

At 1 lb/a, the maximum dosage of DDT employed against forest defoliators, an overall reduction of about 10% of the bird population was indicated in a large 600 sq mi operation against fir tussock moth in Idaho.[63] This dosage applied in Montana did not reduce the juvenile/adult ratio in three species of grouse.[62] In Idaho, prespray and postspray counts of breeding pairs showed a 10% reduction which was no different from that observed at the same time on an unsprayed control plot.[2] No changes in bird population were noted with 1 lb/a applied to coniferous forest in Ontario or to deciduous forest in Maryland and Pennsylvania. Evidence for the safety of 0.5 lb/a was obtained in Illinois and New Jersey.[13] The effects of the single yearly 0.5 lb/a application used in New Brunswick to combat the spruce budworm with DDT were found to be as follows:[88]

Moved into adjoining unsprayed areas: myrtle warbler, magnolia warbler, least flycatcher, olive-backed thrush.

Found in fewer numbers the following year because of destruction of the budworm that had originally attracted high populations: Blackburnian warbler, bay-breasted warbler.

Population level unaffected the following year: Tennessee warbler, black-throated green warbler, black-headed vireo, winter wren.

Populations unusually high on the sprayed area in the year of treatment: brown creeper, black-capped chickadee.

It was concluded by wildlife experts[46] and by the U.S. Fish and Wildlife Service in 1961 that annual sprays of DDT at 1 lb/a or less were without undesirable side effects on birds.

The treatment of an area with DDT at 2 lb/a for four successive years resulted in population reductions in 3 out of the 26 species inhabiting it, namely the American redstart, parula warbler, and red-eyed vireo, which were reduced by 30–45%.[109]

Orchard Spraying. The DDT treatments formerly applied to apple orchards had little effect on bird populations, although there have been occasional deaths of seed-eating pheasants in the Yakima Valley of Washington and the Okanagan Valley of British Columbia.[113]

In a 100-acre apple orchard at New Paltz, New York sprayed annually with DDT, dieldrin, azinphosmethyl, and carbaryl, the robin population was suffering no undue mortality and the clutch sizes were normal. Since the earthworm and soil-arthropod population in the orchard was only one-fifth of that outside, the birds fed in the surrounding area, particularly on unsprayed cherries. Their average brain concentrations in 1966 and 1967 were 2.4 ppm DDE, 1.4 ppm DDT, and 0.6 ppm dieldrin. A single robin found in tremors had a brain concentration of 2.6 ppm dieldrin and 15 ppm t-DDT.[73]

Shade-Tree Spraying. The application of DDT suspensions to American elm shade-trees to control the bark beetle vectors of the Dutch elm disease has been responsible for killing large numbers of American robins, along with myrtle warblers in Atlantic coast towns and bronzed grackles in the Midwest. At doses of 1–5 lb DDT per tree at Urbana, Illinois, the soil contamination amounted to 5–20 lb/a or 5–10 ppm DDT. The earthworms picked up sufficient DDT residues, namely 50–200 ppm of which about one-third had become DDE, that some 100 of them would contain a lethal dose (ca. 3 mg) for a robin.[8] On the campus of Michigan Stage University, the robin population was brought down from an original 370 down to 15, and these raised virtually no broods at all.[136] In Wisconsin towns sprayed against Dutch elm disease, the average mortality rate of robins was 87%, to be compared with the normal 50% or so that die during the season.[57] Here the surviving population amounted to about 20% of the normal, and as the territories became vacant more robins moved into the deathtrap thus created.[68] The substitution of methoxychlor on the Madison campus allowed the robin population to return from 3 pairs up to 29 pairs.[69] No less than 94 species of birds have been found dead or dying in areas treated with DDT for Dutch elm disease control.

Physiological Impact of DDT on Bird Populations

Effects through Food.   The instances of bird kills in woodland sprayed with DDT are due to secondary poisoning by the oral route, not to direct contact poisoning. Although the direct spraying of nestlings in their nest with DDT at 5 lb/a had no effect,[94] feeding them with budworms taken from an area sprayed at 1 lb/a caused 20% mortality, vireos and passerines being especially susceptible.[47] Indeed, DDT-treated grain has successfully controlled pest populations of house sparrows and starlings.

DDT in the diets of pheasant had little or no effect on egg production or fertility, but hatchability and chick survival was reduced at concentrations of 100 ppm or more.[5] With bobwhite quail on feed containing 100 ppm, the egg production was normal, but the fertility and hatchability were reduced, and chick survival was eventually nil.[34] High dietary doses of DDT have reduced sperm production in cockerels[3] and the bald eagle.[85] The decline in reproductive success of woodcock noted in the DDT-sprayed areas of New Brunswick in the early 1960s was thought to have been due to the DDT residues derived from their diet of earthworms, but they had accumulated considerably more heptachlor during their winter sojourn in the southern United States.[140]

Metabolism of DDT.   Birds metabolize DDT principally by dehydrochlorinating it to DDE, but they also dechlorinate it to DDD; experiments with liver slices or homogenates from pigeons showed that only DDE is produced under aerobic conditions, while DDD is produced under anaerobic conditions perhaps by a nonenzymic process.[17] These two alternative routes have been demonstrated also in the brown-headed cowbird and domestic fowl.[121] In pigeons (rock doves) p,p'-DDT, with a half-life of 28 days, is converted mainly to DDE which is scarcely metabolized further and has a half-life of

250 days in the pigeon; there is also some conversion to DDD, which with a half-life of 24 days is further dehydrochlorinated to DDMU and a little DDMS.[7] In cowbirds DDD is found in all tissues of DDT-dosed birds,[122] but DDE is at least three times as abundant (see Table 8.3) and increases greatly with time in the body.[123] In domestic chicks, DDE is converted to no metabolite except a little DBP, but DDD is converted from DDMS through DDNU, DDOH, DDA, and DDM to DBP.[1] Higher levels of DDE are found in the males, as found in wing surveys of mallard and black duck, presumably because this metabolite is partly eliminated from the female in her eggs.[54] Older birds contain higher *t*-DDT residues than younger, as determined by comparing adults with juveniles in two species of North American cuckoos.[51]

DDT Poisoning and the Brain. The minimum content of DDT in the brain at which death occurred was 50 ppm for American robins and 60 ppm for house sparrows,[10] while for female ring-necked pheasants it was 14 ppm.[66] Lethal brain levels of DDE were around 500 ppm in cowbirds, but considerably lower in pigeons; the lethal level of DDD for cowbirds is 65 ppm.[121] The average brain residues of DDT (+ DDD) in birds killed by DDT (Table 8.2) show that the brain of the pheasant is the most susceptible and that of the *Coturnix* the least susceptible, but it should be observed that the standard deviations were very wide.[134] Double-crested cormorants, fed on a diet containing a 3:3:4 mixture of DDE, DDD, and DDT for 9 wk had accumulated 14 ppm DDE, 11 ppm DDD, and 2 ppm DDT in their brains if they survived, while those that died contained an average of 85 ppm DDD, the compound considered the best indicator of the intoxication.[50] DDE is also neurotoxic, American kestrels dying as a result of chronic DDE feeding when they had accumulated 200–300 ppm DDE in the brain, as compared to a brain content of 15 ppm in those that were surviving.[106] The succession of symptoms leading to death from DDT poisoning have been described for birds in general as the following: tail tremors, stumbling gait, then general tremors, immobility, and convulsions.[58]

Lethal Mobilization. High residues of DDT in birds may be mobilized to become lethal if they are starved or thrown into activity, which processes reduce the adipose fat and release the organochlorine into the body circulation to settle in the phospholipid of the nervous system. House sparrows with DDT residues of 800 ppm in the body fat were all right if well fed, but when they starved they died, the DDT mobilization engendering tremors which reduced the fat almost to zero and sent lethal concentrations into nerve and brain.[10,11] In brown-headed cowbirds, fed a diet containing 500 ppm DDT for 8 days during which period half of them died, half of the sur-

Table 8.2.   Average brain residue levels in birds killed by DDT (Van Velzen and Kreitzer, 1975).

	DDT LC_{50} ppm	DDT + DDD in brain, ppm	Original Reference
Ringnecked Pheasant	311	15	66
Blue Jay	415	23	59
Cardinal	535	27	59
Bobwhite Quail	1390	30	59
House Sparrow	415	43	59
Clapper Rail	1747	44	134
American Robin	--	58	123
Bald Eagle	80*	74	124
Brown-headed Cowbird	--	86	122
Japanese Quail	568	104	122

* ppm <u>dry</u> weight of diet in this case

vivors subsequently died during a subsequent 2–93 days on a clean diet.[123] Some of the cowbirds that had successfully survived 2 months on a diet with 40 ppm DDT would be thrown into tremors ending in death if they were unusually disturbed by workers entering their cage.[121]

The effect of starvation in reducing adipose fat and increasing the DDT residues was brought out (Table 8.3) by comparing cockerels placed on a restricted diet with those maintained on a full diet after treatment, and in which the brain residues (mainly DDE) were six times greater and approaching the lethal level.[36] With cowbirds, it was those which died during the 4-day posttreatment period on restricted rations that showed the tremendous increase in the DDT residues in the brain coupled with the great loss in adipose fat, as if the process of mobilization, once triggered by tremors, would go the whole way to lethal concentrations in the brain.[135]

DDT has itself the effect of augmenting the deposition of lipids in the liver; this effect increases at the expense of the adipose-tissue lipid when the DDT-treated bird is starved, and has been observed in the bobwhite quail[53] and in pigeons and coturnix quail.[38]

Residues Found in the Field.   Blue grouse in a Montana forest sprayed at 0.5 lb/a against the spruce budworm were found to contain 46 ppm *t*-DDT in their fatty tissue at the end of the summer, 22 ppm a year later, and 18 ppm in the third season.[100] American robins sampled in a Maine forest sprayed at 1 lb/a contained whole-body residues of 14 ppm *t*-DDT (ca. 70% of which was DDE) in the year of treatment, 4.5 ppm in the following year, and still 3.5 ppm 8 yr later. During the 9-yr period the *t*-DDT residues in the soil ranged from 0.6 to 1.6 ppm, the DDE content rising gradually from 6 to 12%; those in the earthworms ranged from 0.1 to 0.3 ppm, the DDT content rising from 15 to 21% during the period.[35]

In DDT-contaminated natural environments, insectivorous species such as sparrows and plovers contain much more DDE than plant-feeders such as ptarmigan or snail eaters such as ducks.[14] A cormorant colony near Lake Poinsett, South Dakota showed an average brain residue level of 0.09 ppm DDT and 0.56 ppm DDE.[50] Among British birds in the 1960s, fish-eaters such as herons and grebes accumulated more DDE (and dieldrin) than hawks and owls (Fig. 8.1), and much more than plant-feeders such as moorhens or seed-eaters such as wood pigeons.[95] Nevertheless, pheasants accumulated up to 3000 ppm in their body fat in California fields where DDT had been used in rice-planting operations.[65]

Table 8.3. **Effect of food deprivation on birds that have accumulated DDT from the diet (Ecobichon and Saschenbrecker, 1969; Van Velzen et al., 1972).**

	percent Lipid in Tissues		ppm Organochlorine in Brain		
	Adipose	Brain	DDT	DDD	DDE
Cockerels, after 15 wks on 0.25 ml/day and then 20 weeks on full rations or restricted diet					
Full Rations	68.3	6.9	10.3	4.8	29.5
Restricted Diet	14.8	6.9	47.7	21.2	219.4
Cowbirds, after 13 days on 100 ppm preparatory to 4 days on a restricted diet					
Before Restricted Diet	15.7	5.9	1.3	0.6	3.2
Died during the 4 days	3.8	6.3	23.0	8.7	29.6
Survived the 4 days	17.4	6.1	1.1	0.7	1.6
Kept on Full Rations instead	14.4	6.0	1.0	0.8	2.3

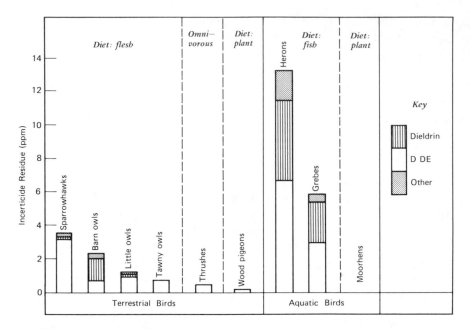

Fig. 8.1. Residues of DDE, dieldrin, and other organochlorines in British birds, contrasting plant eaters, flesh eaters, and fish eaters (from Moore and Walker, 1964).

From residue determinations made on warblers which had collided with television towers in Florida during their spring and fall migrations,[75] it was found that the concentrations of *t*-DDT (including DDD, DDE, and *o,p'*-DDT) in their brains were minuscule (Table 8.4). When the whole-body residues for these warblers, plus two vireo species and the catbird, were plotted against time, a steady decline could be detected in successive years, those in 1973 being one-quarter those in 1964.[74] However, a warbling vireo was found dying in tremors on the edge of a field which had been sprayed with DDT at 0.75 lb/a a month before, and its brain content of *t*-DDT was 460 ppm.[38]

Methoxychlor and PCBs. In contrast to DDT, methoxychlor is rapidly broken down in warm-blooded animals, so that residues are not found in birds when it is sprayed against Dutch elm disease, although present in the earthworms they eat.[70] When methoxychlor is fed to laying hens (Table 8.5), very little is found in their abdominal fat or their eggs, in contrast to its relative dicofol with which the accumulations are considerable.[90] Residues of methoxychlor induced in domestic fowl by feeding very high doses have

Table 8.4. Total DDT residues in brains and adipose fat of warblers migrating through Florida, 1964–1973 (Johnston, 1976).

	Adipose Fat		Brain
	Av. 1964-73	1964 Maximum	Av. 1964-73
Ovenbird	1.7	4.0	0.01
Black-&-White Warbler	2.1	---	0.03
Common Yellowthroat	3.8	6.5	0.06
American Redstart	3.4	10.4	0.17
Northern Waterthrush	8.7	---	0.17
Palm Warbler	4.9	---	0.27

Table 8.5. Extent to which organochlorine insecticides in their diet are accumulated by laying hens in their tissues and eggs, ppm (Cummings et al., 1966, 1967; McCaskey et al., 1968).

	Abdominal Fat	Liver	White Muscle	Eggs
0.45 ppm for 14 weeks				
Lindane	0.7	0.08	--	0.10
DDT	1.6	0.25	--	0.20
Endrin	3.6	0.35	--	0.35
Dieldrin	4.5	0.50	--	0.55
Heptachlor epoxide	5.1	0.61	--	0.48
12 ppm for 5 days				Yolk
Methoxychlor	0.9	--	<0.1	0.3
Isobenzan	10.6	--	0.2	0.5
Chlordane	11.0	--	2.1	3.1
Ovex	13.3	--	0.3	2.4
Dicofol	28.4	--	2.1	14.2

no adverse effect, and are rapidly lost when the birds are placed on clean feed.[102] The sulfonate acaricide ovex also accumulates in hens, as also do the organochlorines isobenzan and chlordane.[90] The cyclodiene derivatives endrin, dieldrin, and heptachlor (epoxide) accumulate to a greater degree than DDT (which is degraded to DDE and DDD which also accumulate), although in all cases a plateau is reached. Lindane, the gamma isomer of BHC, accumulates the least,[30] being the only one of the organochlorines tested to disappear completely after 1 month's withdrawal from exposure to 0.05 ppm in the diet. Hens could tolerate 100 ppm lindane in the diet for a short period without an obvious effect on egg weight or hatchability, although their egg production fell off sharply.[137]

The polychlorinated biphenyls were found to have very low toxicity to quail, the Aroclor series showing the following values (in ppm):

A. 1221	A. 1242	A. 1248	A. 1254	A. 1260
>12,000	>6000	4844	2898	2186

Thus although the toxicity increases with the percentage chlorine in the PCB product (indicated by the last two digits of its number), even those with 54 or 60% chlorine are scarcely toxic.[60] However Aroclor 1254, which was one of the most widely used PCB products, caused embryonic mortality and induced chromosomal alterations in ring doves.[103]

B.   DIRECT EFFECTS OF CYCLODIENE-TYPE INSECTICIDES

Aldrin, Dieldrin, and Heptachlor

The cyclodiene insecticides aldrin and dieldrin have a much higher acute toxicity to birds than toxaphene or lindane; the LD_{50} values (in mg/kg) were as follows:

	Aldrin	Dieldrin	Toxaphene	Lindane
Bobwhite Quail	4	13	90	200
Mourning Dove	13	45	225	375

the quail being up to 30 times more susceptible than mourning doves.[32] Endrin, aldrin, dieldrin, and heptachlor also have a considerable chronic toxicity to birds (Table 8.1). Growing bobwhite quail eventually all die on a diet containing 0.5 ppm aldrin.[34]

Seed Dressings. The practice of coating seed with aldrin, dieldrin, and heptachlor to protect spring-sown wheat against the bulb fly in England resulted in extensive kills of the wood pigeon feeding on the seeds; this the farmers often regarded as a welcome form of pest control, but other birds including pheasants and a few hawks and owls were also killed. For example, on a 1480-acre area in Lincolnshire the kills detected in 1961 were 5668 wood pigeons, 118 stock doves, 89 pheasants, and 59 rooks.[29] Feeding experiments confirmed that it was the cyclodiene seed-dressings that caused the mortalities,[133] and such dressings for spring wheat were, therefore, discontinued. Experience in the Netherlands confirmed the secondary poisoning of the predator birds; corpses of buzzards and long-eared owls predaceous on the wood pigeon, and of the European sparrow hawk predaceous on finches, were found along with the dead wood pigeons soon after the dressed wheat seed was sown in March.[44] Dieldrin residues of the order of 10 ppm and smaller amounts of heptachlor epoxide were found in the livers of these birds, as well as in the barn owl, goshawk, kestrel, and peregrine falcon.[79] It was noted in sharptail grouse that birds sublethally poisoned with dieldrin were unusually susceptible to predation by hawks and coyotes.[92]

Dieldrin. The application of dieldrin at 2 lb/a to control the larvae of the Japanese beetle in fields in the midwest caused many bird deaths, and elimination of bobwhite quail.[24] Applied at 3 lb/a over a large area around Sheldon, Illinois, dieldrin caused about 80% mortality of songbirds, especially robins and meadowlarks, but of pheasants, grackles, and starlings as well[116]; during the summer following the treatment in April, the numbers of birds seen per hour in the cornfields were as follows:

	Horned Lark	Dickcissel	Meadowlark	Grasshopper Sparrow
Untreated	2.04	0.35	0.13	0.09
Treated	0.91	0.19	0.08	0.04

The practice in tropical Africa of spraying tree trunks and vegetation with 5% dieldrin emulsion from knapsack sprayers to control *Glossina* tsetse flies also involves bird mortality. Among eight species found dead within a week of the treatment of an area in South Nyanza, Kenya, brain residues of 15–20 ppm dieldrin were found in the moustached warbler *Melocichla*, the greybacked glasseye *Camaroptera*, and the arrow-marked babbler *Turdoides*, while corpses of kingfishers, turtledoves, and swallows were found in Botswana. Two months after spraying operations along the shores of Lake Victoria in Kenya, dieldrin residues were high in the king-fisher *Ceryle rudis* but not in the cormorant *Phalacrocorax africanus.*[80]

Antimosquito spraying with aldrin or dieldrin at only 0.2 lb/a in California killed some songbirds, ducklings, and redwing blackbirds.[113] Treatment of a wheatfield in Illinois with dieldrin at 0.2 lb/a caused red-winged blackbirds in an adjoining hayfield to abandon their nests; in a con-trol area 2 miles downwind of the application, half of the nests were abandoned where the eggs contained more than 6 ppm dieldrin.[49] Considerable mortality among birds was also caused by soil applications of dieldrin or heptachlor to control the white-fringed beetle in southeastern United States.[121] In May and June 1970 there were deaths of month-old whitefaced ibis in three colonies on the Gulf coast of Texas; since dying birds were found to have brain concentrations of 0.3 ppm DDE and 7.0 ppm dieldrin, it was concluded that they had been poisoned by the invertebrate food their parents had brought them from the aldrin-treated ricefields along the littoral. The symptoms of poisoning observed—muscular weakness, immobility, body tremors, neck arched over back, wing-beat convulsions—appeared to bear this out.[41]

Experiments on *Coturnix* quail showed that dieldrin becomes fatal to birds when the concentrations in the brain reach 4–5 ppm.[128] Even where the gallinaceous birds survived, their reproductivity was reduced. Pheasants fed dieldrin in sublethal concentrations (50 ppm) in the diet showed reductions in egg fertility and egg production which added up to a halving of the reproductive success.[45] A more serious aftereffect was the mortality among the hatched chicks, which was appreciable among pheasants where the mothers had been fed 2 ppm dieldrin, and among bobwhite quail on breed-ing diets containing only 1 ppm dieldrin.[34] *Coturnix* chicks fed sublethal concentrations of endrin, dieldrin, or chlordane progressively lost their avoidance response to a hawk silhouette.[82] When dieldrin was fed to hen pheasants through two generations, the second-generation hens laid eggs with lower fertility and hatchability than the first.[9]

Heptachlor. The inauguration in 1957 of a campaign to eradicate the imported fire ant by means of heptachlor (or alternatively dieldrin) at 2 lb/a led to extensive bird kills, although heptachlor is half as toxic as dieldrin to birds. On four farms in Louisiana, 222 corpses of 28 different species were found after the treatment.[119] In Alabama heptachlor at 2 lb/a significantly reduced the songbird population (DeWitt et al.[105]), while either heptachlor or dieldrin applied in granules at this dosage entirely eliminated the resident bobwhite quail.[23] In Texas, the bird populations decreased by 85%, and the nesting success dropped by 90%. In 1960 the dosage of heptachlor was adjusted downwards to 0.25 lb/acre but there was still a 17% mortality among quail, as compared to 61% with the former 2-lb/a dosage.[83] The sprays or granules of heptachlor were then replaced with baits containing chlordecone (Kepone) or mirex, at area dosages amounting to only 0.01 lb/a. This eradication program, against a type of insect that is very hard to kill, eventually was abandoned, its undesirable side effects having been forcefully reviewed by W. L. Brown (1961) and Rachel Carson (1962).

Although 1 yr was sufficient for virtually complete recovery of bird numbers from applications of heptachlor at 0.25 lb/a in Mississippi,[39] with 2-lb/a applications in Georgia the bobwhite populations had not yet recovered their numbers 3 yr later.[111] The recovery from the seed dressing kills in England was very rapid; a wood pigeon population in East Anglia, which had suffered a 20% reduction in 1961 due to aldrin and heptachlor seed-dressings, recovered to a normal level in the following year when the dressings were suspended.[99] The use of aldrin seed-dressings on rice sown in Louisiana ricefields evidently did not harm purple gallinules or common gallinules, since the egg clutches taken from the treated area were just as large and showed as high a hatchability and hatchling survival as those taken from untreated marshy areas, nor was there any decrease in eggshell thickness in eggs that had a high content of the metabolite dieldrin.[43]

Lethal Residues. Woodcock that had overwintered in fire ant areas treated by heptachlor showed an average increase in whole-body heptachlor residues from 0.3 ppm in 1961 to 7.2 ppm in 1963.[140] When woodcocks were fed on earthworms containing 3 ppm heptachlor, half of them were killed in 38 days; none died when fed on earthworms containing 0.65 ppm for 60 days, but when these woodcocks were put on quarter-rations without heptachlor, half of them succumbed when they had lost 20% of their body weight.[127] With dieldrin, European song-thrushes accumulated whole-body residues to the same concentration as that in the contaminated earthworms on which they were fed; those fed high dieldrin dosages until they died showed an average brain residue of 17 ppm, the premortem loss of adipose fat having raised it to this figure from the critical level of 3 ppm accumulated in the brain from the diet.[72] In birds found dead in an area in Pennsylvania treated at 3 lb/a against the European chafer the average brain levels were 7.5 ppm in the woodcock, 9.6 ppm in the American robin, and 12.5 ppm dieldrin in the starling.[128]

In the 3-yr period 1966–1968, of the 69 bald eagles found dead or dying in the United States and analyzed for residues, 8 of them had 4–10 ppm dieldrin in their brains.[96] By contrast, golden eagles had seldom more than a trace of dieldrin in the brain.[107] In the 7-yr period 1964–1970, of 153 bald eagles analyzed, 15 probably died from dieldrin, 1 from DDE, and 1 from DDT poisoning; a peregrine falcon found dead at Manteo, North Carolina in 1973 contained 5.4 ppm dieldrin in the brain besides 34 ppm DDE and 55 ppm PCBs.[108] Of four ospreys found dead or dying in Connecticut, one had 7.5 ppm dieldrin in the brain.[138]

Other Cyclodiene Insecticides

Endrin is sufficiently toxic to birds that it has been used as a coating to protect seeds from avian attack, or as a direct toxicant in wicks or poly- ethylene polysulfide sticky coatings on buildings. Nevertheless, orchard applications of endrin at 1.25 lb/a designed to control mice killed only the occasional quail or pheasant.[139]

Isobenzan (Telodrin) is perhaps the most toxic of all organochlorines to birds, death supervening in *Coturnix* after a few days on 10 ppm in the diet, when the brain residues reached 1 to 2 ppm (Koeman, 1971[121]). The residues of isobenzan and dieldrin acquired in that area have also resulted in the death of female eider ducks during the period of incubating their eggs when they take little food.[121] Contamination of the Dutch coast with isobenzan and dieldrin from a plant at the mouth of the Rhine River also led to abnormal mortality of chicks of the Sandwich tern in 1964 and 1965,[81] and was a factor in the abrupt decline of the species at that period.[141]

Endosulfan, although of low toxicity to bobwhite quail, has a higher acute toxicity than dieldrin or toxaphene to the mallard duck; the poisoning symptoms are a high carriage, with wings folded over the back and tail pointed downwards, occasional wing tremors, and a goose-stepping ataxia with occasional falls.[131] The chronic toxicity of endosulfan resembles that of lindane and toxaphene in being low (Table 8.1), at least to bobwhite quail.

Toxaphene applied at 1.5 lb/a to control grasshoppers originating from prairie wetlands in North Dakota caused only small losses of ducks and coots (Knedel, 1951[27]), but at 2 lb/a in oil solution it resulted in zero sur- vival of the young of ducks, coots, rails, and terns.[52] Chlordane at 1 lb/a reduced the production of juvenile pintail, shoveler, and blue-winged teal by not less than 50%,[52] while aldrin at only 2 oz/a killed some 25% of the juvenile ducks present (Knedel, 1952[46]). Toxaphene at 300 ppm in the diet has reduced egg production and hatchling viability in pheasants, though not as much as with dieldrin or DDT at lower dosages.[45] Irrigation water from fields treated with toxaphene at 2 lb/a draining into ponds in the Lower Klamath wildlife refuge, California has caused extensive mortality in fish- eating birds[77]; this is described in Chapter 10.

Kepone and Mirex

Chlordecone (Kepone) and mirex are organochlorines of involved ring structure that are highly effective when applied in insect baits, especially against ants and cockroaches. Their acute toxicity, particularly that of mirex (Table 8.7), is very low for birds; but they are extremely persistent in the body, the half-life of mirex in grackles being 28 wk, which is slightly more persistent than DDE. High doses of chlordecone have reduced spermatogenesis and induced female plumage in male pheasants (DeWitt et al., 1963[26]). Residues accumulate in eggs, and a reduction in hatchability and hatchling survival becomes evident when chlordecone at 150 ppm, or mirex at 600 ppm, are fed to the mother hens.[101] In bobwhite quail, however, mirex in the diet at 40 ppm for a year followed by 10 ppm in the second year, although causing 75% mortality, gave no reductions in egg production, hatchability, or chick survival.[55] With bronzed grackles, mirex at 250 ppm in the diet killed 50% of the birds in 38 days, and the lethal level in the brains was 210–458 ppm for grackles, cowbirds, redwings, and starlings.[126] In areas treated with mirex baits in Mississippi, the residues in the adipose fat ranged from 1 to 56 ppm in American robins, 20 to 60 ppm in brown thrashers, and 5 to 104 ppm in blue jays, but the brain residues in the insectivorous birds were never more than 1.5 ppm.[6] Mirex does not show lethal mobilization from fat to brain to any considerable extent.[126]

C. EFFECTS OF ORGANOPHOSPHORUS AND CARBAMATE INSECTICIDES

OP Compounds

A number of these valuable insecticides have high acute toxicity for birds (Tables 8.6, 8.7), although species differ greatly in their susceptibility.[132] Since it is unusual to find residues of these insecticides in birds, the proof of poisoning by these anticholinesterases in incidents of bird kills must rely on the finding of fresh corpses with low brain ChE levels.[125]

Parathion, which was the first OP compound to be widely used, diminished the swallow population in Japanese ricefields when applied to control the rice stem borer, and dead swallows were found.[71] A citrus orchard in South Africa treated with parathion at 7.5 lb/a yielded 800 corpses of 27 species of birds.[20] The use of parathion and other highly toxic OP compounds to control aphids on cole crops in Britain caused severe bird casualties in the early 1950s.[29] Japanese quail fed a diet containing 170 ppm parathion for 5 days would die when their brain ChE levels had diminished to 55–75% of the normal.[86] Mallards feeding on a diet containing 25 ppm parathion showed nearly 50% mortality after a month, and the survivors had lost weight; but mallards confined to ponds to which parathion was applied at 0.5 lb/a every fortnight were unharmed.[78] A diet of 10 ppm parathion over the 90-day egg-laying period of mallards did not affect the

Table 8.6.   **Acute oral toxicities of insecticides to pheasant and other birds: LD_{50} values in mg/kg, 14-day observation period (Tucker and Crabtree, 1970; Tucker and Haegele, 1971).**

	Ring-necked Pheasant	Coturnix Quail	Chukar Partridge	Mallard Duck	Rock Dove (Pigeon)
Organophosphorus insecticides					
Monocrotophos	2.8	3.7	6.5	4.8	2.8
Dicrotophos	3.2	4.3	9.6	4.2	2.0
Demeton	8.2	8.5	15	7.2	8.4
Parathion	12	5.9	24	2.0	2.5
Chlorpyrifos	12	17	61	75	27
Fenthion	18	11	26	5.9	4.6
Temephos	32	84	270	90	50
Methyl demeton	42	84	113	54	15
EPN	53	5.2	14	3.1	5.9
Carbamate insecticides					
Mexacarbate	4.5	3.2	5.2	3.0	65
Propoxur	20	28	24	12	60
Landrin	52	71	60	19	168
Mobam	228	668	237	1130	273
Carbaryl	>2000	2290	--	>2180	2000
Organochlorine insecticides					
Dieldrin	79	70	23	381	27
DDT	1300	840	--	>2240	>4000

production or fertility of eggs, but the shells were significantly thinner.[97] Diets of 8 ppm parathion similarly administered to gray partridge in three successive generations resulted in heavier chicks in each generation, and thicker eggshells than normal in the F_3 generation.[98] Methyl parathion at 0.5 lb/a has caused sickness in birds and some mortality, while applications at 3 lb/a have killed pheasants and sickened many more in this respect being more toxic than parathion itself.[125]

Fenthion, less toxic than parathion to mammals, is equitoxic with it to

Table 8.7. Acute oral toxicities of insecticides to mallard and other birds: LD$_{50}$ values in mg/kg, 14-day observation period (Tucker and Crabtree, 1970).

	Mallard Duck	Ring-necked Pheasant	Other Birds
Organophosphorus insecticides			
Phorate	0.6	7.1	Chukar 13
Phosphamidon	3.0	---	Chukar 9.7, Pigeon 2.5, Dove 3.0
Diazinon	3.5	4.3	---
Tepp	3.6	4.2	Chukar 10
Mevinphos	4.6	1.4	---
Methyl parathion	10	8.2	---
Dimethoate	42	---	---
Naled	52	---	Canada goose 37
Azinphosmethyl	136	75	Chukar 84
Fenitrothion	1190	56	Bobwhite 27
Stirofos	>2000	2000	Chukar >2000
Carbamate insecticides			
Carbofuran	0.4	4.2	Bobwhite 5.0
Methomyl	16	15	---
Aminocarb	22	42	---
Organochlorine insecticides			
Endrin	5.6	1.8	Pigeon 3.5, Sharptail Grouse 1.1
Endosulfan	33	---	---
Aldrin	--	17	Bobwhite quail 6.6
Toxaphene	70	40	Canada goose 85
Mirex	>2400	1500	---
Botanical insecticides			
Nicotine sulfate	587	1600	Coturnix 530, Pigeon >2000
Rotenone	>2000	>1400	---

birds. Both compounds have been used to control the weaverbird *Quelea quelea,* a pest of cereal crops in tropical Africa. Fenthion used as a mosquito larvicide gave the same results as parathion when its hazards were tested on mallards,[78] and when applied at 0.1 lb/a on a wildlife refuge in California it was harmless to pintail and pheasants also.[19] But an aerial application of undiluted fenthion at 1.3 oz/a by the ULV method over 600 ha at Grand Forks, North Dakota at the peak of the spring migration period in May killed many warblers. No less than 453 corpses of 37 species of birds were found, 196 being Tennessee warblers, 88 yellow warblers, and 50 blackpolls; those taken and tested while still warm showed very low ChE activity in the brain. It is probable that the airspray droplets had become concentrated in the thickets along the Red River where the warblers had collected.[117] Similar extensive mortalities have been observed in Louisiana, where insectivorous birds are susceptible to secondary poisoning from insects killed in mosquito-control operations.[125] Occasional bird mortalities were also observed with ULV fenthion applied against mosquitoes in California, and in Ohio where the dead killdeers, redwing blackbirds and spotted sandpipers were found to have a marked brain ChE inhibition.[37] Applied to ricefields in Japan, fenthion ULV sprays killed sparrows.[71]

Chlorpyrifos, effective as a mosquito larvicide at 0.01 lb/a, had no effect on mallards, pintail, and pheasants when applied to a wildlife refuge at 0.10 lb/a,[19] and on a Texas salt marsh treated at 0.025 lb/a the bird counts made on three occasions on the day after the spray were no different for those before it.[87] Its chronic toxicity is about the same as parathion and fenthion, chlorpyrifos in the diet causing domestic chicks to show poor weight gains at 20 ppm, and 75% blood ChE reduction with some mortality at 80 ppm.[16] Of other OP compounds used as mosquito larvicides, fenitrothion at 25 ppm in the diet of mallards caused no mortality and allowed weight gain.[78] Temephos (Abate) showed a chronic LC_{50} of 92 ppm for bobwhite quail in one investigation (Table 8.1) and of 1540 ppm in another. Nevertheless, it was not as harmless as bromophos or stirofos, as shown by the following chronic LC_{50} figures (in ppm) for common North American birds[58]:

	Cardinal	House Sparrow	Blue Jay
Temephos	76	47	30
Bromophos	427	640	223
Stirofos	2835	1000	995

Malathion is probably the safest OP compound for birds. Applied as a mosquito adulticide at 3 oz/a from aircraft by the ULV method, malathion killed no birds, the house sparrow population remaining normal and showing no brain ChE inhibition.[61] When malathion-treated seeds were the only source of food available, house sparrows could be fatally poisoned when their brain ChE had been about four-fifths inhibited, but they ate only one-fifth as much treated seed as they did when it was untreated.[93] Malathion at 500 ppm in the diet of laying hens did not decrease weight gain and only slightly decreased egg population (Table 8.8), while at 15 ppm there was no effect on egg production or hatch, nor on chick survival.[91] A more recent experiment revealed slightly decreased egg production and hatchability at 1 ppm but not at 0.1 ppm, but no decrease in shell thickness even at 10 ppm.[114] When Florida citrus groves were treated with malathion bait-sprays at 0.6 lb/a four times a year to eradicate the Mediterranean fruit fly, there was not a single confirmed death of any bird over the entire state.[15,27] Application of malathion at 1 lb/a to a California forest resulted in no immediate reduction in bird population, although it later declined due to the disappearance of insects as a food supply.[67] When malathion was applied at 2 lb/a to an Ohio forest there was no measurable effect on the population of birds, except that they were unusually quiet for the first 2 days after the spray.[104] Sharptail grouse, for which the malathion LC_{50} was 220 mg/kg, when taken and treated at sublethal levels before being returned to the field displayed exaggerated aggressiveness on the breeding grounds for a day or two.[92]

Monocrotophos (Azodrin), one of the most toxic OP compounds for birds, was applied to control cotton pests over almost a million acres in California in 1967; 22 cases of bird kills were reported, including pheasants, doves, quail, killdeer, meadowlarks, horned larks and blackbirds, as well as sparrows.[12] Monocrotophos has also killed large numbers of birds on cotton in Arizona and in potato fields in Florida. Dicrotophos (Bidrin) has also seriously affected, but not killed, wild birds when applied at 2 oz/a.[125] When laying hens were fed 5 ppm of either of these OP compounds for 3 wk, egg production remained normal, despite a 20–25% decrease in the brain cholinesterase level.[118]

Table 8.8.   Effect of OP compounds on growth and reproduction of laying hens (Ross and Sherman, 1960).

	ppm*	Food Consumption	Weight gain gms	Egg Production Hen-day %
Coumaphos	40	104	162	48
Trichlorfon	60	81	196	29
Diazinon	100	85	-52	36
Malathion	500	94	328	44
Fenchlorphos	100	106	176	47
Control	0	100	260	51

* Dietary concentration for weeks 8-29; weeks 1-7 at lower concentration

Phosphamidon is also very toxic to birds, and when applied at 1 lb/a to a Montana forest as a substitute for DDT it reduced the bird population by 75–80%. Many dead blue grouse were found, with blood ChE reductions of 50–60%, the poisoning being probably due to ingestion of seeds into which this systemic OP had been absorbed[40]; although the bird counts had returned to normal 6 wk after the spray, phosphamidon was discarded as a forest insecticide.[40] The Canadian authorities in New Brunswick had the same experience of bird kills, which sometimes resulted merely from contact of the bird's feet with sprayed twigs.[42] Applied at 1 kg/ha to a 1050-ha Swiss forest to control the larch bud moth, phosphamidon caused 70% reduction in the bird population, 76 corpses of 18 species (principally chaffinches and thrushes) being found over 20 ha; the following year the bird population returned to the normal variety and numbers of species.[115] Diazinon is as dangerous as phosphamidon for ducklings. When laying hens were fed a diet containing 100 ppm diazinon, they lost weight and their egg production was greatly reduced (Table 8.8). Even at 0.1 ppm in the diet of laying hens, diazinon decreased egg production and hatchability and increased embryonic mortality; however, a diet of 10 ppm diazinon for 10 wk did not result in thinning of eggshells.[114]

Azinphosmethyl, the insecticide of choice to control codling moth in apple orchards, is comparable to fenitrothion in its relative safety to birds; although its acute toxicity is higher (Table 8.7), its chronic toxicity is lower both to bobwhite quail (Table 8.1) and to mallards.[78] Bobwhites caged in a large orchard area sprayed from the air six times in the season with azinphosmethyl at rates between 12 oz and 1 lb per acre showed no ill effects (nor did mourning doves or shore birds which frequented a stock tank in the area), and their food supply taken from the sprayed orchards was found to be harmless. However, a 25% reduction in reproductive success was noted in bobwhites fed 5 ppm azinphosmethyl and in pheasants fed at the 10 ppm level.[4] Reduction in growth rate was noted in bobwhite quail fed 20 ppm azinphosmethyl, but there was no reduction in the growth of male *Coturnix* quail even at dietary levels of 500 ppm.[48] Yet in ponds treated six times with azinphosmethyl at 0.4 lb/a twice the dosage rate which kills bass and sunfish, mallards showed no ill effects.[4]

Neurotoxic Symptoms

For acute poisoning with a lethal oral dose of azinphosmethyl as a representative OP compound, the symptoms in ring-necked pheasant and other birds included regurgitation, goose-stepping, ataxia, wing droop, wing spasms, anal sphincter tenesmus, diarrhoea, myasthenia, dyspnea, prostration, and death with opisthotonus or wing-beat convulsions.[131] With chronic poisoning, the most obvious succession of symptoms in birds are a reduction in activity, fluffing of the feathers, followed by a general lethargy.[58] The amount of dietary exposure to azinphosmethyl is best reflected, in bobwhite and Japanese quail, by the degree of reduction of brain cholinesterase.[48] In the ring-necked pheasant, a relatively tolerant species, chronic exposure for 6 wk to 100 ppm in the diet of diazinon, dimethoate, phorate, or methyl demeton induced brain ChE reductions approximating 40% and plasma ChE reductions ranging up to 80%; it was concluded that to establish poisoning with such OP compounds as the cause of death a 90% reduction in brain ChE should be the criterion.[18]

Carbamates

Applications of carbofuran (Furadan) or aldicarb (Temik) can kill birds, although they are safe if used in granules that are then covered with soil.[125] Indeed, methiocarb (Mesurol) is employed to treat seeds and to spray

Table 8.9. Degree of cumulativeness of insecticides as oral poisons for mallards, as derived from the ratio of acute to chronic toxic dose (Matsumura, 1975).

	Chronic MLD* mg/kg/day	$\dfrac{\text{Acute } LD_{50}}{\text{Chronic MLD}}$
DDT	50	45
Dieldrin	1.25	76
Endrin	0.12	45
Parathion	4.5	4.0
Chlorpyrifos	2.5	30
Temephos	2.5	36
Carbaryl	125	17
Propoxur	2.0	2.0
Zectran	1.2	2.4

* The minimum lethal dose, the lowest daily oral intake that kills 1 or 2 out of 6 birds in a 30-day period (Tucker and Crabtree 1970)

cherry and peach orchards to protect them against birds feeding on them. Some carbamates have a high acute toxicity to birds, but they are all characterized by a low chronic toxicity. As compared to the OP insecticide chlorpyrifos, the carbamate mexacarbate showed a very high ratio of the chronic LC_{50} to the acute LD_{50}, as follows[76]:

		Acute, mg/kg	Chronic, ppm
Chlorpyrifos	Coturnix	16	500
	Mallard	75	360
Mexacarbate	Coturnix	3.2	500
	Mallard	3.0	1000

Comparison of the chronic minimum lethal dose (MLD) with the acute dose (Table 8.9) shows that propoxur is among the least cumulative (as contrasted) with dieldrin among the most cumulative) in that the daily administration of one-half of the acute LC_{50} is no more toxic to mallards than the full LC_{50} given in a single dose. Laboratory experiments have shown that high acute doses of carbaryl evoke the following symptoms in birds: ataxia, weakness, salivation, tachypnea, tremors, tetanic paralysis, coma, and convulsions.[131] But applications of carbaryl at 1.25 lb/a to control gypsy moth had no effect on the behavior, condition, reproduction or rearing of young birds in deciduous forests in New York State.[25]

REFERENCES CITED

1. M. B. Abou-Donia and D. B. Menzel. 1968. The metabolism of DDT, DDD, and DDE in the chick. *Biochem. Pharmacol.* **17**:2143–2161.

2. L. Adams, M. G. Hanavan, N. W. Hosley, and D. W. Johnston. 1949. The effects on fish, birds, and mammals of DDT used in the control of forest insects in Idaho and Wyoming. *J. Wildl. Manage.* **13**:245–254.

3. T. F. Albert. 1962. The effects of DDT on the sperm production of the domestic fowl. *Auk* **79**:104–107.

4. C. A. Anderson et al. 1974. Guthion (azinphosmethyl): Organophosphorus insecticide. *Residue Rev.* **51**:123–180.

5. J. A. Azevedo, E. G. Hunt, and L. A. Woods. 1965. Physiological effects of DDT on pheasants. *Calif. Fish Game* **51**:276–293.

6. K. P. Baetcke, J. D. Cain, and W. E. Poe. 1972. Mirex and DDT residues in wildlife and miscellaneous samples in Mississippi. *Pestic. Monit. J.* **6**:14–22.

7. S. Bailey, P. J. Bunyan, B. D. Rennison, and A. Taylor. 1969. The metabolism of DDT, DDD, DDE, and DDMU in the pigeon. *Toxicol. Appl. Pharmacol.* **14**:13–22, 23–32.

8. R. J. Barker. 1958. Notes on some ecological effects of DDT sprayed on elms. *J. Wildl. Manage.* **22**:269–274.

9. W. L. Baxter, R. L. Linder, and R. B. Dahlgren. 1969. Dieldrin effects in two generations of penned hen pheasants. *J. Wildl. Manage.* **33**:96–102.

10. R. F. Bernard. 1963. Studies on the effect of DDT on birds. *Publ. Mich. State Univ. Mus., Biol. Ser.* **2**:155–191.

11. R. F. Bernard. 1966. DDT residues in avian tissues. *J. Appl. Ecol.* **3**(Suppl):193–198.

12. R. van den Bosch. 1969. The toxicity problem—comments by an applied insect ecologist. In *Chemical Fallout.* Ed. M. W. Miller and G. G. Berg. Charles C Thomas. pp. 97–112.

13. A. W. A. Brown. 1951. Insecticides and the balance of animal populations. In *Insect Control by Chemicals.* Wiley. pp. 720–780.

14. N. J. Brown and A. W. A. Brown. 1970. Biological fate of DDT in a sub-arctic environment. *J. Wildl. Manage.* **34**:929–940.

15. W. L. Brown. 1961. Mass insect control programs: Four case histories. *Psyche* **68**:75–109.

16. R. A. Brust, S. Miyazaki and G. C. Hodgson. 1971. Effect of Dursban in the drinking water of chicks. *J. Econ. Entomol.* **64**:1179–1183.

17. P. J. Bunyan, J. M. J. Page, and A. Taylor. 1966. *In vitro* metabolism of *p,p′*-DDT in pigeon liver. *Nature* **210**:1048–1049.

18. P. J. Bunyan, D. M. Jennings, and A. Taylor. 1969. Organophosphorus poisoning, chronic feeding of some common pesticides to pheasants and pigeons. *J. Agric. Food Chem.* **17**:1027–1032.

19. W. E. Burgoyne. 1968. Studies of effects of Dursban and fenthion insecticide on wildlife. *Down to Earth* **24**(2):31–33.

20. W. Buettiker. 1961. Ecological effects of insect control on bird populations. *Int. Union Conserv. Nat., Tech. Meet. Warsaw 1960, Proc.* **8**:48–60.

21. R. Carson. 1962. *Silent Spring.* Houghton Mifflin. pp. 161–167.

22. A. Chappellier and M. Raucourt. 1945. Les rapports entre les traitements arsenicaux antidoryphoriques. *Ann. Epiphyt.* **8**:1–39.

23. S. G. Clawson. 1959. Fire ant eradication—and quail. *Ala. Conserv.* **30**(4):14.

24. S. G. Clawson and M. F. Baker. 1959. Immediate effects of dieldrin and heptachlor on bobwhite. *J. Wildl. Manage.* **23**:215–219.

25. P. F. Connor. 1960. A study of small mammals, birds, and other wildlife in an area sprayed with Sevin. *N.Y. Fish Game J.* **7**:26–32.

26. O. B. Cope. 1971. Interactions between pesticides and wildlife. *Annu. Rev. Entomol.* **16**:325–364.

27. O. B. Cope and P. F. Springer. 1958. Mass control of insects: The effects on fish and wildlife. *Bull. Entomol. Soc. Am.* **4**:52–56.

28. C. Cottam and E. Higgins. 1946. DDT: Its effect on fish and wildlife. *J. Econ. Entomol.* **39**:44–52.

29. S. Cramp. 1973. The effects of pesticides on British wildlife. *Br. Vet. J.* **129**:315–323.

30. J. G. Cummings, K. T. Zee, V. Turner, F. Quinn, and R. E. Cook. 1966. Residues in eggs from low level feeding of five chlorinated hydrocarbon insecticides to hens. *J. Assoc. Off. Anal. Chem.* **49**:354–364.

31. J. G. Cummings, M. Eidelman, V. Turner, D. Reed, and K. T. Zee. 1967. Residues in poultry tissues from low level feeding of five chlorinated hydrocarbon insecticides to hens. *J. Assoc. Off. Anal. Chem.* **50**:418–425.

32. J. H. Dahlen and A. O. Haugen. 1954. The toxicity of certain insecticides to the bob-white quail and mourning dove. *J. Wildl. Manage.* **18**:477–481.

33. P. W. Danckwortt and E. Pfau. 1926. Massenvergiftungen von Tieren durch Arsenbestaubung vom Flugzeug. *Z. Angew. Chem.* **39**:1486–1487.

34. J. B. DeWitt. 1956. Chronic toxicity to quail and pheasants of some chlorinated insecticides. *J. Agric. Food Chem.* **4**:863–866.

35. J. B. Dimond, G. Y. Belyea, R. E. Kadunce, A. S. Getchell, and J. A. Blease. 1970. DDT residues in robins and earthworms associated with contaminated forest soil. *Can. Entomol.* **102**:1722–1730.

36. D. J. Ecobichon and P. W. Saschenbrecker. 1969. The redistribution of stored DDT in cockerels under the influence of food deprivation. *Toxicol. Appl. Pharmacol.* **15**:420–432.

37. J. B. Elder and C. Henderson. 1969. *Field Appraisal of ULV Baytex Mosquito Control Applications on Fish and Wildlife.* U.S. Bur. Sports Fish. Wildl. and Ohio Dept. Nat. Res. Special Report. 74 pp.

38. Environmental Protection Agency. 1975. *DDT: A Review of Scientific and Economic Aspects the Decision to Ban its Use as a Pesticide.* Washington, D.C. Document EPA-540/1-75-022. 300 pp.

39. D. E. Ferguson. 1964. Some ecological effects of heptachlor on birds. *J. Wildl. Manage.* **28**:158–163.

40. R. B. Finley. 1965. Adverse effects on birds of phosphamidon applied to a Montana forest. *J. Wildl. Manage.* **29**:580–591.

41. E. L. Flickinger and D. L. Meeker. 1972. Pesticide mortality of young white-faced ibis in Texas. *Bull. Environ. Contam. Toxicol.* **8**:165–168.

42. C. D. Fowle. 1966. The effects of phosphamidon on birds in New Brunswick forests. *J. Appl. Ecol.* **3**(Suppl.):169–170.

43. J. F. Fowler, L. D. Newsom, J. B. Graves, F. L. Bonner, and P. E. Schilling. 1971. Effect of dieldrin on egg hatchability, chick survival, and egg-shell thickness in purple and common gallinules. *Bull. Environ. Contam. Toxicol.* **6**:495–501.

44. P. Fuchs. 1967. Death of birds caused by application of seed dressings. *Med. Facult. Landbouw. Rijks Univ. Gent.* **32**:855–859.

45. R. E. Genelly and R. L. Rudd. 1956. Effects of DDT, toxaphene, and dieldrin on pheasant reproduction. *Auk* **73**:529–539.

46. J. L. George. 1959. Effect on fish and wildlife of chemical treatments of large areas. *J. Forestry* **57**:250–254.

47. J. L. George and R. T. Mitchell. 1947. The effects of feeding DDT-treated insects to nestling birds. *J. Econ. Entomol.* **40**:782–789.

48. B. J. Gough, L. A. Escurieux, and T. E. Shellenberger. 1967. A comparative study of a phosphorodithioate in Japanese and bobwhite quail. *Toxicol. Appl. Pharmacol.* **10**:12–19.

49. R. R. Graber, S. L. Wunderle, and W. N. Bruce. 1965. Effects of a low-level dieldrin application on a red-winged blackbird population. *Wilson Bull.* **77**:168–174.

50. Y. A. Greichus and M. R. Hannon. 1973. Distribution and biochemical effects of DDT,

DDD, and DDE in penned double-crested cormorants. *Toxicol. Appl. Pharmacol.* **26**:483-494.

51. D. R. J. Grocki and D. W. Johnston. 1974. Chlorinated hydrocarbon residues in North American cuckoos. *Auk* **91**:186-188.

52. W. R. Hanson. 1952. Effects of some herbicides and insecticides on biota of North Dakota marshes. *J. Wildl. Manage.* **16**:299-308.

53. R. J. Hayes. 1972. Effects of DDT on glycogen and lipid levels in bobwhites. *J. Wildl. Manage.* **36**:518-523.

54. R. G. Heath and S. A. Hill. 1974. Nationwide organochlorine and mercury residues in wings of adult mallards and black ducks during the 1969-1970 hunting season. *Pestic. Monit. J.* **7**:153-164.

55. R. G. Heath and J. W. Spann. 1973. Reproduction and related residues in birds fed mirex. In *Pesticides and the Environment: A Continuing Controversy.* Ed. W. B. Diechmann. Intercontinental Med. Book Co., pp. 421-435.

56. R. G. Heath, J. W. Spann, E. F. Hill, and J. F. Kreitzer. 1972. *Comparative Dietary Toxicities of Pesticides to Birds.* U.S. Fish Wildl. Serv., Bur. Sport Fish and Wildl., Special Scientific Report—Wildlife No. 152. 57 pp.

57. J. J. Hickey and L. B. Hunt. 1960. Initial songbird mortality following a Dutch elm disease control program. *J. Wildl. Manage.* **24**:259-265.

58. E. F. Hill. 1971. Toxicity of selected mosquito larvicides to some common avian species. *J. Wildl. Manage.* **35**:757-762.

59. E. F. Hill, W. E. Dale, and J. W. Miles. 1971. DDT intoxication in birds: Subchronic effects and brain residues. *Toxicol. Appl. Pharmacol.* **20**:502-514.

60. E. F. Hill, R. G. Heath, J. W. Spann, and J. D. Williams. 1974. Polychlorinated biphenyl toxicity to Japanese quail as related to degree of chlorination. *Poultry Sci.* **53**:597-604.

61. E. F. Hill, D. A. Eliason, and J. W. Kilpatrick. 1971. Effects of ULV application of malathion on non-target animals. *J. Med. Entomol* **8**:173-179.

62. R. S. Hoffmann, R. G. Janson, and F. Hartkorn. 1958. Effect on grouse populations of DDT spraying for spruce budworm. *J. Wildl. Manage.* **22**:92-93.

63. C. H. Hoffmann and J. P. Linduska. 1949. Some consideration of the biological effects of DDT. *Sci. Monthly* **69**:104-114.

64. N. Hotchkiss and R. H. Pough. 1946. Effect on forest birds of DDT used for gypsy moth control in Pennsylvania. *J. Wildl. Manage.* **10**:202-207.

65. E. G. Hunt. 1966. Biological magnification of pesticides. In *Scientific Aspects of Pest Control.* Nat. Acad. Sciences, Washington Publication 1402. pp. 251-262.

66. E. G. Hunt. 1966. Studies of pheasant–insecticide relationships. *J. Appl. Ecol.* **3**(Suppl.):113-123.

67. E. G. Hunt and J. O. Keith. 1963. *Pesticide-Wildlife Investigations in California, 1962.* Proc. 2nd Ann. Conf. on the Use of Agric. Chems. in Calif. Davis, Calif. 29 pp.

68. L. B. Hunt. 1960. Song bird breeding populations in DDT-sprayed Dutch elm disease communities. *J. Wildl. Manage.* **24**:139-146.

69. L. B. Hunt. 1965. Kinetics of pesticide poisoning in Dutch elm disease control. *U.S. Fish Wildl. Serv. Circ.* **226**:12-13.

70. L. B. Hunt and R. J. Sacho. 1969. Response of robins to DDT and methoxychlor. *J. Wildl. Manage.* **33**:336-345.

71. H. Ishikura. 1962. Impact of pesticide use on the Japanese environment. In *Environmental Toxicology of Pesticides*. Ed. F. Matsumura, G. M. Boush, and T. Misato. Academic. pp. 1–32.

72. D. J. Jefferies and B. N. K. Davis. 1968. Dynamics of dieldrin in soil, earthworms, and song thrushes. *J. Wildl. Manage.* **34**:929–940.

73. E. V. Johnson, G. F. Mack, and D. Q. Thompson. 1976. The effects of orchard pesticide applications on breeding robins. *Wilson Bull.* **88**:16–35.

74. D. W. Johnston. 1974. Decline of DDT residues in migratory songbirds. *Science* **186**:841–842.

75. D. W. Johnston. 1975. Organochlorine pesticide residues in small migratory birds, 1964–73. *Pestic. Monit. J.* **9**:79–88.

76. E. E. Kenaga. 1973. Factors to be considered in the evaluation of the toxicity of pesticides to birds in their environment. *Environ. Qual. Safety* **2**:166–181.

77. J. O. Keith. 1966. Insecticide contaminations in wetland habitats and their effects on fish-eating birds. *J. Appl. Ecol.* **3**(Suppl.):71–85.

78. J. O. Keith and M. S. Mulla. 1966. Relative toxicity of five organophosphorus mosquito larvicides to mallard ducks. *J. Wildl. Manage.* **30**:553–563.

79. J. H. Koeman and H. van Genderen. 1966. Some preliminary notes on residues of chlorinated hydrocarbon insecticides in birds and mammals in the Netherlands. *J. Appl Ecol.* **3**(Suppl.):99–106.

80. J. H. Koeman and J. H. Pennings. 1970. The side-effects and environmental distribution of insecticides used in tsetse control in Africa. *Bull. Environ. Contam. Toxicol.* **5**:164–170.

81. J. H. Koeman, R. C. H. M. Oudejans and E. A. Huisman. 1967. Danger of chlorinated hydrocarbon insecticides in birds' eggs. *Nature* **215**:1094–1096.

82. J. F. Kreitzer and G. H. Heinz. 1974. The effect of sublethal dosages of five pesticides on the avoidance response of *Coturnix* chicks. *Environ. Pollut.* **6**:21–29.

83. J. F. Kreitzer and J. W. Spann. 1968. Mortality among bobwhites confined to a heptachlor contaminated environment. *J. Wildl. Manage.* **32**:874–878.

84. H. Krieg. 1925. Die Bekampfung forstlicher Schadlinge durch Abwurf von Calziumarseniat vom Flugzeug. *Anz. Schadlingsk.* **1**:97–98.

85. L. N. Locke, N. J. Chura, and P. A. Stewart. 1966. Spermatogenesis in bald eagles experimentally fed a diet containing DDT. *Condor* **68**:497–502.

86. J. L. Ludke, E. F. Hill and M. P. Dieter. 1975. Cholinesterase response and related mortality among birds fed cholinesterase inhibitors. *Arch. Environ. Contam. Toxicol.* **3**:1–21.

87. P. D. Ludwig, H. J. Dishburger, J. C. McNeill, W. O. Miller, and J. R. Rice. 1968. Biological effects and persistence of Dursban insecticides in a salt-marsh habitat. *J. Econ. Entomol.* **61**:626–633.

88. D. R. Macdonald and F. E. Webb. 1963. Insecticides and the spruce budworm. In *The Dynamics of Epidemic Spruce Budworm Populations*. Ed. R. F. Morris. Memoirs Entomol. Soc. Can. No. 31. pp. 288–310.

89. F. Matsumura. 1975. Effects of pesticides on wildlife. In *Toxicology of Insecticides*. Plenum. pp. 355–401.

90. T. A. McCaskey, A. R. Stemp, B. J. Liska, and W. J. Stadelmann. 1968. Residues in

egg yolks and tissues from laying hens administered selected chlorinated hydrocarbon insecticides. *Poultry Sci.* 47:564–569.

91. M. W. McDonald, J. F. Dillon, and D. Stewart. 1964. Non-toxicity to poultry of malathion as a grain protectant. *Austral. Vet. J.* 40:358–360.

92. L. C. McEwen and R. L. Brown. 1966. Acute toxicity of dieldrin and malathion to wild sharp-tailed grouse. *J. Wildl. Manage.* 30:604–611.

93. K. N. Mehrotra, Y. P. Beri, S. S. Misra, and A. Phokela. 1967. Physiological effects of malathion on the house sparrow. *Indian J. Exp. Biol.* 5:219–221.

94. R. T. Mitchell. 1946. Effects of DDT spray on eggs and nestlings of birds. *J. Wildl. Manage.* 10:192–194.

95. N. W. Moore and C. H. Walker. 1964. Organic chlorine insecticide residues in wild birds. *Nature* 201:1072–1073.

96. B. M. Mulhern et al. 1970. Organochlorine residues and autopsy data from bald eagles 1966–68. *Pestic. Monit. J.* 4:141–144.

97. H. D. Muller. 1971. Reproductive responses of the mallard duck to subtoxic pesticide ingestion. *Poultry Sci.* 50:1610.

98. H. D. Muller. 1971. Responses of three generations of gray partridge to low level pesticide ingestion. *Poultry Sci.* 50:1610–1611.

99. R. K. Murton and M. Vizoso. 1963. Dressed cereal seed as a hazard to wood pigeons. *Ann. Appl. Biol.* 52:503–517.

100. T. W. Mussehl and R. B. Finley. 1967. Residues of DDT in forest grouse following spruce budworm spraying. *J. Wildl. Manage.* 31:270–287.

101. E. C. Naber and G. W. Ware. 1965. Effect of Kepone and mirex on reproductive performance in the laying hen. *Poultry Sci.* 44:875–880.

102. C. E. Olney, W. E. Donaldson, and T. W. Kerr. 1962. Methoxychlor in eggs and chicken tissues. *J. Econ. Entomol.* 55:477–479.

103. D. B. Peakall, J. L. Lincer, and S. E. Bloom. 1972. Embryonic mortality and chromosomal alterations caused by Aroclor 1254 in ring doves. *Environ. Health Perspect.* 1:103–104.

104. T. J. Peterle and R. H. Giles. 1964. *New Tracer Techniques for Evaluating the Effects of an Insecticide on the Ecology of a Forest Fauna.* Ohio State Univ. Res. Foundation Rep. 1207 (to U.S. Atom. Energy Comm.). 455 pp.

105. D. Pimentel. 1971. *Ecological Effects of Pesticides on Non-target Species.* Exec. Off. President, Off. Sci. Technol. Sup't. Documents, Washington. 220 pp.

106. R. D. Porter and S. N. Wiemeyer. 1972. DDE at low dietary levels kills captive American kestrels. *Bull. Environ. Contam. Toxicol.* 8:193–199.

107. W. L. Reichel, E. Cromartie, T. G. Lamont, B. M. Mulhern, and R. M. Prouty. 1969. Pesticide residues in eagles. *Pestic. Monit. J.* 3:142–144.

108. W. L. Reichel, L. N. Locke, and R. M. Prouty. 1974. Peregrine falcon suspected of pesticide poisoning. *Avian Diseases* 18:478–479.

109. C. S. Robbins, P. F. Springer, and C. G. Webster. 1951. Effects of five-year DDT application on breeding bird populations. *J. Wildl. Manage.* 15:213–216.

110. C. S. Robbins and R. E. Stewart. 1949. Effects of DDT on bird population of scrub forest. *J. Wildl. Manage.* 13:11–18.

111. W. Rosene. 1965. Effects of field applications of heptachlor on bobwhite quail and other animals. *J. Wildl. Manage.* **29**:554–580.

112. E. Ross and M. Sherman. 1960. The effect of selected insecticides on growth and egg production when administered continuously in the feed. *Poultry Sci.* **39**:1203–1211.

113. R. L. Rudd and R. E. Genelly. 1956. *Pesticides: Their Use and Toxicity in Relation to Wildlife.* Calif. Dept. Fish Game, Game Bull. No. 7. 209 pp.

114. E. A. Sauter and E. E. Steele. 1972. The effect of low level pesticide feeding on the fertility and hatchability of chicken eggs. *Poultry Sci.* **51**:71–76.

115. F. Schneider. 1966. Some pesticide-wildlife problems in Switzerland. *J. Appl. Ecol.* **3**(Suppl.):15–20.

116. T. G. Scott, Y. L. Willis, and J. A. Ellis. 1959. Some effects of a field application of dieldrin on wildlife. *J. Wildl. Manage.* **23**:409–427.

117. R. W. Seabloom, G. L. Pearson, L. W. Oring, and J. R. Reilly. 1973. An incident of fenthion mosquito control and subsequent avian mortality. *J. Wildl. Dis.* **9**:18–20.

118. T. E. Shellenberger, G. W. Newell, R. F. Adams, and J. Barbaccia. 1966. Cholinesterase inhibition and toxicological evaluation of two organophosphate pesticides to Japanese quail. *Toxicol. Appl. Pharmacol.* **8**:22–28.

119. R. D. Smith and L. L. Glasgow. 1963. Effects of heptachlor on wildlife in Louisiana. *Ann. Conf. Southern Assoc. Game Fish Comm.* **17**:140–154.

120. J. M. Spiers. 1949. *The Relation of DDT Spraying to the Vertebrate Life of the Forest.* Ontario Dept. Lands Forests, Biol. Bull. No. 2. pp. 141–158.

121. L. F. Stickel. 1973. Pesticide residues in birds and mammals. In *Environmental Pollution by Pesticides.* Ed. C. A. Edwards. Plenum Press, London. pp. 254–312.

122. L. F. Stickel and W. H. Stickel. 1970. Distribution of DDT residues in tissues of birds in relation to mortality, body condition, and time. *Ind. Med. Surg.* **38**:44–53.

123. L. F. Stickel, W. H. Stickel, and R. Christensen. 1966. Residues of DDT in brains and bodies of birds that died on dosages and in survivors. *Science* **151**:1549–1551.

124. L. F. Stickel, N. J. Chura, P. A. Stewart, C. M. Menzie, R. M. Prouty, and W. L. Reichel. 1966. Bald eagle pesticide relations. *Trans. No. Am. Wildl. Nat. Res. Conf.* **31**:190–200.

125. W. H. Stickel. 1974. Effects on wildlife of newer pesticides and other pollutants. *Proc. 53rd Ann. Conf. Western Assoc. of State Game and Fish Comm.* pp. 484–491.

126. W. H. Stickel, J. A. Galyen, R. A. Dryland, and D. L. Hughes. 1973. Toxicity and persistence of mirex in birds. In *Pesticides and the Environment: A Continuing Controversy.* Ed. W. B. Deichmann. Intercontinental Med. Book Co. pp. 437–467.

127. W. H. Stickel, D. W. Hayne, and L. F. Stickel. 1965. Effects of heptachlor-contaminated earthworms on woodcocks. *J. Wildl. Manage.* **29**:132–146.

128. W. H. Stickel, L. F. Stickel, and J. W. Spann. 1969. Tissue residues of dieldrin in relation to mortality in birds and mammals. In *Chemical Fallout.* Ed. M. W. Miller and G. G. Berg. Charles C Thomas. pp. 174–204.

129. E. F. Thomas and A. L. Shealy. 1932. Lead arsenate poisoning in chickens. *J. Agric. Res.* **45**:315–319.

130. I. Tragardh. 1935. The economic possibilities of aircraft dusting against forest insects. *Bull. Entomol. Res.* **26**:487–495.

131. R. K. Tucker and D. G. Crabtree. 1970. *Handbook of Toxicity of Pesticides to Wildlife.* U.S. Fish Wildl. Serv., Bur. Sports Fish Wildl., Resources Publ. No. 84. 131 pp.

132. R. K. Tucker and M. A. Haegele. 1971. Comparative acute oral toxicity of pesticides to six species of birds. *Toxicol. Appl. Pharmacol.* **20**:57–95.

133. E. E. Turtle, A. Taylor, E. N. Wright, R. J. P. Thearle, H. Egan, W. H. Evans, and N. M. Soutar. 1963. The effects on birds of certain chlorinated insecticides used as seed dressings. *J. Sci. Food Agric.* **14**:567–577.

134. A. C. Van Velzen and J. F. Kreitzer. 1975. The toxicity of *p,p′*-DDT to the clapper rail. *J. Wildl. Manage.***39**:305–309.

135. A. C. Van Velzen, W. B. Stiles, and L. F. Stickel. 1972. Lethal mobilization of DDT by cowbirds. *J. Wildl. Manage.* **36**:733–739.

136. G. J. Wallace, W. P. Nickell, and R. F. Bernard. 1961. *Bird Mortality in the Dutch Elm Disease Program in Michigan.* Cranbrook Inst. Sci., Bull. 41. 44 pp.

137. C. C. Whitehead. 1971. The effects of insecticides on production in poultry. *Vet. Rec.* **88**:114–117.

138. S. N. Wiemeyer, P. R. Spitzer, W. C. Krantz, T. G. Lamont, and E. Cromartie. 1975. Effects of environmental pollutants on Connecticut and Maryland ospreys. *J. Wildl. Manage.* **39**:124–139.

139. H. R. Wolfe. 1957. *Orchard Mouse Control with Endrin Sprays.* State Coll. Wash., Extens. Cir. 282. 8 pp.

140. B. S. Wright. 1965. Some effects of heptachlor and DDT on New Brunswick woodcocks. *J. Wildl. Manage.* **29**:172–185.

141. C. F. Wurster. 1969. Chlorinated hydrocarbons and avian reproduction: How are they related? In *Chemical Fallout.* Ed. M. W. Miller and G. G. Berg. Charles C Thomas. pp. 368–389.

9

ORGANOCHLORINE INSECTICIDES AND EGGSHELL THINNING

A. EFFECT OF DIELDRIN, DDT, AND DDE ON REPRODUCTIVE SUCCESS OF RAPTORS

Dieldrin and British Raptors

In 1963 it became evident that the golden eagle in the west of Scotland was suffering from a decline in reproductive success, since only 30% of the eyries were successful and many of the eggs were broken.[57] Restricted only to the region where eagles fed on sheep carrion, this decline was associated with a high egg content (approx. 0.9 ppm) of dieldrin, then being used in sheep dips.[58] At the same time the peregrine falcon, having undergone a population crash in 1956, was also disappearing from Great Britain, only 68 pairs breeding successfully in 1962, as compared to 650 pairs before World War II.[79] Ratcliffe (1967) then compared the eggshells of these two species (and of the European sparrowhawk) from problem areas with those from museum specimens from past years, and found that they were significantly thinner. For a measure of the thickness index, he used the egg weight (mg) divided by the egg length-times-breadth product (mm), and found that the reductions in this index had taken place between 1945 and 1948, the indices for eggs before and after this period being as follows:

	Before	After
Peregrine Falcon	1.84	1.47
Sparrow Hawk	1.45	1.15
Golden Eagle	3.16	2.89

Meanwhile at the Patuxent Biological Station in United States, it was found that the addition of 10 ppm dieldrin to the diet slightly but significantly thinned the eggshells of mallards,[54] although not of ring-necked pheasants.[20] The abandonment of dieldrin sheep-dips in 1966 was followed by a recovery of the peregrine falcon in Britain by 1968. In the golden eagle the egg residues of dieldrin declined to about 0.3 ppm and the proportion of successful eyries rose to nearly 70%,[58] and their eggshell thickness commenced returning towards the normal.[81]

DDT and DDE and American Raptors

Meanwhile in 1963 the 30% annual decline observed in breeding pairs of the fish-eating osprey on the coast of Connecticut was found by Ames (1966) to be associated with a higher content of DDT residues, especially DDE; the Connecticut eggs contained about twice as much DDE as eggs from Maryland where there was considerably less fish contamination and no decline in the osprey population (Table 9.1). Hickey and Anderson (1968) then found that the eggshell weights and thicknesses from herring gull colonies in five different states were inversely correlated with the DDE content of the eggs. The shell thickness at Green Bay, Wisconsin was 11% below normal for a DDE content of 90 ppm in 1967; in 1964, when so much egg breakage had been noted in the herring gull colony,[41] the DDE content of 200 ppm found at that time therefore indicated that the shells had been even thinner. They also found a 20% decrease from the normal in the eggshell thickness of the peregrine falcon (Massachusetts, New Jersey, California) which feeds on birds, and the osprey (New Jersey) and bald eagle (Florida) which feed on fish; a serious population decline had been noted in these two species since 1966.[83] On the other hand, there was no eggshell thinning in the golden eagle, red-tailed hawk and great horned owl, which feed on mammals. In the eggshells of European raptors taken around the Baltic Sea, they found an 8–16% decrease in thickness in the peregrine falcon and European sparrowhawk, and in the osprey and white-tailed eagle which feed on fish.[40] The white-tailed eagles in the Stockholm archipelago were sustaining a greatly reduced hatch of their eggs, in which the DDE content of the lipid extract was almost 1000 ppm.[46]

Table 9.1. Residues (ppm) in eggs of osprey in relation to their fish prey (Ames, 1966).

		DDT	DDD	DDE
Maryland	Atlantic Eel	0.1	0.2	0.1
	Yellow Perch	0.1	0.04	0.1
	Menhaden	Trace	Trace	Trace
	Osprey Eggs	0.07	0.6	2.3
Connecticut	Alewife	1.7	0.2	0.3
	Black-headed Flounder	0.8	0.1	0.4
	Windowpane Flounder	2.0	0.1	0.7
	Osprey Eggs	0.1	0.3	4.7

Falcons. In southern Alberta the prairie falcon *Falco mexicanus,* which was showing a 30% reduction in the number of territories occupied, was laying eggs which on the average were 11% thinner than normal, and contained 16 ppm DDE on a dry-weight basis, equivalent to 3 ppm on the wet-weight basis.[28] The peregrine falcons in the Ungava region in the far north of Quebec, which had a whole-body DDE content of some 13 ppm in 1967, were laying eggs with shells that were 21% thinner than the normal.[4] In the Mackenzie River valley of northwestern Canada, the peregrines contained an average of 280 ppm DDE in their adipose tissue, but did not show any failure in reproduction.[25] In the Yukon River valley of Alaska, the average DDE residues in the peregrines' adipose tissue was 560 ppm in 1967, while their bird prey in that area had whole-body residues ranging from 0.3 to 6.0 ppm DDE[13]; their eggs had a median content of 15 ppm DDE, ranging up to about 100 ppm.[14] In Baja California, the peregrines that preyed on seabirds feeding on fish in the Gulf of California were suffering from a subnormal reproductive success; one cracked egg was taken and showed a residue content of almost 100 ppm DDE, and a one-third reduction in shell thickness.[84] On the other hand, the gyrfalcon which remains as a year-round resident in Arctic regions had an egg DDE content 200 times lower than the peregrine in Alaska.[14]

Other Raptors. Among the accipiter group of hawks, population declines had been noted in Cooper's hawk and the sharp-shinned hawk, but not in the goshawk which feeds on mammals rather than birds, as shown by the census counts of migratory birds at Hawk Mountain, Pennsylvania. Collections made from Oregon, Arizona, and New Mexico as well as from New York and Pennsylvania in 1969 revealed that the eggs of the two bird-feeding hawks had a DDE content 10 times greater than those of the goshawk. In Arizona, eggs of the Cooper's hawk showed thicknesses ranging down to 30% less than normal, and there was a correlation between DDE content and the degree of thinning.[89] Among the buteonine hawks, the rough-legged hawk along the Colville River, Alaska, a resident which feeds on small mammals such as shrews, introduced only 1.2 ppm DDE into its eggs; whereas the migratory peregrine falcon, feeding in Alaska on birds such as willow ptarmigan, shoveller, and mallard, accumulated a DDE residue in their eggs nearly 100 times greater.[55] In the bald eagle, eggs from nonproductive nests in Maine contained much more DDE (and dieldrin) than eggs collected in Wisconsin and Florida from productive and nonproductive nests.[50] By 1969, the eggshell thickness in the Wisconsin, Florida, and Alaska bald eagles was about 15% below normal,[98] and the DDE content in the Maine eggs had a median of 15 ppm, known to be a shell-thinning level for American sparrowhawks.

DDE Content and Shell Thickness

It became clear that DDE was the major shell-thinning factor, since a linear inverse relationship between shell thickness and the DDE content of the egg could be demonstrated not only in the prairie falcon, but also in the double-crested cormorant and the brown pelican (Fig. 9.1), as well as in the peregrine falcon (Fig. 9.2). The correlation in the brown pelican was slightly obscured by the characteristic decrease of DDE which occurred as egg development proceeded towards completion and the yolk lipid decreased by about ten-fold[35]; moreover it is accompanied by a progressive thinning of the shell, amounting to 7% in *Coturnix* quail.[51]

The amounts of DDE in the ovaries of the mother bird, as indicated by the DDE content of her eggs, to produce a given degree of thinning varied according to the species[47]; for example, whereas 20% thinning would correlate with about 12 ppm DDE in the prairie falcon and white pelican, it correlates with about 36 ppm DDE in the gannet and double-crested cormorant, and with approximately 162 ppm in the herring gull and great blue heron (Fig. 9.1). Wherever the residues induced an eggshell thinning more than 10% below the normal thickness, that bird population would be in a decline.[3]

DDE, Shell Thickness, and Hatch Rate

The predominant importance of organochlorine residues and thin eggshells as compared to other environmental conditions as affecting the reproductive success of endangered birds was demonstrated by Wiemeyer et al. (1975), who interchanged eggs of the osprey between Connecticut and Maryland. The Connecticut population already described had declined from 31 active nests in 1961 down to 5 active nests in 1969, and had a 20–30% hatch rate as compared to 80–85% in Maryland. It was found that the Connecticut eggs placed in Maryland nests showed their characteristic reduced hatch, and the Maryland eggs yielded their good hatch rate in Connecticut. The average shell thicknesses of the eggs and their organochlorine residues (in ppm) were as follows:

	Shell Thickness	DDE	Dieldrin	PCBs
Maryland	10% below normal	2.4	0.25	2.6
Connecticut	18% below normal	8.9	0.61	15.0

It was noted that the nest-clutch showed total hatch failure when the average residues exceeded 10 ppm for DDE, 1 ppm for dieldrin, or 15 ppm for PCBs.[100]

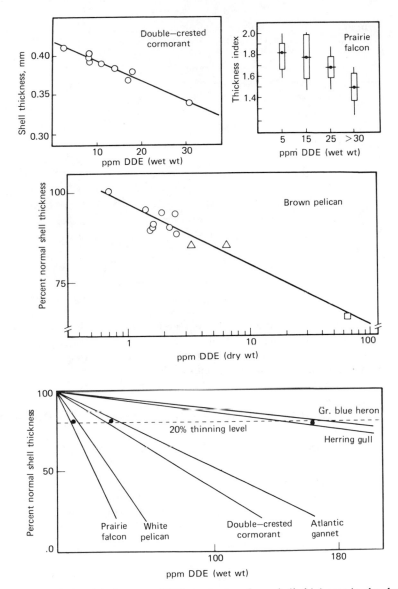

Fig. 9.1. Relationship between DDE content and eggshell thickness in the double-crested cormorant,[3] prairie falcon,[28] brown pelican[8] and six representative species.[47] The graphs for the cormorant and falcon are reprinted with permission from Can. Field-Nat. 83:91–112 and 191–200. 1969.

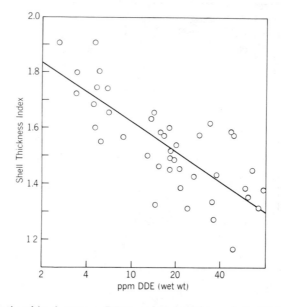

Fig. 9.2. Relationship between DDE content and eggshell thickness index in Alaska peregrine falcons (from Cade et al., 1971).
Reproduced from Science 172 No. 3986:955-7. 1971. Copyright 1971 by the American Association for the Advancement of Science.

B. EFFECT OF DDT RESIDUES ON EGGSHELL FORMATION

Feeding Experiments with DDT and DDE

Laboratory feeding experiments with mallards over a 2-yr period showed that 40 ppm in the diet DDT caused occasional egg-cracking, DDD was without effect, but DDE resulted in frequent cracking, leading to a 40% embryo mortality and a 75% reduction in duckling production.[37] The shell thicknesses for two dietary levels were reported as follows (in mm):

		p,p′-DDT	Tech. DDD	*p,p′* DDE
Control	0.37	—	—	—
10 ppm	—	0.35	0.35	0.33
40 ppm	—	0.33	0.35	0.32

The effect of DDE was greater than that of DDT. In another separate investigation,[94] DDT in the diet at 10 ppm did not result in the thinning of mallard eggshells, but a single oral dose at 1 g/kg resulted in the mallards laying eggs with 25% thinner shells during the next 2 days. A third investigation with *p,p'*-DDT found that 20 ppm in the diet did result in a 20% reduction in the eggshell thickness of mallards.[23] When the American sparrowhawk (*Falco sparverius*) was used as a laboratory indicator of *F. peregrinus,* it was found that whereas shell thinning down to 17% below normal thickness could be induced by a mixture of 5 ppm of DDT and 1 ppm dieldrin in the diet,[74] DDE at 10 ppm did not result in shell thinning until the second year of this diet.[99] In a subsequent investigation with *F. sparverius,* 10 ppm DDE caused a 25% shell-thinning without such a delay.[72] The feeding of 10 ppm DDE also resulted in a 13% shell-thinning in the screech owl,[62] and an 18–29% thinning in black duck with 10% of the eggs cracked[60]; these black duck eggs, which contained 65 ppm DDE, sustained four times as much mortality under the hazards of natural incubation as in an artificial incubator, and produced only 25% as many ducklings as the normal.[59]

Seed Eaters. The feeding of DDT to seed-eating birds did not produce such striking effects, only a 10% decrease in eggshell weight being induced in Asiatic ringdoves[69] and the Japanese quail *Coturnix.*[90] DDE also caused 10% eggshell thinning in ringdoves, accompanied by a reduction in the number of eggs per clutch.[32] In the case of *Coturnix,* the induced thinning was additive to the progressive thinning which occurs with successive layings.[7] In this species, DDE was no more effective than DDT in producing the moderate reductions in shell thickness and shell calcium, or in aggravating the shell breakage by the mother birds.[15] DDT at 30 ppm in the diet did not decrease eggshell thickness in bobwhite quail,[94] and no obvious thinning was noted in ring-necked pheasants fed at dietary levels up to 500 ppm DDT.[42] In the domestic fowl, neither DDT nor DDE, even at 300 ppm in the diet, caused any significant difference in eggshell thickness,[16,22] although in a third investigation DDT at 10 ppm induced an 11% decrease in thickness.[88] In the Bengalese finch the oral ingestion of *p,p'*-DDT at 50–300 μg/day actually induced a 7% increase in shell weight.[43]

Physiological Investigations of Mode of Action

Carbonic Anhydrase. The failure of the shell gland to secrete enough calcium carbonate was first attributed to a lack of carbonate ions due to inhibition of the carbonic anhydrase enzyme necessary to produce them from CO_2. When *Coturnix* quail had been fed on 100 ppm DDT or DDE (and laid eggs with less shell calcium than normal) the carbonic anhydrase activities in the blood and shell gland were as follows:

	Eggshell Ca, %	Carb in Blood, units	Carb in Gland, units
Control	2.58	298	1184
p,p′-DDT	2.37	242	924
p,p′-DDE	2.38	235	663

the effect of DDE being greater than that of DDT.[5] Injection of DDE into Asiatic ringdoves shortly before they are due to lay eggs resulted in a strong depression of carbonic anhydrase in the oviduct and the appearance of thin-shelled eggs.[69] Inhibition of carbonic anhydrase by DDT had been originally demonstrated *in vitro* by Keller (1952); however, neither DDT nor DDE had any effect on the ability of human red cells to hydrate CO_2, and DDT concentrations of at least 500 ppm were required to inhibit bovine red-cell carbonic anhydrase.[24] That the *in vitro* inhibition observed was due to colloidal coprecipitation, a nonspecific artifact, was proved by the isolation of coprecipitates of DDT, DDE, and dieldrin when they were incubated with bovine carbonic anhydrase.[73] When both the carbonic anhydrase and the DDE (or DDE) were kept in suspension by means of DMF and ultrasonication, no inhibition of the enzymes occurred.[61]

Hypothyroidism. A second possible explanation involves the thyroid secretion. The finding that Bengalese finches fed DDT laid smaller eggs with thicker shells was taken by analogy with the domestic fowl to be indicative of induced hyperthyroidism.[43] On the other hand, the feeding of DDT to pigeons, which causes an increase in liver and thyroid weights, in high doses induces a lowering of oxygen consumption and body temperature, two symptoms of hypothyroidism.[45] Hypothyroidism by thyroidectomy in chickens results in thin eggshells.[93] The idea that the sublethal effects of organochlorine insecticides on birds is centered on the thyroid gland offers a unitary theory,[44] but the implication that the shell thinning observed in the field was caused by hypothyroidism is not at present supported by sufficient evidence. Among the several mechanisms that may operate to cause eggshell thinning, thyroid malfunction is probably one of the less important.[19]

Estradiol Deficiency. The third possible explanation also relates to calcium deficiency, but in this case deriving from an upset in reproductive hormones; these hormones promote the deposition of calcium in medullary bone during the breeding season, to be subsequently drawn upon for eggshell production. Pigeons fed estradiol plus 100 ppm DDT developed half as much medullary bone as those fed on estradiol alone.[66] Ringdoves fed 10 ppm DDT showed a lower estradiol content in the blood,[69] which could result in less calcium being stored in medullary bone.[83] The reduction in estradiol was probably due to the effect of DDT in stimulating the microsomal mixed-function oxidases, which are known to metabolize steroids.[17] Pigeons pretreated with 10 ppm DDT and/or 2 ppm dieldrin in their diet for 1 wk yielded livers whose comparative capacities to produce metabolites from steroid hormones *in vitro* were as follows:[68]

	Control	DDT	Dieldrin	DDT + Dieldrin
Testosterone	29	75	111	168
Progesterone	30	78	90	155

This hypothesis is strengthened by findings that whereas dietary DDT does not affect these oxidases in the Japanese quail, it causes substantial increases in the American sparrow hawk.[30] In contradiction, however, DDT is inferior to dieldrin[68] and PCBs as an mfo inducer, and DDE is inferior to DDT.[29] It is of interest that *o,p′*-DDT, which constitutes 10–29% of commercial grades of technical DDT, is the most active of the DDT group of compounds in competing with estradiol at its receptor site in mammals.[53] Injection of *o,p′*-DDT into Japanese quail induced the typical estradiol response in the oviduct, which doubled in weight and thus in glycogen, while *p,p′*-DDT did not induce it.[6] No change in serum Ca was detected,[18] although *o,p′*-DDT in the diet reduced the amount of calcium laid down in the *Coturnix* eggs by 7–15%, which *p,p′*-DDT did also.[7] Since *o,p′*-DDT in the diet of hens has turned out to be only one-tenth as cumulative as *p,p′*-DDT or *p,p′*-DDE,[16] it is considered that this impurity of technical DDT is not likely to be important as an agent of eggshell thinning.[19]

Until recently, the sum of the evidence[96] thus indicated that the residues originating from technical DDT reduced the amount of calcium available for their eggshells because they caused an estradiol deficiency[101]; although the limitation is probably more through lack of carbonate ions than of calcium ions.[19] The distinction was made between DDE, which was considered to decrease the utilization of calcium due to a scarcity of carbonate ions owing to a reduction in carbonic anhydrase activity in the oviduct,[19] and DDT (and more so dieldrin), which apparently reduced the deposition of an available reserve of calcium in medullary bone through increasing the breakdown of the necessary steroid hormones.[70]

ATP-ase Inhibition.   The fourth and most probable explanation relates to a lack of calcium ions being made available from the active-transport process driven by the ATP-ase activity of cell-membrane proteins. The binding of DDT, and still more of DDE, to these proteins has been demonstrated in mammalian tissue.[12] It was thought that such a hypothesis might explain why seed-eating birds, which have a greater calcium absorption efficiency on their calcium-deficient diet, are less susceptible to the shell-thinning effect of DDE.[83] A recent study on the relation of Ca to shell gland itself has shown that there is no reduction in the amount of Ca reaching the shell gland in the blood of Pekin ducks or ringdoves fed diets containing DDE.[71] On the other hand, the presence of as little as 0.2 ppm DDE in the shell gland of the duck strongly inhibits the Ca-ATPase which acts as a calcium pump responsible for active transport of this kation from the blood to the developing shell; this inhibition was also demonstrated *in vitro.*[63] Moreover in the domestic fowl, where DDE does not cause eggshell thinning, it did not inhibit the Ca-ATPase in the shell gland.[26]

C.   AQUATIC BIRDS AND THE EFFECT OF OTHER PESTICIDES

Eggshell Thinning in Aquatic Birds

Pelicans and Cormorants.   Thin eggshells were first observed in double-crested cormorants and white pelicans on the lakes and coasts of Wisconsin, the thinness being correlated with the DDE content.[3] Then colonies of the brown pelican on Anacapa, Santa Barbara, and the Los Coronados Islands off the coast of southern California were discovered in 1968 to have reached the point where virtually no chicks were to be found. On Anacapa Island, off Los Angeles, only five chicks resulted from 1272 attempted nests in 1969. Virtually all the eggs were crushed or perforated, and the shells were on the average 53% thinner than normal for the Pacific brown pelican.[83] As indicators of the contamination of the pelican ovaries, the average DDE content was 72 ppm,[8] and ranged up to 2500 ppm in the lipid extract.[83] The situation on Anacapa Island contrasted with that in three localities on the Gulf of California (Mexico), where there was no eggshell thinning and the total-DDT (mainly DDE) residues in the egg lipids were low, as follows[48:]

	t-DDT, ppm	Shell Thickness, mm
Anacapa Island	1225	0.38
Gulf of California	75	0.56

On the Gulf coast of Florida the brown pelican eggs were of normal thickness in 1969 while those on the Atlantic coast of Florida were slightly thinner (Table 9.2). The two colonies in South Carolina were suffering from low reproductive success, that on Deveaux Bank having sustained a drastic decline in numbers and that on Marsh Island barely holding its own; in 1969 the shell thickness on Deveaux Bank was more than 15% below normal, and the egg DDE content was 6.4 ppm on the average.[8]

The data for the brown pelican on southern United States coasts (Table 9.2) show the close relationship between DDE content and shell thinning, the differences in dieldrin content being immaterial and there being no pronounced differences in mercury content. On Marsh Island in 1971 and 1972, where only 45% of the nests were producing young, no egg-clutches were successful which had an average DDE content exceeding 2.4 ppm or an average dieldrin content exceeding 0.54 ppm; differences in the content of DDT or of PCBs were immaterial.[11]

On the California coast, the double-crested cormorant on Anacapa Island had suffered just as severely from thin eggshells as the brown pelican.[83] In 1972, with the completion of a separate treatment system for the DDT manufacturing plant that had been discharging its wastes through the Los Angeles County sewer, the reproductive success of the brown pelican on Anacapa Island started to return to normal. By 1975 it had almost completely returned to normal on the South Carolina coast.[26]

Sea Birds.   The western herring gull, which feeds on garbage rather than fish, showed no eggshell thinning on the southern California coast. Farther north, at Point Reyes just north of San Francisco, shell thinning and a 50% annual population decline was characterizing the common murre and the ashy petrel.[83] By 1968–1970 the murre's eggs on the Farallon Islands contained about 40 ppm DDE and their shells were 13% thinner than 55 yr

Table 9.2.   **Shell thicknesses and pesticide residues (ppm) in eggs of the brown pelican (Blus, Belisle, and Prouty, 1974; Blus et al., 1975).**

	Thickness, mm	DDE	DDD	DDT	Dieldrin	PCB's	Hg
Florida Keys 1970	0.535	1.02	0.15	0.10	0.06	0.62	0.52
Florida Gulf Coast 1969	0.510	1.37	0.43	0.25	0.10	0.89	0.46
Florida Atlantic Coast 1969	0.494	2.24	0.83	0.40	0.36	2.81	0.37
Louisiana 1973	0.488	1.31	0.19	0.16	0.64	2.89	0.08
South Carolina 1969	0.460	5.24	1.43	0.37	0.94	5.68	0.28
California 1969	0.356	71.70	1.18	1.76	0.09	3.60	0.42

previously.[31,82] The 3–4% annual decline in the Bermuda petrel over the decade before 1968 was associated with a 6 ppm DDE residue in the eggs, although no shell-thickness measurements were taken.[102] A more abrupt decline in hatching rate was noted in the sooty tern on Dry Tortugas, Bahamas in 1969,[85] but in a dwindling colony of common terns in Alberta the eggs which failed to hatch contained no more DDE than those that hatched, the average being 7.6 ppm.[91]

Wading Birds.   The grey heron in eastern England showed a 20% reduction in eggshell thickness, and the average egg residues were 6 ppm DDE and 2.5 ppm dieldrin.[75] In eastern North America, eggshell thinning which ranged up to 20% and was correlated with the DDE content was occurring in the great blue heron in New York and New Jersey.[2] In California, shell thinning to the extents of 10% in the great blue heron and 15% in the common egret was observed in a heronry north of San Francisco; there were significant amounts of DDE in the eggs, but the situation was complicated by PCBs and dieldrin residues in the birds themselves, the latter being acquired in their wintering grounds in the Imperial Valley.[27] On the Canadian prairies the egg content of DDE in the great blue heron was 6–37 ppm, and of dieldrin only about 0.05 ppm.[95] Along the Atlantic coast of the United States in 1972, where egg residues in wading birds did not exceed 4.0 ppm DDE and 4.2 ppm PCBs, the great egret showed no eggshell thinning, nor were there any differences from the pre-1946 thicknesses in two other species of egrets, three species of heron, two species of ibis or the anhinga.[67]

Effects of Other Pesticides

Dieldrin.   The evidence of feeding experiments with dieldrin, mentioned previously, indicates that it significantly reduces the eggshell thickness in mallards[54] but not in ring-necked pheasant.[20] The degree of eggshell thinning was about 6% when the mallards received 10 ppm dieldrin in the diet.[23] In a colony of brown pelicans transferred from Florida to Grand Terre, Louisiana in 1971, the eggshell thickness and the reproductive success in 1973 were subnormal despite a low egg residue content of DDE (1.3 ppm); however since the dieldrin content was high (0.64 ppm) it was considered that this cyclodiene insecticide was responsible, just as it was suspected that endrin contamination might have previously contributed to the final disappearance of native brown pelican populations from Louisiana in 1961.[10]

Polychlorinated Biphenyls.   Although evidence of PCBs causing thin eggshells was obtained by feeding chickens with 100 ppm Aroclor 1254, 1242 or 1260,[64] no eggshell thinning was obtained with Aroclor 1254 in pheasants,[21] mallards or *Coturnix* quail[33] or in bobwhite quail.[38] Residues of PCBs which were present in British birds to about the same extent of DDE, being especially pronounced in grey herons and other fish eaters as well as in bird-feeding raptors, were sufficiently high in 1966–1968 to contribute to breeding failures in sensitive species, though not toxic enough to cause direct bird kills.[77] In North America, a correlation between eggshell thinning and PCB residues was found in the double-crested cormorant but not in the white pelican.[3] In the brown pelican, the Florida eggs had only slight eggshell thinning although the concentrations of PCBs were as high as in the thin-shelled California eggs. However, the following correlation coefficients (negative) between PCBs and other residues, and eggshell thickness and weight, were obtained in eggs from Florida and South Carolina pelican colonies[9]:

	PCBs	DDE	Dieldrin	Mercury
Thickness	.445	.523	.471	.263
Weight	.289	.463	.371	.122

This indicated that dieldrin and the PCBs play a part in the eggshell thinning primarily due to DDE. Moreover Aroclor 1254 and 1262 fed to American kestrels induced a liver microsomal activity that rapidly degraded estradiol.[56]

Mercury. The role of mercury in eggshell thinning is considered to be essentially negligible.[19] When included at 200 ppm in the diet of mallards, the organic mercurial Ceresan M caused no shell thinning where DDT had caused 15% thinning.[34] When given to mallards in a single heavy but sublethal dose, Ceresan M caused no subsequent thinning where DDE had caused severe thinning for weeks after the dose.[33] As is mentioned in the final chapter on fungicides, methylmercury did not reduce eggshell thickness in mallards, pheasants, ringdoves, or American kestrels.

Other Pesticides. The effect of lindane (gamma-HCH) is uncertain, judging by feeding experiments on laying hens; although British experiments found 200 ppm for 1 month not to cause eggshell thinning,[97] American work found significant thinning after 10 wk on 1 ppm lindane.[86] Among the organophosphorus insecticides, neither malathion nor diazinon caused thinning even with 10 ppm for 10 wk.[86] When given to mallards or *Coturnix* quail in a single heavy but sublethal dose, no shell thinning apart from a transitory effect in the first 1–2 days after the dose was caused by heptachlor, toxaphene, chlordecone, parathion, carbaryl, 2,4-D, sodium arsenite, or tetraethyl lead, nor by dieldrin or *o,p*´-DDT.[33]

D. OVERVIEW OF THE SITUATION IN THE 1970s

An extensive study by Anderson and Hickey (1972) of eggshells collected for museums in the United States and Canada up to 1969 (Table 9.3) added further species to the list of predaceous birds in which eggshell thinning had occurred, notably the marsh hawk, common loon and American sparrow hawk, the red-tailed hawk in Montana,[87] the red-shouldered hawk in Texas, and the great horned owl in Florida. It was pointed out that the shell-thinning problem often involved only a fraction of the species' geographic range, and was probably absent from the majority of the bird species in North America. In the American woodcock, in which a decline of reproductive success had been associated with DDT and other organochlorine insecticides, a survey of 10 states in 1969 found that the shells of fresh-laid unembryonated eggs were 10% thinner, whether they had hatched or not.[52]

Eggshell measurements made in Britain by Prestt and Ratcliffe (1972) revealed that thinning had occurred in the merlin, hobby, and kestrel among the Falconidae, and in the rook and carrion crow among the Corvidae. Increased egg breakage was correlated with decreased shell thickness in the peregrine falcon, European sparrow hawk, golden eagle and grey heron. The residues in the eggs of falcons and accipiters were on the average as follows (in ppm):

	DDE	DDD + DDT	Dieldrin	PCBs
Peregrine Falcon	13.8	0.11	0.57	1.1
Merlin	10.5	0.62	2.77	3.0
Kestrel	3.6	0.09	0.23	1.3
Hobby	4.4	0.20	0.20	1.0
Sparrow Hawk	29.5	1.89	4.59	—
Golden Eagle	0.48	0.01	0.34	0

Egg breakage was associated with reproductive failures of the peregrine not only in Britain, but also in France, Switzerland, West and East Germany, and Finland.[39] In Britain, population declines due to organochlorine residues have also occurred in the kestrel, merlin, and barn owl. Populations of the European sparrow hawk (*Accipiter nisus*) had been recovering ever since the discontinuation of dieldrin in sheep-dips in 1966.[76] By 1971, although the sparrow hawks were coming to occupy more territories in southern Scotland, their reproductive success remained low, only 20% of the 113 females in a 500 km² area in Dumfries hatching their entire clutch; the eggs which failed to hatch contained approximately 180 ppm DDE, 35 ppm dieldrin, and 75 ppm PCBs, on the average, in their lipid extract.[65]

Table 9.3. Degrees of eggshell thinning (percent below normal thickness) in populations of predaceous bird species, 1950–1969.

USA and Canada (Anderson and Hickey 1972)

> 20%	10-19%	5-10%	No change
Peregrine falcon	Sharp-shinned hawk	Gyrfalcon	Broad-winged hawk
Prairie falcon	Common loon	Am. Sparrowhawk	Am. Rough-legged hawk
Cooper's hawk	White pelican	Goshawk	Whooping crane
Bald eagle	Herring gull	Red-shouldered hawk	Eastern crow
Marsh hawk	Red-tailed hawk	Golden eagle	
Osprey		Great horned owl	
Brown pelican		Great blue heron	
Double-crested cormorant			
Black-crowned night heron			

Great Britain (Prestt and Radcliffe 1972)

17-21%	10-13%	Approx. 5%	No change
Peregrine falcon	Merlin	Kestrel	Buzzard
Sparrow-hawk	Golden eagle	Hobby	Golden plover
Osprey	Shag	Rook	Greenshank
Grey heron		Carrion crow	Black-headed gull
			Kittiwake
			Guillemot
			Razorbill auk
			Raven

There were some exceptions in the general picture of declines of North American raptors due to the organochlorine insecticides. Cooper's hawk in Arizona, despite the eggshell thinning due to DDE, showed no population reduction in the period 1969 to 1971.[89] The populations of peregrines in Alaska remained steady for 20 years despite the average DDE content along the Colville River being nearly 900 ppm in the egg lipids, with a corresponding reduction of 22% in the eggshell thickness index; however, population decreases were commencing in 1970.[14]

Ospreys on Gardiner's Island on the New York State coast, which had shown a drop from 600 fledglings in 1948 to only 4 in each of the years 1965 and 1966, recovered to 18 in 1973 and 26 in 1974; *t*-DDT residues in unhatched eggs dropped from 12 ppm in the mid-1960s to 3.5 ppm in 1974.[78] Bald eagles and ospreys had been increasing in the Wisconsin–Michigan National Forest ever since 1964 (*Annual Bald Eagle–Osprey Status*, U.S. Forest Service, Milwaukee 1970). Among the total counts of all migrating raptors which had been steadily increasing at Hawk Mountain, Pennsylvania, those of the osprey increased in 1969 and 1970 (*Hawk Mountain Newsletter* No. 43, 1971); this occurred despite the fact that there had been no reduction in DDE and dieldrin levels at that time, as judged by the residues in mallard wings collected throughout the United States between 1966 and 1969.[36] DDT was suspended from general use in the United States in 1970, and aldrin and dieldrin in 1974.

It may be concluded that the declines in reproductive success induced by dieldrin and other cyclodiene insecticides had been due primarily to a general toxic effect on the mother bird, while the declines due to DDT were primarily due to the effect of DDE in decreasing the secretion of the shell gland. The former situation characterized the population declines in Britain, and the latter the declines in the United States.[19] In India, the declines observed in many species of birds were principally due to habitat destruction by the expanding human population, although no investigations of eggshell thinning have been made.[92]

REFERENCES CITED

1. P. L. Ames. 1966. DDT residues in the eggs of the osprey in the northeastern United States and their relation to nesting success. *J. Appl. Ecol.* **3**(Suppl.):87–97.

2. D. W. Anderson and J. J. Hickey. 1972. Eggshell changes in certain North American birds. *Proc. 15th Int. Ornithol. Congr.* 514–540.

3. D. W. Anderson, J. J. Hickey, R. W. Risebrough, D. F. Hughes, and R. E. Christensen. 1969. Significance of chlorinated hydrocarbon residues to breeding pelicans and cormorants. *Can. Field-Nat.* **83**:91–112.

4. D. D. Berger, D. W. Anderson, J. D. Weaver, and R. W. Risebrough. 1970. Shell thinning in eggs of Ungava peregrines. *Can. Field-Nat.* **84**:265–267.

5. J. Bitman, H. C. Cecil, and G. R. Fries. 1970. DDT-induced inhibition of avian shell gland carbonic anhydrase: A mechanism for thin egg-shells. *Science* **168**:594–596.

6. J. Bitman, H. C. Cecil, S. J. Harris, and G. F. Fries. 1968. Estrogenic activity of *o,p′*-DDT in the mammalian uterus and avian oviduct. *Science* **162**:371–372.

7. J. Bitman, H. C. Cecil, S. J. Harris, and G. F. Fries. 1969. DDT induces a decrease in egg-shell calcium. *Nature* **224**:44–46.

8. L. J. Blus, A. A. Belisle, and R. M. Prouty. 1974. Relations of the brown pelican to certain environmental pollutants. *Pestic. Monit. J.* **7**:181–194.

9. L. J. Blus, R. G. Heath, C. D. Gish, A. A. Belisle, and R. M. Prouty. 1971. Egg-shell thinning in the brown pelican. *BioScience* **21**:1213–1215.

10. L. J. Blus, T. Joanen, A. A. Belisle, and R. M. Prouty. 1975. The brown pelican and certain environmental pollutants in Louisiana. *Bull. Environ. Contam. Toxicol.* **13**:646–655.

11. L. J. Blus, B. S. Neely, A. A. Belisle, and R. M. Prouty. 1974. Organochlorine residues in brown pelicans: Relation to reproductive success. *Environ. Pollut.* **7**:81–91.

12. H. Brunnert and F. Matsumura. 1969. Binding of DDT with subcellular fractions of rat brain. *Biochem. J.* **114**:135–139.

13. T. J. Cade, C. M. White, and J. R. Haugh. 1968. Peregrines and pesticides in Alaska. *Condor* **70**:170–178.

14. T. J. Cade, J. L. Lincer, C. M. White, D. G. Roseneau, and L. G. Schwartz. 1971. DDE residues and eggshell changes in Alaska falcons and hawks. *Science* **172**:955–957.

15. H. C. Cecil, J. Bitman, and S. J. Harris. 1971. Effects of dietary DDT and DDE on egg production and egg-shell characteristics of Japanese quail. *Poult. Sci.* **50**:657–659.

16. H. C. Cecil et al. 1972. Dietary *p,p'*-DDT, *o,p'*-DDT, or *p,p'*-DDE and changes in eggshell characteristics and pesticide accumulation in White Leghorns. *Poult. Sci.* **51**:130–139.

17. G. A. H. Conney. 1967. Pharmacological implications of microsomal enzyme induction. *Pharmacol. Rev.* **19**:317–366.

18. A. S. Cooke. 1970. The effect of *o,p'*-DDT on Japanese quail. *Bull. Environ. Contam. Toxicol.* **5**:152–157.

19. A. S. Cooke. 1973. Shell thinning in avian eggs by environmental pollutants. *Environ. Pollut.* **4**:85–102.

20. R. B. Dahlgren and R. L. Linder. 1970. Eggshell thickness in pheasants given dieldrin. *J. Wildl. Manage.* **34**:226–227.

21. R. B. Dahlgren and R. L. Linder. 1971. Effects of polychlorinated biphenyls on pheasant reproduction, behavior, and survival. *J. Wildl. Manage.* **35**:315–319.

22. K. L. Davison and J. I. Sell. 1972. Dieldrin and *p,p'*-DDT effects on egg production and eggshell thickness of chickens. *Bull. Environ. Contam. Toxicol.* **7**:9–18.

23. K. L. Davison and J. L. Sell. 1974. DDT thins shells of eggs from mallard ducks maintained on *ad libitum* or control-feeding regimens. *Arch. Environ. Contam. Toxicol.* **2**:222–232.

24. B. H. Dvorchik, M. Istin, and T. H. Maren. 1971. Does DDT inhibit carbonic anhydrase? *Science* **172**:728–729.

25. J. H. Enderson and D. D. Berger. 1968. Chlorinated hydrocarbon residues in peregrines and their prey species from northern Canada. *Condor* **70**:149–153.

26. Environmental Protection Agency. 1975. *DDT: A Review of Scientific and Economic Aspects of the Decision to Ban Its Use as a Pesticide.* Washington, D.C. Document EPA-540/1-75-022. 300 pp.

27. R. A. Faber, R. W. Risebrough, and H. M. Pratt. 1972. Organochlorines and mercury in common egrets and great blue herons. *Environ. Pollut.* **3**:111–122.

28. R. W. Fyfe, J. Campbell, B. Hayson, and K. Hodson. 1969. Regional population declines and organochlorine insecticides in Canadian prairie falcons. *Can. Field-Nat.* **83**:191–200.

29. J. W. Gillett, T. M. Chan, and L. C. Terriere. 1966. Interactions between DDT analogs and microsomal epoxidase systems. *J. Agric. Food Chem.* **14**:540.

30. J. W. Gillett et al. 1970. Induction of liver microsomal activities in Japanese quail and American sparrowhawk. In *The Biological Impact of Pesticides in the Environment.* Oregon State Univ. Press, Environ. Health Series 1. pp. 59–64.

31. F. Gress, R. W. Risebrough, and F. C. Sibley. 1971. Shell thinning in eggs of the common murre from the Farallon Islands, California. *Condor* **73**:368–369.

32. M. A. Haegele and R. H. Hudson. 1973. DDE effects on reproduction of ring doves. *Environ. Pollut.* **4**:53–57.

33. M. A. Haegele and R. K. Tucker. 1974. Effects of 15 common environmental pollutants on eggshell thickness in mallards and coturnix. *Bull. Environ. Contam. Toxicol.* **11**:98–102.

34. M. A. Haegele, R. K. Tucker, and R. H. Hudson. 1974. Effects of dietary mercury and lead on eggshell thickness in mallards. *Bull. Environ. Contam. Toxicol.* **11**:5–11.

35. W. Hazeltine. 1972. Disagreements on why brown pelican eggs are thin. *Nature* **239**:410–411.

36. R. G. Heath and S. A. Hill. 1974. Nationwide organochlorine and mercury residues in wings of adult mallards and black ducks during the 1969–1970 hunting season. *Pestic. Monit. J.* **7**:153–164.

37. R. G. Heath, J. W. Spann, and J. F. Kreitzer. 1969. Marked DDE impairment of mallard reproduction in controlled studies. *Nature* **224**:47–48.

38. R. G. Heath, J. W. Spann, J. R. Kreitzer, and C. Vance. 1972. Effects of polychlorinated biphenyls on birds. *Proc. 15th Int. Ornithol. Congr.* pp. 475–485.

39. J. J. Hickey (Ed.). 1969. *Peregrine Falcon Populations: Their Biology and Decline.* Univ. Wisconsin Press. 596 pp.

40. J. J. Hickey and D. W. Anderson. 1968. Chlorinated hydrocarbons and egg-shell changes in raptorial and fish-eating birds. *Science* **162**:271–273.

41. J. J. Hickey, J. A. Keith, and F. B. Coon. 1966. An exploration of pesticides in a Lake Michigan ecosystem. *J. Appl. Ecol.* **3**(Suppl.):141–154.

42. E. G. Hunt, J. A. Azevedo, L. A. Woods, and W. T. Castle. 1969. The significance of residues in pheasant tissues resulting from chronic exposures to DDT. In *Chemical Fallout.* Eds. M. W. Miller and G. G. Berg. Charles C Thomas. 335–360.

43. D. J. Jefferies. 1969. Induction of apparent hyperthyroidism in birds fed DDT. *Nature* **222**:578–579.

44. D. J. Jefferies. 1975. The role of the thyroid in the production of sublethal effects by organochlorine insecticides and polychlorinated biphenyls. In *Organochlorine Insecticdes: Persistent Organic Pollutants.* Ed. F. Moriarty. Academic. pp. 131–230.

45. D. J. Jefferies and M. C. French. 1971. Hyper- and hypo-thyroidism in pigeons fed DDT: An explanation for the thin egg-shell phenomenon. *Environ. Pollut.* **1**:235–242.

46. S. Jensen, A. G. Johnels, M. Olsson, and G. Otterlind. 1969. DDT and PCB in marine animals from Swedish waters. *Nature* **224**:247–250.

47. J. A. Keith and I. M. Gruchy. 1972. Residue levels of chemical pollutants in North American birdlife. *Proc. 15th Int. Ornithol. Congr.* pp. 437–454.

48. J. O. Keith, L. A. Woods, and E. G. Hunt. 1970. Reproductive failure in brown pelicans on the Pacific Coast. *Trans. N. Am. Wildl. Nat. Resour. Conf.* **35**:56–63.

49. H. Keller. 1952. Die Bestimmung kleinster Mengen DDT auf enzymanalytischem Wege. *Naturwissenschaften* **39**:109.

50. W. C. Krantz, B. M. Mulhern, G. E. Bagley, A. Sprunt, F. J. Ligas, and W. B. Robertson. 1970. Organochlorine and heavy metal residues in bald eagle eggs. *Pestic. Monit. J.* **4**:136–140.

51. J. F. Kreitzer. 1972. The effect of embryonic development on the thickness of the egg shells of coturnix quail. *Poult. Sci.* **51**:1764–1765.

52. J. F. Kreitzer. 1973. Thickness of the American woodcock eggshell, 1971. *Bull. Environ. Contam. Toxicol.* **9**:281–286.

53. D. Kupfer. 1975. Effects of pesticides and related compounds on steroid metabolism and function. *CRC Crit. Rev. Toxicol.* **4**:83–124.

54. P. N. Lehner and A. Egbert. 1969. Dieldrin and eggshell thickness in ducks. *Nature* **224**:1218–1219.

55. J. L. Lincer, T. J. Cade, and J. M. Devine. 1970. Organochlorine residues in Alaskan peregrine falcons, rough-legged hawks, and their prey. *Can. Field-Nat.* **84**:255–263.

56. J. L. Lincer and D. B. Peakall. 1970. Metabolic effects of polychlorinated biphenyls in the American kestrel. *Nature* **228**:783–784.

57. J. D. Lockie and D. A. Ratcliffe. 1964. Insecticides and Scottish golden eagle eggs. *Br. Birds.* **57**:89–102.

58. J. D. Lockie, D. A. Ratcliffe, and R. Balharry. 1969. Breeding success and organo-chlorine residues in golden eagles in west Scotland. *J. Appl. Ecol.* **6**:381–389.

59. J. R. Longcore and F. B. Samson. 1973. Eggshell breakage by incubating black ducks fed DDE. *J. Wildl. Manage.* **37**:390–394.

60. J. R. Longcore, F. B. Samson, and T. W. Whittendale. 1971. DDE thins eggshells and lowers reproductive success of captive black ducks. *Bull. Environ. Contam. Toxicol.* **6**:485–490.

61. T. H. Maren, E. R. Swenson, D. S. Miller, and W. B. Kinter. 1974. Failure of DDT to inhibit carbonic anhydrase *in vitro* in shell glands of the duck *Anas platyrhynchos. Bull. Mt. Desert Isl. Biol. Lab.* **14**:63–68.

62. M. A. R. McLane and L. C. Hall. 1972. DDE thins screech owl eggshells. *Bull. Environ. Contam. Toxicol.* **8**:65–68.

63. D. S. Miller, A. Seymour, D. Shoemaker, M. H. Winsor, D. B. Peakall, and W. B. Kinter. 1975. Possible enzymatic basis of DDE-induced eggshell thinning in the white Pekin duck, *Anas platyrhynchos. Bull. Mt. Desert Isl. Biol. Lab.* **14**:73–76.

64. Monsanto Company. 1970. *Toxicity, Reproduction, and Residue Study of Aroclor 1242, 1254, and 1260 in White Leghorn Chickens.* Report prepared by Industrial Biotest Laboratory, Northwood, Illinois. 85 pp.

65. I. Newton. 1973. Success of sparrowhawks in an area of pesticide usage. *Bird Study* **20**:1–8.

66. M. I. Oestreicher, D. H. Shuman, and C. F. Wurster. 1971. DDE reduces medullary bone formation in birds. *Nature* **229**:571.

67. H. M. Ohlendorf, E. E. Klaas, and T. E. Kaiser. 1974. Environmental pollution in relation to estuarine birds. In *Survival of Toxic Environments.* Ed. M. A. Q. Khan and J. P. Bederka. Academic. pp. 53–81.

68. D. B. Peakall. 1967. Pesticide-induced breakdown of steroids in birds. *Nature* **216**:505–506.

69. D. B. Peakall. 1970. *p,p′*-DDT: Effect on calcium metabolism and concentration of oestradiol in the blood. *Science* **168**:592–594.

70. D. B. Peakall. 1970. Pesticides and the reproduction of birds. *Sci. Am.* **222**(4):72–78.

71. D. B. Peakall, D. S. Miller, and W. B. Kinter. 1975. Blood calcium levels and the mechanism of DDT-induced eggshell thinning. *Environ. Pollut.* **9**:289–294.

72. D. B. Peakall, J. L. Lincer, R. W. Risebrough, J. B. Pritchard, and W. B. Kinter. 1973. DDE-induced eggshell thinning: Structural and physiological effects in three species. *Comp. Gen. Pharmacol.* **4**:305–313.

73. Y. Pocker, W. M. Beug, and V. R. Ainardi. 1971. Carbonic anhydrase in interactions with DDT, DDE, and dieldrin. *Science* **174**:1336–1339.

74. R. D. Porter and S. N. Wiemeyer. 1969. Dieldrin and DDT: Effects on sparrowhawk eggshells and reproduction. *Science* **165**:199–200.

75. I. Prestt. 1970. Organochlorine pollution of rivers and the heron (*Ardea cinerea*). *Int. Union. Cons. Nature, 11th Tech. Mtg.* **1**:95–102.

76. I. Prestt and D. A. Ratcliffe. 1972. Effects of organochlorine insecticides on European birdlife. *Proc. 15th Int. Ornithol. Congr.* pp. 486–513.

77. I. Prestt, D. J. Jefferies, and N. W. Moore. 1970. Polychlorinated biphenyls in wild birds in Britain and their avian toxicity. *Environ. Pollut.* **1**:3–26.

78. D. Puleston. 1975. Return of the osprey. *Nat. Hist.* **84**:52–59.

79. D. A. Ratcliffe. 1973. The status of the peregrine in Great Britain. *Bird Study* **10**:56–88.

80. D. A. Ratcliffe. 1967. Decrease in eggshell weight in certain birds of prey. *Nature* **215**:208–210.

81. D. A. Ratcliffe. 1970. Changes attributable to pesticides in egg breakage frequency and eggshell thickness in some British birds. *J. Appl. Ecol.* **7**:67–115.

82. R. W. Risebrough. 1969. Chlorinated hydrocarbons in marine ecosystems. In *Chemical Fallout*. Eds. M. W. Miller and G. G. Berg. Charles C Thomas. pp. 5–23.

83. R. W. Risebrough, J. Davis and D. W. Anderson. 1970. Effects of chlorinated hydrocarbons on birds. In *The Biological Impact of Pesticides in the Environment*. Ed. J. W. Gillett. Environ. Health Series 1, Oregon State University. pp. 40–53.

84. R. W. Risebrough, P. Reiche, D. B. Peakall, S. G. Herman, and M. N. Kirven. 1968. Polychlorinated biphenyls in the global ecosystem. *Nature* **220**:1098–1102.

85. W. C. Robertson. 1969. *Sooty Tern Hatch Failure*. Smithsonian Institution Center for Short-Lived Phenomena. Event Notification Report 85-69-657.

86. E. A. Sauter and E. E. Steele. 1972. The effect of low level pesticide feeding on the fertility and hatchability of chicken eggs. *Poult. Sci.* **51**:71–76.

87. J. C. Seidensticker and H. V. Reynolds. 1971. The nesting, reproductive performance, and chlorinated hydrocarbon residues in the red-tailed hawk and great horned owl in south-central Montana. *Wilson Bull.* **83**:408–418.

88. S. I. Smith, C. W. Weber, and B. L. Reid. 1970. Dietary pesticides and contamination of yolks and abdominal fat of laying hens. *Poult. Sci.* **49**:233–236.

89. N. E. R. Snyder, H. A. Snyder, J. L. Lincer, and R. T. Reynolds. 1973. Organochlorines, heavy metals, and the biology of North American accipiters. *BioScience* **23**:300–315.

90. L. F. Stickel and L. I. Rhodes. 1970. The thin eggshell problem. In *The Biological Impact of Pesticides in the Environment*. Oregon State University, Environ. Health Sci. Series No. 1. pp. 31–35.

91. B. Switzer, V. Lewin, and F. H. Wolfe. 1971. Shell thickness, DDE levels in eggs, and reproductive success in common terns in Alberta. *Can. J. Zool.* **49**:69–73.

92. N. S. Talekar. 1971. DDT and reproduction of birds. *Pesticides (Bombay)* **5**(5):11–20.

93. L. W. Taylor and B. R. Burmester. 1940. Effect of thyroidectomy on production, quality and composition of chicken eggs. *Poult. Sci.* **19**:326–331.

94. R. K. Tucker and H. A. Haegele. 1970. Eggshell thinning as influenced by method of DDT exposure. *Bull. Environ. Contam. Toxicol.* **5**:191–194.

95. K. Vermeer and R. T. Reynolds. 1970. Organochlorine residues in aquatic birds in the Canadian prairie provinces. *Can. Field-Nat.* **84**:117–130.

96. G. W. Ware. 1975. Effects of DDT on reproduction in higher animals. *Residue Rev.* **59**:119–140.

97. C. C. Whitehead, J. N. Downie, and J. A. Phillips. 1972. BHC not found to reduce the shell quantity of hen's eggs. *Nature* **239**:411–412.

98. S. N. Wiemeyer, B. M. Mulhern, F. J. Ligas, R. J. Hensel, J. E. Mathisen, F. C. Robards, and S. Postupalsky. 1972. Residues of organochlorine pesticides, polychlorinated biphenyls, and mercury in bald eagle eggs and changes in shell thickness. *Pestic. Monit. J.* **6**:50–53.

99. S. N. Wiemeyer and R. D. Porter. 1970. DDE thins egg-shells of captive American kestrels. *Nature* **227**:737–738.

100. S. N. Wiemeyer, P. R. Spitzer, W. C. Krantz, T. G. Lamont, and E. Cromartie. 1975. Effects of environmental pollutants on Connecticut and Maryland ospreys. *J. Wildl. Manage.* **39**:124–139.

101. C. F. Wurster. 1969. Chlorinated insecticides and avian reproduction: how are they related? In *Chemical Fallout* Eds. M. W. Miller and G. G. Berg. Charles C Thomas. pp. 368–387.

102. C. F. Wurster and D. B. Wingate. 1968. DDT residues and declining reproduction in the Bermuda petrel. *Science* **159**:979–981.

10

INSECTICIDE RESIDUES AND BIOTIC FOOD CHAINS

A. WATER TRANSPORT FROM TREATED TERRAIN

Erosive Runoff

The amount of a field application of insecticide that runs off into streams and lakes depends on the degree of slope of the ground, the fineness of the soil, and the degree of vegetation cover.[154] Water transport of the organochlorines depends on erosive runoff, since they are strongly absorbed to soil particles. While most OP insecticides can leach, they are usually degraded before they have traveled very far. Most of the loss is in the first good rain after the application. The following figures on organochlorine concentrations in runoff and for the total loss during a year (Table 10.1) show that unless the treated ground has been newly ploughed the loss is not great, and less than with most herbicides.

Table 10.1.   Runoff of organochlorine insecticides from different conditions of fields (Pionke and Chesters, 1973).

Ground Cover	Soil Type	Percent Slope	ppm in Runoff	Percent lost to Runoff
Aldrin at 1.5 kg/ha, Kentucky (63)				
Ploughed	Silt	1-2	--	5.2
DDT at 0.73 kg/ha, Maine (47)				
Potatoes	Loam	8	7.0-83	1.6
Endrin at 1.5 kg/ha, Maine (47)				
Potatoes	Loam	8	1.0-49	0.9
Endosulfan at 1.0 kg/ha, Maine (47)				
Potatoes	Loam	8	1.0-19	0.35
Dieldrin at 5.6 kg/ha, Ohio (29)				
Corn	Silt	14	0.4-4.1	0.07

Moreover the contribution of agricultural land to the pollution of creeks and rivers may be considerably less than that of urban areas contributing outfalls through storm sewers; the Red Cedar River in Michigan in the late 1960s received little DDT from the farmland upstream, but the main input came from the waste-water treatment plants of municipalities engaged in annual spraying against Dutch elm disease.[208] Another point as far as river fish are concerned is that the residues that may kill them come in pulses with each good rain; in Twin Bayou, Mississippi, bordered by a 275-acre cottonfield, a heavy rain on August 14 brought down mud containing 1 ppm DDT, which killed bluegill and green sunfish, gizzard shad, and a few *Gambusia.*[48]

A season's treatment of cottonfields consisting of a succession of 12 DDT applications each at 1 lb/a lost a total of 2.8% to runoff during the calendar year, whereas with 12 successive applications of toxaphene at 2 lb/a only 0.4% of the total could be found in the runoff; the water of ponds in the area came to contain 13 ppb DDT and 65 ppb toxaphene as a result of these applications.[19] In Oklahoma, the runoff of endrin from cottonfields resulted in a maximum concentration in drainage ditches of 50 ppt in the water. The runoff from cornfields treated with aldrin at 5 lb/a against soil insects gave ditch concentrations not exceeding 20 ppt, and the maximum content in the water of the Kankakee River was 200 ppt. There were no residues of endrin or dieldrin in the aqueous sediments.[178]

On irrigated land where runoff is assured, aerial applications of DDT at 2 lb/a and of diazinon at 1 lb/a yielded water concentrations (particlefree) peaking at 500 ppm DDT and at 1000 ppm diazinon on the day after the last application.[79] Runoff water in ditches in the Bradford Marsh of Ontario, Canada, a black-soil area devoted to truck farming, when sampled throughout the summer of 1972 showed peaks of parathion at 50 ppt, ethion at 100 ppt, and diazinon an early peak at 1 ppb and a later one at 2 ppb; these three OP insecticides could be detected in every sample from April to October.[69] In New York orchards in 1973, the seven successive sprays of azinphosmethyl each at 0.75 lb/a gave a maximum soil residue of 4.5 ppm by late August, but it had all disappeared by October; soil residues from carbaryl applications at 3 lb/a had decreased to less than 1 ppm within a week of each treatment.[104]

The residues in fish resulting from runoff were measured in a Wisconsin orchard area in 1967 and an Iowa corn-growing area in 1970. The runoff in two creeks embracing a hill with an orchard where dieldrin had been applied not only on the trees but also for rodent control gave the following residues in three species of Wisconsin fish (in ppb)[138]:

	Brook Trout	Muddler	Creek Chub
t-DDT	162	62	72
Dieldrin	11	12	10

The runoff of dieldrin from aldrin-treated cornfields in Iowa gave residues in the edible flesh of 12–175 ppb in the walleyes, crappies, and bluegills, and 0.1–1.6 ppm in the catfish and carp inhabiting the stream which drained the area.[137]

Leaching

The organochlorine insecticides become so tightly bound to the soil particles that they do not leach into the ground water. With the OP insecticides, however, it is a different matter; for example, when a mixture of parathion and aldrin is added to a 60-cm column of soil, parathion percolates through but aldrin does not.[111] The comparative leaching abilities of OP compounds are, for example: phorate, disulfoton > parathion, methyl parathion > ethion, carbophenothion.[102] But the majority of OP insecticides employed against soil insects have no greater leaching mobility than the least mobile herbicides, only thionazin being as leachable as the ureas or triazines (Fig. 10.1).

B. FATE OF RESIDUES IN WATER BODIES

Organochlorine Insecticides

Apolar compounds of this class either reach the aquatic sink adsorbed onto soil particles in the runoff, or when directly applied to water become adsorbed onto the suspended matter. DDT and the cyclodiene insecticides are the organochlorines that are most actively adsorbed. When representa-

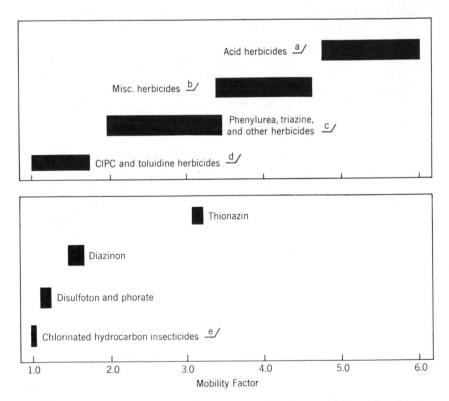

Fig. 10.1. Mobilities of herbicides and insecticides in a subirrigated soil column (from Pionke and Chesters, 1973 after Harris, 1969).
Reprinted with permission from Journal of Agricultural and Food Chemistry; copyright by the American Chemical Society.

tive organochlorines were added to a turbid aqueous suspension, the ratio of adsorbed to free insecticide was 100:1 for *p,p'*-DDT, 10:1 for methoxychlor, and 4:1 for endosulfan.[162]

The importance of anaerobic conditions in promoting the degradation of organochlorine insecticides has been stressed in the review of Pionke and Chesters (1973). Even when the redox potential is decreased by chemical means, aerobic aqueous systems cannot degrade DDT, endrin, or lindane.[109] Raw river water containing colloidal material but kept aerobic did not degrade DDT, dieldrin, endrin, or lindane. On the other hand, it degraded endosulfan and heptachlor completely in 1–2 wk, and converted 80% of added aldrin to dieldrin.[44]

DDT and Methoxychlor. When a pond was treated with DDT at 0.02 ppm, an effective concentration for mosquito control, the DDT disappeared from the water after 3 wk, and was found in the mud for 8 wk after the treatment.[21] Greater amounts of DDT reach the lake bottom where the sedimenting material is composed of finer particles.[7] In waters from six California lakes that were probably oxygen-depleted, DDT was dechlorinated to DDD in amounts which, depending on the plankton content, ranged up to 95%.[132] More than 80% of the microbial isolates from Lake Michigan water were capable of degrading DDT to DDD, and nearly half of them carried the dechlorination further to dichlorodiphenylethane (DDNS); 20% of the isolates produced DDE also.[124] From marine waters, the most active isolates producing DDD came from inshore locations such as Kaneohe Bay on Oahu and the Houston ship canal.[150]

The degradation of DDT in the sediments was similar to that in water. Out of nearly 200 microbial isolates from hydrosoils of Lake Michigan and its tributaries, almost 80% produced DDD and fully 50% produced the further metabolite DDNS, while about one-third of the isolates produced DDE.[124] Of the marine hydrosoils, isolates from the sea floor were weak in degradative capacity, while those taken from bays and estuaries formed DDD with small amounts of DDNS and dichlorodiphenyl-hydroxyethane (DDOH).[150] Incubation of DDT with aquatic sediments in the laboratory, which intensifies the degradation process,[4] resulted in its further conversion to dichlorobenzophenone (DBP), along with minor amounts of the metabolites DDMU, DDMS, DDNU, DDM, and DBH[152] (see Chapter 4).

DDT in sewage sludge is rapidly converted to DDD under the anaerobic conditions there prevailing[78]; dichlorodiphenylchloroethane (DDMS) and DBP appear as further metabolites.[4] On prolonged incubation with the sludge (up to 6 months), DDMU, DDNU, and DDM can also be detected.[152] The toxic metabolite DDCN (dichlorodiphenylacetonitrile) was discovered in sewage sludge and lake sediments in Sweden.[86]

Although none of the bacteria that effect these degradations to DDD (and beyond) in water, sediment and sewage sludge have been identified, they probably belong to genera already known to be active on DDT in soils (e.g., *Aerobacter, Hydrogenomonas*). In sea water, representative marine algae were being found to produce only DDE, six out of eight species tested being active.[18,161] The percentage conversions for three species tested in both investigations were as follows:

	After 14 days[4]	After 24 days[161]
Skeletonema costatum	4.5	4.9
Cyclotella nana	3.9	1.2
Amphidinium carteri	0	2.8

The marine diatom *Cylindrotheca closterium*[98] and the freshwater diatom *Nitzschia*[133] also produce only DDE, and in small amounts; qualifications of these tests are discussed in the review of Johnsen (1976).

When water bodies are treated with DDT, the macrophytes first take up much of the insecticide, and then give it up to the water and thence to the sediments. In a Florida salt marsh ditch treated at 0.2 lb/a, *Ruppia maritima* showed residues of 75 ppm after 3–4 wk, which declined to 9 ppm a month later.[38] In a farm pond treated at 0.02 ppm, the *Potamogeton* contained residues of 30 ppm the day after treatment, which declined to about 4 ppm 2 wk later, at which time the plants started to die presumably from the kerosene in the DDT formulation.[21]

When methoxychlor drains into the water in a model ecosystem (see below), methoxychlor-DDE (dimethoxydiphenyldichloroethylene) is found only in the *Physa* snails.[128] When added to ponds at 0.04 ppm, methoxychlor could not be found in the bottom sediment.[101] *Flavobacterium harrisonii* slowly produced methoxychlor-DDE at a second-order rate constant of 1:1, but this metabolite could not be produced by *Aspergillus, Chlorella,* or *Bacillus subtilis*.[149] The coliform bacterium *Aerobacter aerogenes* also produces this olefin derivative by dehydrochlorinating methoxychlor.[125]

Cyclodiene Derivatives and Lindane.   Aldrin is seldom found in sediments, having been usually all converted to dieldrin, but it is 10 times as persistent as DDT in sewage sludge.[78] While the alga *Dunaliella* can degrade it to aldrin diol,[150] *Chlorella pyrenoidosa* can remove the endomethane group and reduce the methylene linkage in the nonchlorinated half of the naphthalene framework.[45] Dieldrin is quite stable in aquatic environments,

although about 50% of the organisms isolated from Lake Michigan sediment, including *Dunaliella,*[150] were capable of rearranging it to photodieldrin.[123] Algae taken from a brackish fish-pond in Hawaii could degrade
endrin to keto-endrin.[150] Heptachlor is converted by *Chlorella pyrenoidosa*
to the epoxide, which is partly converted to a keto derivative.[45] In sewage
sludge, only 1 day is required for complete conversion of heptachlor at 1
ppm to its epoxide, which in turn is very refractory to further degradation.[78]
Moreover, heptachlor degradation is not suppressed by a decreased reduction potential of the aqueous medium, [109] since apart from its epoxidation it
is mainly hydrolyzed to hydroxychlordene.

Toxaphene can remain in the treated water for some considerable time.
The concentrations (ppb) in the water of two Oregon lakes treated with
toxaphene in 1961 were as follows:[185]

	1961	1962	1963	1964
Deep Oligotrophic Lake	40	2.1	1.2	0.64
Shallow Eutrophic Lake	88	0.63	0.41	0.02

In eight Wisconsin lakes treated with 100 ppb toxaphene, the sediments
continued to show residues between 0.2 and 1 ppm for 3–9 yr after the treatment.[92] In a shallow New Mexico lake, treated at 50 ppb, the toxaphene in
the water had dropped to 1 ppb a month later, a concentration maintained
for the ensuing 9 months; the maximum concentration in the sediments was
only 150 ppb.[95] By contrast, the residues in the aquatic plants 1 wk after the
applications reached a peak of 18 ppm, and remained above 2 ppm for 3 yr.

Lindane (gamma-BHC) is stable in lake water, but the influence of the
mud below it is to precipitate it and cause settling,[111] so that in artificial
impoundments fully two-thirds of it has reached the hydrosoil after 9 days.
As with DDT, so with lindane the finer sediments adsorb more of this
organochlorine from the water.[113] The sediment causes its conversion to
alpha-BHC and a small amount of delta-BHC, the proportion converted in
9 days being 15% of the gamma-BHC under aerobic conditions, and 90%
under anaerobic conditions.[143] This conversion to alpha-BHC, about one-
quarter as toxic as gamma-BHC, was found in sediments from the bottom
of Pearl Harbor, Hawaii, and *Pseudomonas putida* was discovered to effect
the conversion very slowly if NAD was added.[10] Small amounts of gamma-
PCCH were produced from gamma-BHC by *Chlorella pyrenoidosa,*[45] and
by *C. vulgaris* and *Chlamydomonas reinhardtii* (Sweeney, 1968[189]). Lindane
at 1 ppm in sewage sludge is completely degraded within a day, under conditions where 10 days were required for DDT and 40 days for DDD; the
conversion proceeded faster under anaerobic conditions.[78]

Organophosphorus and Carbamate Insecticides

OP Compounds. Most of the organophosphorus insecticides, with the notable exception of diazinon which degrades in acid, are hydrolyzed in water at alkaline pH's.[154] In raw river water taken from the Little Miami river, Ohio, the percentages of the OP compound remaining undegraded after 2 wk at room temperature, during which time the pH rose from 7.3 to 8.0, were as follows[44]:

Malathion	10	Parathion	30
Fenthion	10	Ethion	75
Carbophenothion	10	Dimethoate	85
Methyl parathion	10	Monocrotophos	100

Among carbamate insecticides, the percentages remaining undegraded after 1-wk exposure were the following:

Methiocarb	0	Propoxur	50
Carbaryl	5	Aminocarb	60
Mexacarbate	15		

Parathion applied to ponds at 1 lb/a decreases from an initial 450 ppb in the water down to about 3 ppm in 2 wk, while the average concentrations in the hydrosoil 3 wk after the application are about 30 ppb. At 0.1 lb/a, the dosage employed to control mosquito larvae, 0.01 ppm parathion was found in the water 4 days after treatment and none remained in the hydrosoil.[139] Sediments from an eutrophic alkaline lake were more active than those from an oligotrophic acid lake in degrading parathion, mainly to amino-parathion, the rates being 0.74 and 0.28% per day, respectively.[56] In alkaline water this biological reduction is reinforced by chemical hydrolysis, which produces diethylphosphorothioic acid (DEPTA) and *p*-nitrophenol.[147] Among the organisms isolated from polluted water, *Bacillus subtilis* is the most active in degrading parathion to aminoparathion; *Aspergillus oryzae* is partly effective, while *Penicillium, Mucor, Proteus,* and *Saccharomyces* are inactive. The unicellular alga *Chlorella pyrenoidosa* also degraded parathion to aminoparathion, with no *p*-nitrophenol or *p*-aminophenol appearing.[120] Seepage of parathion from a canal draining sprayed citrus groves contaminated shallow wells supplying water to a Florida town for several months in 1962–1963, although the concentrations did not exceed 1 ppb. Temporary breakdowns of a waste treatment facility in an Alabama parathion-producing plant in 1961 and again in 1966 allowed sufficient quantities to reach and pass through the municipal sewage system to kill fish along 28 miles of Choccolocco Creek and out into the Coosa River.[144]

In lake water held in the dark, methyl parathion was all degraded in less than 7 months, whereas 0.2% of the parathion was still remaining after 1 yr. Methyl parathion and fenitrothion are rapidly degraded by *B. subtilis* through reduction of the *p*-nitro group, while malathion, fenthion, and diazinon are more slowly inactivated by this bacterium.[206]

Azinphosmethyl, like parathion, can percolate through 60 cm of soil.[111] When employed as an experimental piscicide at 1 ppm, azinphosmethyl disappears from the water within 2 wk of the treatment.[129] Its degradation follows first-order kinetics, and while stable on the acid side of neutrality, its half-life at pH 8.6 in sterile water at 25°C is 28 days.[76]

Malathion, quite stable in neutral and acid waters, is readily degraded by aquatic bacteria such as *Pseudomonas, Xanthomonas, Comamonas,* and *Flavobacterium,* which hydrolyze off one of the ethyl groups of the succinyl-ester side chain.[148] The product of this biological hydrolysis is the malathion β-monoacid, along with some diethyl maleate,[209] viz.:

Whereas the product of chemical hydrolysis is the malathion α-monoacid and diethyl fumarate, viz.:

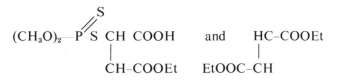

although the malathion metabolite excreted by the rat was conclusively identified by NMR assessment as the α-monoacid.[32] Bacteria from river water produce in addition small amounts of dimethylphosphorodithioic acid (DMPDTA) and malathion dicarboxylic acid.[148] Although malathion can inhibit sewage microorganisms, when the sludge is in high concentration these microorganisms are stimulated by the malathion and become adapted to hydrolyze it.[157]

Chlorpyrifos applied to ponds at 1 lb/a decreases from an initial 220 ppb in the water down to about 3 ppb in 2 wk, thus resembling parathion; however, the concentration in the hydrosoil was still 250 ppb 2 wk after the application, following a maximum 520 ppb a week earlier. As with parathion, quite high residues (20–30 ppm) were reached in the aquatic vegetation, but they decreased to less than 1 ppm after 2 wk.[85] Chlorpyrifos applied to a Texas salt marsh at 0.05 lb/a, which produced a maximum concentration of 8 ppb an hour after application, had disappeared from the water 1 wk later, and from that time on could not be found in the hydrosoil.[115] In polluted waters chlorpyrifos is quite rapidly hydrolyzed, and the particles of organic matter adsorb it from the water.[171]

Carbamates. Carbaryl, applied to Oregon estuaries to control the glass shrimp which are pests of Pacific oyster beds, had a half-life in sea water of 38 days at 8°C, producing the hydrolytic product 1-naphthol, which is adsorbed on the bottom mud along with some intact carbaryl.[97] This initial hydrolysis is chemical in nature, the bacteria being incapable of it.[83,149] The 1-naphthol produced is readily degraded by aquatic bacteria such as *Pseudomonas, Bacillus, Brevibacterium,* and *Flavobacterium,* which use it as a source of carbon, but did not degrade carbaryl itself.[149] *Flavobacterium* degrades the 1-naphthol to *o*-hydroxycinnamic acid and salicylic acid.[83] After 2 wk in sea water, 80% of the 1-naphthol still remains in the water if it has been sterilized, as compared with only 2% if the water had not been sterilized of microorganisms.[106]

The carbamate Zectran (mexacarbate) often employed in forest spraying was degraded by *Pseudomonas* and *Trichoderma viride* to dimethyl-aminoxylenol, the N-methylcarbamic acid moiety being removed by hydrolysis; four other metabolites were identified among those produced by several microbial isolates from forest soils.[11] Similarly with carbofuran (Furadan), the principal degradation product rapidly produced in water of a model ecosystem was carbofuran phenol and its 3-hydroxy derivative, as well as the 3-keto, 3-hydroxy, and N-hydroxymethyl derivatives of carbofuran itself.[207]

Although not a carbamate but an aliphatic ester, the juvenile-hormone mimic methoprene (Altosid) should be mentioned here because of its use against mosquito larvae; its half-life in pond water is approximately 30 hr, the breakdown being entirely due to microorganisms. It undergoes hydrolysis and O-demethylation, and 7-methoxycitronellic acid is among the metabolites.[172]

C. FATE OF RESIDUES IN THE BIOTA

Uptake of Insecticides from Water

Microorganisms. The algae and bacteria in the water are very efficient concentrators of insecticides, their small size and consequently high surface-to-mass ratio making for rapid and thorough adsorption. Exposed to 1 ppb of DDT or methoxychlor, *Aerobacter aerogenes* and *Bacillus subtilis* concentrated them from the water by 1140–3400 times within half an hour, and adsorbed just as much DDT when they had been killed by autoclaving.[89] Within 4 hr, suspensions of *Agrobacterium tumefaciens* could concentrate all of the DDT and nine-tenths of the dieldrin to which they had been exposed in 25 ml of water, while *Trichoderma viride* and *Streptomyces* absorbed about three-quarters of each of these organochlorines.[30]

The freshwater algae *Microcystis, Anabaena, Scenedesmus,* and *Oedogonium* each concentrated DDT from 1 ppm in the water to 130–270 ppm in their bodies within 1 wk, and did the same for aldrin, dieldrin, and endrin.[186] The uptake of DDT by *Chlorella* was completed within 15 seconds.[177] The marine diatom *Cylindrotheca closterium* concentrated DDT by 190 times, while degrading it to DDE.[98] The accumulations by certain algae and protozoa exposed to 1 ppm of DDT or parathion for 7 days were as follows:

		DDT	Parathion
Anacystis nidulans	(blue-green)	849	50
Scenedesmus obliquus	(green)	626	72
Euglena gracilis	(flagellate)	99	62
Paramecium bursaria	(ciliate)	264	94

These figures appear high because they are expressed in ppm per *dry* weight of the microorganisms.[57] When exposed to the ^{14}C-labeled insecticide in a model ecosystem, the number of times by which *Oedogonium cardiacum* algae concentrated the radioactivity from the water after 33 days were as follows:

Carbaryl[169]	127	Aldrin[126]	1790
Parathion[169]	144	Dieldrin[169]	2210
Mirex[127]	610	Endrin[127]	4530

Arthropods. When aquatic arthropods were exposed to DDT in water concentrations between 50 and 100 ppt for a 3-day period, they achieved the following biomagnifications[90]:

Daphnia magna	114,000	*Gammarus fasciatus*	20,600
Onconectes nais	2,900	*Hexagenia bilineata*	32,600
Palaemonetes kadakiensis	5,000	*Chironomus* larvae	47,800

Larvae of the midge *Chironomus tentans* took up as much *p,p′*-DDE when they were dead as when alive, the uptake being simply proportional to their cuticular area.[40]

Fish.   When the Atlantic croaker was exposed to 1 ppb DDT in salt water for 2 wk, it concentrated it by 12,000 times, and concentrated 0.1 ppb by 40,000 times.[66] Brown trout exposed to 2.3 ppb and given DDT-free food for 3 wk concentrated this insecticide 3000 times in their flesh.[80] Such uptake can occur very fast, exposure of Atlantic salmon fingerlings to 1 ppm DDT for 1 hr resulting in 30 ppm of DDT residues in the liver.[156] The top-minnow *Fundulus heteroclitus* concentrated parathion by 80 times when exposed to 120 ppb parathion in runoff water for 2 days,[131] and the mosquitofish *Gambusia affinis* came to contain 15 ppm parathion after only 4 hr of exposure to 20 ppb in a pond treated at 0.1 lb/a.[139]

Relation to Food Chain.   The direct accumulation of the organochlorine from the water can in certain cases make the additional uptake from the food irrelevant. Reticulated sculpins exposed to dieldrin in laboratory aquatic took up the same amount of the insecticide whether or not the *Tubifex* worms on which they were fed were contaminated with dieldrin.[31] Smallmouth bass fingerlings in artificial pools treated with DDT took up as much of this organochlorine when the aquatic invertebrates inhabiting those pools were replaced as a fish-food source by brine shrimp separately reared free of DDT. Since the pool invertebrates were found to lose their DDT content in parallel with the decline in the DDT in the water, Hamelink, Waybrant, and Ball (1971) proposed that uptake depended on the exchange equilibrium between the tissues and the surrounding water, the source of gain and loss in the fish being the gills, as well as cloacal excretion. They argued that the 172,000-fold biomagnification they found for DDT in the smallmouth bass, as compared to the 19,000-fold average accumulation in the invertebrates and the 7000-fold uptake in the algae, derived from an exchange equilibrium where the extent of the accumulation depended on the fat content of the fish and the liposolubility of the insecticide.

Accordingly there should be an inverse relationship between the water solubility of the organochlorine and its accumulation in fish (Table 10.2). Indeed, among eight species of salt-water fish sampled from San Francisco Bay in 1970, their ranking on the basis of their *t*-DDT content turned out to be the same as their ranking by lipid content; moreover there was a positive correlation between the DDT residues and the fat content among the individuals in each of the species, except dwarf perch.[41] A similar correlation between uptake and fat content characteristics of the species was found in the various kinds of fish in Lake Michigan[159] and in Lake Poinsett, South Dakota.[65]

When the various tissues were compared in goldfish, no correlation was found between lipid content and DDT (or dieldrin) uptake, and the half-life of these organochlorines in the adipose tissue was no greater than in other

Table 10.2. Inverse relationship of pesticide water-solubility to accumulation in fish (Hamelink, Waybrant, and Ball, 1971).

	Solubility in Water ppm	Maximum Accumulation whole fish
Lindane	10	100 x
Toxaphene	3	10,000 x
Dieldrin	0.25	10,000 x
DDT	.0012-.037	100,000 - 1 million x
2,4-D	725	150 x

tissues.[58] When exposure of the goldfish to ^{14}C-dieldrin was followed by exposure to unlabeled dieldrin, the turnover in the tissues was complete in 2 wk. Grzenda, Taylor, and Paris (1972) have, therefore, proposed a receptor-complex theory, which concludes that there is a readily mobilized fraction in the lipid with which the organochlorine forms a charge-transfer complex.

To settle the question whether DDT uptake was through the food chain or through exchange-equilibrium uptake, Macek and Korn (1970) made a precise comparison under continuous-flow rather than static conditions. Brook trout were exposed for 120 days (*i*) in water containing 3 ppt DDT with clean food and (*ii*) in clean water with food containing 3 ppm DDT. They were found to take up whole-body residues of only 0.026 ppm from the contaminated water and clean food, whereas from the clean water and contaminated food they accumulated *t*-DDT residues amounting to 1.9 ppm. Macek (1969) concluded that the fish of Lake Michigan in 1967 could not have directly absorbed the DDT residues amounting to 3–10 ppm from water containing 3 ppt (the level actually present in the lake) during the 2–5 yr they had been in the lake, but rather that the bulk of the residues had accumulated through the food chain.

When a single mosquitofish was kept for 3 days in water containing an organochlorine insecticide at 1–3 ppb, the uptake by exchange equilibrium amounted to considerably less than that accumulated when food chain organisms, including its mosquito prey, were present (Table 10.3); this difference was very great with DDT and DDE, but with endrin and lindane the differential was only twofold, and with mirex the fish took up more insecticide in the absence of the food chain.[127]

The situation in the field, where the contamination is deposited in the sediments and then mobilized by bottom browsers such as ostracods or

Table 10.3.   Uptake of pesticides from water by northern silversides in presence or absence of mosquito larvae in the food chain (Matsumura and Benezet, 1973).

	μg on Sand	ppb in whole Fish	
		Larvae absent	Larvae present
DDT	1.8	458	337
Gamma-HCH	1.5	2904	1080
Zectran	1.1	213	76
TCD Dioxin	1.6	2	708

mosquito larvae, can be mimicked in the laboratory by coating a sand hydrosoil with the pesticide. In such experiments it was found that the presence of such aquatic invertebrates as *Daphnia,* ostracods, and *Culex* larvae would actually reduce the amount taken up by fish, in this case the northern silversides, by competing for the restricted amounts of pesticide available under these confined conditions (Table 10.3). Only with the pesticide contaminant TCDD was the amount in the fish greater when *Culex* was present than when absent, because only these mosquito larvae could mobilize this organochlorine from the sand grains of the hydrosoil.[122]

Food Chains

In his review on the bioconcentration aspects of DDT, Bevenue (1976) quotes the statement of the Administrator of the U.S. Environmental Protection Agency in his cancellation of DDT in 1972, in which he concludes that DDT is an "uncontrollable chemical" because of its "persistence and biomagnification in the food chain."

Aquatic Food Chains.   The first fully developed example of the poisoning of birds through the food chain emerged when DDD was applied to Clear Lake, California in 1957. The recreation amenities of this 75 sq mi warm shallow eutrophic lake were severely handicapped by its producing large numbers of the Clear Lake gnat, *Chaoborus astictopus.* DDD was, therefore, chosen to control the aquatic larvae of this pest species, and treatment of the entire lake in 1949 with an average concentration of 14 ppb achieved 99% control without killing any of its considerable fish population. It was not necessary to repeat the application for 5 yr, and the second application in 1954 achieved 99% control at a dosage of 20 ppb. It became necessary to reapply 3 yr after that, and this time the control achieved at 20 ppb was less satisfactory, the *Chaoborus* larvae being more tolerant to DDD than before.

About 1000 pairs of the western grebe *Aechmophorus occidentalis* bred on Clear Lake in the years before 1949; although the species characteristically winters on the Pacific coast and breeds on inland lakes, many western grebes visited Clear Lake in winter to feed on the fish. After the first application in 1949 there was no breeding on the lake, and indeed not for the next 17 yr. After the second application in 1954, some of the winter-visiting grebes died. When the deaths of winter-visiting grebes were repeated after the 1957 application, the corpses were analyzed[84] and their visceral adipose tissue was found to contain DDD residues of the order of 1600 ppm (0.16% of the wet weight). Likewise the visceral fat of the fish contained the following concentrations of DDD (ppm):

| Bluegills | 125–254 | Brown bullhead | 342–2500 |
| Blackfish | 700–983 | Largemouth bass | 1550–1700 |

It was, therefore, evident that the DDD that had relieved the lakeside cottagers from the gnat pest, while insufficient to kill the fish, had accumulated in them enough to kill the fish-eating western grebe by secondary poisoning. The source of the residues in the fish was the abundant phytoplankton, which was found to contain 5 ppm DDD. Since the residues were so much higher in the predaceous largemouth bass than in the plankton-feeding blackfish, it was concluded that the DDD had accumulated in an aquatic food chain consisting of plankton → herbivore fish → predaceous fish → fish-eating birds (Fig. 10.2). Thus the water concentration of 0.02 ppm was magnified by 80,000 times in the visceral fat of bass and grebes.[167]

Some grebes eventually returned to breed in 1967, 10 yr after the third and last application; the 150 pairs laid about 500 eggs of which 200 hatched,

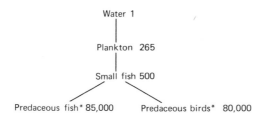

Water 1

Plankton 265

Small fish 500

Predaceous fish* 85,000   Predaceous birds* 80,000

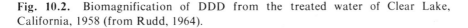

* Residues in visceral fat; whole—body residues for plankton and small fish.

Fig. 10.2. Biomagnification of DDD from the treated water of Clear Lake, California, 1958 (from Rudd, 1964).

and from which 50 juveniles survived. The DDD residues in nestlings amounted to 600 ppm in the fat extracted from the yolk sac, of which about half consisted of the chloroethylene and chloroethane metabolites.[75] The 60 tons of DDD applied in the three lake treatments, reinforced by the DDD produced from bacterial decomposition of the surprisingly small amount of DDT in the runoff from orchard lands around Clear Lake, had resulted in the lake sediments containing 0.5–1.0 ppm DDD (plus some DDE) in the top 5 inches.[168] By 1969, after 12 yr had elapsed since the last DDD application, more than 50% of the breeding pairs raised young; by the late summer of 1971 there were over 600 grebes on the lake, of which one-third were juveniles raised there. Although the higher residues in the predaceous bass could be largely due to exchange-equilibrium magnification operating over their longer life,[136] the food chain concept is quite valid, and afforded at the time a means of comprehending the eventual environmental impact of the persistent and stable organochlorine insecticides.

A clear example of an aquatic food chain culminating in birds was obtained by a survey of the effect of runoff from the orchard areas of Door county, Wisconsin on the biota of Green Bay and Lake Michigan.[77] In these areas 30 tons of DDT had been applied annually, along with 15 tons of DDD and 15 tons of methoxychlor, during the preceding 15 yr. Residue analyses demonstrated the role of the amphipod shrimp *Pontoporeia affinis* in concentrating the DDT to a level 40 times that in the sediments; these in turn served as food for the fish and for the old-squaw duck in the bay, which magnified the residue 10 times more (Fig. 10.3). While the ring-billed gull, feeding on terrestrial insects as well as fish, further concentrated the residues by five times, the herring gull feeding almost entirely on fish concentrated it 20 times again. The total magnification of 10,000 times in

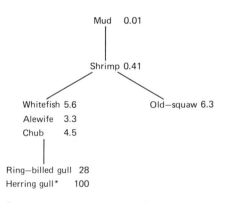

Mud   0.01

Shrimp 0.41

Whitefish 5.6   Old—squaw 6.3
Alewife   3.3
Chub   4.5

Ring—billed gull 28
Herring gull*   100

*Breast muscle for birds, muscle for whitefish;
 whole—body residues for others.

Fig. 10.3. Biomagnification of DDT in Green Bay, Wisconsin, 1964: *t*-DDT residues in ppm (from Hickey, Keith, and Coon, 1966).

Table 10.4. Biomagnification of DDT in Green Bay, Wisconsin, 1964: residues of DDT and metabolites in ppm (Hickey, Keith, and Coon, 1966).

		DDT	DDD	DDE
Bottom Mud (50' depth)		.006	.003	.002
Shrimp (Pontoporeia)		.14	.03	.24
Whitefish (Coregonus)	Muscle	1.8	0.8	3.0
Old-Squaw Duck	Muscle	0.8	0.7	4.8
	Fat	34	9	94
	Brain	0.2	0.1	1.4
Herring Gull	Muscle	14.1	4.8	79.9
	Fat	390	125	1925
	Brain	3.9	1.1	15.8
	Brain	3.9		

the breast muscle of herring gulls as compared to the sediments was accompanied by extensive dehydrochlorination of the DDT residues (Table 10.4), so that the eggs contained 200 ppm DDE. The Green Bay colony produced less than half as many young as the normal production for the herring gull, since one-third of the eggs did not produce hatchlings. It was observed that the eggshells were frequently damaged, appearing at the time as if the residue-containing mother gulls were careless in incubating them.[99]

Another example of an ecosystem with a long history of DDT application was a Long Island salt marsh which had been sprayed annually at 0.2 lb/a for mosquito control for a number of years. The *t*-DDT residues in the salt marsh, in ppm wet-weight on the whole-body basis, were found to be as follows:

Water	0.0005	Silversides and minnows	0.23–0.94
Plankton	0.04	Needlefish (*Strongylura*)	2.07
Bay shrimp	0.16	Mergansers, cormorants, gulls	20–35

One ring-billed gull showed a whole-body residue of 75 ppm *t*-DDT.[205]

Terrestrial Food Chains.   The study described in Chapter 8 of the residues resulting from the application of DDT to control Dutch elm disease vectors also revealed a food chain (Table 10.5). The earthworms magnified the *t*-DDT residues by 10 times as they dragged the DDT-contaminated leaves down into the DDT-contaminated soil. These residues are biomagnified yet again by 10 times in the brains of the robins which prey on the earthworms. In the process most of the DDT becomes converted to DDE.[8]

In a broadleaved Pennsylvania forest sprayed once with DDT at 0.5 lb/a, the residues in the streams 3 months later were 10 ppb in the sediment, 0.2 ppm in crayfish, and 0.6 ppm in suckers and brook trout.[34] In Missouri cornfields which had been treated with aldrin at 1 lb/a annually for the preceding 17 yr, the average residues were as follows (ppm):

Soil	0.03	White-footed mice	1.0
Earthworms and carabids	1.0–1.5	Garter snakes	12.4

Nearly all of the residue was in the form of the epoxidation product dieldrin.[103] Small mammals inhabiting Mississippi fields treated with mirex at 0.85 lb/a showed the following whole-body residue (in ppm)[202]:

Herbivores	(*Sigmodon, Pitymys*)	0.01
Omnivores	(*Peromyscus, Reithrodontomys, Oryzomys*)	0.21
Predators	(*Blarina brevicauda*)	1.00

As has been already seen in Chapter 7, organochlorine residues in a terrestrial environment were much less than in an aquatic environment, except for the suburban areas where exceedingly high dosages of DDT had been applied for Dutch elm disease control.

On the tundra bordering Hudson Bay, where an area around Fort Churchill, Manitoba had received DDT treatments at 0.2 lb/a for 12 suc-

Table 10.5. Biomagnification of DDT under elm shade-trees at Urbana, Illinois: residues in ppm* (Barker, 1958).

Soil (upper 2")		5.9 - 9.5	0.7 - 5.4
Leaves after fall		20 - 28	1 - 4
Earthworms		86	33
American Robins	Liver	19	744
	Brain	90	252

*Dosage 2 lb/tree, giving 1.3 lb/acre on soil, 140 ppm

on leaves

cessive years for mosquito control, and where the terrain is characterized by snow-melt pools, the residues (in ppm) were as follows[23]:

Pool sediments	0.4	Soil	0.09
Aquatic invertebrates	0.3–0.5	Terrestrial plants	0.2–0.11
Fish	0.7–3.6	Lemmings and squirrels	0.5–0.7

Among the abundant summer population of birds, the relation of residues in the adipose tissue to their diet was as follows:

Willow ptarmigan (seeds)	3–4	Ducks and plovers (invertebrates)	20–52
Sparrows (seeds and insects)	11–17	Gulls and terns (fish)	56–64

In the arctic tern, in which the residues in the fatty tissue ranged up to 188 ppm, fully 80% of the *t*-DDT was DDE.

Food Chain Residues in Untreated Areas. By the close of the 1960s, residues had become a problem even in areas which had not been treated at all. Samples of northern pike (*Esox lucius*) and mussels (*Mytilus edulis*) taken in 1967 and 1968 in untreated areas of North Atlantic countries revealed 0.01–0.3 ppm in the soft tissue of the mussels, and 0.02–2 ppm in the lateral muscle of the pike; the highest residues in the pike were in Italy, and the highest in mussels were in *Modiolus demissus* in the United States.[81] In Lake Poinsett, South Dakota, a 12-sq mi body of water used for recreation and surrounded by agricultural land, the biota of the lake contained the following *t*-DDT residues (in ppb)[65]:

Water	0.08	Crayfish	2.0
Sediment	2.2	Aquatic insects	919
Zooplankton and algae	5.0	Fish	100

The fish residues were highest in pike, bass, carp, and catfish, and lowest in bluegills, shiners, yellow perch, and walleye. Other organochlorines found in the biota were heptachlor and its epoxide, aldrin and dieldrin, and toxaphene. The source of the contamination was the runoff from the fields which deposited the insecticides in the lake sediment.

It has been observed in lakes on the prairies that where invertebrates are scarce, organochlorine residues fail to enter the food chain that leads to fish. In Lake Michigan the food chain allowed Coho salmon, introduced from hatcheries at the age of 16 months, to take up DDT residues to the extent of 4 ppm in their first year in the lake when they weigh 3 lb, and accumulate 12 ppm in the second year when they grow to 12 lb weight.[160] The final stage in the food chain, namely fish to birds, was demonstrated by the side effects of treating Big Bear Lake, California with 0.1 ppm toxaphene as a piscicide to eliminate populations of rough fish (two spp. of goldfish). The moribund fish contained 3–21 ppm toxaphene in their flesh, and as a result many fish-eating birds—ducks, terns, gulls, eared grebes, and white pelicans—died of secondary poisoning. As judged by the residues found in the adipose tissue of surviving birds, the victims must have contained toxaphene residues amounting to at least 13 ppm in terns and 1700 ppm in pelicans.[167]

Model Ecosystems

A means of following the fate of an insecticide as it reaches the aquatic sink, and of obtaining a comparative assessment of its biomagnification or biodegradability, is offered by the model ecosystem of Metcalf (1974). A sand bed in an all-glass aquarium measuring $25 \times 30 \times 45$ cm is graded so that half is a pool, while the other half is sown to sorghum (*S. halepense*), which when it has grown to 10 cm height is treated with 5 mg of the radio-labeled insecticide under investigation. It is then artificially infested with 10 large salt-marsh caterpillars (*Estigmene acraea*), and the water is colonized with a filamentous green alga (*Oedogonium cardiacum*), and sometimes with the water weed *Elodea*. The aquatic invertebrates added were water fleas (*Daphnia magna*) and snails (*Physa* sp.), and sometimes the crab *Uca manelensis*. Some 300 larvae of the mosquito *Culex pipiens* are introduced 27 days later. At the end of 30 days, a single mosquitofish (*Gambusia affinis*) is added, and finally all the elements of the ecosystem are assayed for radioactivity (after chromatographic separation) 3 days later, that is, 33 days after its initial treatment and colonization.[128]

Organochlorines. With DDT (Table 10.6), the *Gambusia* was found to have accumulated 18.6 ppm of the original insecticide, representing an Ecological Magnification (EM) of 84,000 times the 0.22 ppb present in the water; the *Gambusia* contained only 0.85 ppm of polar metabolites of DDT as compared to 56.7 ppm of apolar material, principally DDT, DDD, and DDE, thus indicating a Biodegradability Index (BI) of 0.015 for DDT. The month-old ecosystem as a whole had produced in the organisms about twice as much DDE as the DDT remaining.[128] *Tilapia* differed from *Gambusia* and *Lepomis* sunfish in producing more DDD than DDT.[158] When the metabolites DDE and DDD were themselves tested in a model ecosystem, their biomagnification and biodegradability were similar to that of DDT. But methoxychlor[96] showed an EM figure of only 1550 and a BI as high as

Table 10.6. Accumulation and conversion of DDT and its
relatives to their metabolites in snails, mosquito
larvae, and mosquitofish in a model ecosystem, ppm
(Metcalf, Sangha, and Kapoor, 1971).

	Water	Physa	Culex	Gambusia
DDT	.00022	7.6	1.8	18.6
DDD	.00012	1.6	0.4	5.3
DDE	.00026	12.0	5.2	29.2
Polar Metabolites	.00320	1.0	1.5	0.85
DDD	.0004	3.3	3.4	33
Apolar Metabolites	---	1.2	---	3.6
Polar Metabolites	.0056	1.1	---	2.0
DDE	.0053	103.5	159.5	145.0
Polar Metabolites	.0027	18.1	9.4	4.8
Methoxychlor	.00011	13.2	---	0.17
Apolar Metabolites	.00028	1.7	---	---
Polar Metabolites	.00125	0.8	---	0.16

0.94 for *Gambusia* (Table 10.7); the principal metabolite found in *Gambusia, Tilapia* and *Lepomis* was the 2-hydroxy-1,1-diphenyltrichloroethane, with some of the 2,2-dihydroxy metabolite,[158] and these were excreted into the water.

Dieldrin and endrin were even less biodegradable than DDT, but their biomagnification figure in *Gambusia* was similar to that for methoxychlor. However the biomagnification of dieldrin in the *Physa* snails was 115,000 times that in the water, the highest figure recorded in the model ecosystem.[170] Lindane and mirex accumulated less than methoxychlor, lindane being more and mirex less biodegradable than DDT. The PCBs were intermediate between dieldrin and DDE, the EMs for trichlorobiphenyl being approximately 6000, and for tetra- and pentachlorobiphenyl approximately 12,000.[169]

OP Compounds. Among the organophosphorus compounds tested, parathion accumulated as residues in *Gambusia* but not in the invertebrates; *p*-nitrophenol also was present in *Gambusia* only.[169] Chlorpyrifos showed about the same accumulation but with less degradation, while chlorpyrifos-

Table 10.7. Bioaccumulation and biodegradation of insecticides after 33 days in a model ecosystem: data from *Gambusia affinis* added for the last 3 days (Metcalf et al., 1973).

	Ecological Magnification	Exchange-Equilibrium[a]	Biodegradability Index
	Ratio Gambusia/Water	Magnification	Ratio Polar/Apolar
DDT	84,500	344	0.015
DDD	83,500	---	0.054
DDE	27,400	217	0.032
Dieldrin	2,700	---	0.0018
Methoxychlor[b]	1,550	---	0.94
Endrin	1,340	680	0.009
Lindane	560	166	0.091
Parathion[c]	335	---	29.8
Chlorpyrifos[d]	314	---	1.02
Mirex	219	530	0.0145
Chlorpyrifosmethyl[d]	95	---	3.05
Carbofuran	0	---	~∞
Carbaryl[c]	0	---	~∞

[a] 3 days in insecticide-treated water without other ecosystem organisms

[b] from Metcalf, Sangha and Kapoor (1971)

[c] from Sanborn (1974)

[d] from Metcalf (1974)

methyl accumulated less and degraded more than its ethyl analog; in both cases the metabolites were principally the leaving group 3,5,6-trichloropyridinol, with small amounts of other polar compounds.[126] The phosphoroamidothioate Orthene did not itself accumulate in the ecosystem animals, although unknown metabolites were found in them.

Carbamates. None of the carbamate insecticides tested proved to be present as residues in the ecosystem fauna, and thus their EM is zero and their BI approaches infinity (Table 10.7). Residues of Bux and its metabolites were present in the alga *Oedogonium* and the waterweed *Elodea,* but were absent from the animals in the system. Two unknown metabolites of carbaryl were found in the snail and crab.[169] The highly toxic carbofuran did not accumulate in any of the organisms, the principal metabolite being 7-hydroxy-dimethyldihydrocarbofuran (carbofuran phenol) found abundantly in the snail and in small amounts in the mosquitofish along with polar conjugates.[126]

D. RESIDUES IN RIVERS, LAKES, AND ESTUARIES

Insecticides in Streams and Rivers

By 1964 the organochlorines could be detected in most United States rivers; out of 56 rivers sampled, dieldrin was found in 39, DDT in 25, and endrin in 22 of them.[192] In 1966–1968, rivers in the western half of the United States yielded 72 samples positive for DDT, as compared to 16 for dieldrin, 10 for lindane, 3 for endrin, 1 for aldrin, and 0 for heptachlor and/or its epoxide.[121] Of 118 American rivers surveyed in 1968, 63 were positive for some organochlorine, dieldrin being the most frequent; the maximum concentrations found were 316 ppm for DDT (Beaulieu River, Fla.), 840 ppt DDD (Kansas River, Kan.) and 407 ppt for dieldrin (Tombigbee River, Miss.).[110] By the end of the 1960s, the concentration of DDT residues in the lower Rio Grande was 10–20 ppt. The record contamination of a stream was in Cypress Creek, Tennessee, where the sediments contained 0.3% aldrin, 0.9% dieldrin, 1.0% endrin, and 3% chlordane; the samples were taken downstream of an insecticide-manufacturing plant.[9]

In Big Creek, draining DDT-treated tobacco lands in Norfolk county, Ontario, the concentrations found in the water in 1970 never exceeded 70 ppt. The annual treatments had been discontinued after the season of 1969, when the soil contained 3–5 ppm DDT. The maximum concentration in the stream sediments was 441 ppb, and the highest residue in fish (creek chub and suckers) was 1 ppm.[130]

In Flint Creek, draining 15,000 acres of cotton in northern Alabama, the average water concentrations in 1963 were 1 ppt DDT and 78 ppt toxaphene, while the mud contained 0.4 ppm DDT and no detectable amounts of toxaphene; the BHC residues were 26 ppt in the water and 10 ppb in the mud.[145] Residues in the *Hexagenia* mayfly nymphs so valuable as fish food were about 0.2 ppm of DDT and of DDE, and about 0.5 ppm toxaphene. Residues in the edible portions of the stream fish were as follows (in ppm)[58]:

	DDT	DDE	Toxaphene	BHC
Largemouth Bass	0.2	0.07	0.8	0.01
Green Sunfish	0.5	0.1	0.3	rare
Gizzard Shad	0.4	0.1	1.3	0.04

Organochlorine residues in British rivers reached their maximum in 1965, the peak amounts being 908 ppt for DDT, 390 ppt for BHC, and 2940 ppt for dieldrin; many rivers were negative, and the average figures were very much less.[43] A survey of OP residues made in six English rivers in 1968–1970 revealed the presence of diazinon in five and malathion in two of them, the amounts not exceeding 70 ppt. Phorate, chlorfenvinphos, and demeton residues were each found in one of the rivers, while carbophenothion was found in the Chelmer river in Essex in amounts up to 8 ppb.[22] In Mill Creek, draining orchard areas in Michigan, azinphosmethyl can be detected in occasional surges after heavy rain or when the harvested apples are washed.

Organochlorines in the Great Lakes

Lake Michigan is a bypass from the Great Lakes drainage system to the St. Lawrence River and the sea, and would be a complete cul-de-sac but for a tiny drainage southward through the Chicago River. Two of its bays receive runoff from orchard areas, namely Green Bay in Wisconsin which already in 1964 had 14 ppb *t*-DDT in its bottom sediments,[77] and Grand Traverse Bay in Michigan in which sediment concentrations of 1–5 ppm lindane, heptachlor and parathion were found in 1968.[108]

In 1967, on the first return of Coho salmon, which had been planted and matured in the lake, to the rivers of southern Michigan to spawn, a heavy mortality was observed among the sac-fry obtained from the eggs of these salmon.[91] Consequently, residue determinations were made by Leland, Bruce, and Shrimp (1973) on the sediments in the southern half of the lake in 1969 and 1970 (Table 10.8). The average level of *t*-DDT was 18.6 ppb throughout that vast area, being higher on the east side and highest in the

Table 10.8.   Residues of DDT and dieldrin in the sediments of the southern part of Lake Michigan 1969–1970: median values in ppb (Leland, Bruce, and Shimp, 1973).

	Top 2 cm	2-6 cm	6-12 cm
Dieldrin	2.0	Trace	Trace
p,p'-DDT	9.3	3.8	3.0
p,p'-DDD	3.0	0.5	Not det.
p,p'-DDE	2.2	0.8	0.6
o,p'-DDT	1.2	0.7	Trace
t-DDT	18.5	6.3	3.4

center of the lake (Fig. 10.4). On the west side, much of the DDT had been converted to DDE; on the east side, where the sediments had a high content of clay and organic matter, DDD was the predominant compound as a result of such anaerobic and reducing conditions. This input of DDT residue into the lake was due not only to orchard spraying, but more importantly to Dutch elm disease control operations and to moth-proofing establishments[159] where the contamination reached the rivers through the tiled drains of municipalities. The lake sediments also contained residues of dieldrin at an average of 2 ppb, restricted to the upper layer (Table 10.8); the main source of these residues would be the aldrin applied to cornfields to control *Diabrotica* rootworms.

Although at that time the concentrations in the lake water were only 3 ppt *t*-DDT and 1 ppt dieldrin, some 80% of the commercial catch of Coho salmon, chub and whitefish had residues exceeding the 5 ppm tolerance limit for DDT, and the same proportion exceeded the 0.3 ppm tolerance for dieldrin.[201] When a comparison was made with the fish of the other Great Lakes, the *t*-DDT residues were present in them in the following order of concentrations (Table 10.9): Michigan > Huron > Ontario > Erie > Superior. With respect to the residues of dieldrin (Table 10.10), the fish of Lake Michigan again showed the highest and Lake Superior the lowest concentrations. In 1970 the average residues of *t*-DDT in the principal game fish, for Lake Michigan (southern half) in comparison with other lakes, were as follows[160]:

Coho Salmon		Lake Trout	
Lake Michigan	14 ppm	Lake Michigan	19 ppm
Lake Erie	2.2	Lake Superior	4.4

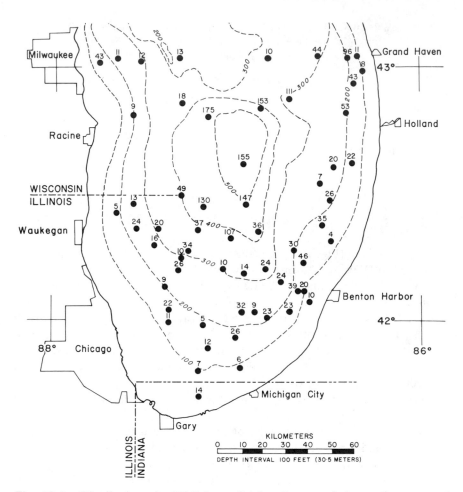

Fig. 10.4. Distribution of *t*-DDT in surficial sediments of the southern part of Lake Michigan, 1969–1970, in ppb. (from Leland, Bruce, and Shimp, 1973).
Reprinted with permission from Environmental Science and Technology, copyright by the American Chemical Society.

About half of the *t*-DDT was DDE, and about one-tenth was DDD. By 1971 virtually all the Coho salmon and lake trout carried residues exceeding the 5 ppm tolerance limit for *t*-DDT.

They also exceeded the tolerance limit for PCBs, the average whole-body residues in Lake Michigan fish ranging from 2.7 ppm in smelt to 15 ppm in lake trout, as compared to the average *t*-DDT residues which ranged from 1

Table 10.9. *t*-DDT residues in fish in the Great Lakes, 1967–1968: whole-body, ppm (Reinert, 1970).

	Superior	Huron	Erie	Ontario	Michigan
Alewife	0.7	2.4	1.6	2.0	3.9
American Smelt	0.3	0.8	1.1	1.6	2.3
Bloater (Chub)	1.1	3.6	---	---	8.6
Lake Herring	1.4	---	---	3.5	6.7
Yellow Perch	---	1.6	0.9	2.1	3.2

ppm in suckers to 16 ppm in rainbow trout.[187] The PCB profile resembled that of the mixture sold as Aroclor 1254.

When PCBs were investigated in the fish of Lake Ontario, the residues in northern pike and long-nosed gar were identified and quantitated in ppm as follows:[94]

Aroclor 1242	0.9–3.7	Aroclor 1260	0.4–1.1
Aroclor 1254	1.0–2.3	Mirex	0.02–0.05

The source of the PCB contamination was considered to have been the atmosphere.

This was also thought to be the origin of the mirex found in Lake Ontario, but there was at least one formulating plant bordering the lake which was using mirex in the manufacture of ant traps. In Mississippi, where mirex granules were applied to control the fire ant, localities which had never been treated showed residues not very much less than the treated areas, the overall averages in the fauna of ponds and estuaries being as

Table 10.10. Dieldrin residues in fish in the Great Lakes, 1967–1968: whole-body, ppm (Reinert, 1970).

	Superior	Huron	Erie	Ontario	Michigan
Alewife	.05	.05	.14	.06	.11
American Smelt	.02	.04	.04	.10	.06
Bloater	.03	.11	--	--	.23
Lake Herring	.02	--	--	.07	.20
Yellow Perch	--	.03	.05	.05	.08

follows (in ppm)[141]:

Molluscs	0.15	Crustacea	0.44
Fish	0.26	Annelida	0.63
Insects	0.29		

This dispersal of the mirex applied to the land and its accumulation in water bodies is of concern because it is scarcely biodegradable and its toxicity to young crayfish, for example, is such that its LC_{50} is less than 0.1 ppb.[114]

Insecticides in Estuaries

By the late 1960s, insecticide contamination brought by the rivers to the Gulf of Mexico resulted in residues ranging from 0.01 to 0.2 ppm *t*-DDT in the estuarine fish, while DDT residues in the Atlantic oyster ranged up to 11 ppm.[26] The major input occurred when the rivers were in flood and brought down particles in suspension.[28] On the Pacific coast, the Dungeness crab off San Francisco contained 2.4 ppm DDE in 1968, and its productivity was decreasing.[134] In Pensacola Bay the *t*-DDT in croaker and pinfish rose from 0.05 ppm in the spring to 0.4 ppm in the fall from the contributions of crop treatments in the hinterland.[67] The decline in sea trout in the Laguna Madre of Texas, described in Chapter 6, was associated with increased organo-chlorine residues in the menhaden on which they feed. The national monitoring program conducted on molluscs (mussels, clams and oysters) around the shores of the United States from 1965 to 1972 found the highest average residues to be 5.4 ppm for *t*-DDT (Florida) and 2.8 ppm for PCBs; there was an instance of 54 ppm toxaphene residues in a Georgia estuary.[27] The residue level in molluscs reached their peak in 1968, and showed a sharp decline in 1971, as shown by the following frequency distribution of samples according to their DDT content:

	<10 ppb	10–100 ppb	>100 ppb
1967	9	90	20
1971	87	26	2

By 1968–1969 the average residues of endrin and dieldrin in the Atlantic oyster off the mouth of the Mississippi River were, respectively, 1.3 and < 1.0 ppb; this was much less than those found in the years 1964–1966 which had been preceded by a 10-yr period of heavy loads of these cotton insecticides reaching the river.[166]

In the Thames estuary of England, *t*-DDT residues in mussels and oysters were more moderate than in the southeastern United States, seldom exceeding 0.1 ppm.[155] Nevertheless the residues in the pink shrimp ranged up to a concentration which was just below that found in shrimps killed by long-term exposure to DDT; the water dosage which was thus fatal was 0.2 ppb, and this amount of DDT has been found in Texas river-water[121] flowing into shrimp nursery areas. The deliberate addition of carbaryl to Yaquina Bay, Oregon to control the crustacean and molluscan pests of the Pacific oyster did not pose a residue problem, since it was more than 90% hydrolyzed in 10 days.[97]

In Biscayne Bay, Florida, surface slicks formed from oily organic matter emitted from canal outflows constitute areas of concentration of the lipophilic organochlorines on which the little fish come to feed, and the big fish come to feed on the little fish. In the bay, a water concentration of 0.1 ppt *t*-DDT is converted to a slick concentration of 10 ppb, mainly DDE. Slicks escaping into the Florida current to contribute to ocean contamination contained 60 ppt DDE at the end of the 1960s.[173] Oil slicks from petroleum-derived hydrocarbons are also efficient concentrators,[72] while the surface microlayer of the open sea has a higher concentration than the exceedingly low content in the water beneath it.[14]

E. GLOBAL CONTAMINATION WITH INSECTICIDES

Insecticides in the Atmosphere

This subject has been recently reviewed by Wheatley (1973) and by Spencer (1975). During the 1960s, concentrations of aldrin, chlordane, and toxaphene at approximately 1–10 ng/m³ were detected in the air of agricultural communities of the United States.[182] In 1965, the air over central London[2] contained DDT at 12 ng/m³. In 1967, the maximum concentrations of DDT in the air over agricultural areas in Florida and Mississippi reached levels close to 1000 ng/m³ (Table 10.11). The concentration of the *o,p'*-DDT present in technical DDT was also high, as was DDE which is a photolytic metabolite of DDT[135] and is less nonvolatile than DDT. At the nine United States stations sampled in 1967, the maximum concentration of other insecticides at any station in ng/m³ were as follows (Barney, 1970[153]):

BHC	22	Endrin	58
Heptachlor	19	Parathion	465
Aldrin	8	Malathion	2
Dieldrin	30		

Table 10.11. Maximum concentrations (ng/m³) of DDT residues in air over rural and urban communities during 1967 (Stanley et al., 1971).

	DDT	DDE	o,p'-DDT*
Orlando, Fla.	1560	131	510
Stoneville, Miss.	950	47	252
Dotham, Ala.	177	13	92
Riverside, Cal.	24	11	6.2
Baltimore, Md.	20	2.4	3.0
Fresno, Cal.	11	6.4	5.5
Buffalo, N.Y.	11	--	2.9
Salt Lake City, Ut.	8.6	--	1.4
Iowa City, Iowa	2.7	3.7	2.1

* including o,p'-DDE

Air-Borne Drift of Spray. Vapor concentrations of parathion, azinphosmethyl, malathion, and carbaryl have been quite high in orchards immediately after spray operations. Insecticidal fogs dispersed against adult mosquitoes left between 1 and 10 ng/m³ malathion in the air.[182] The pesticide losses to the air are greater with mist-blowers than with aircraft,[190] and the off-target losses with aircraft spraying exceeded 50% on alfalfa[191] and approached 75% on cornfields.[79]

Evaporation from Soil and Water. The vapor pressure of DDT is only 1.5 × 10⁻⁷ mm Hg, yet the average DDT content of air at the end of the 1960s was 1 ng/m³,[155] and thus the world's atmosphere in the early 1970s was estimated to contain some 8000 metric tons of DDT.[50]

The loss of DDT from the land is faster the greater the moisture content of the soil, and the same has been observed for other organochlorines and other pesticides.[61] The water film on the soil particles prevents the tight adsorption of the pesticide, and provides a "wick effect" as the water rises to replace what is evaporated at the soil surface, the DDT tending to accumulate at the water–air interface.[71] Determinations of loss from English soils indicated that losses of DDT from moist soil would be of the order of 1–2 lb/a-yr, the rate being 0.17 lb/a for each summer month.[113] However, determinations by American workers,[49,60] presumably made on dry soils, have indicated that the losses were of the order of 0.1 lb/a-yr. When the evaporated DDT was trapped, it was found that 80–160 days after application to the soil at 40 ppm the air concentration 10 cm above its surface was 15–50 ng/m^3; with DDD the amounts evaporated were greater.[199]

At the surface of the water, the low water solubility of DDT gives it a high activity coefficient, so that the ratio of contaminant to water in the vapor phase is much higher than that in the water. Thus the half-lives in shallow water (1 m depth) of the organochlorines and PCBs are the following:

	Vapor Pressure, mm Hg	Water solubility, ppm	Half-life
DDT	1.5×10^{-7}	1.2×10^{-3}	3.7 days
Lindane	9.4×10^{-6}	7.3	289 days
Dieldrin	1×10^{-7}	0.25	723 days
Aroclor 1254	7.7×10^{-5}	1.2×10^{-2}	1.2 min

Thus lindane, despite its high volatility, has a longer persistence in exposed water than DDT, and the PCB Aroclor 1254 takes to the air readily because of its low water solubility.[119] The PCBs that have been employed in commerce as constituents of paints, plastics, paper products, and casing waxes, besides transformers, condensers and heat exchange systems, are readily transported throughout the world by volatilization.[55]

Air-Borne Dust. Another source of loss of DDT and other nonvolatile organochlorines to the air would be windblow of soil particles. Atmospheric dust deposited by a rainstorm at Cincinnati, Ohio on 26 January 1965 contained the following concentrations (ppm) of pesticides:

DDT + DDE	0.6	Heptachlor epoxide	0.004
Chlordane	0.5	Dieldrin	0.003
Ronnel	0.2	2,4-D	0.003

this material had originated from a dust storm in West Texas.[33] Yet in the air above Bermuda, less than 4% of the organochlorine content was on the dust particles.[14]

Organochlorines in Rain. It is estimated that the time constant for DDT in air is 4.3 yr,[204] before it is returned to the continents and oceans in rain. During this time, it is exposed to ultraviolet light at wavelengths that have proved capable of photodegrading it to DDE and to trichlorobiphenyls.[135] In 1964, English rain contained DDT at 3 ng/liter (ppt) in the countryside,[195] and 400 ng/l DDT plus 70 ng/liter DDE in central London.[1] The average *t*-DDT content for three areas in Ohio at this time was 187 ng/liter, 18 ng/liter of which was DDE.[193] In 1966, the average for six localities in the British Isles was 80 ng/liter, of which 20 ng/liter was DDE and 14 ng/liter DDD; at this time rain residues of BHC and dieldrin as well as DDT were found on the Shetland Islands.[183] By 1971 the average DDT content of the rain falling on the Hawaiian Islands was only 4 ng/liter.[13]

Organochlorines in the Oceans

Water and Sediments. Concentrations of DDT amounting to 0.15–0.5 ppt in the surface water were found in the Sargasso Sea[14] and 1 ppt in the North Sea[155]; the 2–6 ppt found in Pacific waters off California was associated with particulate suspensoids.[37] However, in a transect made across the North Atlantic all the water samples were negative for DDT although they showed 2–94 ppt of PCBs; and the ocean sediments which contained about 1 ppb of PCBs revealed no trace of DDT residues.[73]

Nevertheless, judging from the rain falling on islands (4 ppt on Hawaii, 80 ppt on Britain) and from the snow falling on the Antarctic coast in 1966 (40 ppt in the melt water[151]), the oceans must have received some 12,000 tons of DDT each year from the precipitation. This may be compared to the 200 tons/yr calculated for the input of DDT from the world's rivers, based on data for the Mississippi.[142] Evidence for the disappearance of added DDT was obtained when samples of sea water were treated with 3 ppb DDT and sealed; after 58 days the concentration had declined to 0.23 ppb. In a similar experiment, 0.7 ppb DDE declined to 0.035 ppb DDE in 20 days.[200] That DDT was in fact reaching the open ocean was first proved by analysis of the fish oils obtained from the western Pacific fishery of Japan in which the residues ranged from 1 to 10 ppm *t*-DDT; the situation was similar for the western North Atlantic fishery (U.S. Senate Committee on Government Operations, 1964[77]).

Marine Mammals. The most telling evidence of marine contamination comes from the large and long-lived aquatic mammals (Table 10.12), in whose blubber DDT could be found in quite high concentrations, often accompanied by dieldrin. In harbor porpoises taken from the northwest Atlantic in 1970, the lipid extracted from the blubber contained 29–119 ppm DDT, 26–122 ppm DDE, and 13–48 ppm DDD.[51] In the sea lions on the islands off the California coast the *t*-DDT residues in the blubber

Table 10.12.   Organochlorine residues in blubber of marine mammals, 1966–1973: average figures (ppm).

		t-DDT	Dieldrin	Ref.
California Sea-Lion	E. Pacific, S. California	911	----	(107)
Harbor Seal	Bay of Fundy, Canada	65.4	0.18	(52)
Ringed Seal	Gulf of Bothnia, Sweden	63.0	----	(87)
Harbor Porpoise	N. Atlantic, Scotland	42.7	9.9	(82)
Pilot Whale	St. Lucia, Lesser Antilles	16.2	0.03	(53)
Fur Seal*	N. Pacific, Pribiloff Islands	15.9	0.06	(5)
Grey Seal	N. Atlantic, Scotland	14.5	0.79	(82)
Harp Seal	N. Atlantic, Scotland	11.2	0.07	(82)
Sperm Whale	E. Pacific, California	5.8	0.02	(203)
Long-snouted Dolphin	St. Lucia, Lesser Antilles	3.5	0.03	(53)
White Whale	Arctic Ocean, Canada	2.25	----	(3)
Grey Whale	Behring Sea	0.23	0.06	(203)
Weddell Seal	Antarctic	0.06	----	(20)

*   Seal pups

exceeded 900 ppm, 94% of it being DDE.[39,107] In confined waters, ringed seals and harbor seals from the Swedish coast of the Baltic Sea contained 120–130 ppm *t*-DDT in the blubber lipid.[87] About the same level of contamination was found in harbor seals in the Bay of Fundy, whose blubber contained 65 ppm *t*-DDT which was slightly more than half DDE.[52] In open waters in the eastern Pacific, the grey whale which migrates to the Behring Sea to feed on amphipods had much lower residues than the sperm whale which is often found in California waters. The sea otter (*Enhydra lutris*) off the California coast contained an average level of 11 ppm *t*-DDT in its body fat, and one individual had 9 ppm in its brain.[174]

DDT residues have been found in the white whale in the Arctic Ocean off the mouth of the Mackenzie River, and in Weddell seals in the Ross Sea off Antarctica. Dieldrin residues, also ubiquitous, were highest in harbor porpoises off the west coast of Scotland. PCB residues greatly exceeded DDT residues in harbor seals in the Gulf of Maine where they amounted to 28–240 ppm in the blubber; but they were only about 1 ppm in pilot whales off the West Indies.[53] Mercury levels in liver tissue were 2–51 ppm in harbor seals in the Bay of Fundy[52] and 19–157 ppm in pilot whales off St. Lucia,[53] and 74–170 ppm in California sea lions off Oregon.[24]

The t-DDT levels reaching the brain in Nova Scotian harbor seals amounted to only 0.5 ppm, and even in sea lions on the California islands they never exceeded 14 ppm,[107] still well below the level that causes mortality in mammals. Deaths of sea lions off the Oregon coast in the early 1970s were probably due to leptospirosis, and the levels of DDE in the brain and PCBs in the adipose tissue were no greater in sick than in healthy animals[24]; nor was there any difference in t-DDT residue content between dead and live sea lions on California's Santa Barbara Channel Islands.[107] However, the increasing numbers of premature stillbirths on these islands was probably associated with DDT residues, since they were eight times higher in females with such troubles than in females which gave birth to normal sea lion pups.[39]

Marine Food Chains. The concentration of organochlorines from ocean waters by the biota is illustrated by data on PCBs in mid-Atlantic, where the residues in the water never exceeded 1 ppt and those in the sediments were about 1 ppb; here the various elements in the ocean biota were found to contain the following residue levels (in ppb)[73]:

Bottom invertebrates	1	Plankton	200
Mid-depth fish	10	Sea birds	1200
Upper-level fish	50	Sea mammals	3000

It is probable that the phytoplankton removes all the DDT in sea water except that adsorbed to the finest suspended particles.[37] Thus some of it would enter the food chain, but probably most sinks to the bottom sediment.[155] The DDT residue levels accumulated by phytoplankton were 0.2–0.5 ppb in the open Atlantic (mainly in *Sargassum*),[74] 4 ppb off the west coast of Scotland,[198] and 20–470 ppb off the California coast where the levels in 1969 were three times higher than in 1955.[36] Zooplankton samples often contained about 10 times more than the phytoplankton.[74,198] From other published data, Portmann (1975) in his review was able to assemble the following residue tabulation (in ppb) for t-DDT residues in ocean biota:

Phytoplankton	0.2–0.5	Fish	0.6–330
Zooplankton	0.01–9.5	Birds	600–3100
Molluscs	8–100	Mammals	10–20,000

The data for molluscs came from mussels and oysters on the Atlantic and North Sea coasts of Britain, while those for fish came from herring and cod in the North Sea (being 10 times greater in the herring), and from flyingfish in the open Atlantic in which the highest residue was 4 ppb. The residues of DDT and PCBs in cod liver oil from the Baltic Sea were high enough to cause it to be banned for sale in some countries. The data for birds came from the guillemot in the Irish Sea.

Ocean-Going Birds. The far-ranging birds that feed on ocean fish have provided the most telling evidence of the global contamination with organochlorines. Glaucous gulls on Bear Island in the Barents Sea 250 miles north of Scandanavia, where the surrounding water comes from the western North Atlantic, contained on the average 17 ppm DDE and 24 ppm PCBs in their livers; one gull found in convulsions carried a residue load of 67 ppm DDE and 311 ppm PCBs, and this in a species which does not migrate farther south.[16] Adelie penguins off Cape Crozier, Antarctica contained 24–152 ppb *t*-DDT in their body fat[175]; in chinstrap penguins, traces of DDT, traces of DDT, dieldrin, and BHC were found in Antarctic waters and similar DDT residues were found in the South Orkney islands.[62,184] It should be noted that analyses made at this time would not distinguish between DDT and certain of the PCB compounds, although it has been stated that the Adelie penguins contained no PCB.[163]

By 1974, residues of the livers of fish-eating birds in the South Atlantic were of the order of 100 ppb for *t*-DDT and for PCBs. Birds frequenting British waters showed DDE levels of several ppm, and PCBs some 10 times that amount, while in the Mediterranean the PCB figure was lower and the DDE higher.[17] One of the many uses of these chlorinated biphenyls is as an additive to the antifouling paint applied to boat hulls. PCBs in fish oil are undesirable, since they caused an epidemic of a chloracne-type disease in Japan, with four fatalities, when a leak from a heat-exchange system contaminated rice oil.[105]

Residues in Seas. Residues in the bird fauna of land-surrounded seas were at first higher than in the ocean, until corrective action was taken at the close of the 1960s. In 1965, on the islands off the North Sea coast close to the border of England and Scotland, fish-eating birds had accumulated the following residues (in ppm)[165]:

	Liver		Eggs	
	DDE	Dieldrin	DDE	Dieldrin
Cormorant (*Phalacrocorax carbo*)	4.14	0.19	3.8	1.22
Sandwich Tern	—	—	0.75	0.17
Lesser Black-backed Gull	0.62	0.26	0.63	0.43

The shag (*Phalacrocorax aristotelis*) was sufficiently abundant on the North Sea coast that a residue monitoring program could be inaugurated in 1964, and by 1971 it was evident that the residues of DDE in their eggs had reached their peak in 1968, and those of dieldrin (HEOD) in 1966, with a steady decline thereafter.[35] Photodieldrin, a toxic photoconversion product of HEOD, was not found in the shag's eggs.[164]

In the Baltic Sea in 1968, residues of *t*-DDT and PCBs were found to be as follows (in ppm fresh wt)[87]:

	Mussel (*Mytilus edulis*)	Herring	Guillemot (eggs)	Grey Seal
t-DDT	0.02	0.68	40	36
PCBs	0.084	0.27	16	6.1

The PCB concentrations in the herring (*Clupea harengus*) and predators was less than half those of *t*-DDT. In Nova Scotia waters, however, the PCB residues in the herring were approximately 0.5 ppm as compared to 0.25 ppm for *t*-DDT.[210] It is possible that PCBs may have been involved in the secondary poisoning of sea birds in the Irish Sea in 1969 and of terns off Long Island in 1971, since in both cases the PCB residues were unusually high.

Residues along the California Coast.   On the shores of the ocean, the California coast was a special case, with high DDT residues due not only to agricultural runoff and domestic use, but also to effluents from a manufacturing plant. In 1967 residues of 2.4 ppm DDE (3 ppm *t*-DDT) were found in the king crab *Limulus* 5 miles offshore in the Santa Barbara Channel, and of 4 ppm *t*-DDT in the Pacific oyster in the Hedionda lagoon farther south.[134] In 1970 the *t*-DDT residues 1 mile off San Diego were, on the average, 4.3 ppm in the red abalone, 6.5 ppm in the sheepshead, and 27 ppm in sand bass.[140] DDT residues in the sand crab *Eremita analoga* pinpointed the source of the greatest contamination. This was White's Point, the outfall of the Los Angeles County Sewer, where the *t*-DDT residues in the crab reached 8 ppm, in contrast to the usual 0.1 ppm residues found along the entire length of coast from San Francisco south to the Mexican border.[25] The concentrations of DDT in the ocean sediment 4 km out from its mouth were of the order of 10 ppm, rising to 1% directly at the outfall. The discharge from the DDT manufacturing plant into this sewer had been 200–500 kg/yr.[55]

In 1971, the DDT effluent was recycled and thus reduced from 10–15 lb/

day down to 1 oz/day, or approximately 10 lb/yr.[176] By 1973, although the DDT residues around the Los Angeles outfall had been steadily declining, they were still increasing in the sea lions of the Santa Barbara Channel Islands.[118] Far up the coast in a northerly direction, the hake (*Merluccius productus*) off Cape Foulweather, Oregon contained an average of 0.28 ppm *t*-DDT, of which DDT constituted half and DDE a quarter.[181] Far south in the Gulf of California, the high ratio of DDT to PCBs (10:1) which characterized the Pacific coast still held good in that remote region.[163]

Mathematical Treatment of Ecological Data

There is a need for predicting the behavior of an insecticide introduced into the environment, and for bringing together the quantitative results obtained for each compound into a predictive system. Components may be assembled with their cause and effect equations, a systems diagram can arrange them in their several relationships, and a digital computer can calculate the expected results from given inputs.[54]

Computer Models. After the insecticide has been applied to terrain, its rate of loss may be predicted from first-order kinetics; Walker (1974) has developed a simulation model (though primarily suitable for herbicides) that takes into account the fluctuations in temperature and moisture content at the soil surface under field conditions. Bailey, Swank, and Nicholson (1974) have provided systems diagrams to predict the runoff rate for successive rainfall events superimposed upon the continuous background attenuation of the soil content of insecticide, along with the positions of the rate constants and equilibrium constants to be determined; the visual model also is designed to predict sediment loading, and thus much of the input into water bodies.

The uptake and subsequent loss of residue by aquatic organisms in treated or contaminated water bodies have been characterized, on the basis of first-order kinetics for both uptake and loss, by the mathematical model of Eberhardt, Meeks, and Peterle (1971). A summary has been provided by Kenaga (1972) of the type of data required to determine the bioconcentration potential of a pesticide. Harrison and co-workers (1970) have developed equations for food chains, and calculated the length of time it takes for the input to reach its final expression in the highest trophic level.

Woodwell, Craig, and Johnson (1971) concluded that at the end of the 1960s the world's biota contained in their bodies about 5400 metric tons of DDT, less than one-thirtieth of the annual production at that time; the partition among the various organisms was estimated to be as follows (in metric tons):

Land	Wild animals	9	Crops	1400
	Domestic animals	170	Forests and grass	181
	Humans	300	Swamps	24
	Birds	0.2	Lakes and streams	0.4
Sea	Fish	650	Open-ocean algae	100
	Marine mammals	55	Littoral algae	300
	Invertebrates	302	Attached algae	2000

These authors calculated that agricultural soils at that time contained DDT at an average level of 1.5 lb/a, and that its half-life in the soil was 5 yr. They concluded that the main pesticide dynamics of DDT was its transfer from the soil to the oceans and eventually to the abyss. Nisbet and Sarofim (1972) evaluated this transport process, principally through the atmosphere and rain, for the considerably more volatile PCBs.

Predictions. The most serious implication of the systems model of Harrison et al. (1970), which balanced the inputs (surface, atmospheric, and water) with outputs (in the same three categories) for the state of Wisconsin, was that the length of time it takes to reach equilibrium in the biota depends on their life spans, which with eagles may be as much as 80 yr. It was pointed out by Bloom and Menzel (1971) that this model did not take into account the metabolism and excretion of DDT that the organisms bring about; it should however be noted that DDE as a terminal metabolite is far from an alleviation, although the route through DDD to polar metabolites such as DDA is indeed a detoxication.

Eventualities. In the event, it turned out that the DDT residues in Lake Michigan fish peaked in 1971 (Table 10.13). The residues of DDT and dieldrin in starlings and the wings of mallards had been falling since 1969

Table 10.13. DDT residues in Lake Michigan fish, 1969–1974: *t*-DDT in the whole fish, ppm (Reinert in Environmental Protection Agency, 1975).

	1969	1970	1971	1972	1973	1974
Lake Trout	9.9	19.2	13.0	11.3	10.0	---
American Bloater	---	9.8	6.2	4.3	2.1	1.3
Coho Salmon	11.8	14.0	9.8	7.2	4.5	---

Table 10.14. **Organochlorine residues in birds in the United States, 1969–1974: nationwide averages, ppm (White, 1976; White and Heath, 1976).**

	1967-8	1969	1970	1972	1974
Starlings, whole-body					
DDE	1.64	----	0.84	0.79	0.62
Dieldrin	0.14	----	0.12	0.10	0.06
PCB's	----	----	0.66	0.42	0.11
Mallard, wings					
DDE	----	1.03	----	0.44	----
Dieldrin	----	0.05	----	0.02	----
PCB's	----	1.29	----	1.24	----
Black Duck, wings					
DDE	----	1.32	----	0.35	----
Dieldrin	----	0.14	----	0.02	----
PCB's	----	1.37	----	1.36	----

(Table 10.14) although a drop in PCB content was not evident until 1974. DDT residues in small birds in Florida had been declining since 1964.[93] On the other hand, both DDT and PCB residues in the great skua of the ocean were still increasing in 1974.[17] It must be remembered that the much-needed use of DDT for malaria control and for agriculture in the tropics and subtropics is continuing, despite an abrupt termination of input of this insecticide in the North Temperate zone.[55]

REFERENCES CITED

1. D. C. Abbott, R. B. Harrison, J. O'G. Tatton, and J. Thomson. 1965. Organochlorine pesticides in the atmospheric environment. *Nature* **208**:1317–1318.
2. D. C. Abbott, R. B. Harrison, J. O'G. Tatton, and J. Thomson. 1966. Organochlorine pesticides in the atmosphere. *Nature* **211**:259–261.
3. R. F. Addison and P. F. Brodic. 1973. Occurrence of DDT residues in beluga whales from the Mackenzie delta. *J. Fish Res. Bd. Can.* **30**:1733–1736.
4. E. S. Albone, G. Eglinton, N. C. Evans, J. M. Hunter, and M. M. Rhead. 1972. Fate of DDT in Severn estuary sediments. *Environ. Sci. Technol.* **6**:914–919.

5.   R. E. Anas and A. J. Wilson. 1970. Organochlorine pesticides in fur seals. *Pestic. Monit. J.* **3**:199–200.

6.   G. W. Bailey, R. R. Swank, and H. P. Nicholson. 1974. Predicting pesticide run-off from agricultural land: A conceptual model. *J. Environ. Qual.* **3**:95–102.

7.   T. E. Bailey and J. R. Hannum. 1967. Distribution of pesticides in California. *J. Sanit. Eng. Div., Am. Soc. Civil Engineers, Proc.* **93**:27–43.

8.   R. J. Barker. 1958. Notes on some ecological effects of DDT sprayed on elms. *J. Wildl. Manage.* **22**:269–274.

9.   W. F. Barthel, J. C. Hawthorne, J. H. Ford, G. C. Bolton, L. L. McDowell, E. H. Grissinger, and D. A. Parsons. 1970. Pesticide residues in sediments of the lower Mississippi River and its tributaries. *Pestic. Monit. J.* **3**:8–66.

10. H. J. Benezet and F. Matsumura. 1973. Isomerization of γ-BHC to α-BHC in the environment. *Nature* **243**:480–481.

11. H. J. Benezet and F. Matsumura. 1974; Factors influencing the metabolism of mexacarbate by microorganisms. *J. Agric. Food Chem.* **22**:427–430.

12. A. Bevenue. 1976. The "bioconcentration" aspects of DDT in the environment. *Residue Rev.* **61**:37–112.

13. A. Bevenue, J. W. Hylin, Y. Kawano, and T. W. Kelley. 1972. Organochlorine pesticide residues in water, sediment, algae, and fish, Hawaii, 1970–71. *Pestic. Monit. J.* **6**:56–60.

14. T. F. Bidleman and C. E. Olney. 1974. Chlorinated hydrocarbons in the Sargasso atmosphere and surface water. *Science* **183**:516–518.

15. S. G. Bloom and D. B. Menzel. 1971. Decay time of DDT. *Science* **172**:213.

16. J. A. Bogan and W. R. P. Bourne. 1972. Organochlorine levels in Atlantic seabirds. *Nature* **240**:358.

17. W. R. P. Bourne. 1975. Discussion at close of J. E. Portmann's paper. *Proc. Royal Soc. London. B* **189**:303–304.

18. G. W. Bowes. 1972. Uptake and metabolism of DDT by marine phytoplankton and its effect on growth and chloroplast electron transport. *Plant Physiol.* **49**:172–176.

19. J. R. Bradley, T. J. Sheets, and M. D. Jackson. 1972. DDT and toxaphene movement in surface water from cotton plots. *J. Environ. Qual.* **1**:102–105.

20. H. V. Brewerton. 1969. DDT in fats of Antarctic animals. *New Zealand J. Sci.* **12**:194–199.

21. W. R. Bridges, B. J. Kallman, and A. K. Andrews. 1963. Persistence of DDT and its metabolites in a farm pond. *Trans. Am. Fish. Soc.* **92**:421–427.

22. G. T. Brooks. 1972. Pesticides in Britain. In *Environmental Toxicology of Pesticides.* Eds. F. Matsumura, G. M. Boush, and T. Misato. Academic. pp. 61–114.

23. N. J. Brown and A. W. A. Brown. 1970. Biological fate of DDT in a sub-arctic environment. *J. Wildl. Manage.* **34**:929–940.

24. D. R. Buhler, R. R. Claeys, and B. R. Mate. 1975. Heavy metal and chlorinated hydrocarbon residues in California sea lions (*Zalophus californianus*). *J. Fish. Res. Bd. Can.* **32**:2391–2397.

25. R. Burnett. 1971. DDT residues: Distribution of concentrations in *Emerita analoga* along coastal California. *Science* **174**:606–608.

26. P. A. Butler. 1969. Monitoring pesticide pollution. *BioScience* **19**:889–891.

27. P. A. Butler. 1973. Organochlorine residues in estuarine mollusks, 1965–72: National monitoring program. *Pestic. Monit. J.* **6**:238–246.

28. P. A. Butler, R. Childress, and A. J. Wilson. 1972. The association of DDT residues with losses in marine productivity. In *Marine Pollution and Sea Life*. Ed. M. Ruivo. Fishing News (Books) Ltd. pp. 262–266.

29. J. H. Caro and A. W. Taylor. 1971. Pathways of loss of dieldrin from soils under field conditions. *J. Agric. Food Chem.* **19**:379–384.

30. C. I. Chacko and J. L. Lockwood. 1967. Accumulation of DDT and dieldrin by microorganisms. *Can. J. Microbiol.* **13**:1123–1126.

31. G. G. Chadwick and R. W. Brocksen. 1969. Accumulation of dieldrin by fish and selected fish-food organisms. *J. Wildl. Manage.* **33**:693–700.

32. P. R. Chen, W. P. Tucker, and W. C. Dauterman. 1969. Structure of biologically produced malathion monoacid. *J. Agric. Food Chem.* **17**:86–90.

33. J. M. Cohen and C. Pinkerton. 1966. Widespread translocation of pesticides by air transport and rain-out. *Adv. Chem. Ser.* No. 60. pp. 163–176.

34. H. Cole, D. Barry, D. E. H. Frear, and A. Bradford. 1967. DDT levels in fish and stream sediments after aerial spray application in northern Pennsylvania. *Bull. Environ. Contam. Toxicol.* **2**:127–146.

35. J. C. Coulson, I. R. Deans, G. R. Potts, J. Robinson, and A. N. Crabtree. 1972. Changes in organochlorine contamination of the marine environment of eastern Britain monitored by shag eggs. *Nature* **236**:454–456.

36. J. L. Cox. 1970. DDT residues in marine phytoplankton: Increase from 1955 to 1969. *Science* **170**:71–73.

37. J. L. Cox. 1971. DDT residues in sea water and particulate matter in the California Current system. *Fishery Bull.* **69**:443–450.

38. R. A. Croker and A. J. Wilson. 1965. Kinetics and effects of DDT in a tidal marsh ditch. *Trans. Am. Fish. Soc.* **94**:152–159.

39. R. L. DeLong, W. G. Gilmartin, and J. G. Simpson. 1973. Premature births in California sea lions; association with high organochlorine pollutant residue levels. *Science* **181**:1168–1169.

40. S. K. Derr and M. J. Zabik. 1972. Biologically active compounds in the aquatic environment: Studies on the uptake of DDE by the aquatic midge *Chironomus tentans*. *Arch. Environ. Contam.* **2**:151–164.

41. R. D. Earnest and P. E. Benville. 1971. Correlation of DDT and lipid levels in San Francisco Bay fish. *Pestic. Monit. J.* **5**:235–241.

42. L. L. Eberhardt, R. L. Meeks, and T. J. Peterle. 1971. Food chain model for DDT kinetics in a freshwater marsh. *Nature* **230**:60–62.

43. C. A. Edwards (Ed.). 1973. *Persistent Pesticides in the Environment*. 2nd edition. CRC Press, Cleveland, Ohio, 170 pp.

44. J. W. Eichelberger and J. J. Lichtenberg. 1971. Persistence of pesticides in river water. *Environ. Sci. Technol.* **5**:541–544.

45. E. Elsner, D. Bieniek, W. Klein, and F. Korte. 1972. Verteilung und Umwandlung von Aldrin, Heptachlor, und Lindan in der Grunalge *Chlorella pyrenoidosa*. *Chemosphere* **1**:247–250.

46. Environmental Protection Agency. 1975. *DDT: A Review of Scientific and Economic Aspects of the Decision to Ban Its Use as a Pesticide*. Document EPA-540/1-75-022. 300 pp.

47. E. Epstein and W. J. Grant. 1968. Chlorinated insecticides in runoff water as affected by crop rotation. *Proc. Soil Sci. Soc. Am.* **32**:423–426.

48. D. E. Ferguson, J. L. Ludke, J. P. Wood, and J. W. Prather. 1965. The effect of mud on the bioactivity of pesticides on fishes. *J. Miss. Acad. Sci.* **11**:219–228.

49. V. H. Freed, R. Haque, and D. Schmedding. 1972. Vaporization and environmental contamination by DDT. *Chemosphere* **1**:61–65.

50. R. A. E. Galley. 1974. The atmospheric distribution of organochlorine insecticides. *Chem. Ind.* 2 March. 179–181.

51. D. E. Gaskin, M. Holdrinet, and R. Frank. 1971. Organochlorine pesticide residues in harbour porpoises from the Bay of Fundy region. *Nature* **233**:499–500.

52. D. E. Gaskin et al. 1973. Mercury, DDT, dieldrin, and PCB in harbour seals from the Bay of Fundy and Gulf of Maine. *J. Fish. Res. Bd. Can.* **30**:471–475.

53. D. E. Gaskin et al. 1974. Mercury, DDT, dieldrin, and PCB in two species of Odontoceti from St. Lucia, Lesser Antilles. *J. Fish. Res. Bd. Can.* **31**:1235–1239.

54. J. W. Gillett, J. Hill, A. W. Jarvinen, and W. P. Schoor. 1974. *A Conceptual Model for the Movement of Pesticides through the Environment.* US Environmental Protection Agency, Washington. Document EPA-660/3-74-024. 80 pp.

55. E. D. Goldberg. 1975. Synthetic organic halides in the sea. *Proc. Royal Soc. London B* **189**:277–289.

56. D. A. Graetz, G. Chesters, T. C. Daniel, L. W. Newland, and G. B. Lee. 1970. Parathion degradation in lake sediments. *J. Water Pollut. Control Fed.* **42**:R76–94.

57. W. W. Gregory, J. K. Reed, and L. E. Priester. 1969. Accumulation of parathion and DDT by some algae and protozoa. *J. Protozool.* **16**:69–71.

58. A. R. Grzenda, G. J. Lauer, and H. P. Nicholson. 1964. Water pollution by insecticides in an agricultural river basin. II. The zooplankton, bottom fauna, and fish. *Limnol. Oceanogr.* **9**:318–323.

59. A. R. Grzenda, W. J. Taylor, and D. F. Paris. 1972. The elimination and turnover of ^{14}C-dieldrin by different goldfish tissues. *Trans. Am. Fish. Soc.* **101**:686–690.

60. W. D. Guenzi and W. E. Beard. 1970. Volatilization of lindane and DDT from soils. *Soil Sci. Soc. Am. Proc.* **34**:443–447.

61. W. D. Guenzi and W. E. Beard. 1974. Volatilization of pesticides. In *Pesticides in Soil and Water.* Ed. W. D. Guenzi. *Soil Sci. Soc. Am.* Madison, Wis. pp. 107–122.

62. D. L. Gunn. 1972. Dilemmas in conservation for applied biologists. *Ann. Appl. Biol.* **72**:105–127.

63. C. T. Haan. 1971. Movement of pesticides by run-off and erosion. *Trans. Am. Soc. Agric. Engineers* **14**:445–449.

64. H. L. Hamelink, R. C. Waybrant, and R. C. Ball. 1971. A proposal: Exchange equilibria control the degree chlorinated hydrocarbons are magnified in lentic environments. *Trans. Am. Fish. Soc.* **100**:207–214.

65. M. R. Hannon, Y. A. Greichus, R. L. Applegate, and A. C. Fox. 1970. Ecological distribution of pesticides in Lake Poinsett, South Dakota. *Trans. Am. Fish. Soc.* **99**:496–500.

66. D. J. Hansen. 1966. *Annual Report of the Gulf Breeze Biological Laboratory, 1965.* U.S. Bur. Comm. Fish. Circ. 247. pp. 10–11.

67. D. J. Hansen and A. J. Wilson. 1970. Significance of DDT residues from the estuary near Pensacola, Fla. *Pestic. Monit. J.* **4**:51–56.

68. C. I. Harris. 1969. Movement of pesticides in soil. *J. Agric. Food Chem.* **17**:80–82.

69. C. R. Harris and J. R. W. Miles. 1975. Pesticide residues in the Great Lakes region of Canada. *Residue Rev.* **57**:27–79.

70. H. L. Harrison, O. L. Loucks, J. W. Mitchell, D. F. Parkhurst, C. R. Tracy, D. G. Watts, and V. J. Yannacone. 1970. Systems studies of DDT transport. *Science* **170**:503–508.

71. G. S. Hartley. 1969. Evaporation of pesticides. In *Pesticide Formulations Research: Physical and Colloidal Chemical Aspects.* Ed. R. F. Gould. Am. Chem. Soc. Advan. Chem. Series **86**:115–134.

72. R. Hartung and G. W. Klinger. 1970. Concentration of DDT by sediment polluting oils. *Environ. Sci. Technol.* **4**:407–10.

73. G. R. Harvey. 1969. DDT and PCB in the Atlantic. *Oceanus* **18**:19–23.

74. G. R. Harvey, V. J. Bowen, R. H. Backus, and G. D. Grice. 1973. Chlorinated hydrocarbons in open ocean Atlantic organisms. In *The Changing Chemistry of the Oceans.* Eds. D. Green and D. Jagner. Wiley. pp. 177–186.

75. S. G. Herman, R. L. Garrett, and R. L. Rudd. 1969. Pesticides and the western grebe. In *Chemical Fallout—Current Research on Persistent Pesticides.* Eds. M. W. Miller and G. G. Berg. Charles C Thomas. pp. 24–53.

76. B. Heuer, B. Yaron, and Y. Birk. 1974. Guthion half-life in aqueous solutions and on glass surfaces. *Bull. Environ. Contam. Toxicol.* **11**:532–537.

77. J. J. Hickey, J. A. Keith, and F. B. Coon. 1966. An exploration of pesticides in a Lake Michigan ecosystem. *J. Appl. Ecol.* **3**:(Suppl.):141–154.

78. D. W. Hill and P. L. McCarty. 1967. Anaerobic degradation of selected chlorinated hydrocarbon pesticides. *J. Water Pollut. Control Fed.* **39**:1259–1277.

79. E. Hindin, D. S. May, and G. H. Dunstan. 1966. Distribution of insecticides sprayed by airplane on an irrigated corn plot. In *Organic Pesticides in the Environment.* Am. Chem. Soc., Advan. Chem. Series **60**:132–145.

80. A. V. Holden. 1962. A study of the absorption of C^{14}-labelled DDT from water by fish. *Ann. Appl. Biol.* **50**:466–477.

81. A. V. Holden. 1970. International cooperative study of organochlorine pesticide residues in terrestrial and aquatic wildlife. *Pestic. Monit. J.* **4**:117–135.

82. A. V. Holden and K. Marsden. 1967. Organochlorine pesticides in seals and porpoises. *Nature* **216**:1274–1276.

83. L. Hughes. 1971. *A Study of the Fate of Carbaryl Insecticide in Surface Waters.* Ph.D. Dissertation, Purdue University.

84. E. G. Hunt and A. I. Bischoff. 1960. Inimical effects on wildlife of periodic DDD applications to Clear Lake. *Calif. Fish Game* **46**:91–106.

85. S. H. Hurlbert, M. S. Mulla, J. O. Keith, W. E. Westlake, and M. E. Duesch. 1970. Biological effects and persistence of Dursban in freshwater ponds. *J. Econ. Entomol.* **63**:43–52.

86. S. Jensen, R. Gothe, and M. O. Kindstedt. 1972. Bis-(*p*-chlorophenyl)acetonitrile (DDCN), a new DDT derivative formed in anaerobic digested sewage sludge and lake sediment. *Nature* **240**:421–422.

87. S. Jensen, A. G. Johnels, M. Olsson, and G. Otterlind. 1969. DDT and PCB in marine animals from Swedish waters. *Nature* **224**:247–250.

88. R. E. Johnsen. 1976. DDT metabolism in microbial systems. *Residue Rev.* **61**:1–28.

89. B. T. Johnson and J. O. Kennedy. 1973. Biomagnification of *p,p'*-DDT and methoxychlor by bacteria. *Appl. Microbiol.* **26**:66–71.

90. B. T. Johnson, C. R. Saunders, H. O. Sanders, and R. S. Campbell. 1971. Biological

magnification and degradation of DDT and aldrin by freshwater invertebrates. *J. Fish. Res. Bd. Can.* **28**:705–709.

91. H. E. Johnson and R. C. Ball. 1972. Organic pesticide pollution in an aquatic environment. In *Fate of Organic Pesticides in the Aquatic Environment*. Am. Chem. Soc., Adv. Chem. Series **111**:1–10.

92. W. D. Johnson, G. F. Lee, and D. Spyridakis. 1966. Persistence of toxaphene in treated lakes. *Internat. J. Air Water Pollut.* **10**:555–560.

93. D. W. Johnston. 1975. Organochlorine pesticide residues in migratory songbirds. *Science* **186**:841–842.

94. K. L. E. Kaiser. 1974. Mirex: An unrecognized contaminant of fishes from Lake Ontario. *Science* **185**:523–525.

95. B. J. Kallman, O. B. Cope, and R. J. Navarre. 1962. Distribution and detoxication of toxaphene in Clayton Lake, New Mexico. *Trans. Am. Fish. Soc.* **91**:14–22.

96. I. P. Kapoor, R. L. Metcalf, R. F. Nystrom, and G. K. Sangha. 1970. Comparative metabolism of methoxychlor, methiochlor, and DDT in mouse, insects, and in a model ecosystem. *J. Agric. Food Chem.* **18**:1145–1152.

97. J. F. Karinen, J. G. Lamberton, N. E. Stewart, and L. C. Terriere. 1967. Persistence of carbaryl in the marine estuarine environment. *J. Agric. Food Chem.* **15**:148–156.

98. J. E. Keil and L. E. Priester. 1969. DDT uptake and metabolism by a marine diatom. *Bull. Environ. Contam. Toxicol.* **4**:169–173.

99. J. A. Keith. 1966. Reproduction in a population of herring gulls (*Larus argentatus*) contaminated by DDT. *J. Appl. Ecol.* **3**(Suppl.):57–70.

100. E. E. Kenaga. 1972. Guidelines for environmental study of pesticides: Determination of bioconcentration potential. *Residue Rev.* **44**:73–113.

101. H. D. Kennedy, L. F. Eller, and D. F. Walsh. 1970. *Chronic Effects of Methoxychlor on Bluegills and Aquatic Invertebrates.* U.S. Dept. Interior, Bureau Sport Fish. Wildl. Tech. Paper 53. 18 pp.

102. P. H. King and P. L. McCarty. 1968. A chromatographic model for predicting pesticide migration in soils. *Soil Sci.* **106**:248–261.

103. L. J. Korschgen. 1970. Soil–foodchain–pesticide–wildlife relationships in aldrin-treated fields. *J. Wildl. Manage.* **34**:186–199.

104. R. J. Kuhr, A. C. Davis, and J. B. Bourke. 1974. Dissipation of Guthion, Sevin, Polyram, Phygon, and Systox from apple orchard soils. *Bull. Environ. Contam. Toxicol.* **11**:224–230.

105. M. Kuratsune, T. Yoshimura, J. Matsuzaka, and A. Yamaguchi. 1972. Epidemiologic study on Yusho, a poisoning caused by polychlorinated biphenyls. *Environ. Health Perspect.* **1**:119–128.

106. J. G. Lamberton and R. R. Claeys. 1970. Degradation of 1-naphthol in sea water. *J. Agric. Food Chem.* **18**:92–96.

107. B. J. LeBoeuf and M. L. Bonnell. 1971. DDT in California sea lions. *Nature* **234**:108–109.

108. H. V. Leland, W. N. Bruce, and N. F. Shimp. 1973. Chlorinated hydrocarbon insecticides in sediments of southern Lake Michigan. *Environ. Sci. Technol.* **7**:833–838.

109. W. O. Leshniowsky, P. R. Dugan, R. M. Pfister, J. I. Frea, and C. I. Randles. 1970. Adsorption of chlorinated hydrocarbon pesticides by microbial floc and lake sediment and its ecological implications. *Internat. Assoc. Great Lakes Res., Proc. 13th Conf.* pp. 611–618.

110. J. J. Lichtenberg, J. W. Eichelberger, R. C. Pressman, and J. E. Longbottom. 1970. Pesticides in the surface waters of the United States—a 5-year summary, 1964–68. *Pestic. Monit. J.* **4**:71–86.

111. E. P. Lichtenstein, K. R. Schulz, R. F. Skrentny, and Y. Tsukano. 1966. Toxicity and fate of insecticide residues in water. *Arch. Environ. Health* **12**:199–212.

112. C. P. Lloyd-Jones. 1971. Evaporation of DDT. *Nature* **229**:65–66.

113. E. G. Lotse, D. A. Graetz, G. Chesters, G. B. Lee, and L. W. Newland. 1968. Lindane adsorption by lake sediments. *Environ. Sci. Technol.* **2**:353–357.

114. J. L. Ludke, M. T. Finley, and C. Lusk. 1971. Toxicity of mirex to crayfish, *Procambarus blandingi*. *Bull Environ. Contam. Toxicol.* **6**:89–96.

115. P. D. Ludwig, H. J. Dishburger, J. C. McNeill, W. O. Miller, and J. R. Rice. 1968. Biological effects and persistence of Dursban in a salt-marsh habitat. *J. Econ. Entomol.* **61**:626–633.

116. K. J. Macek. 1969. Biological magnification of pesticide residues in food chains. In *The Biological Impact of Pesticides in the Environment*. Oregon State University, Environ. Health Sciences Series No. 1. pp. 17–21.

117. K. J. Macek and S. Korn. 1970. Significance of the food chain in DDT accumulation by fish. *J. Fish Res. Bd. Can.* **27**:1496–1498.

118. J. S. MacGregor. 1974. Changes in the amount and proportions of DDT, DDE, and DDD in the marine environment of southern California. *Fish. Bull.* **72**:275–293.

119. D. Mackay and A. W. Wolkoff. 1973. Rate of evaporation of low-solubility contaminants from water bodies to atmosphere. *Environ. Sci. Technol.* **7**:611–614.

120. M. Mackiewicz, K. H. Deubert, H. B. Gunner, and B. M. Zuckerman. 1969. Study of parathion degradation using gnotobiotic techniques. *J. Agric. Food Chem.* **17**:129–130.

121. D. B. Manigold and J. A. Schulze. 1969. Pesticides in selected western streams: A progress report. *Pestic. Monit. J.* **3**:124–135.

122. F. Matsumura and H. J. Benezet. 1973. Studies of the bioaccumulation and microbial degradation of TCDD. *Environ. Health Perspect., Exp. Issue* **5**:253–258.

123. F. Matsumura, K. C. Patil, and G. M. Boush. 1970. Formation of photodieldrin by microorganisms. *Science* **170**:1206–1207.

124. F. Matsumura, K. C. Patil, and G. M. Boush. 1971. DDT metabolized by microorganisms from Lake Michigan. *Nature* **230**:325–326.

125. J. L. Mendel, A. K. Klein, J. T. Chen, and M. S. Walton. 1967. Metabolism of DDT and some other chorinated organic compounds by *Aerobacter aerogenes*. *J. Assoc. Official Anal. Chemists* **50**:897–903.

126. R. L. Metcalf. 1974. A laboratory model ecosystem to evaluate compounds producing biological magnification. *Essays Toxicol.* **5**:17–38.

127. R. L. Metcalf, I. P. Kapoor, P. Y., Lu, C. K. Schuth, and P. Sherman. 1973. Model ecosystem studies of the environmental fate of six organochlorine pesticides. *Environ. Health Perspect. Exp. Issue* **4**:35–44.

128. R. L. Metcalf, G. K. Sangha, and I. P. Kapoor. 1971. Model ecosystem for the evaluation of pesticide biodegradability and ecological magnification. *Environ. Sci. Technol.* **5**:709–713.

129. F. P. Meyer. 1965. The experimental use of Guthion as a selective fish eradicator. *Trans. Am. Fish. Soc.* **94**:203–209.

130. J. R. W. Miles and C. R. Harris. 1970. Insecticide residues in a stream and controlled

drainage system in agricultural areas of southwestern Ontario. *Pestic. Monit. J.* **5**:289–294.

131. C. W. Miller, B. M. Zuckerman, and A. J. Charig. 1966. Water translocation of diazinon and parathion off a model cranberry bog. *Trans. Am. Fish. Soc.* **95**:345–349.

132. R. P. Miskus, D. P. Blair, and J. E. Casida. 1965. Conversion of DDT to DDD by bovine rumen fluid, lake water, and reduced porphyrins. *J. Agric. Food Chem.* **13**:481–483.

133. S. Miyazaki and A. J. Thorsteinson. 1972. Metabolism of DDT by freshwater diatoms. *Bull. Environ. Contam. Toxicol.* **8**:81–83.

134. J. C. Modin. 1969. Chlorinated hydrocarbon pesticides in California bays and estuaries. *Pestic. Monit. J.* **3**:1–7.

135. K. W. Moilanen and D. G. Crosby. 1973. *Vapor-Phase Decomposition of p,p´-DDT and Its Relatives.* 165th Nat. Mtg. Am. Chem. Soc. (reviewed by W. F. Spencer, *Residue Rev.* **59**, p. 111).

136. F. Moriarity. 1972. The effects of pesticides on wildlife: Exposure and residues. *New Scientist* **53**:594–596.

137. R. L. Morris and L. G. Johnson. 1971. Dieldrin levels in fish from Iowa streams. *Pestic. Monit. J.* **5**:12–16.

138. R. T. Moubry, J. M. Helm, and G. R. Myrdal. 1968. Chlorinated pesticide residues in an aquatic environment located adjacent to a commercial orchard. *Pestic. Monit. J.* **1**:(4):27–29.

139. M. S. Mulla, J. O. Keith, and F. A. Gunther. 1966. Persistence and biological effects of parathion residues in waterfowl habitats. *J. Econ. Entomol.* **59**:1085–1090.

140. T. O. Munson. 1972. Chlorinated hydrocarbon residues in marine animals of southern California. *Bull. Environ. Contam. Toxicol.* **7**:223–228.

141. S. M. Naqvi and A. A. de la Cruz. 1973. Mirex incorporation in the environment: residues in non-target organisms. *Pestic. Monit. J.* **7**:104–111.

142. National Academy of Sciences. 1971. *Chlorinated Hydrocarbons in the Marine Environment.* Persistent Insecticides Monitoring Panel. Committee on Oceanography, E. D. Goldberg, Chairman. Washington, 42 pp.

143. L. W. Newland, G. Chesters, and G. B. Lee. 1969. Degradation of γ-BHC simulated lake impoundments as affected by aeration. *J. Water Poll. Control Fed.* **41**, R174–188.

144. H. P. Nicholson. 1969. Occurrence and significance of pesticide residues in water. *J. Washington Acad. Sci.* **59**:77–85.

145. H. P. Nicholson, A. R. Grzenda, G. J. Lauer, W. S. Cox, and J. I. Teasley. 1964. Water pollution by insecticides in an agricultural river basin. I. Occurrence of insecticides in river and treated municipal water. *Limnol. Oceanogr.* **9**:310–317.

146. I. C. T. Nisbet and A. F. Sarofim. 1972. Rates and routes of transport of PCBs in the environment. *Environ. Health Perspect.* **1**:21–38.

147. D. F. Paris and D. L. Lewis. 1973. Chemical and microbial degradation of ten selected pesticides in aquatic systems. *Residue Rev.* **45**:95–124.

148. D. F. Paris, D. L. Lewis, and N. L. Wolfe. 1975. Rates of degradation of malathion by bacteria isolated from an aquatic system. *Environ. Sci. Technol.* **9**:135–138.

149. D. F. Paris, D. L. Lewis, J. T. Barnett, and G. L. Baughman. 1975. *Microbial Degradation and Accumulation of Pesticides in Aquatic Systems.* U.S. Environ. Prot. Agency, Ecol. Res. Series EPA-660/3-75-007. 45 pp.

150. K. C. Patil, F. Matsumura, and G. M. Boush. 1972. Metabolic transformations of DDT, dieldrin, aldrin, and endrin by marine microorganisms. *Environ. Sci. Technol.* **6**:629–632.

151. T. J. Peterle. 1969. DDT in Antarctic snow. *Nature* **224**:620.

152. F. K. Pfaender and M. Alexander. 1972. Extensive microbial degradation of DDT *in vitro* and DDT metabolism by natural communities. *J. Agric. Food Chem.* **20**:842–846.

153. D. Pimentel. 1971. *Ecological Effects of Pesticides on Nontarget Species.* Exec. Off. President, Off. Sci. Technol. Sup't. Documents, Washington. 220 pp.

154. H. B. Pionke and G. Chesters. 1973. Pesticide–sediment–water interactions. *J. Environ. Quality* **2**:29–45.

155. J. E. Portmann. 1975. The bioaccumulation and effects of organochlorine pesticides in marine animals. *Proc. Royal Soc. London B* **189**:291–304.

156. F. H. Premdas and J. M. Anderson. 1963. The uptake and detoxification of C^{14}-labelled DDT in Atlantic salmon, *Salmo salar. J. Fish. Res. Bd. Can.* **20**:827–836.

157. C. W. Randall and R. A. Lauderdale. 1967. Biodegradation of malathion. *J. Sanit. Eng. Div., Am. Soc. Civil Engineers* **93**:145–146.

158. K. A. Reinbold, I. P. Kapoor, W. F. Childers, W. N. Bruce, and R. L. Metcalf. 1971. Comparative uptake and biodegradability of DDT and methoxychlor by aquatic organisms. *Bull. Ill. Nat. History Surv.* **30**:405–415.

159. R. E. Reinert. 1970. Pesticide concentrations in Great Lakes fish. *Pestic. Monit. J.* **3**:233–240.

160. R. E. Reinert and H. L. Bergman. 1973. Residues of DDT in lake trout and Coho salmon from the Great Lakes. *J. Fish. Res. Bd. Can.* **31**:191–199.

161. C. P. Rice and H. C. Sikka. 1973. Uptake and metabolism of DDT by six species of marine algae. *J. Agric. Food Chem.* **21**:148–152.

162. E. M. Richardson and E. Epstein. 1971. Retention of three insecticides on different soil particles suspended in water. *Proc. Soil Sci. Am.* **35**:884–887.

163. R. W. Risebrough, P. Reiche, D. B. Peakall, S. G. Herman, and M. N. Kirven. 1968. Polychlorinated biphenyls in the global ecosystem. *Nature* **220**:1098–1102.

164. J. Robinson, A. Richardson, B. Bush, and K. E. Elgar. 1966. A photoisomerization product of dieldrin. *Bull. Environ. Contam. Toxicol.* **1**:127–132.

165. J. Robinson, A. Richardson, A. N. Crabtree, J. C. Coulson, and G. R. Potts. 1967. Organochlorine residues in marine organisms. *Nature* **214**:1307–1311.

166. D. R. Rowe, L. W. Canter, P. J. Snyder, and J. W. Mason. 1971. Dieldrin and endrin concentrations in a Louisiana estuary. *Pestic. Monit. J.* **4**:177–183.

167. R. L. Rudd. 1964. *Pesticides and the Living Landscape.* Univ. Wisconsin Press. 320 pp.

168. R. L. Rudd and S. G. Herman. 1972. Ecosystem transferral of pesticide residues in an aquatic environment. In *Environmental Toxicology of Pesticides.* Eds. F. Matsumura, G. M. Boush, and T. Misato. Academic. pp. 471–485.

169. J. R. Sanborn. 1974. *The Fate of Select Pesticides in the Aquatic Environment.* U.S. Environ. Prot. Agency, Ecol. Research Series EPA-660/3-74-025. 83 pp.

170. J. R. Sanborn and C. C. Yu. 1973. The fate of dieldrin in a model ecosystem. *Bull. Environ. Contam. Toxicol.* **10**:340–346.

171. C. H. Schaefer and E. F. Dupras. 1970. Factors affecting the stability of Dursban in polluted waters. *J. Econ. Entomol.* **63**:701–705.

172. D. A. Schooley, B. J. Bergot, L. L. Dunham, and J. B. Siddall. 1975. Environmental degradation of the insect growth regulator methoprene. II. Metabolism by aquatic microorganisms. *J. Agric. Food Chem.* **23**:293–298.

173. D. B. Seba and E. F. Corcoran. 1969. Surface slicks as concentrators of pesticides in the marine environment. *Pestic. Monit. J.* **3**:190–193.

174. S. B. Shaw. 1971. Chlorinated hydrocarbon pesticides in California sea otters and harbor seals. *Calif. Fish Game.* **57**:290–294.

175. W. J. L. Sladen, C. M. Menzie, and W. L. Reichel. 1966. DDT residues in Adelie penguins and a crab-eater seal from Antarctica. *Nature* **210**:670–678.

176. M. Sobelman. 1973. DDT and pelicans. *Nature* **241**:225.

177. A. Sodergren. 1968. Uptake and accumulation of C^{14}-DDT by *Chlorella* sp. *Oikos* **19**:126.

178. B. I. Sparr, W. G. Appleby, D. M. DeVries, J. V. Osmun, J. M. McBride, and G. L. Foster. 1966. Insecticide residues in waterways from agricultural use. In *Organic Pesticides in the Environment.* Am. Chem. Soc., Adv. Chem. Series **60**:146–162.

179. W. F. Spencer. 1975. Movement of DDT and its derivatives into the atmosphere. *Residue Rev.* **59**:91–117.

180. C. W. Stanley, J. E. Barney, M. R. Helton, and A. R. Yobs. 1971. Measurements of atmospheric levels of pesticides. *Environ. Sci. Technol.* **5**:430–435.

181. V. F. Stout. 1968. Pesticide levels in fish of the northeast Pacific. *Bull. Environ. Contam. Toxicol.* **3**:240–246.

182. E. C. Tabor. 1966. Contamination of urban air through the use of insecticides. *Trans. N.Y. Acad. Sci.* **28**:569–578.

183. K. R. Tarrant and J.O'G. Tatton. 1968. Organochlorine pesticides in rainwater in the British Isles. *Nature* **219**:725–727.

184. J. O'G. Tatton and J. H. A. Ruzicka. 1967. Organochlorine pesticides in Antarctica. *Nature* **212**:346–348.

185. L. C. Terriere, U. Kiigemagi, A. R. Gerlach, and R. L. Borovicka. 1966. The persistence of toxaphene in lake water and its uptake by aquatic plants and animals. *J. Agric. Food Chem.* **14**:66–69.

186. B. D. Vance and W. Drummond. 1969. Biological concentration of pesticide by algae. *J. Am. Water Works Assoc.* **61**:360–362.

187. G. C. Veith. 1975. Baseline concentrations of polychlorinated biphenyls and DDT in Lake Michigan fish, 1971. *Pestic. Monit. J.* **9**:21–29.

188. A. Walker. 1974. A simulation model for prediction of herbicide persistence. *J. Environ. Quality* **3**:396–400.

189. G. W. Ware and C. C. Roan. 1970. Interaction of pesticides with aquatic microorganisms and plankton. *Residue Rev.* **33**:15–45.

190. G. W. Ware, E. J. Apple, W. P. Cahill, P. D. Gerhardt, and K. R. Frost. 1969. Pesticide drift. II. Mist-blower vs. aerial application of sprays. *J. Econ. Entomol.* **62**:844–846.

191. G. W. Ware, W. P. Cahill, P. D. Gerhardt, and J. M. Witt. 1970. Pesticide drift. IV. On-target deposits from aerial application of pesticides. *J. Econ. Entomol.* **63**:1982–1983.

192. L. Weaver, C. G. Gunnerson, A. W. Breidenbach, and J. J. Lichtenberg. 1965. Chlorinated hydrocarbon pesticides in major U.S. river basins. *U.S. Public Health Rep.* **80**:481–493.

193. S. R. Weibel, R. B. Weidner, J. M. Cohen, and A. G. Christianson. 1966. Pesticides and other contaminants in rainfall and runoff. *J. Am. Water Works Assoc.* **48**:1075–1084.

194. G. A. Wheatley. 1973. Pesticides in the atmosphere. In *Environmental Pollution by Pesticides*. Ed. C. A. Edwards. Plenum. pp. 365–408.

195. G. A. Wheatley and J. A. Hardman. 1965. Indications of the presence of organochlorine insecticides in rainwater in central England. *Nature* **207**:486–487.

196. D. H. White. 1976. Nationwide residues of organochlorines in starlings, 1974. *Pestic. Monit. J.* **10**:10–17.

197. D. H. White and R. G. Heath. 1976. Nationwide residues of organochlorines in wings of adult mallards and black ducks, 1972–73. *Pestic. Monit. J.* **9**:176–185.

198. R. Williams and A. V. Holden. 1973. Organochlorine residues from plankton. *Marine Pollut. Bull.* **4**:109–111.

199. G. H. Willis, J. F. Parr, and S. Smith. 1971. Volatilization of soil-applied DDT and DDD from flooded and non-flooded plots. *Pestic. Monit. J.* **4**:204–208.

200. A. J. Wilson, J. Forester, and J. Knight. 1969. *Chemical Assays.* U.S. Dept. Interior Circ. 335, Gulf Breeze Lab p. 20.

201. D. R. Winter. 1970. *Pesticide Residues and Their Implications in the Upper Great Lakes.* Paper presented to Am. Assoc. Adv. Sci., 29 December.

202. J. L. Wolfe and B. R. Norment. 1973. Accumulation of mirex residues in selected organisms after an aerial treatment. *Pestic. Monit. J.* **7**:112–116.

203. A. A. Wolman and A. J. Wilson. 1970. Occurrence of pesticides in whales. *Pestic. Monit. J.* **4**:8–9.

204. G. M. Woodwell, P. P. Craig, and H. A. Johnson. 1971. DDT in the biosphere: Where does it go? *Science* **174**:1101–1107.

205. G. M. Woodwell, C. F. Wurster, and P. A. Isaacson. 1967. DDT residues in an east coast estuary: A case of biological concentration of a persistent pesticide. *Science* **156**:821–824.

206. M. Yasuno, S. Hirakoso, M. Sasa, and M. Uchida. 1965. Inactivation of some organophosphorus insecticides by bacteria in polluted water. *Jap. J. Exp. Med.* **35**:545–563.

207. C. C. Yu, G. M. Booth, D. J. Hansen, and J. R. Larsen. 1974. Fate of carbofuran in a model ecosystem. *J. Agric. Food Chem.* **22**:431–434.

208. M. J. Zabik, B. E. Pape, and J. W. Bedford. 1971. Effect of urban and agricultural pesticide use on residue levels in the Red Cedar River. *Pestic. Monit. J.* **5**:301–308.

209. R. G. Zepp, N. L. Wolfe, J. A. Gordon, R. C. Finche, and G. L. Baughman. 1975. *Chemical and Photochemical Transformations of Selected Pesticides in Aquatic Systems:* U.S. Environ. Protection Agency, Ecol. Res. Ser. EPA-600/3-76-067.

210. V. Zitko. 1971. Polychlorinated biphenyls and organochlorine pesticides in some freshwater and marine fish. *Bull. Environ. Contam. Toxicol.* **6**:464–468.

11

HERBICIDES: PERSISTENCE AND PLANT ECOSYSTEM EFFECTS

A. HERBICIDE EFFECTS ON THE PLANT ECOSYSTEM

The essential role of herbicides in crop production is to protect monocultures, and to prevent the bared arable land from being overgrown by plant cover natural to the site and climate. The various chemicals employed have a certain selectivity from one seed-plant group to another, certain being effective against grassy weeds in broadleaf crops, and others effective against broadleaf weeds in cereal crops (Table 11.1). The time of application, whether preplanting, preemergence, or postemergence (foliar), offers another means of exaggerating the differential between crop and weed. Finally the physical characteristic of the compound that decides the depth to which it penetrates may protect deep-rooted crops against shallow-rooted weeds (e.g., simazine, diuron), or shallow-rooted crops against deep-rooted weeds (e.g. picloram, dalapon). The rationale and procedures of weed control may be found in the books by Klingman and Ashton (1975) and by Crafts (1975).

Long-Term Effects of Single Applications

Grassland. The ecological effect of a selective herbicide such as 2,4-D has been investigated on rangeland in Colorado, where it was applied as the butyl ester at 3 lb/a to control pocket gophers. The reduction in flowering plants (forbs) that provide the seeds that are the gophers' food supply was found to be fairly permanent. The result of single sprayings on Black

Table 11.1. Some widely used herbicides grouped according to their principal weed target, with modes of application indicated.

Broadleaf Weeds	Grasses	Both (selective)	All (non-selective)
2,4-D (3)	Triallate (2,3)	Simazine (1,2)	Atrazine
MCPA (3)	Propanil (3)	Atrazine (2)	Fenuron
Silvex (3)	TCA (2,3)	Diuron (2)	Monuron
2,4,5-T (3)	Dalapon (3)	Linuron (2,3)	Diuron
Endothall (1,2)	DCPA (2,3)	Chlorpropham (2,3)	TCA
Dicamba (3)		Alachlor (1,2)	2,3-6-TBA
Picloram (3)		Dichlobenil (2)	Prefix
		Dinoseb (2,3)	
		Amitrole (3)	
		Trifluralin (1,2)	
		Diphenamid (1,2)	
		Fenac (2)	

(1) Preplanting, (2) preemergence, (3) postemergence, applications

Mesa[73] show the following percentage compositions of forbs as against grasses:

	Prespray 1959	Postspray 1959	1964
South Crystal Creek	77	9	44
Myers Gulch	63	9	10

Detailed analyses of the plant species involved were made on Grand Mesa in 1951, and again in 1960 after 2,4-D had been applied between 1955 and 1958.[75] The results (Table 11.2) show a particular increase in slender wheatgrass (*Agropyron trachycaulum*), and decrease in *Agoseris* and American vetch. The application of a nonselective herbicide, sodium cacodylate at 7 lb/a, to a fescue meadow in Tennessee resulted in a general

Table 11.2.   Floral change induced by 2,4-D ester (3 lb/a) on a Colorado upland meadow; frequency of occurrence of each species* (Turner, 1969).

	1951 before	1960[†] after		1951 before	1960[†] after
Forbs			Grasses		
Cinquefoils	100	71	Letterman Needlegrass	99	96
Common Dandelion	100	71	Sedges	83	65
Agoseris	97	26	Slender Wheatgrass	47	80
Orange Sneezeweed	80	17	Trisetums	47	60
Douglas Knotweed	72	88	Bluegrass	44	28
American Vetch	47	2	Brome Grasses	19	28
Aspen Peavine	46	26	Alpine timothy	9	6
Lupines	43	10	Alpine fescue	2	4
Penstemons	30	26	Shrubs		
Aspen Fleabane	22	4	Silver Sagebrush	10	3
Prairiesmoke Sieversia	11	6			

* on 12.5-ft^2 plots spaced at 18 to the acre

† 2-5 years after the spray was applied

reduction from which all species recovered sooner or later, fescue being first.[50] An unexpected consequence of this nonpersistent herbicide was a stimulation of bacterial and fungal decomposition of the soil litter, an important sink for nutrients.

Forest. The outstanding example of ecosystem disruption by single or sporadic applications of herbicides is the military airspraying conducted in South Vietnam during the decade 1962–1972, mainly with an equal mixture of 2,4-D and 2,4,5-T (Agent Orange) at 25 lb/a, but also with a 4:1 mixture of 2,4-D and picloram (Agent White) at 7.5 lb/a. More than 11,000 sq km of semidecidous Dipterocarp forest was treated, often twice. A single spraying would partially, and a double spraying wholly, revert it to the initial grass–sedge stage of the grass–shrub–forest succession. However, sufficient tree seeds survived to germinate and grow into secondary forest, provided that the seedlings were not overwhelmed by an invasion of bamboo species. It is considered that this period of reduced species diversity could involve a serious leakage from the pool of soil nutrients to be lost in runoff.[81] Nearly 1100 sq km of mangrove swamp, one-third of all the mangrove in South Vietnam, was also treated. Fingers of mangrove regeneration could be observed from the air 6 yr after a treatment, and it was estimated that the original condition would return after 20 yr.[74] However, the visiting commission of the U.S. National Academy of Sciences (1974) was of the opinion that "it may take well over a 100 years for the mangrove area to be reforested."

Effects of Continued Applications

Field Crop Weeds. The continuing use of one herbicide season after season may promote the incidence of certain species of weeds rather than others. The use of 2,4-D to control broadleaf weeds in sugar cane in Hawaii led to infestations of foxtail grasses (*Setaria*) and crab grasses (*Digitaria*).[30] Applications of atrazine in corn in Michigan favored crab grass (*D. sanguinalis*) and witch grass (*Panicum capillare*), and led to infestations of fall panicum (*P. dichotomiflorum*) and green foxtail (*S. viridis*).[46]

Data have been reported for spring wheat plots in Saskatchewan to which 2,4-D amine or 2,4-D ester had been applied every spring from 1947 to 1972, or MCPA every spring from 1953 to 1972, the dosage having been 1.5 lb/a (Table 11.3). When the weeds that were found emerging before each annual application between 1968 and 1972 were compared with those on check plots, it was clear that the MCPA regime had reduced the average weed regeneration by 69%, and the 2,4-D ester and amine by 83% and 86%, respectively. Of the individual species, stinkweed, which had been originally the most prevalent species, was also the one showing the most reduction, while redroot pigweed that had been the least prevalent showed the least reduction (Table 11.3). During the 20-yr period no weed species completely disappeared under the herbicide regime, and three species (wild buckwheat, Canada thistle, and perennial sowthistle) made their first appearance.[51] Canada thistle cannot be satisfactorily controlled with 2,4-D, and dicamba has to be substituted to control this species.

Orchard Weeds. The application of herbicides to the ground cover of orchards eventually results in almost pure stands of grasses, as happens where silvex or 2,4,5-T ester or amine has been applied for six successive years.[66] Nevertheless compounds designed to control broadleaved weeds do

Table 11.3. **Percent reductions of weeds in wheatfields treated annually for 20 yr: final prespray infestations compared to untreated checks (McCurdy and Molberg, 1974).**

Weed Species[*]	2,4-D amine	2,4-D ester	MCPA[†]
Stinkweed	97	94	98
(Thlaspi arvense)			
Russian Thistle	88	58	35
(Salsola pestifer)			
Lamb's Quarters	90	85	86
(Chenopodium album)			
Wild Buckwheat	32	54	51
(Polygonum conrolvulus)			
Wild Tomato	52	53	23
(Solanum triflorum)			
Redroot Pigweed	55	15	30
(Amaranthus retroflexus)	___	___	___
Total of all weeds	86	83	69

[*] Listed in order of relative abundance

[†] MCPA amine in some years, MCPA ester in others

encourage certain broadleaved species to become abundant in the orchard after successive annual applications, the important genera being:

Rumex; after 2,4,5-T, simazine, diuron, terbacil
Plantago; after 2,4,5-T amine, diuron, monuron
Polygonum; after silvex, simazine, diuron
Convolvulus; after simazine, diuron, terbacil
Rubus; after simazine, diuron, terbacil, dichlobenil
Cerastium; after dalapon, amitrole

In orchards, Canada thistle has followed general terbacil treatments in Michigan,[57] and dandelion has followed simazine treatments in Massachusetts. Among pest grasses, yellow foxtail, fall panicum, and witch grass have followed simazine or dichlobenil in Michigan, with quack grass after simazine in Massachusetts and crab grass after dichlobenil in North Carolina.

Effect on Pathogenic Fungi. An indirect effect of herbicides on the plants which they are designed to protect is their promotion of certain fungal diseases; it has been most commonly noted in the damping-off fungus *Rhizoctonia solani* attacking various crops under herbicide treatment. Not infrequently, one herbicide will promote the incidence of the disease, while another suppresses it. Records of these occurrences have been collected by Katan and Eshel (1973), and the more important examples are as follows:

Disease promoted
Rhizoctonia solani	Damping-off, cotton	Trifluralin
Helminthosporium sativum	Seedling disease, barley	Maleic hydrazide
Fusarium oxysporum	Wilt disease, tomato	MH, Dalapon
Alternaria solani	Early blight, tomato	2,4-D

Disease suppressed
Cercosporella herpotrichoides	Foot rot, wheat	Diuron
Fusarium oxysporum	Wilt disease, tomato	Propham, TCA
Alternaria solani	Early blight, tomato	MH, Dalapon

Thus it may be seen that on tomato the herbicides MH and dalapon promote wilt disease but suppress early blight. The hormone-type herbicide 2,4-D has also promoted tobacco mosaic virus infection of tobacco, cucumber and cotton.[9,42] On the other hand, many herbicides increase the population of the fungus *Trichoderma viride*[1]; this species suppresses many pathogenic fungi in the soil, as is described more fully in Chapter 13.

Plant Resistance to Herbicides

Preexisting Natural Tolerance. The especial tolerance of fall panicum and green foxtail to atrazine in cornfields was due to these weeds absorbing less of the herbicide than the crop to be protected,[46] but the situation could be corrected by a postemergence spray with another triazine, namely cyanazine, to which they did not show such tolerance. This characteristic had not been developed by the continued use of atrazine, since fall panicum from untreated areas was found to be no more susceptible than the treated problem areas in New Jersey.[69]

Corn (*Zea mays*) is resistant in itself to simazine and atrazine primarily because it can detoxify them, conjugating the triazine to glutathione by means of an S-transferase enzyme present in the leaves.[67] In addition, corn roots contain a "sweet substance" which nonenzymically catalyzes the hydroxylation of simazine and atrazine by nucleophilic attack on the 2-Cl substituent (Fig. 11.1). This compound, identified as 2,4-dihydroxy-3-keto-7-methoxy-1,4-benzoxazine (MBOA), exists either free or as its 2-glucoside.[27,60] Cereals that are less resistant than corn respond to sublethal levels of simazine by taking up more water and nitrate, thus gaining more plant weight and total protein.[59]

The reason that 2,4-D kills broadleaf weeds and not grasses again depends on detoxication; excised roots of alfalfa, soybean, and peanut, in contrast to those of oats, barley, and corn, were found incapable of hydroxylating the benzene ring of 2,4-D.[80] Red currants (*Ribes sativum*), characteristically resistant to 2,4-D, decarboxylate 50% of the herbicide applied to the leaves in 1 wk, while black currants (*R. nigrum*) that are susceptible detoxified only 2% of the applied herbicide.[49] The effectiveness of 2,4-D to delay fruit and leaf abscission of apples does not extend to the McIntosh variety, whose leaves can break down one-third of the applied compound to CO_2 in the same time that a susceptible variety detoxifies only 0.6%.[15]

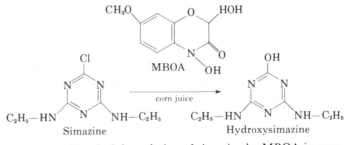

Fig. 11.1   Chemical degradation of simazine by MBOA in corn.

Susceptibility Differences between Ecotypes. It has long been known that different populations ("ecotypes") of weed species differ in their susceptibility to herbicides. First studied in the effect of dalapon on grassy weeds, a wide range of susceptibility was found for 9 lines of yellow foxtail (*Setaria lutescens*) and 16 lines of giant foxtail (*S. faberii*) from different locations in Maryland.[64] Of five strains of Johnson grass (*Sorghum halepense*) collected from different localities in Arizona, one was much more susceptible to dalapon than the others[26]; similar variations between populations in their susceptibility to dalapon have been reported for quack grass, Bermuda grass, and barnyard grass, and to other herbicides by barnyard grass and hoary cress.

The marked variation in susceptibility to amitrole shown by Canada thistle (*Cirsium arvense*) was studied in 20 different populations in Norway; it had no consistent geographical pattern, and was evidently correlated not with the uptake of amitrole but with its detoxication to an unknown metabolite.[16] Of 10 ecotypes of Canada thistle collected in four states of northwestern United States, one from Ada County, Idaho was especially tolerant to amitrole, and it was among those most tolerant to 2,4-D also.[38] When the reaction of field bindweed (*Convolvulus arvensis*) to 2,4-D was studied in 51 strains from the United States and Canada, the most resistant one (from New Mexico) showed 83% weight gain after the treatment, while the most susceptible one (from Canada) showed an 87% weight loss.[79]

A wide range of susceptibility to the carbamate herbicide propham was found in wild oats (*Avena fatua*) when 63 strains from the northwestern United States and Canada were assessed; those strains with nondormant seed tended to be the more resistant.[63] Lines obtained from 214 seed collections of wild oats along a 60-mile length of the Red River Valley, North Dakota showed a normal frequency curve of susceptibility to the thiocarbamate herbicides diallate and triallate which ranged from almost complete resistance to almost complete susceptibility to a single diagnostic dose[40]; such variation could provide the background for the development of resistant populations by the selective effect of these herbicides.

Development of Resistant Populations. While the preexistence of refractory ecotypes provides a potential source of populations resistant to herbicides, there are three examples of herbicide-resistant strains having developed in response to selection pressure in the sense that they have become increasingly difficult to control with each successive application of that herbicide (Table 11.4). This has been reported for the use of 2,4-D against the fireweed *Erechtites hieracifolia* on certain sugar plantations in Hawaii,[30] and for certain populations of Canada thistle resisting control by MCPA in Norway.[16] The clearest case of resistance developing to continued

Table 11.4.   Examples of weed populations resistant to herbicides.

Emergence of Resistant Populations

Erechtites hieracifolia	2,4-D	Hawaii 1955
(Hawaiian fireweed)		
Cirsium arvense	MCPA	Norway 1973
(Canada thistle)		
Senecio vulgaris	Atrazine	Washington state 1968
(Common groundsel)		
Poa annua	Metoxuron	France 1974
(Annual bluegrass)		

Existence of Resistant Ecotypes

Sorghum halepense	Dalapon	Arizona 1963
(Johnson grass)		
Convolvulus arvensis	2,4-D	New Mexico 1963
(Field bindweed)		
Avena fatua	Propham	North Dakota 1963
(Wild oats)		
Setaria lutescens	Dalapon	Maryland 1960
(Yellow foxtail)		
Cirsium arvense	Amitrole	Idaho 1970
(Canada thistle)		

herbicide operations comes from a tree nursery near Puyallup, Washington. Here an infestation of common groundsel (*Senecio vulgaris*), having been treated with atrazine or simazine once or twice every year since 1958, noticeably resisted control with either of these triazines in 1968, although it was not cross-resistant to dichlobenil or the urea herbicides.[62] The resistance was confirmed by comparing the Puyallup strain with a normal strain from the Willamette Valley, Oregon, and it was also shown to four other triazine herbicides.[58]

Resistance has been induced experimentally by applying the urea herbicide metoxuron to the annual bluegrass *Poa annua*.[21] A series of 14 applications made over a 5-yr period to a population on a market-garden path at Montpellier, France at intervals corresponding to the maturing and seeding of the grass resulted in a strain showing 33% survival to a dosage (0.3 kg/ha) that had allowed only 3% survival in this and two other ecotypes of this species. This resistance induced in the field population did not start to increase materially until after the eighth application. Resistance to siduron shown by certain genotypes of foxtail barley has proved to be due to three genes, each dominant, all of which are required for the character to be expressed[65]; atrazine tolerance in flax is also due to additive genes, while the occasional atrazine susceptibility found in corn derives from a single mutant gene.

B. PERSISTENCE OF HERBICIDE RESIDUES IN THE SOIL

Disappearance and Degradation

Unlike some of the persistent pesticides, residues of herbicides do not build up from one year to the next, since at the dosages used on croplands very few persist in the soil for more than 12 months. Of the compounds used as selective insecticides (Table 11.5), 2,4-D, MCPA, the bipyridylium compounds and the organic arsenicals are the least persistent, and so are picloram and trifluralin (Fig. 11.2). With atrazine applied at 2 lb/a to silty clay loam in Pennsylvania in the spring, 70% of the original herbicide remained in the soil 1 month later, 20% after 4 months, and in the following year 15% after 11 months and 5% after 16 months.[22] A cornfield treated with simazine cannot be planted to susceptible crops in the next year. Corn soils in the dry land of western Nebraska, that had received atrazine at 4 lb/a for the previous 3 yr, suppressed the growth of soybeans planted in them by 15-30%; but there was no atrazine carry-over where these soils were irrigated.[8] For total vegetation control on roadsides and industrial sites, simazine and monuron at higher dosages may be herbicidal for 2 yr. The organic herbicides picloram and fenac are as persistent as the inorganic arsenite, borate, and chlorate, and thus may be used for total vegetation control.

In characterizing the rate of loss of herbicide residues, the "half-life" concept has been useful, since it is somewhat independent of the dosage. As points on the disappearance curve, the DT_{50} and DT_{90} are taken as relative disappearance times, and allow compounds to be compared for their longevity with figures derived from the literature (Table 11.6). The longevity of course is inverse to temperature, the half-lives in months for certain triazines, ureas, and carbamates[19] being as follows:

	15°C	30°C		15°C	30°C		15°C	30°C
Atrazine	6.0	2.0	Diuron	7.0	5.5	Chlorpropham	3.0	1.5
Ametryn	6.0	4.5	Fenuron	4.5	2.2	Propham	0.4	0.2

Table 11.5. **Persistence of herbicidal activity in temperate-zone soils at the usual application rates (Klingman and Ashton, 1975).***

1 Month or Less (Temporary Effects)	1-3 Months (Early Season Control)	3-12 Months (Full Season Control)	Over 12 Months (Total Vegetation Control)
Amitrole	Bentazon	Alachlor	Arsenite
AMS	Butachlor	Ametryn	Borate
Barban	Butylate	Atrazine	Bromacil
Cacodylic acid	CDAA	Benefin	Chlorate
Chloroxuron	CDEC	Bromoxynil	Fenac
Dalapon	Chloramben	Chlorobromuron	Picloram
2,4-D	Chlorpropham	Cyprazine	Tebuthiuron
2,4-DB	Cycloate	DCPA	Terbacil
Dinoseb (DNBP)	Diallate	Dicamba	2,4,6-TBA
Diquat	2,4-DEP	Dichlobenil	
DSMA	Diphenamid	Diuron	
Endothall	EPTC	Fenuron	
Fluorodifen	Mecoprop	Fluometuron	
Glyphosate	Naptalam	Linuron	
Metham	Pebulate	Metobromuron	
MCPA	PCP	Metribuzin	
MH	Propachlor	Monuron	
Molinate	Pyrazon	Norea	
MSMA	Siduron	Oryzalin	
Nitrofen	Silvex	Prometryn	
Paraquat	TCA	Propazine	
Phenmedipham	Triallate	Simazine	
Propanil	2,4,5-T	Terbutol	
Propham	Vernolate	Trifluralin	

* A table of references on the longevity of specific dosages and concentrations may be found in Alexander (1969).

The highly persistent picloram would have a DT_{50} of 5 months under Montana conditions as compared to 1.5 months under Texas conditions.[24]

The rate of disappearance of the residues is proportional to the concentration remaining, being expressed by the equation $-dc/dt = Kc^n$, where c = concentration, t = time, K = the rate constant, and n = the exponent; in first-order kinetics $n = 1$, and in second-order kinetics the

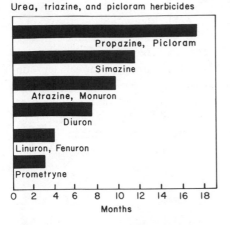

Fig. 11.2. Persistence of herbicides in soils: time taken for a compound applied to soil at the normal dosage to decrease by more than 75% (from Kearney et al., 1969).

Table 11.6. Average 50% and 90% disappearance times (in days) for herbicides (Hamaker, 1972).

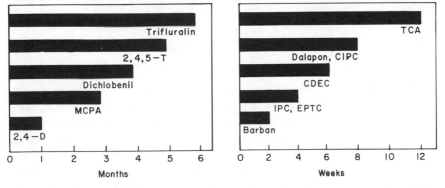

	DT_{50}	DT_{90}		DT_{50}	DT_{90}		DT_{50}	DT_{90}
Atrazine	130	284	Chlorpropham	43	53	2,4-D	17	26
Simazine	105	360	Chloramben	34	65	Dalapon	15	27
Fenuron	69	156	Linuron	24	169	Propham	8	22

exponent is some other value. The matter is discussed fully by Freed and Haque (1973) and by Haque and Ash (1974), and equations have been developed in great detail by Hamaker (1972) to cover the various environmental conditions and to be suitable for computer treatment.

Most herbicides undergo a certain amount of purely chemical degradation in soil, but the half-life exclusively for the slow chemical process ranges from 1 yr with chlorpropham to 10 yr for diuron, with atrazine intermediate.[28] Dichlobenil is slowly hydrolyzed in soil to 2,6-dichlorobenzamide as a nonbiological process, its half-life being more than 5 months.[6] The relative importance and role of biological degradation are described in the next chapter.

Disappearance and Translocation

Additional factors governing the disappearance of residues include volatilization, photodecomposition, adsorption and leaching (Fig. 11.3) as explained in detail in the reviews of Helling, Kearney, and Alexander (1971) and of Weber and Weed (1974), and summarized in the book by Klingman and Ashton (1975).

Volatilization.   This takes place mainly at the time of application, and compounds characterized by volatility such as EPTC and trifluralin are customarily tilled into the soil when applied as preemergent herbicides in order to avoid this volatilization. Applied in aqueous solution to the soil surface, EPTC is 70% lost in the first day if the soil is moist, 45% if it is dry; other thiocarbamate herbicides are less volatile, in this descending order: vernolate > pebulate > molinate.[20] When the trifluralin which evaporated from a sprayed sandy loam surface was trapped, the yield in the first 100 hr was 4.5% of that applied.[4] Applied in emulsion to clay loam, chlorpropham is 15% lost, and propham 30% lost, in the first day; this happened when the soil was moist, but when it was dry the losses were only 1 and 2%, respectively.[56] The esters of 2,4-D (but not the salts or the acid) are sufficiently volatile that as much as 10% of the application may be volatilized in the first 24 hr,[18] enough to damage susceptible crops such as tomatoes by vapor action. Under intense sunlight, the soil surface may become hot enough to volatilize herbicides of low vapor pressure, such as atrazine. Simazine has a lower vapor pressure than atrazine, while prometryn and prometone are more volatile triazines.[41] The amounts of triazine herbicides volatilized increase by about 50% for a 10° temperature increase from 35°C to 45°C in the soil.[43] Residues of atrazine and 2,4-D PGBE ester have been found in rain a mile away from fields treated with these herbicides 3 wk before.[10]

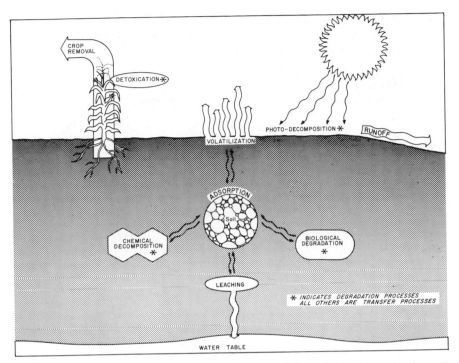

Fig. 11.3. Processes influencing the behavior and fate of herbicides in the soil (from Weber and Weed, 1974).
Reprinted from Pesticides in Soil and Water, edited by W. B. Guenzi, page 229, 1974 by permission of the Soil Science Society of America.

Average rates of loss from fields have been quoted by Haque and Ash (1974) to be as follows:

Vapor index 1, loss of <0.1 kg/ha-yr   Dalapon, MCPA, 2,4-D, 2,4,5-T
        2   0.2–3.0   Propanil, Trifluralin
        3   7–14   Alachlor

Once the herbicide has percolated into the soil, volatilization is a minor factor, especially where the soil is poorly aerated.

The role of moisture in promoting the loss of volatile herbicides is exemplified by EPTC, which is very persistent in dry soil, but if applied to a soil of 35% moisture content, three-quarters of the dose has volatilized by the time the soil has dried.[17] Trifluralin, a crystalline solid with a vapor pressure (at 25.5°C) of only 2×10^{-4} mm Hg in this form, when in aqueous solution shows a pressure of 126 mm Hg at 80°C.[4] However in their review of pesticide volatilization, Spencer, Farmer, and Cliath (1973) conclude that there is no firm evidence for an enhancement of volatility due to codistillation with water, but rather it is that the soil moisture displaces the pesticide from the soil particles so that their volatility can take effect. Indeed soil moisture, while increasing the volatilization of atrazine, decreases that of simazine.[43]

Photodecomposition. This may take place when the herbicide is in vapor form, or when it is dissolved in water. The effect of sunlight in inactivating simazine and atrazine has been frequently reported, but degradation due to photolysis has been proved only for paraquat and diquat deposited on or evaporating from leaf surfaces. A number of herbicides are readily photolysable in aqueous solution when irradiated with ultraviolet light in the laboratory, including DNOC, dinoseb, trifluralin, monuron, picloram, and diphenamid,[13] but of these only dinoseb and trifluralin have an absorption maximum (375 nm wavelength) in that part of the solar ultraviolet that reaches the earth, the maximum for others being below 250 nm and thus well below the lower limit (292 nm) of the ultraviolet that reaches the earth.[47] When riboflavin is present in the water, it acts as a sensitizer to permit the rapid photolysis of amitrole, and even of 2,4-D, silvex, and 2,4,5-T. The breakdown products of photolysis are usually found to be much the same as those which are produced by chemical degradation in unsterilized soil, a notable exception being the production of chloroform from TCA. The degree to which photodecomposition is a factor in residue disappearance is difficult to establish, but it is probably minor except for herbicides such as amitrole which are not readily degraded by soil microorganisms. Trifluralin, on the other hand, is not photodecomposed when irradiated under quartz plates, volatility being the main source of loss from soils.[36] Atrazine and simazine are photodecomposed when irradiated in water but not at wavelengths exceeding 300 nm being hydroxylated to 2-hydroxyatrazine.[55] However, when irradiated on aluminum planchets both these triazines were considerably photolysed even at 311 nm wavelength.[41]

In field experiments at 7200 feet altitude in Wyoming, where ultraviolet insolation is strong, exposure of certain triazine and urea herbicides for 25 days caused a 10–45% loss in sunlight compared to 2–16% loss in the shade; the ranking in order of loss was atrazine > fenuron > diuron > simazine > monuron. During this experiment the soil temperature in the sun was 120°F on the average; when subsequent experiments were made at the height of summer, when the soil temperatures were between 140 and 180°F, the loss in the shade eventually came to be as great as that in the sun, showing that volatilization could mask photolytic effects.[11] In summer sunlight in the San

Joaquin Valley, California, the breakdown of monuron in aqueous solution was about 6% in 2 wk, the N-dimethyl groups being first oxidized to aldehyde groups, followed by dealkylation, as also occurs in the degradation by soil microorganisms.[14] Picloram in aqueous solution is 30% degraded in 6 days of Texas sunlight, the rate being half that at 254 nm wavelength, its pyridine nucleus being destroyed and Cl being liberated.[23] Diquat is rapidly photolyzed in aqueous solution, sunlight degrading 70% of it in 3 wk, the end products being picolinamide and picolinic acid.[68] At the other extreme, chloramben loses no herbicidal activity to insolation when sprayed on dry soil, although the loss is 12% on wet soil.[39]

Adsorption. Binding of herbicides to particles is more marked in clay than in sandy soils, since there is a greater ratio of surface to volume. The montmorillonite clays characteristic of the northern temperate zone are much more adsorptive than the kaolinite clays of the subtropics and tropics.[47] The greatest adsorption of herbicides is onto organic matter, which may exceed 15% in much soils; humic acid is the principal adsorbent, as shown by exact tests with 2,4-D.[78] Particles that are anionic in charge vigorously adsorb the kationic herbicides which replace the Ca, Mg or H ions on the particle surface, and the clays that have high base-exchange capacity are the most adsorptive; the amount of surface area that the soil mineral offers is also important (Table 11.7). On the average for all nonmuck soils examined, the highly kationic herbicide paraquat has an equilibrium constant in the Freundlich equation of 353, as compared to 38 for silvex and 0.55 for fenuron, the most soluble of the herbicides tested.[25] Triazine, urea, carbamate, and anilide herbicides, which are alkaline compounds, are readily adsorbed onto the acid form of montmorillonite (Table 11.8), while com-

Table 11.7. **Surface characteristics of soil components and minerals as absorbents (Bailey and White, 1964).**

	Cation-exchange Capacity meq/100 g	Surface Area in^2/g
Organic Matter	200-400	500-800
Vermiculite	100-150	600-800
Montmorillonite	80-150	600-800
Illite	10-40	65-100
Kaolinite	3-15	7-30
Oxides and Hydroxides	2-6	100-800

**Table 11.8.   Amounts of herbicide adsorbed (micromoles/g)
on the two forms of montmorillonite clay (Bailey,
White, and Rothberg, 1968).**

	Na Salt	Acid		Na Salt	Acid
Atrazine	3.8	22.7	Fenuron	102	381
Propazine	0.4	2.7	Monuron	1.3	17.9
Chlorpropham	4.1	8.0	Diuron	1.5	6.8
Propham	0.2	0.9	Propanil	7.6	11.1

pounds such as picloram, amiben, 2,4-D, and 2,4,5-T, which are acidic, are not adsorbed by montmorillonite.[3]

Herbicides in the adsorbed state are thereby sequestered from degradation by microorganisms; this was demonstrated by adding montmorillonite to a soil and preventing the breakdown of diquat added thereto, an effect not achieved with the less-adsorptive kaolinite.[76] The adsorption of dichlobenil onto soil organic matter prevents it from being hydrolyzed to 2,6-dichlorobenzamide.[6] Incidentally, adsorbed herbicides are thereby prevented from exerting side effects on soil flora and fauna. With organic matter as adsorbents, however, the degradation of atrazine and linuron was not negatively correlated with the degree of adsorption, but was directly correlated to the activity of microorganisms in the soils.[29]

Whereas the adsorption on organic matter is not materially affected by temperature, the adsorption onto soil minerals is much greater at low temperatures than high, as the following percentages adsorbed[33] demonstrate:

	Muck soil (pH 5.6)		Bentonite (pH 8.5)	
	0°C	50°C	0°C	50°C
Monuron	45	48	47	8
Simazine	44	43	93	36
Atrazine	42	40	41	12

While no relationship can be discerned between the water solubility of the herbicide and its liability to be adsorbed, the pH of the soil moisture is very important. More of the herbicide is adsorbed in an acid than an alkaline soil, as the following percentages as indicated by Kd values[52] demonstrate for atrazine at three different soil temperatures:

Soil pH	10°C	20°C	30°C
3.9	12.9	12.6	10.4
5.3	8.0	8.2	6.8
8.0	6.2	6.3	4.4

It is seen that soil pH is certainly no less important than soil temperature in deciding the amount adsorbed. Indeed, a selection of 25 different soils ranked on the basis of decreasing pH showed essentially the same ranking for increased adsorption of atrazine and simazine.[72] Soil moisture may also increase the phytotoxicity of the herbicide by enhancing its desorption; this happens with EPTC, atrazine, and chloramben, but there is no effect with propachlor and alachor, and the effect is the opposite with trifluralin and picloram.[71]

A highly soluble chemical, if not tightly adsorbed, will be leached more easily. Judging herbicides by their heat of solution (in kcal/mole) as a qualitative indicator of adsorption (Table 11.9), it can be seen that leaching is roughly the inverse of adsorption. Paraquat and diquat are immediately adsorbed and do not leach. Simazine and other triazines, monuron and most urea herbicides, and chlorpropham remain in the surface layer of the soil longer, while adsorption and decomposition prevents any significant undecomposed residues penetrating more than 15 cm deep. Simazine is more strongly adsorbed to soil particles than propazine and atrazine, while prometone and prometryn are the most strongly adsorbed of the triazine herbicides.[34] Soil particles of montmorillonite clay convert atrazine and propazine to the keto form of the protonated hydroxy derivative, which is

Table 11.9.   Adsorption and leaching characteristics of representative herbicides (Freed and Haque, 1973).

	Adsorption kcal/mole	Soil Adsorption	Leaching		Adsorption kcal/mole	Soil Adsorption	Leaching
Simazine	9.0	Large	Little	Chlorpropham	4.9	Interm'te	Mod'te
2,4,5-T	8.9	"	"	Fenuron	3.9	"	"
Fenac	6.7	"	"	Dichlobenil	2.8	Small	"
2,4-D	6.1	"	Mod'te	Amiben	2.8	Interm'te	Much
Monuron	6.0	"	Little	2,3,6-TBA	1.6	Small	Much

Table 11.10.   **Effect of rainfall on the penetration of herbicides in a loam soil (Haque and Freed, 1974).**

	Rainfall ins.	Maximum Penetration ins.	Depth of Max. Conc'n ins.
Monuron	1	1.75	0.25
	3	1.75	0.25
2,3,6-TBA	3	12	8
Atrazine	12	12	7
Simazine	12	7	1

held tightly to the clay,[61] while atrazine itself may be strongly adsorbed to the insoluble humic acid of the soil's organic content.[48]

Leaching.   The depth to which herbicides are leached naturally depends on the rainfall, as shown experimentally by watering the treated soil (Table 11.10). For a rainfall of 150 mm, alachlor, propanil, and trifluralin penetrate about 10 cm, while MCPA, 2,4-D, and 2,4,5-T applied as acids penetrate not more than 20 cm, and dalapon-Na leaches to a depth of more

Table 11.11.   **Percentage of *s*-triazines from a model plant–soil system* 5 months after treatment at 2–4 kg/ha (Best and Weber, 1974).**

	Atrazine		Prometryn	
	pH 5.5	pH 7.5	pH 5.5	pH 7.5
Topsoil Extract[†]	71.7	65.4	82.5	54.1
Topsoil Fixed	13.8	14.8	4.1	20.0
Subsoil Extract[†]	1.6	5.5	1.5	8.1
Leachate	0.1	0.2	0.1	2.5
Respired as CO_2	0.8	0.9	0.1	0.5
Removed by Plants	1.4	3.9	0.5	1.4
Total Recovery	89.4	90.6	88.8	86.6

* Corn, cotton and soybean planted in silt loam adjusted for pH

† Extracted with methanol and then with glacial acetic acid

than 50 cm.[31] The relative mobilities in soil water of phenoxy acids, triazines, ureas and other herbicides have been already displayed (Fig. 10.1). Whereas the maximum depth to which monuron can penetrate is equivalent to half the rainfall, TCA can penetrate to a depth two to three times the rainfall.[35] Atrazine leaches to deeper soil levels than monuron or simazine.[7] Experiments with radiolabeled atrazine have shown that losses from leaching are very small, and that little of the triazine can be removed by harvesting the crop plants (Table 11.11); loss from volatilization was too low to be measured.[5] Thus the main source of loss is molecular degradation, which is mainly chemical in the case of atrazine, and mainly microbiological in the case of prometryn.

Although the water-soluble salts of 2,4-D do leach, its esters do not. Sodium borate, highly water soluble, leaches freely and thus can attack deep-rooted plants. Sodium pentachlorophenate, dalapon-Na, and TCA-Na leach faster than the microbial decomposition can detoxify the leachate. Dalapon shows so low an adsorbability, even in muck soils, that there is a danger of its being leached to locations where its effect is not desired.[44]

REFERENCES CITED

1. M. Alexander. 1969. Microbial degradation and biological effects of pesticides in soil. In *Soil Biology: Reviews of Research.* UNESCO, Paris (Place de Fontenoy, 75 Paris - 7e). pp. 209–240.

2. G. W. Bailey and J. L. White. 1964. Review of adsorption and desorption of organic pesticides by soil colloids. *J. Agric. Food Chem.* **12:**324–332.

3. G. W. Bailey, J. L. White, and T. Rothberg. 1968. Adsorption of organic herbicides by montmorillonite. *Soil Sci. Soc. Am. Proc.* **32:**222–234.

4. C. E. Bardsley, K. E. Savage, and J. C. Walker. 1968. Trifluralin as influenced by concentration, time, soil moisture content, and placement. *Agron. J.* **60:**89–92.

5. J. A. Best and J. B. Weber. 1974. Disappearance of *s*-triazines as affected by soil pH using a balance-sheet approach. *Weed Sci.* **22:**364–373.

6. G. G. Briggs and J. E. Dawson. 1970. Hydrolysis of 2,6-dichlorobenzonitrile in soils. *J. Agric. Food Chem.* **18:**97–99.

7. O. C. Burnside, C. R. Fenster, and G. A. Wicks. 1963. Dissipation and leaching of monuron, simazine and atrazine in Nebraska soils. *Weeds* **11:**209–213.

8. O. C. Burnside, C. R. Fenster, and G. A. Wicks. 1971. Soil persistence of repeated annual applications of atrazine. *Weed Sci.* **19:**290–293.

9. P. C. Cheo. 1969. Effect of 2,4-dichlorophenoxyacetic acid on tobacco mosaic virus infection. *Phytopathology* **59:**243–244.

10. J. M. Cohen and C. Pinkerton. 1966. Widespread translocation of pesticides by air transport and rainout. Am. Chem. Soc., Wash. D.C., Adv. Chem. Ser. **60:**163–176.

11. R. D. Comes and F. L. Timmons. 1965. Effect of sunlight on the phytotoxicity of some phenylurea and triazine herbicides on a soil surface. *Weeds* **13:**81–87.

12. A. S. Crafts. 1975. *Modern Weed Control.* Univ. California Press, Berkeley. 440 pp.

13. D. G. Crosby and M-Y Li. 1969. Herbicide photodecomposition. In *Degradation of Herbicides.* Eds. P. C. Kearney and D. D. Kaufman. Dekker. pp. 321–363.

14. D. G. Crosby and C. S. Tang. 1969. Photodecomposition of monuron. *J. Agric. Food Chem.* **17**:1041–1044.

15. L. J. Edgerton and M. B. Hoffman. 1961. Fluorine substitution affects decarboxylation of 2,4-D in apple. *Science* **134**:341–342.

16. L. C. Erickson and K. Lund-Hoie. 1974. Canada thistle distribution and varieties in Norway and their reactions to ¹⁴C-amitrole. *Res. Norweg. Agric.* **25**:615–624.

17. S. C. Fang, P. Theisen, and V. H. Freed. 1961. Effects of water evaporation, temperature, and rates of application on the retention of EPTC in various soils. *Weeds* **9**:569–574.

18. V. H. Freed. 1970. Global distribution of pesticides. In *The Biological Impact of Pesticides in the Environment.* Ed. J. W. Gillett. Environ. Health Series No. 1, Oregon State Univ., Corvallis. pp. 1–10.

19. V. H. Freed and R. Haque. 1973. Adsorption, movement and distribution of pesticides in soils. In *Pesticide Formulations.* Ed. W. Van Valkenburg. Dekker. pp. 441–459.

20. R. A. Gray and A. J. Weierich. 1965. Factors affecting the vapor loss of EPTC from soils. *Weeds* **13**:141–147.

21. P. Grignac. 1974. Selection d'un biotype de paturin annuel resistant au metoxuron par repetition de traitements herbicides. *C.R. Acad. Agric. France* **60**:401–408.

22. J. K. Hall, M. Pawlus, and E. R. Higgins. 1972. Losses of atrazine in runoff water and soil sediment. *J. Environ. Quality* **1**:172–176.

23. R. C. Hall, C. S. Giam, and M. G. Merkle. 1968. The photolytic degradation of picloram. *Weed Res.* **8**:292–297.

24. J. W. Hamaker. 1972. Decomposition: Quantitative aspects. In *Organic Chemicals in the Soil Environment.* Eds. C. A. I. Goring and J. W. Hamaker. Dekker. pp. 255–340.

25. J. W. Hamaker and J. M. Thompson. 1972. Adsorption. In *Organic Chemicals in the Soil Environment.* Eds. C. A. I. Goring and J. W. Hamaker. Dekker. pp. 51–143.

26. K. C. Hamilton and H. Tucker. 1964. Response of selected and random plantings of Johnsongrass to dalapon. *Weeds* **12**:220–221.

27. R. H. Hamilton and D. E. Moreland. 1962. Simazine: Degradation by corn seedlings. *Science* **135**:373–374.

28. R. J. Hance. 1969. Further observations on the decomposition of herbicides in soil. *J. Sci. Food Agric.* **20**:144–145.

29. R. J. Hance. 1974. Soil organic matter and the adsorption and decomposition of the herbicides atrazine and linuron. *Soil Biol. Biochem.* **6**:39–42.

30. N. S. Hanson. 1956. Dalapon for control of grasses on Hawaiian sugar-cane land. *Down to Earth* **12**(2):2–5.

31. R. Haque and N. Ash. 1974. Factors affecting the behavior of chemicals in the environment. In *Survival in Toxic Environments.* Eds. M. A. Q. Khan and J. P. Bederka. Academic. pp. 357–371.

32. R. Haque and V. H. Freed. 1974. Behavior of pesticides in the environment: Environmental chemodynamics. *Residue Rev.* **52**:89–116.

33. C. I. Harris and G. F. Warren. 1964. Adsorption and desorption of herbicides by soils. *Weeds* **12**:120–126.

34. C. I. Harris, D. D. Kaufman, T. J. Sheets, R. G. Nash, and P. C. Kearney. 1968. Behavior and fate of s-triazines in soils. *Adv. Pest. Control Res.* 8:1–55.

35. G. S. Hartley. 1960. Physicochemical aspects of the availability of herbicides in soils. In *Herbicides and the Soil.* Eds. E. K. Woodford and J. R. Sagar. Blackwell Scientific Publs., Oxford. pp. 63–78.

36. E. R. Hein and J. V. Parochetti. 1970. Residual herbicidal activity of trifluralin, benefin, and nitralin as affected by volatility and ultra-violet light. *Weed Sci. Soc. Am. Abstr. 1970,* No. 151, pp. 78–79.

37. C. S. Helling, P. C. Kearney, and M. Alexander. 1971. Behavior of pesticides in soils. *Adv. Agron.* 23:147–240.

38. J. M. Hodgson. 1970. The response of Canada thistle ecotypes to 2,4-D, amitrole and intensive cultivation. *Weed Sci.* 18:253–255.

39. A. R. Isensee, J. R. Plimmer, and B. C. Turner. Effect of light on the herbicidal activity of some amiben derivatives. *Weed Sci.* 17:520–523.

40. R. Jacobsohn and R. N. Andersen. 1968. Differential responses of wild oat lines to diallate, triallate, and barban. *Weed Sci.* 16:491–494.

41. L. S. Jordan, W. J. Farmer, J. R. Goodin, and B. E. Day. 1970. Volatilization and non-biological degradation of triazine herbicides *in vitro* and in soils. *Residue Rev.* 32:267–286.

42. J. Katan and Y. Eshel. 1972. Interactions between herbicides and plant pathogens. *Residue Rev.* 25:25–44.

43. P. C. Kearney, T. J. Sheets, and J. W. Smith. 1964. Volatility of seven s-triazines. *Weeds* 12:83–87.

44. P. C. Kearney, C. I. Harris, D. D. Kaufman, and T. J. Sheets. 1965. Behavior and fate of chlorinated aliphatic acids in soil. *Adv. Pest Control Res.* 6:1–30.

45. P. C. Kearney, E. A. Woolson, J. R. Plimmer, and A. R. Isensee. 1969. Decontamination of pesticides in soils. *Residue Rev.* 29:137–149.

46. A. D. Kern, W. F. Meggitt, and D. Penner. 1975. Uptake, movement, and metabolism of cyanazine in fall panicum, green foxtail, and corn. *Weed Sci.* 23–277–282.

47. G. C. Klingman and F. M. Ashton. 1975. *Weed Science: Principles and Practice.* Wiley-Interscience. 431 pp.

48. G. C. Li and G. T. Felbeck. 1972. A study of the mechanism of atrazine adsorption by humic acid from muck soil. *Soil Sci.* 113:140–148.

49. L. C. Luckwill and C. P. Lloyd-Jones. 1960. 2,4-Dichlorophenoxyacetic acid in leaves of red and of black currant. *Ann. Appl. Biol.* 48:613–625.

50. C. R. Malone. 1972. Effects of a non-selective arsenical herbicide on plant biomass and community structure in a fescue meadow. *Ecology* 53:507–512.

51. E. V. McCurdy and E. S. Molberg. 1974. Effects of the continuous use of 2,4-D and MCPA on spring wheat production and weed populations. *Can. J. Plant Sci.* 54:241–245.

52. M. D. McGlamery and F. W. Slife. 1966. The adsorption and desorption of atrazine as affected by pH, temperature and concentration. *Weed Sci.* 14:237–239.

53. R. L. Metcalf. 1966. Metabolism and fate of pesticides in plants and animals. In *Scientific Aspects of Pest Control.* Nat. Acad. Sci., Washington, Publ. 1402. pp. 230–250.

54. National Academy of Sciences. 1974. Defoliation in South Vietnam. *Nat. Acad. Sci. News Report* 24(3-4), pp. 1, 4–11.

55. B. E. Pape and M. J. Zabik. 1970. Photochemistry of selected *s*-triazine herbicides. *J. Agric. Food Chem.* **18**:202–207.

56. J. V. Parochetti and G. F. Warren. 1966. Vapor losses of IPC and CIPC. *Weeds* **14**:281–285.

57. A. R. Putnam. 1969. The consequences of repeated herbicide applications in new fruit plantings. *Ann. Rep. Mich. Hort. Soc.* **99**:64–65.

58. S. R. Radosevich and A. P. Appleby. 1973. Relative susceptibility of two common groundsel (*Senecio vulgaris*) biotypes to six *s*-triazines. *Agron. J.* **65**:553–555.

59. S. K. Ries and V. Wert. 1972. Simazine-induced nitrate absorption related to plant protein content. *Weed Sci.* **20**:569–572.

60. W. Roth and E. Knuesli. 1961. Beitrag zur Kenntnis der Resistenzphanomene einzelner Pflanzen gegenuber den phytotoxischen Wirkstoff Simazin. *Experientia* **17**:312–313.

61. J. D. Russell et al. 1968. Mode of chemical degradation of *s*-triazines by montmorillonite. *Science* **160**:1340–1342.

62. G. R. Ryan. 1970. Resistance of common groundsel to simazine and atrazine. *Weed Sci.* **18**:614–616.

63. D. J. Rydrych and C. I. Seeley. 1964. Effect of IPC on selections of wild oats. *Weeds* **12**:265–267.

64. P. W. Santelmann and J. A. Meade. 1961. Variation in dalapon susceptibility within the species *Setaria lutescens* and *S. faberii*. *Weeds* **9**:406–410.

65. A. B. Schooler, A. R. Bell, and J. D. Nalewaja. 1972. Inheritance of siduron tolerance in foxtail barley. *Weed Sci.* **20**:167–169.

66. O. E. Schubert. 1972. Plant cover changes following herbicide applications in orchards. *Weed Sci.* **20**:124–127.

67. R. H. Shimabukuro, D. S. Frear, H. R. Swanson, and W. C. Walsh. 1971. Glutathione conjugation, an enzymatic basis for atrazine resistance in corn. *Plant Physiol.* **47**:10–14.

68. A. E. Smith and J. Grove. 1969. Photochemical degradation of diquat in dilute aqueous solution and on silica-gel. *J. Agric. Food Chem.* **17**:609–613.

69. W. F. Smith and R. D. Ilnicki. 1972. Effect of atrazine on fall panicum collected in New Jersey. *Weed Soc. Am. Abst. 1972 Meet.* p. 48.

70. W. F. Spencer, W. J. Farmer, and M. M. Cliath. 1973. Pesticide volatilization. *Residue Rev.* **49**:1–47.

71. R. L. Stickler, E. L. Knake, and T. D. Hinesley. 1969. Soil moisture and effectiveness of preemergence herbicides. *Weed Sci.* **17**:257–259.

72. R. E. Talbert and O. H. Fletchall. 1965. The adsorption of some *s*-triazines in soils. *Weeds* **13**:46–51.

73. H. P. Tietjen, C. H. Halvorson, P. L. Hegdal, and A. M. Johnson. 1967. 2,4-D herbicide, vegetation and pocket gopher relationships on Black Mesa, Colorado. *Ecology* **48**:635–643.

74. F. H. Tschirley. 1969. The ecological consequences of the defoliation program in Vietnam. *Science* **163**:779–786.

75. G. T. Turner. 1969. Responses of mountain grassland vegetation to gopher control, reduced grazing, and herbicide. *J. Range Manage.* **22**:377–383.

76. J. B. Weber and H. D. Coble. 1968. Microbial decomposition of diquat adsorbed on montmorillonite and kaolinite clays. *J. Agric. Food Chem.* **16**:475–478.

77. J. B. Weber and S. B. Weed. 1974. Effects of soil on the biological activity of pesticides. In *Pesticides in Soil and Water*. Ed. W. D. Guenzi. Soil Sci. Soc. Am., Madison, Wisconsin. pp. 223–256.

78. R. L. Wershaw, P. J. Burcar, and M. C. Goldberg. 1969. Interaction of pesticides with natural organic materials. *Environ. Sci. Technol.* **3:**271–273.

79. J. W. Whitworth. 1964. The reaction of strains of field bindweed to 2,4-D. *Weeds* **12:**57–58.

80. M. Wilcox, D. E. Moreland, and G. C. Klingman. 1963. Aryl hydroxylation of phenoxyaliphatic acids by excised roots. *Physiol. Plant.* **16:**565:571.

81. G. M. Woodwell. 1970. Effects of pollution on the structure and physiology of ecosystems. *Science* **168:**429–433.

12

HERBICIDES: EFFECTS ON INVERTEBRATE AND VERTEBRATE FAUNA

A. EFFECT OF HERBICIDES ON INVERTEBRATE FAUNA

Collembola and Mites

As early as 1952, it was found in experiments in northern Germany that an application of 2,4-D sodium salt had no effect on soil populations of Collembola and mites.[6] When either 2,4-D amine or MCPA acid were applied at 10 times the normal dosage to lawn turf in Argentina, the numbers of mites, Collembola and other soil fauna were found to be unchanged both 6 days and 4 months later.[53] Neither MCPA nor the sodium salt of 2,4-D caused any significant reduction in Collembola populations when applied to loam soil in western Germany.[8] A barley field (three plots) in southern England, which had been treated with the sodium salt of MCPA at 2 lb/a for 10 of the 13 preceding years, when compared with three control plots infested with chickweed, charlock, bindweed, and knotgrass,[18] showed the following average numbers per square meter throughout the year:

	Mites	Collembola	Other Insects	Other Arthropods
MCPA-treated	16,519	6410	1181	3359
Untreated	15,374	6263	1107	2837

Here one-fifth of the mites were predaceous Mesostigmata and nearly all the rest were hyphae-eating Cryptostigmata, and one-half of the springtails were surface-feeding Entomobryids and the other half were subterranean Podurids. The lack of a difference in the MCPA-treated field indicates that these soil fauna had reached a successful equilibrium with the farming practice, besides being helped by the remnants of undestroyed chickweed along with a grass population higher than the untreated plots.

By contrast, barley fields newly treated with MCPA or MCPB were found to have about one-half the soil fauna (exclusive of mites) that untreated fields had; examples of the figures obtained were as follows[61]:

	Numbers/m^3		Biomass, kg/ha	
	Treated	Untreated	Treated	Untreated
MCPA (Crux Easton, Hants)	138	254	1.4	2.1
MCPB (Silwood Park, Berks)	443	792	4.0	10.7

Since Collembola were the most important group, constituting 42% of the population numbers, and because they accounted for 30% of the invertebrates detectable in the crops of partridge chicks inhabiting the fields, it was concluded that these and other herbicides were causatively associated with the general decline that has been observed in the gray partridge in Britain.

Studies in Nova Scotia[24] demonstrated that the effect of herbicides on the soil fauna mainly resulted from the alterations in ground cover that they induced. Four different herbicides were applied at normal rates to grassland composed of browntop (*Agrostis tenuis*), sweet vernal grass (*Anthoxanthum odoratum*) and timothy (*Phleum pratense*). The results (Table 12.1) show that it was where the herbicide severely reduced the vegetative ground cover (atrazine and monuron) that the soil populations had decreased, while where the herbicide replaced the killed grasses with broadleaved plants (dalapon and TCA) the fauna was somewhat richer than before.

Table 12.1. Effects of four herbicides on macroflora and soil fauna of grassland sampled the year after application (Fox, 1964).

	Untreated	Atrazine 8 lb/a	Dalapon 20 lb/a	Monuron 10 lb/a	TCA 80 lb/a
Ground cover in percent					
Grasses	96	1	1	1	2
Clovers	0.5	0	2	0	1
Dandelion	1.5	18	60	1	11
Field Daisy	0.5	0	22	2	66
Sorrel	0.5	1	2	21	0
Yarrow	0	0	1	0	4
Bare or Mossy	1	80	12	75	16
Average numbers per sample					
Mites*	21	8	38	5	30
Springtails*	33	13	38	12	39
Wireworms[†]	1.6	0.3	1.7	0.2	1.8
Millipedes[†]	0.7	0.7	1.2	0.3	1.3
Earthworms[†]	4.4	2.6	5.0	2.5	2.2

* Samples 2.5" diam. x 3" deep   † Samples 6" square by 6" deep

Simazine applied at 2–3 kg/ha resulted in a slight reduction in the soil mites, but an increase in the total numbers of Collembola.[62] In Britain it was found that simazine applications at 2–4 lb/a reduced the total numbers of these soil invertebrates by 13–50%, predatory mites and Isotomid springtails being particularly suppressed.[21] Applied to fallow ground twice annually at 1.5 lb/a, simazine treatments repeated for 6 yr resulted in the following invertebrate numbers (per core 15 cm deep and 5 cm diameter) in the treated and control plots[22]:

		Treated	Control
Mites	Prostigmata	3.4	7.7
	Mesostigmata	0.2	2.3
Springtails	Isotomidae	0	1.2
	Onychiuridae	1.8	1.8

This contrasted with the lack of suppressive effect from treatments with MCPA at 1.5 lb/a or with linuron at 0.75 lb/a. In addition, nontarget dipterous and coleopterous larvae in the soil are reduced by simazine applications.[21]

DNOC when used as a herbicide promptly reduced the soil populations of springtails[20,21]; applied at 5 kg/ha to a loam soil in western Germany, it caused a 12% reduction over the 5 months following the application.[8] But whereas a DNOC treatment at 6 kg/ha applied to winter rye in eastern Germany resulted in an initial 50% reduction in Collembola and soil mites, after 5 months had elapsed the populations were considerably greater than before the application (Table 12.2); *Folsomia* was the most susceptible, *Tullbergia* recovered its numbers more quickly, and *Hypogastrura* became common instead of being a rarity. Even when treatments were applied at 1200 kg/ha for potato defoliation, higher soil populations than before were noted 9 months later; this increased population of mites and springtails was attributed to the treatments having induced a richer soil content of bacteria, fungi, and nematodes, which serve as food for these arthropods.[34]

Collembola. When dalapon was applied at 17 kg/ha to clear turf for reseeding in Ireland, the populations of Collembola over the following 10 months were reduced (Fig. 12.1), an effect attributable to the sward being opened up. When paraquat was applied at 11 kg/ha, an even larger reduction was obtained.[17] Neither dalapon nor 2,4,5-T were found to reduce Collembolan populations in the Netherlands.[20] In the Irish field trials, soil mites were reduced more than the springtails, although there was

Table 12.2. **Recovery and proliferation of springtail and soil mite populations 4 and 11 months after applications of DNOC and 1200 kg/ha, respectively: numbers per 2 liters of soil (Karg, 1964).**

Collembola	6 kg/ha, 4 mths		1200 kg/ha, 11 mths	
	Unspr.*	DNOC	Unspr.	DNOC
Tullbergia krausbaueri	445	739	284	553
Folsomia fimetaria	128	122	142	247
Hypogastrura succinea	2	10	--	--
Hypogastrura assimilis	--	--	--	211
Isotoma notabilis	11	8	19	13
Isotoma viridis	2	5	7	2
Entomobryidae spp.	3	2	1	1
Sminthuridae spp.	8	33	2	4
Mites				
Brachychthonius spp.	20	55	38	10
Oppia ninus	3	46	1	--
Tyrophagus spp.	15	19	48	115
Tarsonemus spp.	27	21	4	6
Microtydeus spp.[†]	543	731	114	117
Scutacaridae spp.	4	125	17	4
Pyemotidae spp.	98	65	16	76
Parasitiformes spp.	38	63	23	93

* The plots to be sprayed had populations very similar to the unsprayed plots

[†] Plus Coccotydeus spp.

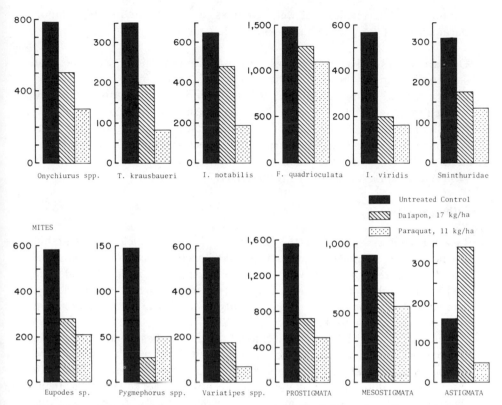

Fig. 12.1. Effect of dalapon and paraquat applied to Irish turf on the populations of Collembola and mites: total numbers sampled over a 10-month posttreatment period (from Curry, 1970). T = *Tullbergia,* I = *Isotoma,* F = *Folsomia.*

one species, *Minunthozetes semirufus,* which increased after the dalapon treatment.

From the review of Edwards and Thompson (1975) of tests of herbicides on soil arthropods, it may be concluded that only simazine and DNOC, and possibly atrazine, have a direct lethal effect on Collembola, viz.:

Without Effect	Suppressive on Vegetated Plots	Suppressive on All Plots
2,4-D	Dalapon	Simazine
2,4,5-T	Paraquat	DNOC
MCPA	Pyrazon	
Linuron	Monuron	
Triallate	Atrazine	

Among those herbicides that were suppressive on vegetated plots, dalapon paraquat and TCA proved harmless on fallow plots.

Orchard Mites. The predaceous Phytoseiid mites, along with the pest Tetranychid spider mites on which they prey, overwinter in the ground cover until the following summer. When the herbicides that are commonly applied to the ground cover of orchards were tested by the slide-dip method (Table 12.3), it was found that dalapon, paraquat, terbacil, and 2,4-D alkanolamine salt were toxic to mites at levels below the recommended herbicidal dosage, and the first two were more toxic to the predator *Amblyseius fallacis* than to its prey *Tetranychus urticae*.[54]

Other Arthropods

Insect Predators. Tests of the direct toxicity of herbicides to Carabid predators have been performed on the ground beetles *Bembidion femoratum* and *B. ustulatum*. They showed that DNOC, simazine, amitrole, dalapon, and sodium chlorate were virtually harmless at normal dosage levels. Diquat and pyrazon caused some mortality, while 2,4-D and chlorpropham were definitely toxic to these predators.[46]

The toxicity of 2,4-D to Coccinellid predators has been a factor in promoting an abundance of aphids. In New Brunswick, outbreaks of *Rhopalosiphum padi* and other aphid species occurred on oats treated with 2,4-D amine at 0.5 lb/a because after the postemergence spray the fields carried about one-third as many ladybeetles (mainly *Hippodamia 13-punctata*) as untreated fields.[2] Previous laboratory experiments had revealed that 2,4-D sprayed at this dosage caused 70–75% mortality to the three species of Coccinellids involved, and that the survivors showed a reduced growth rate.[1]

Table 12.3. **Toxicity of herbicides to *Tetranychus urticae* and**
**   *Amblyseius fallacis* (Rock and Yeargan, 1973).**

	T. urticae	A. fallacis
Paraquat	1.5	0.012
2,4-D alkanolamine	0.9	0.2
Dalapon	2.6	0.3
Terbacil	0.9	0.5
Diuron	5.8	9.5
Simazine	5.6	11.1

Secondary Effects through the Plants. Area application of a herbicide such as 2,4-D may have secondary effects leading to certain insect pest problems. The increase in the proportion of grasses on terrain in Nova Scotia treated with 2,4-D amine for that purpose resulted in an infestation of the wireworm *Agriotes sputator* in the underlying soil.[24] In Saskatchewan, 2,4-D applications increased wireworm damage for another reason, by delaying germination and seedling growth.[25] The plant-stimulatory effect of 2,4-D at sublethal dosages resulted in a fivefold increase of the fecundity of the pea aphid when fed on broad bean plants dipped in 4 ppm of the herbicide, and this was associated with an increase in the concentration of some of the amino acids.[44] Treatment of rice plants with 2,4-D resulted in a 40% increase in the size that the larvae of the rice stem borer (*Chilo suppressalis*) attained; since the same result was not obtained by feeding them a mixture of rice stems and 2,4-D, the increased larval growth was concluded to have been due to the higher level of nitrogenous compounds induced by the herbicide in the growing plant.[32]

Honeybees. 2,4-D and its esters if applied in a water carrier are relatively harmless to honeybees, although there were some losses during large-scale applications in New Zealand. Silvex, MCPA, 2,4,5-T, amitrole, TCA, and picloram are nontoxic to bees, while endothall has some slight toxicity. The phenoxy esters and picloram can be toxic if sprayed with oil carriers. The arsenical herbicides MSMA and cacodylic acid, as well as paraquat and DNOC, are definitely toxic to honeybees,[45] although the risk posed by cacodylic acid and paraquat in the field is low as compared with insecticides.[3] Herbicides can cause indirect losses to bees due to the destruction of nectar-bearing plants, but the use of 2,4-D in roadside weed and brush control has increased bee forage by encouraging the growth of sweet clover.[7] There is no information on the effect of herbicides on hymenopterous parasites of pest insects, but presumably the general picture resembles that found for honeybees.

Other Invertebrates

Earthworms. Some herbicides are directly toxic to earthworms, while others have virtually no direct effect; for example, immersion experiments with 100 ppm concentrations showed monuron to be fatal, but 2,4-D to be harmless.[43] Applied to grassland in Nova Scotia, TCA, atrazine, and monuron caused a reduction of the earthworm population, but only with TCA was it independent of a general reduction in vegetative cover.[24] TCA applied at normal rates to cultivated fields in Russia, on the other hand, has no discernible effect on earthworms or ants, nor did dalapon, 2,4-D sodium salt, or simazine.[31] Although soil treated with simazine at 12 lb/a did not kill any *Lumbricus* or *Helodrilus* earthworms inhabiting it for a month (DeVries, 1962[51]), British experiments with two annual sprays of simazine at 1.5 lb/a in each successive year suppressed the population level of the small Enchytraeid worms, the reduction in Lumbricids being not significant.[22] Application in the Netherlands of propham at 3.4 lb/a, chlorpropham at 1.8 lb/a, or DNOC at 3.6 lb/a caused 30–40% mortality in *Lumbricus castaneus* but had no effect on *Allolobophora caliginosa*.[20] In Britain, paraquat treatments to allow clean culture and slit-seeding of wheat had the effect on inducing higher populations of earthworms than in ploughed wheatfields.[22]

Nematodes.   Soil nematodes withstand immersion in 5 ppm 2,4-D, but 50 ppm suppresses their reproduction.[67] Applications of 2,4-D to vegetable seed beds were found to control the rootknot nematode *Meloidogyne*[13]; while DCPA had the effect of suppressing this pest on onions, trifluralin promoted its attacks on tomato and alfalfa.[4] Treatment of red clover and oat plants with 2,4-D made them susceptible to *Ditylenchus dipsaci,* and transformed the tissues of red clover and the calluses of alfalfa into suitable media for the reproduction of *Aphelenchoides ritzemabosi*.[66] Simazine applied to cornfields at 2–3 kg/ha in Germany did not significantly affect the numbers of soil nematodes.[62] Application of amitrole at 5 lb/a to colonial bentgrass killed nearly 50% of the nematode *Anguia agrostis* forming galls in the flower heads, while dalapon at the same dosage killed nearly 95% of them.[16] Orchard soils sprayed with a mixture of amitrole and simazine (the mixture Domatol) at 6 kg/ha experienced a reduction of about 75% in their nematode population, and a transitory reduction of about two-thirds of their rotifer population.[69]

B.   EFFECT OF HERBICIDES ON WILDLIFE AND LIVESTOCK

Nearly all the compounds developed as herbicides have a very low toxicity to warm-blooded animals (Table 12.4). The lowest LD_{50} and LC_{50} values reported, namely for 2,4,5-T and diquat, are not as high as those for most insecticides. Nevertheless, the accidental ingestion of paraquat concentrates has cost more than 100 human lives over the years, the symptoms being a progressive fibrosis and epithelial proliferation of the lungs.[37] The phenoxy herbicides are scarcely toxic to mammals, the acute oral LD_{50} of 2,4-D being 800 mg/kg for the European rabbit and mule deer, and that of 2,4,5-T being approximately 500 mg/kg for livestock in general.[55] However, even if their application in the field may have no direct toxic effect on wildlife, herbicides may reduce wildlife populations by altering the plant ecosystem,

Table 12.4. **Oral toxicity of herbicides to game birds and laboratory rats: acute LD$_{50}$ and chronic LC$_{50}$ values.**

	Mallard		Pheasant		Coturnix	Rat
	mg/kg*	ppm[†]	mg/kg*	ppm[†]	ppm[†]	mg/kg[‡]
2,4-D	2000	>5000	472	>5000	>5000	666
2,4,5-T	--	>5000	--	1875	>5000	300
Silvex	>2000	--	--	4000	>5000	1070
Dalapon	--	>5000	--	>5000	>5000	8450
Simazine	--	>5000	--	>5000	>5000	>5000
Atrazine	>2000	>5000	--	>5000	--	3080
Monuron	--	>5000	--	4500	>5000	3600
Diuron	>2000	>5000	--	>5000	>5000	3400
Diquat	564	>5000	--	3750	1500	420
Amitrole	>2000	>5000	--	>5000	>5000	5000
Fenac	--	>5000	--	>5000	>5000	1780
Trifluralin	>2000	--	>2000	--	--	>10,000
Picloram	>2000	>5000	>2000	>5000	--	8200
Dichlobenil	>2000	--	1189	1750	--	3160

* Single oral dose in capsules, 2-week observation period

    (Tucker and Crabtree 1970)

[†] Treated feed for 5 days, clean feed for 3 days

    (Heath et al. 1972)

[‡] From USDI 1970, House et al. 1967, and WSA 1967 <u>in</u> Pimentel 1971

particularly when they are used to eliminate the cover afforded by untended hedgerows and ditch banks. Conversely, they may be so employed as to improve plant and tree communities as a wildlife habitat or forage source.

Weed-Tree Control in Forest and Game Management

Forest Management. Aerial spraying of 2,4,5-T at 1–2 lb/a from the ethyl or isopropyl ester has been widely employed to release pine and spruce from overtopping hardwoods such as scrub oak; the application is made after the new coniferous needles have hardened.[5] This treatment fortunately does not kill the blueberries, elderberries, and holly shrubs valuable as wildlife food; although it may eliminate den-trees, severe effects on opossums and squirrels may be eliminated by operating in a 50-acre checkerboard pattern.[47] However, the sprayed areas do not provide preferred feeding ground for mobile game species such as wild turkeys.

Studies in west-central Alabama[14] undertook to measure the quality of wildlife forage left when forest land was prepared for planting to a crop of loblolly pine. The rating chosen was the average of the individual ratings of the species present, ranging from poor for bracken to excellent for *Rubus,* oak, and flowering dogwood. The cheapest method of site preparation was found to be aerial spraying with 2,4,5-T either at 2 lb/a followed by broadcast burning, or by 2,4,5-T at 6 lb/a alone, but both methods leave an understory cover with a forage rating inferior to untreated sites or sites submitted to mechanical clearing. The addition of picloram to the 2,4,5-T ester formulation resulted simultaneously in a better stocking density and height growth of the planted pine, and also in a forage quality that is scarcely inferior to that of untreated sites.

Game Management.   2,4,5-T at 2 lb/a has also been employed to prevent trees from encroaching on the abandoned fields that frequently provide deer pasturage in forest reserves, or for creating new openings in the forest. An opening of 20–40 acres in a 50-foot strip a quarter to half a mile long is best for deer; while 1.5–2 acres suffice for rabbits and ruffed grouse, the sharptail grouse requires areas of 1 sq mi.[56] The 2,4,5-T formulations, applied in Michigan as a broadcast airspray, or better by ground sprays or by adding it to girdles or frills cut in the trees, result in producing trunk shoots, berry bushes, forbs, and grasses that are apparently ideal for deer, rabbits and grouse.[27] In Minnesota, the foliage regrowth of the tree-shrub mountain maple stimulated by 2,4-D and 2,4,5-T was readily eaten by deer.[40] Application of 2,4-D at 2 lb/a in ester formulation to various types of forest, whether aspen, jack pine, or red oak, yielded in each case an increase in choice browse species such as choke cherry, dogwood, and juneberry, and resulted in a fourfold increase in the population of whitetail deer.[39]

Possible Hazards.   Although there are reports of vegetation treated with 2,4-D being repellent to cattle and cottontail rabbits when they have a choice of untreated forage,[51] deer showed no preference for either treated or untreated shrub growth.[38] The residues of 2,4,5-T taken up by blacktail deer after 1.5 months' pasturage on an area in Oregon sprayed at 2 lb/a were all below 20 ppb in their tissues, while 400 ppb was found in the stomach and 40 ppb in the faeces.[49] Ammonium sulfamate (ammate) is sometimes substituted to avoid crop damage from airspray drift, and deer fed ammate-treated foliage apparently made greater weight gains (Haugen, 1951[59]).

One technical grade of 2,4,5-T was found to be teratogenic in that its administration to female rats and mice for 9 days induced a 20–50% frequency of malformations such as cleft palate in their foetuses[15]; this product contained 30 ppm of the dioxin TCDD (2,3,7,8-tetrachlorodibenzodioxin).[14] TCDD itself has been shown to cause kidney abnormalities as well as cleft palate in rat embryos, being teratogenic when applied in high concentration at a crucial time in their development.[48] This dioxin is also highly toxic, the acute oral LC_{50} levels (for a 45-day observation period) being 0.03 mg/kg for rats, and 0.001 mg/kg for guinea pigs.[58] In 1971 grades of 2,4,5-T containing less than 0.5 ppm TCDD, were considered acceptable for use, and subsequent production practices ensured that the dioxin content never exceeded 0.1 ppm.[71] Amitrole, also employed for brush elimination on forest land, had induced an enlargement of the thyroid in severely dosed rats; although this is reversible, it may occasionally lead to carcinoma.[50]

Brush Control in Range Management

2,4-D formulations have been employed to control sagebrush, particularly *Artemisia tridentata*, to improve cattle range in western United States, including at least 300,000 acres in the state of Colorado.[57] These treatments produced little faunal change in Colorado, apart from a small decrease in deer use in some areas (Anderson, 1960[51]). Treatment of an area in eastern Idaho with 2,4-D at 2 lb/a from the isopropyl ester removed the *Artemisia* but left the bitterbrush (*Purshia tridentata*) which is a valuable forage species for livestock and big game alike.[10] The clearing of chamise brush (*Adenostoma fasciculatum*), by making openings with fire followed by herbicides to suppress sprouts and seedlings, resulted in increases in quail, grouse, hares, rabbits, and blacktail deer; the average deer population per square mile was thereby increased from 20 up to 75 and the production of young was increased from 0.7 up to 1.3 fawns per doe.[9]

Another pest of rangeland in Colorado is the pocket gopher (*Thomomys talpoides*), which on Grand Mesa feeds on broadleaved forbs such as *Taraxacum* and *Penstemon*; treatment with 2,4-D at 3 lb/a from the butyl ester decreased the forbs by 83% without decreasing the total herbaceous ground cover; this resulted in the diet of the gophers becoming one-half grass, and a reduction of the pocket gopher population of 87% by the following year.[35] On the other hand, *Peromyscus* deermice and *Microtus* voles were not reduced.[64] Applications of 2,4-D ester at 2 lb/a on four occasions to Idaho highlands during a 10-yr period reduced pocket gopher mounds by 93%, and winter casts by 94%, since they eliminated fleshy-rooted plants such as *Claytonia, Erythronium* and *Lomatium* which serve as a food source in the spring.[30] 2,4-D applications at 2 or 3 lb/a in western Colorado reduced populations of the least chipmunk as well as the pocket gopher, and montane vole populations were higher for as long as the resultant increase in grass cover persisted.[33] A concise review of the beneficial effects, as well as the possible hazards, of the use of 2,4-D and 2,4,5-T in range and forest management has been contributed by Schroeder (1972).

Vegetation Control under Power Lines

Herbicidal treatment of a power line right-of-way through upland oak country in central Pennsylvania was found to produce the same increase in wildlife numbers as maintenance by cutting. Broadcast treatment of mixed 2,4-D and 2,4,5-T esters or of ammate resulted in grass-forb clearings that were attractive for young wild turkeys seeking their insect food.[11] The cleared strip climaxed as a bracken–sedge–goldenrod–blueberry type interspersed with sweetfern (*Comptonia peregrina*). This led to a population density of whitetail deer of 1 per 9 acres on the right-of-way, as compared to the typical upland oak density of 1 per 60 acres. The bracken which provides early summer forage, and the goldenrod, loosestrife, and sorrel eaten in light summer, were found to be high in protein and fat content, and so were the blueberries and sweetfern resorted to at the end of winter. The herbicide treatment applied in 1953 and 1954 did not need to be repeated until 1966. Meanwhile a considerable population of ruffed grouse was established within 60 yards of the right-of-way, and cottontails and wood-chuck had become established on the right-of-way itself.[12]

Applications of 2,4,5-T (PGBE ester) to power line right-of-way overgrown with sassafras trees in southeastern Michigan had no adverse effects on the 11 species of birds and 11 mammalian species frequenting them over the succeeding 4 yr, and the browse was improved by the diversification contributed from the growth of other trees and shrubs.[28]

Hazards to Grazing Animals through the Plants

The use of 2,4-D and 2,4,5-T may offer unanticipated hazards to livestock which eat the weeds exposed to sublethal doses of the herbicide. Ragwort, toxic but at the same time repellent to cattle, may have its sugar content increased by 2,4-D contamination to the point that it becomes attractive to them.[70] An increase in HCN content is another unexpected hazard arising from the use of phenoxy herbicides; applied at 1 lb/a to Sudan grass (*Sorghum vulgare sudanense*), 2,4-D has produced a 35% increase and 2,4,5-T a 70% increase in hydrocyanic acid.[63] A more direct hazard is involved in the use of sodium arsenite as a herbicide, since it is attractive to grazing animals.[47]

The nitrate content of plants which have received sublethal doses of 2,4-D or 2,4,5-T may be dangerously increased in some weed species, for example, a 47% increase in *Eupatorium maculatum* with 2,4-D, a 36% increase in *Impatiens biflora* with 2,4,5-T.[26] This nitrate is converted to toxic nitrite in the rumen of ruminants. Nitrate increases caused by 2,4-D that could be hazardous to cattle have been found in pigweed, lambsquarters, smartweed, Canada thistle, and sugar beet.[51] The death of 40% of a herd of 600 reindeer in the spring of 1970, which had fed on coniferous vegetation sprayed with a mixture of 2,4-D and 2,4,5-T the year before (Lundholm 1970[51]), was probably not due to the herbicide residues in it (which were high), but to its nitrate content; the herd was evidently weakened after the winter in the field, since reindeer from an experimental farm were not poisoned by the same vegetation.

Effect of 2,4,5-T and other Phenoxys on Gallinaceous Birds

The toxicity of 2,4,5-T to game birds is not quite as low as that of 2,4-D (Table 12.4). Nevertheless the chronic LC_{50} of 2,4,5-T as the butoxyethanol ester to pheasant, mallard, and coturnix quail is comfortably in excess of 3000 ppm,[29] and that for the free acid to bobwhite chicks and mallard ducklings is slightly lower.[36] When 2,4,5-T free acid (containing 0.6 ppm TCDD) was fed at 50 ppm in the diet to adult bobwhites for 18 wk, the number of eggs they laid and of young surviving was no less than the untreated controls.[36] It has, however, been reported that fertile eggs of pheasant and grey partridge sprayed with 2,4-D amine in water at 0.5 lb/a showed about 75% mortality of embryos by the nineteenth day of incubation, with the survivors developing malformations particularly in the gonadal region,[41] and that fertile hen's eggs dipped in a water-based formulation containing 4000 ppm 2,4,5-T showed 25% mortality.[42]

Therefore a large-scale experiment was performed by applying sprays either of an equal mixture of 2,4-D and 2,4,5-T as the isooctyl esters at 10 lb/a, or of 4 parts 2,4-D plus 1 part picloram, to egg-clutches of chickens and pheasants in simulated nests. It was found that the embryo mortalities were not in excess of the normal 4%, despite penetration into the egg of 0.5–1.0 ppm of phenoxy residues, while chick survival was normal despite some persistence of the herbicide in their tissues.[60] When domestic hens were pastured for 14 days on grass sprayed with MCPA, 2,4-D and 2,4,5-T at normal dosages, the spray being repeated on every one of these days, there were no signs of ill health and no loss in weight, only the birds consumed more water. The egg production was reduced for a 3-wk period after the spraying began, the reductions amounting to some 25% with MCPA, and 15% with 2,4-D and 2,4,5-T, but the hatchability of these eggs was normal.[19]

On the other hand, the use of MCPA and other herbicides on wheatfields in West Sussex, England had led to a decrease in chick survival of grey partridge since 1955.[52] This was attributed to the 50% reduction in soil arthropods, already described,[61] leading to the death of chicks during cold spring periods before a crop of cereal aphids can materialize to save them.[52] It also should be noted that one of the consequences of military defoliation of mangrove swamp in South Vietnam, with a 50:50 mixture of 2,4-D and 2,4,5-T at 25 lb/a, was a pronounced scarcity of birds.[68]

REFERENCES CITED

1. J. B. Adams. 1960. Effects of spraying 2,4-D amine on coccinellid larvae. *Can. J. Zool.* **38**:285–288.

2. J. B. Adams and M. E. Drew. 1965. Aphid populations in herbicide-treated oat fields. *Can. J. Zool.* **43**:789–794.

3. L. D. Anderson and E. L. Atkins. 1968. Pesticide usage in relation to beekeeping. *Annu. Rev. Entomol.* **13**:213–238.

4. J. L. Anderson and G. D. Griffin. 1972. Interactions of DCPA and trifluralin with seedling infection with root-knot nematode. *Weed Sci. Soc. Am. Abstr.* p. 5.

5. J. L. Arend. 1959. Airplane application of herbicides for releasing conifers. *J. Forestry* **57**:738–749.

6. F. G. von Baudissin. 1952. Die Wirkung von Pflanzenschutzmitteln auf Collembolen und Milben in verschiedenen Boden. *Zool. Jahrb. (Okol)* **81**:47–90.

7. H. P. Beilmann. 1950. Weed killers and bee pasture. *Am. Bee J.* **90**:542–543.

8. H. Bieringer. 1968. Quantitative untersuchungen an die Boden Collembolen des Hohenheimer Dauerherbizidversuches. *Z. Pflanzenkr. Pflanzenpathol. Pflanzenschutz, Sonderheft* **4**:157–162.

9. H. H. Biswell, R. D. Taber, D. W. Hendrick, and A. M. Schultz. 1952. Management of chamise brushlands for game in North Coast region of California. *Calif. Fish Game* **38**:453–484.

10. J. P. Blaisdell and W. F. Mueggler. 1956. Effect of 2,4-D on forbs and shrubs associated with big sagebrush. *J. Range Manage.* **9**:38–40.

11. W. C. Bramble and W. R. Byrnes. 1958. Use of a power line right-of-way by game after chemical brush control. *Penn. Game News* **29**:17–25.

12. W. C. Bramble and W. R. Byrnes. 1972. *A Long-Term Ecological Study of Game Food and Cover on a Sprayed Utility Right-of-way.* Purdue Univ., Indiana Agric. Exp. Stn. Bull. No 885. 20 pp.

13. D. S. Burgis and J. R. Beckenbach. 1948. Herbicides for control of weeds in vegetable seed-beds also control rootknot. *Proc. Am. Soc. Hort. Sci.* **52**:461–463.

14. M. C. Carter, J. W. Martin, J. E. Kennamer, and M. K. Causey. 1975. Impact of chemical and mechanical site preparation on wildlife habitat. *Down to Earth* **31**(2): 14–18.

15. K. D. Courtney, D. W. Gaylor, M. D. Hogan, H. L. Falk, R. R. Bates, and I. Mitchell. 1970. Teratogenic evaluation of 2,4,5-T. *Science* **168**:864-866.

16. W. D. Courtney, D. V. Peabody, and H. M. Austenson. 1962. Effect of herbicides on nematodes in bentgrass. *Plant Dis. Rptr.* **46**:256-257.

17. J. P. Curry. 1970. The effects of the herbicides paraquat and dalapon on the soil fauna. *Pedobiologia* **10**:329-361.

18. B. N. K. Davis. 1965. The immediate and long-term effects of the herbicide MCPA on soil arthropods. *Bull. Entomol. Res.* **56**:357-366.

19. N. Dobson. 1954. Chemical sprays and poultry. *Agriculture (J. Min Agric. U.K.)* **61**:415-418.

20. J. van der Drift. 1963. The influence of biocides on the soil fauna. *Tijdschr. Planteziekt.* **69**:188-199.

21. C. A. Edwards. 1964. Effects of pesticide residues on soil invertebrates and plants. *Br. Ecol. Soc. Symp.* **5**:239-261.

22. C. A. Edwards. 1970. Effects of herbicides on the soil fauna. *10th Br. Weed Contr. Conf. Proc.* **3**:1052-1057.

23. C. A. Edwards and A. R. Thompson. 1973. Pesticides and the soil fauna. *Residue Rev.* **45**:1-79.

24. C. J. S. Fox. 1964. The effects of five herbicides on the numbers of certain invertebrate animals in grassland soil. *Can. J. Plant Sci.* **44**:405-409.

25. W. B. Fox. 1948. 2,4-D as a factor in increasing wireworm damage of wheat. *Sci. Agric.* **28**:423-424.

26. P. A. Frank and B. H. Grigsby. 1957. Effects of herbicidal sprays on nitrate accumulation in certain weed species. *Weeds* **5**:206-217.

27. L. W. Gysel. 1957. Effects of different methods of releasing pine on wildlife food and cover. *Down to Earth* **13**(2):2-3.

28. L. W. Gysel. 1962. Vegetation changes and animal use of a line right-of-way after application of a herbicide. *Down to Earth* **18**(1):7-10.

29. R. G. Heath, J. W. Spann, E. F. Hill, and J. F. Kreitzer. 1972. *Comparative Dietary Toxicities of Pesticides to Birds.* U.S. Bur. Sport Fish Wildl., Special Scientific Report, Wildlife No. 152. 57 pp.

30. A. C. Hull. 1971. Effect of spraying with 2,4-D upon abundance of pocket gophers in Franklin basin, Idaho. *J. Range Manage.* **23**:137-145.

31. A. M. Ilijin. 1969. The toxic effect of herbicides upon ants and earthworms. *Zool. Zh.* **48**:141-143.

32. S. Ishii and C. Hirano. 1963. Growth responses of larvae of the rice stem borer to rice plants treated with 2,4-D. *Entomol. Exp. Applic.* **6**:257-262.

33. D. R. Johnson and R. M. Hansen. 1969. Effects of range treatment with 2,4-D on rodent populations. *J. Wildl. Manage.* **33**:125-132.

34. W. Karg. 1964. Untersuchungen uber die Wirkung von DNOC auf die Mikroarthropoda des Bodens. *Pedobiologia* **4**:138-157.

35. J. O. Keith, R. M. Hansen, and A. L. Ward. 1959. Effect of 2,4-D on abundance and foods of pocket gophers. *J. Wildl. Manage.* **23**:137-145.

36. E. E. Kenaga. 1975. The evaluation of the safety of 2,4,5-T to birds in areas treated for vegetation control. *Residue Rev.* **59**:1-19.

37. R. D. Kimbrough. 1974. Toxic effects of the herbicide paraquat. *CHEST* **65**:655 675.

38. L. W. Krefting and H. L. Hansen. 1963. Use of phytocides to improve deer habitat in Minnesota. *Southern Weed Conf., Proc.* **16**:209–216.

39. L. W. Krefting and H. L. Hansen. 1969. Increasing browse for deer by aerial applications of 2,4-D. *J. Wildl. Manage.* **33**:784–790.

40. L. W. Krefting, H. L. Hansen, and M. H. Stenlund. 1956. Stimulating regrowth of mountain maple for deer browse by herbicides, cutting, and fire. *J. Wildl. Manage.* **20**:434–441.

41. Y. Lutz-Ostertag and H. Lutz. 1970. Action nefaste de l'herbicide 2,4-D sur le developpement embryonnaire et la fecondite de gibier a plume. *C.R. Acad. Sci. Ser. D* **271**:2418–2421.

42. Y. Lutz-Ostertag and R. Didier. 1971. 2,4,5-T et sterilite embryonnaire. *C.R. Soc. Biol.* **165**:2364–2366.

43. L. W. Martin and S. C. Wiggans. 1959. *The Tolerance of Earthworms to Certain Insecticides, Herbicides, and Fertilizers.* Okla. State Univ. Exp. Stn., Proc. Ser. P-334.

44. R. C. Maxwell and R. F. Harwood. 1960. Increased reproduction of pea aphids on broad beans treated with 2,4-D. *Ann. Entomol. Soc. Am.* **53**:199–205.

45. J. O. Moffett, H. L. Morton, and R. H. Macdonald. 1972. Toxicity of some herbicidal sprays to honeybees. *J. Econ. Entomol.* **65**:32–36.

46. G. Muller. 1971. Laboruntersuchungen zur Wirkung von Herbiziden auf Carabiden. *Arch. Pflanzenschutz* **7**:351–364.

47. National Academy of Sciences. 1968. *Weed Control* (Vol. 2 of *Plant and Animal Pest Control*). Nat. Acad. Sci. (Washington) Publ. 1597. 471 pp.

48. D. Neubert, P. Zens, A. Rothenwallner, and H. J. Merker. 1973. A survey of the embryotoxic effects of TCDD in mammalian species. *Environ. Health Perspect.* **5**:67–79.

49. M. Newton and L. A. Norris. 1968. Herbicide residues in blacktail deer from forests treated with 2,4,5-T and atrazine. *Proc. Western Soc. Weed Sci.* **22**:32–34.

50. L. A. Norris. 1971. Chemical brush control: Assessing the hazard. *J. Forestry* **69**:715–720.

51. D. Pimentel. 1971. *Ecological Effects of Pesticides on Non-Target Species.* Office Sci. Technol., Exec. Office Pres., Gov't. Printing Office, Washington, D.C. 220 pp.

52. G. R. Potts. 1970. The effects of the use of herbicides in cereals on the feeding ecology of partridges. *Proc. 10th Br. Weed Control Conf.* pp. 299–302.

53. E. H. Rapoport and G. Cangioli. 1963. Herbicides and the soil fauna. *Pedobiologia* **2**:235–238.

54. G. C. Rock and D. R. Yeargan. 1973. Toxicity of apple orchard herbicides and growth regulating chemicals to *Neoseiulus fallacis* and twospotted spider mite. *J. Econ. Entomol.* **66**:1342–1343.

55. V. A. Rowe and T. A. Hymas. 1954. Summary of toxicological information on 2,4-D and 2,4,5-T type herbicides. *Amer. J. Veter. Res.* **15**:622–629.

56. L. C. Ruch. 1957. Creating and maintaining wildlife openings in wooded areas by means of a herbicide. *Down to Earth* **12**(4), 2–3, 16.

57. M. H. Schroeder. 1972. Relationships of range and forest management with herbicides on wildlife. *Proc. W. Soc. Weed Sci.* **25**:11–14.

58. B. A. Schwertz et al. 1973. Toxicology of chlorinated dibenzo-*p*-dioxins. *Environ. Health Perspect.* **5**:87–99.

59. E. L. Shafer. 1965. *Deer Browsing of Hardwoods in the Northeast.* U.S. Forest Serv. Res. Paper NE-33. 37 pp.

60. J. Somers, E. T. Moran, B. S. Reinhart, and G. Stephenson. 1974. Effect of external application of pesticides to the fertile egg on hatching and early chick performance. *Bull. Environ. Contam. Toxicol.* **11**:33–38, 339–342.

61. T. R. E. Southwood and D. J. Cross. 1969. The ecology of the partridge: Breeding success and the abundance of insects in natural habitats. *J. Anim. Ecol.* **38**:497–509.

62. K. Steinbrenner, F. Naglitsch and I. Schlicht. 1960. Die Einfluss der Herbizide Simazin auf die Bodenmikroorganismen und die Bodenfauna. *Albrecht Thaer Arch.* **4**:611–631.

63. C. R. Swanson and W. C. Shaw. 1954. The effect of 2,4-D on the hydrocyanic acid and nitrate content of Sudan grass. *Agron. J.* **46**:418–421.

64. H. P. Tietjen, C. H. Halvorson, P. L. Hegdal, and A. M. Johnson. 1967. 2,4-D herbicide, vegetation and pocket gopher relationships on Black Mesa, Colorado. *Ecology* **48**:635–643.

65. R. K. Tucker and D. G. Crabtree. 1970. *Handbook of Toxicity of Pesticides to Wildlife.* U.S. Dept. Interior, Fish and Wildl. Serv., Resource Publ. No. 84. 131 pp.

66. J. M. Webster. 1967. Some effects of 2,4-D herbicide on nematode-infested cereals. *Plant Pathol.* **16**:23–26.

67. J. M. Webster and D. Lowe. 1966. The effect of 2,4-D on the host–parasite relationships of some plant-parasitic nematodes. *Parasitology* **56**:313–322.

68. A. H. Westing. 1971. Ecological effects of military defoliation on the forests of South Vietnam. *Bioscience* **21**:893–898.

69. E. Wilcke. 1968. Unkrautbekampfung in Obstbau: Nebenwirkung auf die Bodenfauna. *Z. Pflanzenkr. Pflanzenpathol. Pflanzenschutz, Sonderheft* **4**:163 167.

70. C. J. Willard. 1950. Indirect effects of herbicides. *N. Central Weed Contr. Conf., Proc.* **7**:110–112.

71. J. G. Wilson. 1973. Teratological potential of 2,4,5-T. *Down to Earth* **28**(4):14–17.

13

HERBICIDES AND THE SOIL MICROFLORA

A. BREAKDOWN OF HERBICIDES BY SOIL MICROORGANISMS

The contribution of microorganisms to the total breakdown of herbicides in soils may be gauged by assessing the breakdown in a soil that has been autoclaved as compared with that in the original condition. The breakdown of 2,4-D and dichlobenil is more than five times greater in the unautoclaved soil, and the breakdown of MCPA and simazine is more than two to five times greater where the microorganisms have not been inactivated[81]; on the other hand, with trifluralin the autoclaved soil is as active as the unautoclaved. The relative liability of various herbicides to microbial breakdown was assessed by adding the radiolabeled compounds to a silt loam and measuring the cpm of radioactive CO_2 obtained from the sterilized soil[150]; these were as follows:

Amitrole	110,000	Simazine	350
Amiben	6,500	Atrazine	125
Fenac	1,500	Propazine	10

The unusually high yield of CO_2 from amitrole was due to the microorganisms rapidly attacking the fragments initially produced by abiotic agents.

In the review of microbial degradation in the soil presented by Alexander (1969), 15 herbicides are tabulated with the names of the microorganisms identified as degrading them. The review of Kaufman and Kearney (1970) lists 42 species of bacteria and fungi degrading simazine, atrazine, prometryn and nine other triazine herbicides. Species belonging to 16 genera of bacteria, 2 of actinomycetes and 13 of fungi may now be tabulated for their ability to degrade 20 of the commonly used herbicides (Tables 13.1, 13.2). It is possible that some microorganismal enzymes (e.g., dehydrogenases, ureases) which are known to be free in the soil[188] may degrade herbicides. The most important biochemical reactions degrading herbicides in the soil fall under the heading of dealkylation, either O–R cleavage as for the phenoxy acids and dicamba, or N–R cleavage as for triazines, urea herbicides, dinitrotoluidines, and bipyridylium compounds.[126] Aromatic hydroxylation is exceptionally important to prepare herbicide molecules possessing benzene rings for their final degradation.[43] The metabolites produced by microorganismal action have been reviewed concisely by Kearney (1966) and Kearney, Kaufman, and Alexander (1967), and covered in detail in the book by Kearney and Kaufman (1969) and the reviews by Helling, Kearney, and Alexander (1971), Meikle (1972), and Kaufman (1974). Abstracts of the original papers may be found in the compilations of Menzie (1969, 1974).

Table 13.1. Soil organisms that degrade herbicides of simpler molecular construction. The references are numbered.

BACTERIA	MCPA	2,4-D	2,4,5-T	MCPB	Dalapon	TCA	PCP	Endothall	Dichlobenil	Allyl Alcohol
Achromobacter spp.	196	196	21							
Agrobacterium sp.					103	105				
Alcaligenes sp.					113					
Arthrobacter globiformis		9								
Arthrobacter spp.	148	148			103	103		185		
Azotobacter sp.										106
Brevibacterium sp.			97*							
Corynebacterium sp.	110	110					38			
Flavobacterium spp.	24	196			202					
Micrococcus sp.					149					
Mycoplana sp.	198	198	198							
Pseudomonas putida										104
P. dehalogenans					95	95				
P. cruciviae		220	220				220			

Pseudomonas spp.	59	59		94	132	2		104
Sporocytophaga congregata		110						
ACTINOMYCETES								
Nocardia spp.		110	215	94	95			104
Streptomyces spp.		30		94				
FUNGI								
Aspergillus niger	58	58	58					
Aspergillus sp.				94				
Cephaloasca fragrans						44		
Fusarium sp.							168	
Geotrichum sp.							168	
Penicillium spp.				113		44	168	
Trichoderma viride				113	103			
Trichoderma spp.						44	168	106

*Substrate 2, 3, 6—TBA

Table 13.2. Soil organisms that degrade herbicides with nitrogen in the molecule. The references are numbered.

	Simazine	Atrazine	Chlorpropham	Paraquat	Diphenamid	Propanil	Monuron	Picloram	Trifluralin	DNOC
BACTERIA										
Aerobacter aerogenes				206						
Achromobacter sp.	37		119	226						
Agrobacterium spp.	122		119	206				227		
Arthrobacter spp.			39			31		227		
Bacillus spp.				206				227		
Bacillus sphaericus							213*			
Sarcina sp.				14			92			
Clostridium pasteurianum				14						
Corynebacterium spp.				14						79
Mycobacterium sp.			154							
Flavobacterium spp.			119					227		
Pseudomonas spp.	28		119	206						201
Xanthomonas sp.							92			
ACTINOMYCETES										
Nocardia spp.				161						
Streptomyces spp.	28			161				227		

FUNGI

Aspergillus fumigacus	28	118						
A. niger	159	159				27*	227	66
A. repens	159	159				227	227	
Aspergillus spp.	118			134		92	227	
Botrytis alii							227	
Cephalosporium acremonium	159	159						
Cladosporium herbarum	159	159						
Cylindrocarpon spp.	77							
Fusarium moniliforme	28	118						
Fusarium spp.	41	41	154		18			
Geotrichum sp.	41	41						
Helminthosporium pedicellatum	159						227	
Penicillium cyclopium	159	159						
Penicillium spp.	28	118	154		26	92	227	
Rhizopus spp.	28	118						
Stachybotrys atra	28							
Trichoderma viride	28	118		134				

* Linuron

Adaptation of the Soil Population

When phenoxy herbicides are first applied at normal rates, 2,4-D persists in the soil for about 2 wk, MCPA for about 1 month, and 2,4,5-T for about 2 months. Over a range of soils in Hawaii, the rate of 2,4-D breakdown was found to be proportional to the number of aerobic bacteria present.[1] The breakdown was slow at first, but the rate picked up as those bacteria that could utilize the herbicide as a carbon source increased their numbers, the bacterial population thus becoming adapted to the herbicide. With phenoxy herbicides, the length of the preliminary latent period before the soil became "enriched" with adapted microorganisms was in the order 2,4-D < MCPA < 2,4,5-T, so that the times taken for detoxication to be effected in laboratory soil perfusion experiments were 14, 70, and 270 days, respectively.[8] Among the organisms cultured from the enriched soil, *Arthrobacter globiformis* could grow on a substrate of 2,4-D as the sole carbon source, and its inoculation into new soil with 2,4-D resulted in an immediate breakdown of the herbicide.[9] In a similar way, a *Pseudomonas* was later isolated that could utilize MCPA as the sole carbon source.[68]

The latent periods observed for each of the phenoxy herbicides, and the organisms subsequently isolated, were much the same for any soil sample. The adapted strains isolated are found to have the enzymes that degrade the phenoxyacetic acids, but to lose their adaptation in laboratory culture without these substrates.[11] The adapted types may persist in the soil for 12 months after they have been induced by 2,4-D, even a very small dose having proved sufficient.[212] Thus the consequence of repeated spraying with 2,4-D (or the phenoxy compound MCPA) is that certain species in the soil microflora become adapted to break down the herbicide. The half-life of 2,4-D in Saskatchewan wheatfield soils, once as long as 80 days, has shortened to 14 days or less after a quarter-century of its application to the soil and its microflora.[45]

Whereas the microflora in soil perfused with 2,4-D is now adapted to degrade MCPA also, but not 2,4,5,-T, a perfusion with MCPA adapts it to 2,4-D and 2,4,5-T as well.[10] This duality was also revealed in *Achromobacter,* where a strain adapted to MCPA was found to lose this capacity if grown on 2,4-D instead.[197] In *Flavobacterium peregrinum,* where conditioning with 2,4-D results in the strain metabolizing 2,4-D but not MCPA, a strain conditioned with MCPA was no different in that it could now metabolize 2,4-D but not MCPA.[198] Thus MCPA could be characterized as an inducer but not a substrate in *Flavobacterium*; in *Achromobacter,* the primary metabolite 2,4-dichlorophenol was inducer of both 2,4-D degrading and MCPA-degrading enzyme systems.[197]

The phenomena of lag period and enrichment have been noted in the breakdown of dalapon.[113] Lag periods have also been detected in the soil degradation of picloram, atrazine and diuron; however, they are not as universal an occurrence as the initial experiments with 2,4-D had indicated.[81] In contrast to the phenoxy compounds, with the urea herbicides (fenuron, monuron, diuron, etc.) the latent period was very long before adaptation occurred, and the resulting breakdown was very slow.[184]

Soils treated with the comparatively simple molecules that constitute the present herbicides have yielded a considerable number of bacteria and fungi that can degrade them, including the actinomycetes in the genera *Streptomyces* and *Nocardia,* and the yeast *Lipomyces starkeyi.*[14] It is perhaps significant that whereas bacteria are the main group found metabolizing the phenoxyalkanoic and benzoic acid derivatives (Table 13.1), the fungi figure more importantly among the microorganisms breaking down the herbicides which have nitrogen in the molecule and are more persistent (Table 13.2).

Agents and Pathways of Herbicide Degradation

Phenoxy Acids. 2,4-D is a herbicide whose breakdown in soil is solely the result of biological activity , since the process is completely inhibited by sodium azide.[8] The first product of 2,4-D breakdown by *Arthrobacter, Achromobacter, Pseudomonas, Flavobacterium,* or *Nocardia* is 2,4-dichlorophenol (Fig. 13.1), followed by 3,5-dichlorocatechol.[55,197,203] Enzyme preparations from *Arthrobacter* indicate a pathway through these two metabolites[147,204] to 2,4-dichloromuconic acid, while *Achromobacter* converts the 3,5-dichlorocatechol to the semialdehyde of hydroxydichloromuconic acid.[98] The final path to CO_2 lies through chloromaleylacetic acid, identified by Fernley as a 2,4-D metabolite in 1959, which *Arthrobacter* enzymes degrade through maleylacetic acid to succinic acid.[52] By a similar initial cleavage of the phenyl ether linkage, *Arthrobacter* and *Flavobacterium* break MCPA down to 2-methyl-4-chlorophenol,[24] and the corresponding catechol has been found in *Pseudomonas.*[68] A small amount of the 6-hydroxy derivative is produced by *Pseudomonas* from 2,4-D and MCPA.[55]

The fungus *Aspergillus niger,* on the other hand, first hydroxylates 2,4-D in the 5 position on the phenyl ring, and MCPA likewise.[58] With MCPA and MCPB, the preliminary step is the beta-oxidation of the side chain, carried out by *A. niger* and the actinomycete *Nocardia opaca.*[215] The esters of 2,4-D, the form in which they are usually applied to water, are first hydrolyzed by microorganisms before the active compound is degraded, which proceeds so thoroughly that the 2,4-dichlorophenol is not accumulated.[3]

*s-***Triazines.** Although much simple hydrolysis takes place in acid soils, triazine herbicides may be degraded by microorganismal action.[138] The principal degradation product of atrazine (Fig. 13.2) to be found in soil is

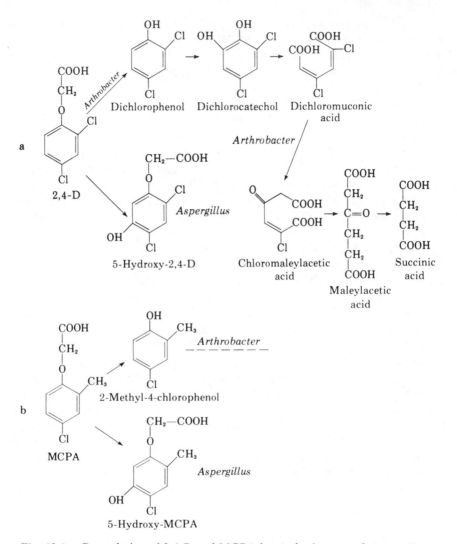

Fig. 13.1. Degradation of 2,4-D and MCPA by *Arthrobacter* and *Aspergillus*.

hydroxyatrazine,[87] a nonphytotoxic product of chemical hydrolysis rather than of microbial action,[6] and it too is relatively stable against breakdown by microorganisms. Simazine and propazine, like atrazine, are converted to their hydroxy forms almost as rapidly in soil treated with the microbial inhibitor sodium azide as in untreated soil.[87] All three triazines when added to soils as radiolabeled compounds yield the hydroxy metabolite, which comes to constitute about 50% of the extractable material within 2 months,

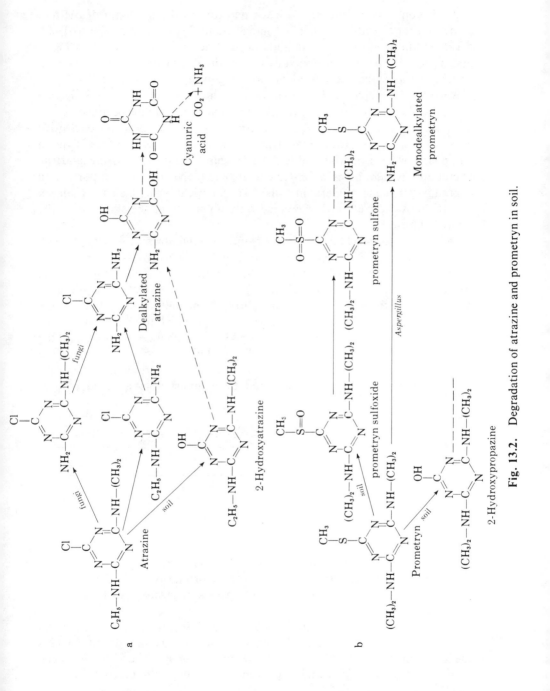

Fig. 13.2. Degradation of atrazine and prometryn in soil.

and this conversion increases with temperature.[88] In a silt loam of pH 6.6, some 40% of the atrazine applied had become hydrolyzed to the hydroxy derivative in 4 wk, whereas in soils around neutrality it was only 10% so hydrolyzed.[186] The further breakdown of hydroxyatrazine proceeds faster in anaerobic rather than aerobic soils, resulting in the production of CO_2.[85]

Simazine, atrazine, and prometryn can, however, be metabolized by a considerable number of soil fungi.[159] The main metabolite produced from simazine by *Aspergillus fumigatus* cultures is the monodesethyl derivative 2-chloro-4-amino-6-ethylamino-*s*-triazine, although a doubly desethylated compound was also produced.[122] The fungus *Trichoderma viride* and the bacterium *Pseudomonas aureofaciens* appear to be the most important in degrading simazine in German soils.[28] In American soils, a *Penicillium* sp. was the most active, but it degraded only about 3% of the simazine in the treated culture.[41]

In contrast to the chloro-*s*-triazines simazine and atrazine, in which much of the herbicide is first chemically hydrolyzed to the hydroxy-triazine before being biologically dealkylated, the methylthio-*s*-triazine prometryn is straightway metabolized biologically.[120] This less persistent herbicide used on cotton is oxidized in soil to the sulfoxide and sulfone as well as to the hydroxy derivative.[171] When prometryn was applied to a silt loam, the production of hydroxypropazine (Fig. 13.2) commenced to outstrip the sulfoxide or sulfone 2 months after the application[22]; 5 months after the application the main metabolite was the hydroxylated product, just as with atrazine, whether or not the soil had been adjusted to an acid pH, as follows:

	Atrazine		Prometryn	
	pH 5.5	pH 7.5	pH 5.5	pH 7.5
Unaltered	11	35	42	8
Hydroxylated	33	10	24	17
Dealkylated	1.2	4.5	1.0	0.3

When prometryn and related triazines were exposed to pure cultures of *Aspergillus fumigatus*, degradation occurred primarily by N-dealkylation, as with simazine[121]; at least four other species of *Aspergillus* break down prometryn.[91]

Atrazine, in which one of the N substituents is isopropyl, was dealkylated by a number of fungi, although hydroxyatrazine was also produced[118] and is the predominant metabolite with *Fusarium roseum*.[41] The CO_2 produced early in simazine and atrazine degradation may be associated with the desalkylation.[138] Soils of pH 6.7 produce CO_2 from the desethylation of atrazine more than twice as fast as those at neutrality,[186] and this biological decomposition is doubled by a temperature increase from 10 to 30°C.[155] Soil bacteria and fungi, although capable of deethylating atrazine, are evidently unable to cleave its triazine ring unless it is first chemically converted to hydroxyatrazine.[187] Cyanuric acid has been postulated as an intermediate; in soil it is degraded to CO_2 10 times as fast as the chlorodiaminotriazine metabolite and 100 times as fast as atrazine, while the soil fungus *Stachybotrys chartarum* can degrade all of it to CO_2 in 8 wk.[221] In any case, ring cleavage of *s*-triazines is evidently a slow process.[120]

Bipyridyliums. The bipyridyl herbicides such as paraquat and diquat become very tightly absorbed to soil particles; thus although they degrade slowly their residues are biologically inactive. They are also persistent in hydrosoils, an application of diquat or paraquat to ponds at 0.3 lb/a leaving bottom residues which range up to 1.7 and 7.9 ppm, respectively.[64] The fungus *Neocosmospora vasinfecta* could reduce paraquat to the free radical,[65] an essential stage in the intoxication process of this herbicide in higher plants. Microorganisms isolated from soil that could break down paraquat included the bacteria *Clostridium pasteurianum* and *Corynebacterium fascians*, which metabolized one-third of it in 3 wk, and the yeast *Lipomyces starkeyi* which could utilize it as the sole source of nitrogen.[14] Paraquat could also serve as the sole source of N for *Aerobacter aerogenes*, *Agrobacterium tumefaciens*, *Pseudomonas fluorescens*, and *Bacillus cereus*.[206] An isolate of unidentified bacterium produced two metabolites, one being similar to the desmethyl derivative of the dipyridinium ion arising out of the free radical, and the other being similar to the carboxy-mono-pyridinium ion.[65] This 1-methyl-4-carboxypyridinium ion (Fig. 13.3) was detected as a metabolic product of *Streptomyces, Nocardia* and nine other actinomycete isolates from soil.[161] Washed suspensions of a species of *Achromobacter* released methylamine from this moiety, and after decarboxylation to CO_2 the remaining five carbon atoms gave rise to succinate and formate.[226]

Ureas and Uracils. Diuron disappears from nonsterilized soil about five times as fast as from sterilized soil.[93] Soils in which the activity of microorganisms has been enriched by the addition of glucose break down diuron more than 10 times as fast as atrazine, and the rate of breakdown is tripled by a soil temperature increase from 10°C up to 30°C.[155] Of the *Aspergillus* fungi in the soil, *A. niger* is more active, and *A. tamarii* less, than *A. sydowi*, in degrading diuron, monuron and other urea herbicides.[160] *Bacillus sphaericus* isolated from soil produces 3,4-dichloroaniline (DCA)

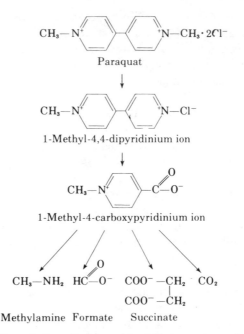

Fig. 13.3.   Degradation of paraquat by actinomycetes and bacteria.

by hydrolyzing linuron at the amide bond (Fig. 13.4), the side chain yielding CO_2 and dimethyl-hydroxylamine[53]; this bacterium also hydrolyzes linuron[213] and metobromuron but is unable to degrade fluometuron or diuron.[53] The bromoaniline released from metobromuron is acetylated to the bromoacetanilide,[208] thus avoiding the condensation to TCAB (see below). Out of 94 species of soil microorganisms (81 fungi, 13 bacteria) no less than 92 decomposed monuron, diuron and linuron to the chloranilines MCA or DCA, by various processes including demethylation, demethoxylation, hydroxylation and decarboxylation of the side chain; moreover 78 of them metabolized the chloranilines produced.[182] The MCA released from monuron may be oxidized (e.g., by *Fusarium oxysporum*) to 4-nitrobenzene or condensed to 4,4′-dichloroazobenzene.[123] Microorganisms in cotton soils degrade diuron to 3,4-dichloroaniline (DCA) by stepwise demethylation followed by hydrolysis of the amide[47]; the DCA subsequently disappears, but without the appearance of TCAB.[20] The rates of biodegradation of the urea herbicides in soil—fenuron > monuron > diuron > neburon—are the reverse of the order of their molecular weight and their adsorption rates on soil.[69]

When the uracil herbicides bromacil and terbacil are applied to soil, the *sec*-butyl or *tert*-butyl substituents are hydroxylated rather than removed.[67]

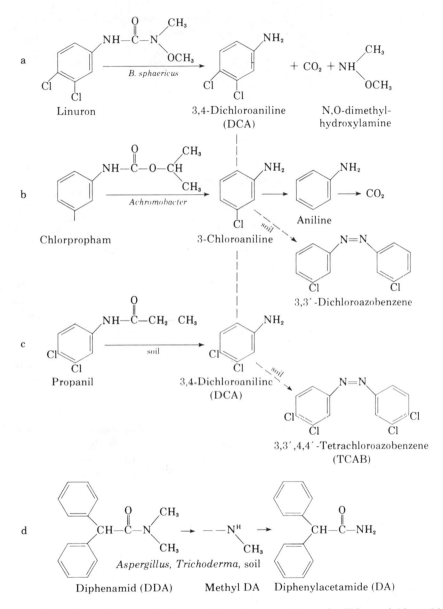

Fig. 13.4. Degradation of (a) urea, (b) carbamate, (c) acylanilide, and (d) amide herbicides in soil: linuron, chlorpropham, propanil, and diphenamid.

Table 13.3.   Degradation of benzene-ring structure by microorganisms: percent breakdown after 8 days' exposure to cultures (McClure, 1974).

	Arthrobacter	Mycobacterium	Fusarium	Penicillium
Propham	100	100	100	100
Chlorpropham	14	0*	27	20
Swep	0	0	0	0
Propanil	5	10	11	5

* Some hydrolysis of the side-chain occurred

Carbamates.   The carbanilate herbicide chlorpropham (CIPC) is hydrolyzed by *Achromobacter, Agrobacterium* and *Flavobacterium* to leave 3-chloroaniline[119]; *Pseudomonas* degrades it to 3-chloroaniline as an end product,[127] but *Arthrobacter* and *Achromobacter* can dechlorinate this breakdown product to aniline, which in turn yields CO_2.[39] Propham (IPC) would produce aniline directly. Whereas aniline is rapidly broken down in soils, 3-chloroaniline persists (Fig. 13.4) and is somewhat toxic to soil microorganisms. A mixed culture habituated to propham also degrades chlorpropham and swep at an accelerated pace[153] but it cannot handle fenuron. Pure isolates of two bacterial and two fungal species obtained from this culture (Table 13.3) proved capable of breaking the benzene rings of propham, chlorpropham and propanil, but not of swep.[154] Another fate of 3-chloroaniline is to dimerize into 3,3′-dichloroazobenzene, which was detected in a chlorpropham-treated sandy loam, probably due to microorganismal action.[17] The carbanilate herbicide swep yields 3,4-dichloroaniline (DCA) which in turn produces 3,3′, 4,4′-tetrachloroazobenzene.[16] This compound (TCAB) is an undesirable pollutant, since the azobenzenes as a class are very stable, and some of them (e.g., 4-dimethylaminoazobenzene) are carcinogenic. The initial hydrolysis of chlorpropham, which often does not persist long enough to give season-long control for this very reason, can be delayed by combining this herbicide with a dimethylcarbamate insecticide such as carbaryl, an inhibitor of the hydrolyzing enzyme.[91]

Thiocarbamates.   The thiocarbamate herbicide EPTC (Eptam), which is readily volatilized from soil, owes about two-thirds of its degradation to microorganisms. Nothing is known about the metabolites produced, beyond the fact that inactivation is well in advance of CO_2 production. Judging by the fate of EPTC in plants and animals, the first step is hydrolysis to ethylmercaptan and dipropylamine (Fig. 13.5), which are later broken down.[56] Butylate (Sutan) was degraded by cultures of several soil fungi within 3–7 days, producing several amines.[76] Diallate has a longer period of phytotoxicity in soil, possibly due to the appearance of allyl alcohol as a secondary product; this would have been by transthiolation and dechlorination of the 2,3-dichloroallyl mercaptan arising from the initial hydrolysis of the herbicide.[116] Having a half-life of 2–4 wk, its breakdown is mainly due to microorganisms; among the fungi tested, *Penicillium janthinellum* and *Phoma eupyrena* were active, and *Trichoderma harzianum* degraded diallate to CO_2 while using it as a carbon source.[5] Triallate, which also owes the bulk of its degradation to microorganisms, was found to be broken down by species of *Penicillium,* which at the same time sequester it by adsorption.[46]

Acylanilides and Amides. Propanil is hydrolyzed to 3,4-dichloroaniline (DCA) by cells and extracts of *Fusarium solani,* in the same way that the related acylanilide Karsil is hydrolyzed by *Penicillium.*[183] *Penicillium piscarium* converts propanil to DCA and *Geotrichum candidum* converts DCA to TCAB, thus preventing it from intoxicating the *Penicillium; Geotrichum* could not break down propanil.[26] Thus propanil is degraded in soils to DCA

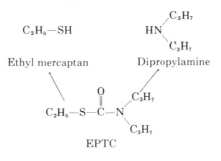

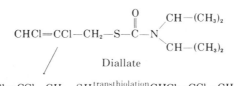

Fig. 13.5. Degradation products of thiocarbamate herbicides in soil: EPTC and triallate.

(Fig. 13.4), which in a sandy loam is largely converted into TCAB.[15] About one-third of the amount of propanil applied to soil is converted after 3 wk into DCA and TCAB, much of these metabolites becoming unextractable, presumably due to adsorption onto organic matter.[145] TCAB has been detected in ricefields 3 yr after they had been routinely treated with propanil.[133] The amide herbicide diphenamid is progressively demethylated in soil to the acetamides MDA and DA,[73] and these compounds have been produced by the soil fungi *Aspergillus candidus* and *Trichoderma viride*.[134] With the amide alachlor, degradation in a sandy loam soil removes the methoxymethyl substituent from the amide nitrogen.[86]

Dinitroanilines.   The preemergent herbicide trifluralin evidently owes its persistence to a virtual immunity from microorganismal breakdown, since unautoclaved soil has been found to be no more active than autoclaved.[173] In soil, this N-substituted dinitrotoluidine is stripped of its two propyl substituents and reduced to the diaminotoluidine (Fig. 13.6). The reduction of the nitro groups would be expected to precede the dealkylation under anaerobic conditions, and vice versa under aerobic conditions. *Aspergillus*

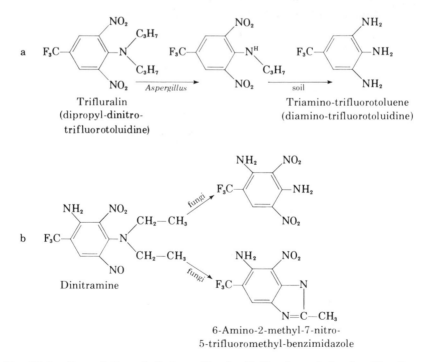

Fig. 13.6.   Degradation of dinitroaniline herbicides by soil fungi: trifluralin and dinitramine.

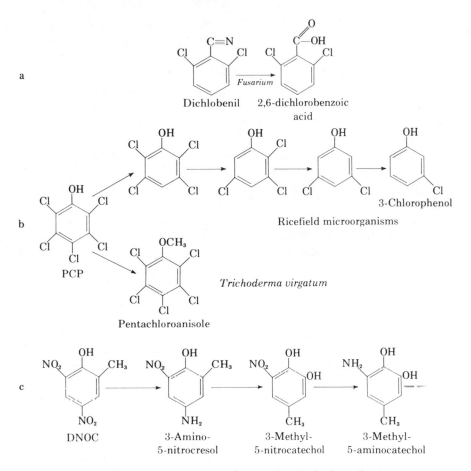

Fig. 13.7. Degradation of benzoic and phenolic herbicides by soil microorganisms: dichlobenil, PCP, and DNOC.

niger can remove one of the propyl groups, though not actively, while *Fusarium* or *Trichoderma* are inactive.[66] The dinitrotoluidine herbicide dinitramine is similarly stripped of its N-alkyl substituents by *Aspergillus fumigatus* and *Poecilomyces* fungi, but at the same time they can cyclize it to a benzimidazole derivative.[141] As a group, the dinitroaniline herbicides are degraded more rapidly in anaerobic than aerobic soils.[90]

Benzoics and Phenols. Soil treated with dichlobenil, a persistent herbicide, comes to contain equal parts of dichlobenil and 2,6-dichlorobenzoic acid[192]; this conversion was demonstrated in four genera of fungi, *Fusarium* being particularly active.[168] The dichlorobenzoic acid (Fig. 13.7), along with three minor unknowns, accounted for 95% of the metabolites when dichlobenil was applied to sandy soil.[210] Dicamba, a dichlorinated benzoic acid derivative, is quite rapidly broken down by unknown soil microorganisms, possibly by decarboxylation.[200]

PCP is progressively dechlorinated by the microorganisms in ricefield soils (Fig. 13.7) to produce three tetraphenols, two trichlorophenols, and three chlorophenols.[99] A saprophytic coryneform bacterium has been isolated that can use PCP as a sole source of carbon and energy, rapidly producing CO_2 from it.[38] The fungi *Trichoderma, Cephaloasca,* and *Penicillium* methylate the phenolic group of PCP to form pentachloroanisole.[44] The dinitro herbicide DNOC quite rapidly disappears from soil, eventually yielding nitrate from the action of *Pseudomonas* and *Corynebacterium* (*Arthrobacter*) *simplex*[79]; the more persistent relative DNBP (dinoseb) is also degraded by the latter bacterium, but very little nitrite appears. The initial metabolite, 3-amino-5-nitrocresol, was identified in cultures of *Pseudomonas* and *Rhizobium* spp.[83] A *Pseudomonas* isolated from garden soil perfused with DNOC was found to produce, like *C. simplex,* 3-methyl-5-nitrocatechol, 3-methyl-5-amino-catechol, and 2,3,5-trihdroxy-toluene before the benzene ring opened.[201]

Aliphatics.   Among the simple chlorinated aliphatic acids, dalapon is oxidatively dechlorinated by *Arthrobacter* to pyruvic acid, perhaps through 2-chloro-2-hydroxypropionic acid,[130] the latter being incorporated as alanine into the bacteria[19]; 2-chloropropionate was identified among the metabolites produced by *Alternaria.*[63] Soils vary greatly in the rate of dalapon decomposition by their microflora, which depends on a good moisture content.[131] Trichloroacetic acid (TCA) is much more refractory to biological degradation than dalapon.[102] Two species of *Arthrobacter* could degrade TCA to oxalate,[71] and a representative *Pseudomonas* from soil incubated with ^{14}C-labeled TCA released the chlorine, the radioactive carbon going into amino acids such as serine incorporated into the bacteria.[132] A similar breakdown with evolution of chloride was shown with *P. dehalogenans* and a species of the actinomycete *Nocardia.*[95] Allyl alcohol, a soil herbicide and fungicide, rapidly disappears by microorganismal degradation, acrylic and propionic acids probably being intermediates.[107]

Organophosphorus Compounds.   Two phosphoroamidothioates developed as insecticides and acaricides have proved effective in eliminating crab grass and other weeds in turf. One is bensulide and the other is DMPA (Zytron) originally developed as fly larvicide. DMPA is hydrolyzed by soil microorganisms such as *Aspergillus clavatus* to 2,4-dichlorophenol, which is slightly fungitoxic; but at the normal application levels (15 lb/a producing 5–10 ppm) this metabolite is also broken down by *Aspergillus.* The O-methyl-N-isopropyl-phosphorothioic acid left by this hydrolysis is actually stimulatory to *Penicillium*.[60] Glyphosate (phosphonomethylglycine) is a successful crab grass herbicide that is completely degraded within a month principally by microorganisms taking it to CO_2, the phosphonate moiety being strongly adsorbed to the clay minerals in the soil.[194]

Other Organics. The nonvolatile aminopyridine herbicide picloram is very persistent in soils, under either aerobic or anaerobic conditions, being dependent on microorganisms for its breakdown.[84] It may be degraded by a number of soil microorganisms (Table 13.2) when placed in nutrient culture, but its role as a carbon source is negligible.[227] The only metabolite detected in picloram-treated soil is 6-hydroxypicloram (Fig. 13.8), and there is no evidence of its being decarboxylated to 4-amino-2,3,5-trichloropyridine as it is in plants. The initial breakdown in soil releases a high proportion of inorganic chloride, and the hydroxypicloram never accumulates to more than 1% of the added picloram, despite being almost as stable; it is, therefore, considered that the breakdown occurs by the pyridine ring of picloram being cleaved by a dioxygenase, such as that implicated in the degradation of vitamin B_6 and nicotinic acid.[157] Pyrazon, a sugarbeet herbicide with a phenyl ring attached to a pyridazine nucleus, is rapidly dephenylated in mineral soils but is quite persistent in muck soils, the metabolite in either case being 5-amino-4-chloro-pyridazinone.[49,190]

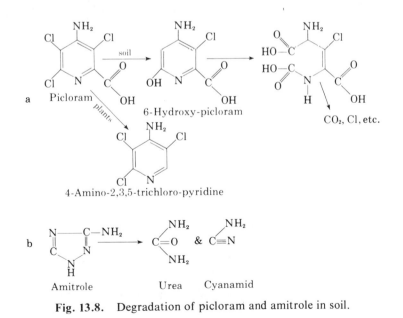

Fig. 13.8. Degradation of picloram and amitrole in soil.

Endothall is quite rapidly converted to CO_2 by hydrosoil, and an *Arthrobacter* was isolated that could use it as the sole source of carbon, incorporating its fragments into glutamic, aspartic, and citric acids[185]; *A. globiformis* can degrade endothall but not its phthalic moiety.[108] The triazole herbicide amitrole is readily broken down in soils into urea, cyanamid and nitrogen (Fig. 13.8), but the initial ring cleavage is carried out abiotically by free radicals, riboflavin, and light. Although microorganisms are active on the cleavage fragments, the initial cleavage is by agents not inhibited by the cytochrome oxidase inhibitor potassium azide, and is not increased by the restoration of microorganisms to autoclaved soil.[124]

Inorganics.   Among the arsenical herbicides, MAA (methanearsonic acid) and cacodylic acid (dimethylarsinic acid) have fates similar to that of sodium arsenite. In aquatic sediments they are completely converted to dimethylarsine,[101] being themselves successive steps on a pathway of reduction and methylation (Fig. 13.9) by the bacteria *Methanobacterium* and *Desulfovibrio*,[152] and by *Scopulariopsis brevicaulis*.[117] Three species of fungi isolated from sewage, namely *Candida humicola, Gliocladium roseum,* and a species of *Penicillium,* were active in producing trimethylarsine gas from arsenate, arsenite, and monosodium or disodium methanearsonate.[42]

Under aerobic conditions in soils, however, slightly more of the applied cacodylic acid is oxidatively converted to CO_2 and arsenate than is reduced to dimethylarsine or other volatile alkylarsines,[222] while the inorganic arsenicals are inactivated by forming salts with the Fe and Al provided by the free oxides in the soil.[223] Four soil organisms—a fungus, a bacterium, and two actinomycetes—were isolated which converted MSMA (the sodium

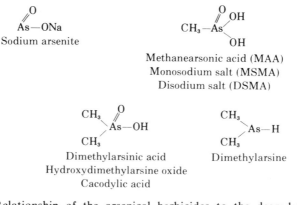

Fig. 13.9.   Relationship of the arsenical herbicides to the degradation product dimethylarsine.

salt of MAA) into CO_2 and inorganic arsenate.[211] This monosodium methanearsonate can be applied annually to soils at dosages up to 25 kg/ha without any accumulation of extra elemental arsenic occurring.[179] Chlorate, applied as the sodium salt for total vegetation control, is slowly reduced by soil microorganisms, apparently to the chloride.[109]

B. EFFECT OF HERBICIDES ON SOIL MICROORGANISMS

Susceptibilities of the General Population and the Nitrifiers

Compounds applied to the soil as selective herbicides have caused either no change in the total microorganism flora, or a reduction that proved transitory. Even with the so-called soil sterilants, any reduction that may occur will eventually be followed by recovery. Often the initial reduction is followed by a net increase in microorganisms, as microorganismal genotypes become prevalent that break down the herbicides. Notable temporary reductions of certain species that often respond with a subsequent surge have been produced by field application of amitrole, PCP, DNBP, chlorpropham, and EPTC.[162]

Temporary reductions in the nitrifying bacteria of the genera *Nitrobacter, Nitrosomonas,* and *Nitrosococcus* are induced by many herbicides, while the nodule-inhabiting *Rhizobium* species are unaffected[51] and the nitrogen-fixing *Azotobacter* may actually be stimulated. Since ammonium is a fertilizer for plants, and nitrates and nitrites can be toxic to herbivores, a restraining of nitrification may actually be beneficial in order to achieve the right balance of the three nutrients for a given crop[74]; there are agricultural chemicals, such as nitrapyrin (N-Serve) now available to moderate the biological conversion of ammonia fertilizers.

An overview of the action of herbicides on soil microflora by Audus (1970) has pinpointed certain of them which have a material effect at field rates. Dinoseb and dalapon can affect total bacterial numbers for 2–3 months, while PCP and EPTC may cause a temporary reduction; on the other hand, endothall can be utilized by bacteria to increase their numbers. Nitrogen fixation by *Azotobacter* may be decreased by PCP, DNOC, and propham, while the phenoxy herbicides can suppress legume nodulation if not the *Rhizobium* bacteria. Ammonification is decreased only by PCP applications in ricefields, while it is increased by endothall, EPTC and propham. Nitrification is liable to be inhibited by field applications of the phenolics PCP, DNOC, and dinoseb, as well as by sodium chlorate, TCA, dalapon and trifluralin; propanil affects nitrification and may cause the accumulation of nitrite. Among the range of sensitivities of fungi to herbicides, it may be generalized that certain pathogenic species such as *Sclerotium rolfsii* and *Pythium* spp. are especially susceptible to most herbicides, while *Trichoderma* and *Penicillium* are resistant. Although most pathogenic *Fusarium* species are susceptible, *F. oxysporum* is not.[12]

To show differences in susceptibility between groups or species of microorganisms it takes experiments with culture media or with solutions percolated through soil, under conditions much more drastic than in normal agronomic practice. A very rough rule-of-thumb is that every 1 lb/a applied puts a concentration of 1 ppm of the herbicide in the top layer of soil,[136,178] or at most 2 ppm.[61] Thus it is rare that soil concentrations of herbicides in cropland exceed 3 ppm. Published assessments tabulated by Alexander (1969) list the following herbicides that have inhibited nitrification at the parts-per-million stated in parentheses:

PCP	(5)	Chlorpropham	(12)
DNBP	(10)	Monuron	(25)

On the other hand, no inhibition was shown by the concentrations cited, some very considerable, of the following herbicides:

Propham	(25)	EPTC	(50)
TCA	(29)	Dalapon	(150)
2,4-D	(50)	Arsenite	(500)

The majority of the bacteria in the soil are not as sensitive as the nitrifying species.

Effects of herbicides on algae are perhaps academic where field crops are concerned, but they must be important in rice culture. The algae are very susceptible to herbicides, 0.5 ppm atrazine completely inhibiting the growth of *Chlamydomonas reinhardii,* while the nitrogen-fixing *Tolypothrix tenuis* is inhibited by concentrations of 2,4-D and trifluralin lower than those required to achieve weed control.[91]

Effects on Fungi and Bacteria

Phenoxy Acids.   With 2,4-D acid, amines or esters applied at the practical dosages between 0.25 and 4 lb/a, a total of 27 investigations found that they had no effect on the total number of microorganisms, or on their activity as judged by the oxygen consumed or the CO_2 produced.[25,61,170] Where some initial inhibition in microorganismal activity was detectable, as in light soils, it was transitory.[4,62] With artificial laboratory experiments it was found that aerobic bacteria were inhibited at lower concentrations than facultative anaerobes,[225] gram-positive species were more susceptible to 2,4-D and 2,4,5-T than the gram-negative,[164] and spore formers more susceptible than non–spore formers.[195]

With respect to nitrification after normal field applications of 2,4-D, the review of Fletcher (1960) reveals that there was either no effect detectable (three investigations) or a transitory reduction (three investigations). In percolation experiments, the total two-stage nitrifying conversion is inhibited when the 2,4-D concentration in the percolate reaches 50 ppm, but after some time the nitrification is resumed.[189] With 2,4-D mixed in the soil, the ammonifying organisms are not inhibited at concentrations more than twice those that inhibit the nitrifying organisms.[111] At a 2,4-D concentration of 100 ppm in the soil, nitrification is about 10% decreased, but ammonification is somewhat stimulated.[144] It takes over 500 ppm in the soil to inhibit *Azotobacter,* and in none of the 11 investigations reviewed by Fletcher (1960) were these nitrogen-fixing bacteria suppressed by field applications of 2,4-D; in fact they were stimulated in three of them. The various species of *Rhizobium* that inhabit the root nodules of legumes are not affected by operational dosages of 2,4-D,[61] but the formation of root nodules in legumes such as *Phaseolus vulgaris* is inhibited at dosages as low as 0.01 lb/a.[169]

At unusually high application rates[176] or in soil percolates,[100] 2,4-D may increase the number of bacteria due to a suppressive effect on soil protozoa. Application of 2,4-D to potato plants in unsterilized soil led to the appearance of an unidentifiable actinomycete with fungistatic properties.[214] It has been reported that normal application rates of 2,4-D[4] and above-normal application rates of 2,4-D and 2,4,5-T[164] enhance the growth of fungi in soil. *Aspergillus niger* is one of the species most resistant to 2,4-D and is benefited by its presence,[33] although in culture solution this species is inhibited at a 2,4-D concentration one-fifth that of a concentration of picloram that was harmless.[7] *Trichoderma viride* and a species of *Penicillium* are also resistant to 2,4-D, and to MCPA.[137]

There are three investigations in which MCPA applications were found not to suppress soil microorganisms, and six investigations in which normal rates of 2,4,5-T were without adverse effect.[61] Among other phenoxy compounds, 2,4-DP and 2,4-DB did not reduce the microorganismal populations,[216] while silvex did not affect *Streptomyces* even at 50 lb/a[29] and did not reduce nitrification at field rates.[80]

s-Triazines. Simazine, atrazine, propazine, prometone, and other aminotriazines applied at normal field rates did not reduce total microfloral activity as judged from the CO_2 production by the treated soil.[25,78,172] Soil that has been treated with simazine continues its normal CO_2 production, and yields vigorous isolates of *Streptomyces* actinomycetes and *Aspergillus* and *Penicillium* fungi.[32] Applications of simazine at 2–3 kg/ha were without effect on the total numbers of bacteria or of fungi, although the actinomycetes were somewhat reduced.[199] Sometimes an acceleration of nitrification is observed with simazine and atrazine (Balicka, 1959[170]; Tsvetkova, 1966[112]) and an increase in the numbers of the nitrogen-fixing *Azotobacter.* Soil applications of simazine at 6 kg/ha also stimulate both *Nitrosococcus* and *Nitrosomonas.*[78,199] When simazine is percolated through soil at 6 ppm, it inhibits the growth of *Nitrobacter* but not of *Nitrosomonas,*[57] and a similar result was obtained with an Indian soil treated with 5 ppm simazine.[163] This may present a problem, since *Nitrobacter* is necessary to oxidize (to nitrate) the enhanced nitrite production resulting from the stimulation of the ammonifying and nitrifying species just mentioned.[89] Applications of simazine favored those soil bacteria that were able to degrade it, such as *Empedobacter, Achromobacter, Microbacterium, Bacterium*[37] and *Arthrobacter.*[112] Soil fungi also are unaffected in total numbers by normal application rates of simazine,[54] although there may be shifts in the relative proportions of species.[78]

Atrazine in laboratory experiments did not inhibit *Arthrobacter, Corynebacterium,* and *Pseudomonas flavescens,* but it did reduce the initial growth rate of *Bacillus subtilis* colonies.[193] On fungi, atrazine inhibited the pathogenic *Sclerotium rolfsii* at 20 ppm, but stimulated the growth of *Trichoderma viride* at levels as high as 80 ppm. Other laboratory experiments have confirmed the stimulatory effect of atrazine on *T. viride*;[41,177] although field application at 4 lb/a actually reduced the numbers of useful inhibitor of root-rot fungi,[89] the net result is a suppression of the pathogenic *S. rolfsii* which is so much more susceptible to atrazine.[180] Of 29 species of fungi tested, *Epicoccum nigrum* was the only species to be stimulated at all concentrations ranging up to 140 ppm.[177] But it should be noted that *Fusarium roseum* grew more rapidly than normal in 10 ppm atrazine.[41] Moreover, *F. oxysporum* joins *T. lignorum* and *Penicillium* as the fungi known to be the most resistant to triazines as well as PCP.[209]

Prometryn and ametryn were triazines that inhibited *Nitrobacter* but not *Nitrosomonas,* at the unduly high concentration of 100 ppm in the soil.[50] One triazine is sufficiently toxic to pathogenic fungi to be marketed as a fungicide for use *inter alia* on turf and onions; this anilino–triazine, called anilazine (Dyrene), is readily broken down by soil bacteria,[89] but is more toxic than any of the triazine or other herbicides to bluegill sunfish.[170]

Urea Compounds.   Monuron and diuron, applied as selective herbicides at 0.5–2 lb/a, have shown no adverse effect on the soil microflora in practice,[61,93] although in soil percolation experiments the considerably water-soluble monuron has shown itself more active than other urea herbicides and triazines in reducing the numbers of *Nitrosomonas* and *Azotobacter*, and particularly *Nitrobacter*.[34] Diuron resembled simazine in the effect that 5 ppm admixed into Oregon soils had on the CO_2 production, which was to reduce it by about one-quarter in the first month, with recovery starting in the second month.[36] It resembled simazine in being nontoxic and sometimes growth-promoting for species of fungi such as *Trichoderma viride* and *Penicillium*[72] which, being antagonistic to the growth of pathogenic *Fusarium* species, are thus helpful against fungal infections in the soil.[114] Monuron proved to be a strong inhibitor of nitrifying bacteria at high concentrations in soil percolates,[174] but when perfused at 5 ppm neither monuron nor diuron or linuron decreased the biological oxidation of NH_4 to NO_3.[40] Since field applications at 1–5 lb/a have suppressed soil nitrifiers,[167] it is evident that the 50 lb/a employed when these herbicides are used in soil sterilants should severely suppress nitrification. At these concentrations the DMU (dichlorophenyl-methylurea) produced by diuron would inhibit *Nitrobacter*, while the further metabolite 3,4-DCA (dichloroaniline) would inhibit *Nitrosomonas* (Fig. 13.10). Some of the urea herbicides (e.g., diuron, linuron) are quite potent inhibitors of denitrifying bacteria while others (e.g., monuron, fenuron) are not; DCA and the 3- and 4-chloroanilines, produced as metabolites of urea, amide, and carbamate herbicides, are also inhibitory.[23]

Propanil and its immediate metabolite 3,4-DCA is an inhibitor of *Nitrosomonas* and not *Nitrobacter*.[40] It is the most active in this regard of a number of herbicides in the following series: propanil > picloram > paraquat > dichlobenil, the reduction in nitrate not being accompanied by an accumulation of nitrate.[48]

Carbamates and Thiocarbamates. Of these compounds, the most herbicidal such as propham (IPC) tend to be the most inhibitory for soil nitrification.[174] Thus whereas propham at field rates produces a transitory suppression of CO_2 production and of *Azotobacter*,[12] it is the nitrification

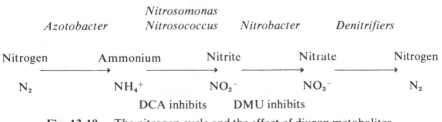

Fig. 13.10. The nitrogen cycle and the effect of diuron metabolites.

process which is much more susceptible.[61] Chlorpropham (CIPC), a weaker nitrification inhibitor in soil percolates,[80] definitely suppressed ammonifying and nitrifying bacteria at excessive dosages of 12–15 lb/a[70]; at the normal application rate of 2–8 lb/a chlorpropham either showed some suppression,[167] particularly of *Nitrobacter,* or none at all (Balicka, 1969[170]). While tending to increase *Azotobacter,* chlorpropham reduced cellulose decomposition in the soil. The thiocarbamate herbicide EPTC similarly did not depress nitrification at the normal rates of 2–5 lb/a, but did impair cellulose decomposition. EPTC resembled diuron in inhibiting CO_2 production by about one-quarter when added to Oregon soils at 5 ppm[36]; it also reduced the numbers of bacteria in alfalfa plots for 2 months[175] and suppressed nitrite oxidation by *Nitrobacter.*[219] Triallate was inhibitory to some species of soil bacteria, particularly the gram-positive species.[46] The thiolcarbamate pebulate (Tillam) resembled atrazine in not inhibiting *Arthrobacter* or *Pseudomonas,* but in suppressing the initial multiplication of *B. subtilis.*[193]

Bipyridyliums.   Paraquat applied to Oregon soils caused a short-term reduction in the bacterial and fungal populations.[205] Both diquat[217] and paraquat[218] differ from other herbicides in being more toxic to *Trichoderma viride* and *Rhizopus stolonifer* than to *Fusarium culmorum* and *Aspergillus niger.* Thus applications to potatoes favor infestations of *F. culmorum* rather than *T. viride* to infect haulms, and these bipyridyls encourage *A. niger* rather than *R. stolonifer* to colonize wheat chaff.

Selective Organics.   Picloram, applied against deep-rooted plants, did not affect either microfloral numbers nor nitrification even with soil treatments at 100 ppm,[75] although it inhibited nitrification when added to tropical soils at this level.[50] At 1 ppm in Oregon soils, all picloram really did was to increase the numbers of *Trichoderma* and *Aspergillus* fungi at the expense of *Mucor* and *Penicillium.*[207] Dalapon has seriously reduced bacterial numbers in alfalfa plots[175]; the nitrifying bacteria are the most susceptible, but recover from the suppression within 3 wk.[224] Moreover, dalapon has been applied for grass control without effect on total bacterial numbers at dosages up to 40 lb/a, and actually stimulated them at 70 lb/a.[224] The nitrogen-fixing *Azotobacter* were particularly resistant to dalapon,[151] although *A. chroococcum* was reduced by applications in Europe.[179]

Amitrole, though not affecting *Azotobacter,* is one of the more inhibitory herbicides as regards nitrification and even general CO_2 production[181]; moreover, it inhibits the proliferation of the microorganisms which degrade dalapon.[115] At the usual application levels of 2–4 lb/a, amitrole caused a 2-wk suppression of nitrification,[167] and the nitrate levels in prairie brown soils had not returned to normal until 1 yr after the application.[35] Endothall at the practical level of 1–8 lb/a did not reduce the general microfloral CO_2 production,[140] and in soil perfusion it stimulated the biological oxidation of NH_4 to NO_3.[48]

Maleic hydrazide, when applied as a suppressant of plant growth at 300 lb/a, is entirely inactive against *Azotobacter* and other soil bacteria[165]; even when the soil contains 1000 ppm MH there is only slight suppression of the microflora. The phosphoroamidothioate herbicide DMPA had no effect on *Streptomyces, Thiobacillus, Rhizobium,* and *Azotobacter* at soil concentrations many times that which occur in practice; soil treated for seven successive years at 15 lb/a had about half as many bacteria, and two-thirds as many actinomycetes and fungi, as untreated soil.[60] Trifluralin reduces the populations of nitrifiers in the soil, but does not affect *Rhizobium*.[51,82]

Nonselective Organics. TCA, used as a soil sterilant particularly against grasses at 10–100 lb/a, not surprisingly reduces soil nitrification[167] and general microorganismal activity[139,140]; what is surprising is that no irreparable change has been produced by these heavy applications.[61] Indeed, high rates of TCA have often been observed to have no deleterious effect on bacteria, actinomycetes and fungi.[96] The general contact herbicide dinoseb (DNBP) is fairly toxic to soil bacteria and to fungi (especially *Sclerotium rolfsii*), even if used at a selective dosage of 3 lb/a. On the other hand, the related dinitro compound DNOC, which has been withdrawn from use as a herbicide in the United States, reduced the microflora only temporarily or not at all when applied at 4–10 lb/a[61]; moreover, all species of *Rhizobium* tested were resistant to it.[81] PCP (pentachlorophenol), generally applied as the sodium salt at 4–16 lb/a, is usually characterized by an immediate depressant effect on the microflora, followed by a stimulation[140] associated with those bacterial species that stain with methylene blue.[61] PCP also inhibits nitrogen fixation, ammonification, and nitrification, but the bacteria responsible may recover their numbers as they begin to break down this rather biostable molecule.[12] Pentachlorophenol has a beneficial effect in that, like dinoseb and picloram, it suppresses the pathogenic fungi *Sclerotium rolfsii, S. bataticola,* and *Rhizoctonia*.[13]

Inorganics. Sodium arsenite, applied at 100–200 lb/a as a soil sterilant on industrial sites, has proved to be comparatively nontoxic to nitrifying organisms.[142,174] It is now largely replaced by the arsenical sodium cacodylate as a nonselective herbicide. Sodium chlorate, the nonselective herbicide most widely used as a semi-permanent sterilant at 50–400 lb/a, does not affect the total microfloral activity even at the top dosage.[166] Laboratory experiments prove that high concentrations of sodium chlorate can inhibit the ammonium–nitrite conversion[191] and particularly the nitrite–nitrate oxidation.[143] The lower dosage rates cause a temporary reduction in nitrification, and with higher dosage rates this reduction is not repaired until the next growing season.[142,166]

Effects on Algae

The action of organic herbicides on the soil algae has been scarcely studied, but it is already evident that the compounds differ widely in their effect and the algal species differ widely in their response. At concentrations in liquid culture ranging from 1 to 10 ppm, the effect was often inhibitory ($\times$), and sometimes stimulatory ($+$) or lacking ($-$), as shown by the following results of one investigation[146]:

		Atrazine	Metobromuron	Diphenamid
Chlamydomonas	*reinhardii*	xx	xx	+
Chlam.	*eugametos*	–	xx	–
Chlorella	*vulgaris*	x	+	x
Chlor.	*pyrenoidosa*	x	x	–

The inhibitory effect of the urea herbicide, as well as of atrazine, is not unexpected since both are inhibitors of photosynthesis. Amitrole also markedly suppresses algae in soil, but they recover their numbers in about 3 months.[135]

REFERENCES CITED

1. E. K. Akamine. 1951. Persistence of 2,4-D toxicity in Hawaiian soils. *Botan. Gaz.* **112**:312–319.

2. M. Alexander. 1969. Microbial degradation and biological effects of pesticides in soil. In *Soil Biology: Reviews of Research.* UNESCO, Paris (Place de Fontenoy 75, Paris-7e). pp. 209–240.

3. O. M. Aly and S. D. Faust. 1964. Studies on the fate of 2,4-D and ester derivatives in natural surface water. *J. Agric. Food Chem.* **12**:541–546.

4. G. R. Anderson and C. O. Baker. 1950. Some effects of 2,4-D in representative Idaho soils. *Agron. J.* **42**:456–458.

5. J. P. E. Anderson and K. H. Domsch. 1976. Microbial degradation of diallate in soils and by pure cultures of soil microorganisms. *Arch. Environ. Contam. Toxicol.* **4**:1–7.

6. D. E. Armstrong, G. Chesters, and R. F. Harris. 1967. Atrazine hydrolysis in soil. *Soil Sci. Soc. Am., Proc.* **31**:61–66.

7. W. R. Arnold, P. W. Santelmann, and J. Q. Lynd. 1966. Picloram and 2,4-D effects with *Aspergillus niger* proliferation. *Weeds* **14**:89–90.

8. L. J. Audus. 1951. The biological detoxication of hormone herbicides in soil. *Plant and Soil* **3**:170–192.

9. L. J. Audus. 1951. The biological detoxication of 2,4-D in soils; isolation of an effective organism. *Nature* **166**:356.

10. L. J. Audus. 1952. The decomposition of 2,4-D and MCPA in the soil. *J. Sci. Food Agric.* **3**:268–274.

11. L. J. Audus. 1960. Microbiological breakdown of herbicides in soils. In *Herbicides and the Soil*. Eds. E. K. Woodford and G. R. Sagar. Blackwell Scientific Publications, Oxford. pp. 1–9.

12. L. J. Audus. 1970. The action of herbicides on the microflora of the soil. *Proc. 10th Br. Weed Control Conf.* **3**:1036–1051.

13. D. C. Bain. 1961. Effect of various herbicides on some soil fungi in cultures. *Plant Disease Rep.* **45**:814–817.

14. B. C. Baldwin, M. F. Bray, and M. J. Geoghegan. 1966. The microbial decomposition of paraquat. *Biochem J.* **101**:15P.

15. R. Bartha and A. Pramer. 1967. Pesticide transformation to aniline and azo compounds in soil. *Science* **156**:1617–1618.

16. R. Bartha and A. Pramer. 1969. Transformation of the herbicide swep in soil. *Bull. Environ. Contam. Toxicol.* **4**:240–245.

17. R. Bartha, H. A. B. Linke, and A. Pramer. 1968. Pesticide transformations: Production of chloroazobenzenes from chloroanilines. *Science* **161**:582–583.

18. R. Bartha, H. A. B. Linke, and A. Pramer. 1969. Umwandlung von Unkrautbekampfungsmitteln zu Azoverbindungen durch Bodenmikroorganismen. *Umschau* **69**:182–183.

19. M. L. Beall, P. C. Kearney, and D. D. Kaufman. 1964. Comparative metabolism of 1-C^{14} and 2-C^{14} labeled dalapon in soil. *Weed Sci. Soc. Am. Abstr.* Feb. 10–13, pp. 12–13.

20. I. J. Belasco and H. L. Pease. 1969. Investigation of diuron- and linuron-treated soils for 3,3′, 4,4′-tetrachloroazobenzene. *J. Agric. Food Chem.* **17**:1414–1417.

21. G. R. Bell. 1957. Some morphological and biochemical characteristics of a soil bacterium which decomposes 2,4-dichlorophenoxyacetic acid. *Can. J. Microbiol.* **3**:821–840.

22. J. A. Best and J. B. Weber. 1974. Disappearance of *s*-triazines as affected by soil pH using a balance-sheet approach. *Weed. Sci.* **22**:364–373.

23. J. M. Bollag and C. L. Nash. 1974. Effect of chemical structure of phenylureas and anilines on the denitrification process. *Bull. Environ. Contam. Toxicol.* **12**:241–248.

24. J. M. Bollag, C. S. Helling, and M. Alexander. 1967. Metabolism of MCPA by soil bacteria. *Appl. Microbiol.* **15**:1393–1398.

25. W. B. Bollen. 1961. Interactions between pesticides and soil microorganisms. *Annu. Rev. Microbiol.* **15**:69–92.

26. L. M. Bordeleau and R. Bartha. 1971. Ecology of herbicide transformation: Synergism of two soil fungi. *Soil Biol. Chem.* **3**:281–284.

27. H. Borner, H. Burgemeister, and M. Schroeder. 1969. Untersuchungen uber Aufnahme, Verteilung, und Abbau von Harnstoffherbiziden durch Kulturpflanzen, Unkrauter, and Mikroorganismen. *Z. Pflanzenkrankh. Pflanzenschutz* **76**:385–395.

28. H. Bortels, E. Fricke, and R. Schneider. 1967. Simazinversetzung durch Mikroorganismen verschiedener Boden. *Nachrichenbl. Deutsch. Pflanzenschutzd.* **19**:101–105.

29. H. C. Bounds and A. R. Colmer. 1964. Resistance of *Streptomyces* to herbicides. *Sugar Bull.* **42**:274–276.

30. H. C. Bounds and A. R. Colmer. 1965. Detoxification of some herbicides by *Streptomyces*. *Weeds* **13**:249–252.

31. W. D. Burge. 1972. Microbial populations hydrolyzing propanil and accumulation of DCA and TCAB in soils. *Soil Biol. Biochem.* **4**:379–386.

32. O. C. Burnside, E. L. Schmidt, and R. Behrens. 1961. Dissipation of simazine from the soil. *Weeds* **9**:477–484.

33. L. Calancea and G. Illyes. 1954. The effect of 2,4-D on the production of citric acid by *Aspergillus niger*. See *Chem. Abstr.* **52,** item 12068.

34. J. C. Caseley and L. C. Luckwill. 1965. The effect of some residual herbicides on soil nitrifying bacteria. *Long Ashton Agric. Hort. Res. Stn., Univ. Bristol, Ann. Rep. 1964.* pp. 78–86.

35. P. Chandra. 1964. Herbicidal effects on certain microbial activities in some brown soils of Saskatchewan. *Weed Res.* **4**:54–63.

36. P. Chandra, W. R. Furtick, and W. B. Bollen. 1960. The effects of four herbicides on microorganisms in nine Oregon soils. *Weeds* **8**:589–598.

37. M. Charpentier and J. Pochon. 1962. Bacteries telluriques cultivant sur amino-triazine (simazine). *Ann. Inst. Pasteur, Paris* **102**:501.

38. J. P. Chu and E. J. Kirsch. 1972. Metabolism of pentachlorophenol by an axenic bacterial culture. *Ann. Appl. Microbiol.* **23**:1033–1035.

39. G. C. Clark and S. J. L. Wright. 1970. Degradation of the herbicide isopropyl-N-phenylcarbamate by *Arthrobacter* and *Achromobacter* spp. from soil. *Soil Biol. Biochem.* **2**:217–226.

40. C. T. Corke and F. R. Thompson. 1970. Effects of some phenylamide herbicides and their degradation products on soil nitrification. *Can. J. Microbiol.* **16**:567–571.

41. R. W. Couch, J. V. Gramlich, D. E. Davis, and H. H. Funderburk. 1965. The metabolism of atrazine and simazine by soil fungi. *Southern Weed Conf., Proc.* **18**:623–631.

42. D. P. Cox and M. Alexander. 1973. Production of trimethylarsine gas from various arsenic compounds by three sewage fungi. *Bull. Environ. Contam. Toxicol.* **9**:84–88.

43. D. G. Crosby. 1973. The fate of pesticides in the environment. *Annu. Rev. Plant Physiol.* **24**:467–492.

44. A. J. Cserjesi and E. L. Johnson. 1972. Methylation of pentachlorophenol by *Trichoderma virgatum*. *Can. J. Microbiol.* **18**:45–49.

45. D. R. Cullimore. 1971. Interaction between herbicides and soil microorganisms. *Residue Rev.* **35**:65–80.

46. D. R. Cullimore and A. E. Smith. 1972. Initial studies on the breakdown of triallate. *Bull. Environ. Contam. Toxicol.* **7**:36–42.

47. R. L. Dalton, A. W. Evans, and R. C. Rhodes. 1966. Disappearance of diuron from cotton field soils. *Weeds* **14**:31–33.

48. A. C. Debona and L. J. Audus. 1970. Studies on the effects of herbicides on soil nitrification. *Weed Res.* **10**:250–263.

49. N. Drescher and T. F. Burger. 1970. Microbiological dephenylation of pyrazon in soil. *Bull. Environ. Contam. Toxicol.* **5**:79–84.

50. H. D. Dubey. 1969. Effect of picloram, diuron, ametryne, and prometryne on nitrification in some tropical soils. *Soil Sci. Soc. Am. Proc.* **33**:893–896.

51. E. P. Dunigan, J. P. Frey, L. D. Allen, and A. McMahon. 1972. Herbicidal effects on the nodulation of *Glycine max. Agron J.* **64**:806–808.

52. J. M. Duxbury, J. M. Tiedje, M. Alexander, and J. E. Dawson. 1970. 2,4-D metabolism: Enzymatic conversion of chloromaleylacetic acid to succinic acid. *J. Agric. Food Chem.* **18**:199–201.

53. G. Engelhardt, P. R. Wallnofer, and R. Plapp. 1972. Identification of dimethylhydroxylamine as a microbial degradation product of linuron. *Appl. Microbiol.* **23**:664–666.

54. C. F. Eno. 1962. The effect of simazine and atrazine on certain of the soil microflora and their metabolic processes. *Soil Crop. Sci. Soc. Fla., Proc.* **22**:49–56.

55. W. C. Evans, J. K. Gaunt, and J. J. Davies. 1961. The metabolism of chlorophenoxyacetic acid herbicides by soil microorganisms. *Proc. 5th Internat. Congr. Biochem., Moscow.* pp. 306–307.

56. S. C. Fang. 1969. Thiolcarbamates. In *Degradation of Herbicides.* Eds. P. C. Kearney and D. D. Kaufman. Dekker. pp. 147–164.

57. F. H. Farmer, R. E. Benoit, and W. E. Chappell. 1965. Simazine, its effect on nitrification and its decomposition by soil organisms. *N. E. Weed Contr., Proc.* **19**:350–354.

58. J. K. Faulkner and D. Woodcock. 1965. Metabolism of 2,4-D and MCPA by *Aspergillus niger. J. Chem. Soc.* **1965**:1187–1191.

59. H. N. Fernley and W. C. Evans. 1959. Metabolism of 2,4 dichlorophenoxyacetic acid by soil *Pseudomonas. Biochem. J.* **73**:22P.

60. M. L. Fields and D. D. Hemphill. 1966. Effect of Zytron and its degradation products on soil microorganisms. *Appl. Microbiol.* **14**:724–731.

61. W. W. Fletcher. 1960. The effect of herbicides on soil microorganisms. In *Herbicides and the Soil.* Eds. E. K. Woodford and G. R. Sagar. Blackwell Scientific Publications, Oxford. pp. 20–62.

62. O. Flieg. 1952. Uber das Verhalten von 2,4-D in Boden hinsichtlich mikrobieller Wirkungen, Beweglichkeit, und Abbau. *Mitt. Biol. Zentralanst. Berlin* **74**:133–135.

63. C. L. Foy. 1961. Penetration and translocation of Cl^{36} and C^{14} labeled dalapon. *Plant Physiol.* **36**:688–697.

64. H. H. Funderburk. 1969. Diquat and paraquat. In *Degradation of Herbicides.* Eds. P. C. Kearney and D. D. Kaufman. Dekker. pp. 282–298.

65. H. H. Funderburk and G. A. Bozarth. 1967. Review of the metabolism and decomposition of diquat and paraquat. *J. Agric. Food Chem.* **15**:563–567.

66. H. H. Funderburk, D. P. Schultz, N. S. Negi, R. Rodriguez-Kabana, and E. A. Curl. 1967. Metabolism of trifluralin by soil microorganisms and higher plants. *Southern Weed Contr. Conf., Proc.* **20**:389.

67. J. A. Gardiner, R. C. Rhodes, J. B. Adams, and E. J. Sobodzenski. 1969. Synthesis and studies with labeled bromacil and terbacil. *J. Agric. Food Chem.* **17**:980–986.

68. J. K. Gaunt and W. C. Evans. 1971. Metabolism of 4-chloro-2-methylphenoxyacetate by a soil Pseudomonad. *Biochem. J.* **122**:519–526, 533–542.

69. H. Geissbuhler. 1969. The substituted ureas. In *Degradation of Herbicides*. Eds. P. C. Kearney and D. D. Kaufman. Dekker. pp. 1–50.

70. I. A. Geller and E. G. Khariton. 1961. The effect of herbicides on soil microflora. *Microbiology* **30**:494–499.

71. C. G. Gemell and H. L. Jensen. 1964. Some studies on trichloroacetate-decomposing soil bacteria. *Arch. Microbiol.* **48**:386–392.

72. V. D. Goguadze. 1968. The effect of herbicides on the soil microflora. *Khimiya sel'Khoz.* **4**(9):41–45. (*Weed Abstr.* **17**:451.)

73. T. Golab, J. V. Gramlich, and G. W. Probst. 1968. Studies on the fate of diphenamid in soils. *Abstr. 155th Mtg. Am. Chem. Soc.* A-50.

74. C. A. I. Goring. 1970. Nutrient cycling implications of pesticides. In *Pesticides in the Soil*. International Symposium, Mich. State Univ., East Lansing. pp. 51–57.

75. C. A. I. Goring, J. D. Griffith, F. C. O'Mella, H. H. Scott, and C. R. Youngson. 1967. The effect of Tordon on microorganisms and soil biological processes. *Down to Earth* **22**(4):14–17.

76. R. A. Gray. 1971. Behavior, persistence, and degradation of carbamate and thiocarbamate pesticides in the environment. *Proc. Calif. Weed Contr. Conf.* **1971**:128–141.

77. J. Guillemat. 1960. Interactions entre la simazine et la mycoflore du sol. *Compt. Rend. Acad. Sci. (Paris)* **250**:1343.

78. J. Guillemat, M. Charpentier, P. Tardieux, and J. Pochon. 1960. Interactions entre une chloro-amino-triazine herbicide et la microflore fongique et bacterienne du sol. *Ann. Epiphyt* **11**:261–290.

79. K. Gundersen and H. L. Jensen. 1956. A soil bacterium decomposing organic nitro-compounds. *Acta. Agric. Scand.* **6**:100–114.

80. M. G. Hale, F. H. Hulcher, and W. E. Chappell. 1957. The effect of several herbicides on nitrification in a field soil under laboratory conditions. *Weeds* **5**:331–341.

81. J. W. Hamaker. 1972. Decomposition: Quantitative aspects. In *Organic Chemicals in the Soil Environment*. Eds. C. A. I. Goring and J. W. Hamaker. Dekker. pp. 255–340.

82. Y. A. Hamdi and M. S. Tewfik. 1969. Effect of trifluralin on nitrogen fixation in *Rhizobium* and *Azotobacter* and on nitrification. *Acta Microbiol. Polon. (Ser. b)* **1**:53–57.

83. Y. A. Hamdi and M. S. Tewfik. 1970. Degradation of 3,5-dinitro-*o*-cresol by *Rhizobium* and *Azotobacter* spp. *Soil Biol. Biochem.* **2**:163–166.

84. R. J. Hance. 1967. Decomposition of herbicides in the soil by non-biological chemical processes. *J. Sci. Food Agric.* **18**:544–547.

85. R. J. Hance and G. Chesters. 1969. The fate of hydroxyatrazine in a soil and a lake sediment. *Soil Biol. Biochem.* **1**:309–315.

86. R. S. Hargrove and M. G. Merkle. 1971. The loss of alachlor from soil. *Weed Sci.* **19**:652–654.

87. C. I. Harris. 1965. Hydroxysimazine in soil. *Weed Res.* **5**:275–276.

88. C. I. Harris. 1967. Fate of 2-chloro-*s*-triazines in soil. *J. Agric. Food Chem.* **15**:157–162.

89. C. I. Harris, D. D. Kaufman, T. J. Sheets, R. G. Nash, and P. C. Kearney. 1968. Behavior and fate of *s*-triazines in soils. *Adv. Pest Contr. Res.* **8**:1–55.

90. C. S. Helling. 1976. Dinitroaniline herbicides in soils. *J. Environ. Qual.* **5**:1–14.

91. C. S. Helling, P. C. Kearney and M. Alexander. 1971. Behavior of pesticides in soils. *Adv. Agron.* **23**:147–240.

92. G. D. Hill and J. W. McGahen. 1955. Further studies on soil relationships of the substituted urea herbicides for pre-emergence weed control. *Southern Weed Sci. Soc. Proc.* **8**:284–293.

93. G. D. Hill, J. W. McGahen, H. M. Baker, D. W. Finnerty, and C. W. Bingeman. 1955. The fate of substituted urea herbicides in agricultural soils. *Agron. J.* **47**:93–104.

94. P. Hirsch and M. Alexander. 1960. Microbial decomposition of halogenated propionic and acetic acids. *Can. J. Microbiol.* **6**:241–249.

95. P. Hirsch and R. Stellmach-Helwig. 1962. On the question of the breakdown of dalapon and TCA by rumen and soil bacteria. *Weed Abstr.* **11**:107. See also *Zentr. Bacteriol. Parasitenk. Infektionskr. Hyg.* **114**:683–686.

96. M. E. Hoover and A. R. Colmer. 1953. The action of some herbicides on the microflora of a sugarcane soil. *Proc. Louisiana Acad. Sci.* **16**:21–27.

97. R. S. Horvath. 1971. Cometabolism of the herbicide 2,3,6-trichlorobenzoate. *J. Agric. Food Chem.* **19**:291–293.

98. R. S. Horvath and M. Alexander. 1970. Cometabolism, a technique for the accumulation of biological products. *Can. J. Microbiol.* **15**:1131–1132.

99. A. Ide, Y. Niki, F. Sakamoto, I. Watanabe, and H. Watanabe. 1972. Decomposition of pentachlorophenol in paddy soil. *Agric. Biol. Chem.* **36**:1937–1944.

100. A. J. Ilijin. 1962. Concerning the effect of the herbicide 2,4-D on soil microorganisms. *Microbiology* **30**:1050–1051.

100. A. M. Ilijin. 1962. Concerning the effect of the herbicide 2,4-D on soil microorganisms. Distribution of alkyl arsenicals in model ecosystems. *Environ. Sci. Technol.* **7**:841–845.

102. H. L. Jensen. 1957. Decomposition of chloro-organic acids by fungi. *Nature (Lond.)* **180**:1416.

103. H. L. Jensen. 1957. Decomposition of chloro-substituted aliphatic acids by soil bacteria. *Can. J. Microbiol.* **3**:151–164.

104. H. L. Jensen. 1959. Allyl alcohol as a nutrient for microorganisms. *Nature (Lond.)* **183**:903.

105. H. L. Jensen. 1960. Decomposition of chloroacetates and chloropropionates by bacteria. *Acta Agric. Scand.* **10**:83–103.

106. H. L. Jensen. 1961. Biologisk sonderdeling af ufkrudsmidler i jordbunden. II: Allylalkohol. *Tidsskr. Planteavl.* **65**:185–198.

107. H. L. Jensen. 1961. Some aspects of biological allyl alcohol dissimilation. *Acta Agric. Scand.* **11**:54–62.

108. H. L. Jensen. 1964. Studies on soil bacteria (*Arthrobacter globiformis*) capable of decomposing the herbicide endothal. *Acta Agric. Scand.* **14**:193–207.

109. S. T. Jensen and S. Larsen. 1957. Reduktion og udvaskning af chlorat i jordbunden. *Tidsskr. Planteavl.* **61**:103–118.

110. H. L. Jensen and H. I. Petersen. 1952. Decomposition of hormone herbicides by bacteria. *Acta Agric. Scand.* **2**:215–231.

111. L. W. Jones. 1956. *Effect of Some Pesticides on Microbial Activities of the Soil.* Utah Agric. Exp. Stn. Bull. 390. 17 pp.

112. P. Kaiser, J. J. Pochon, and R. Cassini. 1970. Influence of triazines on soil microorganisms. *Residue Rev.* **32**:211–233.

113. D. D. Kaufman. 1964. Microbial degradation of 2,2-dichloropropionic acid in five soils. *Can. J. Microbiol.* **10**:843–852.

114. D. D. Kaufman. 1964. Effect of triazine and phenylurea herbicides on soil fungi and soybean-cropped soil. *Phytopathology* **54**:897.

115. D. D. Kaufman. 1966. Microbial degradation of herbicide combinations: Amitrole and dalapon. *Weeds* **14**:130–134.

116. D. D. Kaufman. 1967. Degradation of carbamate herbicides in soil. *J. Agric. Food Chem.* **15**:582–591.

117. D. D. Kaufman. 1974. Degradation of pesticides by soil microorganisms. In *Pesticides in Soil and Water.* Ed. W. D. Guenzi. Soil Sci. Soc. Am., Madison, Wis. pp. 133–202.

118. D. D. Kaufman and J. Blake. 1970. Degradation of atrazine by soil fungi. *Soil Biol. Biochem.* **2**:73–80.

119. D. D. Kaufman and P. C. Kearney. 1965. Microbial degradation of isopropyl-N-(3-chlorophenyl)-carbamate. *Appl. Microbiol.* **13**:443–446.

120. D. D. Kaufman and P. C. Kearney. 1970. Microbial degradation of s-triazine herbicides. *Residue Rev.* **32**:235–265.

121. D. D. Kaufman and J. R. Plimmer. 1971. Effect of chemical structure on biodegradation of s-triazine herbicides. *Weed Sci. Soc. Am. Abstr. 1971,* No. 35, p. 18.

122. D. D. Kaufman, P. C. Kearney, and T. J. Sheets. 1965. Microbial degradation of simazine. *J. Agric. Food Chem.* **13**:238–242.

123. D. D. Kaufman, J. R. Plimmer, and U. I. Klingebiel. 1973. Microbial oxidation of 4-chloraniline. *J. Agric. Food Chem.* **21**:127–132.

124. D. D. Kaufman, J. R. Plimmer, P. C. Kearney, J. Blake, and F. S. Guardia. 1968. Chemical versus microbial decomposition of amitrole in soil. *Weed Sci.* **16**:266–272.

125. P. C. Kearney. 1966. Metabolism of herbicides in soil. In *Organic Pesticides in the Environment.* Am. Chem. Soc., Adv. Chem. Ser. 60. pp. 250–270.

126. P. C. Kearney and C. S. Helling. 1969. Reactions of pesticides in soils. *Residue Rev.* **25**:25–44.

127. P. C. Kearney and D. D. Kaufman. 1965. Enzyme from soil bacterium hydrolyzes phenyl carbamate herbicides. *Science* **147**:740–741.

128. P. C. Kearney and D. D. Kaufman (Eds.). 1969. *Degradation of Herbicides.* Dekker. 394 pp.

129. P. C. Kearney, D. D. Kaufman, and M. Alexander. 1967. Biochemistry of herbicide decomposition in soils. In *Soil Biochemistry.* Eds. A. D. McLaren and G. H. Peterson. Dekker. pp. 318–342.

130. P. C. Kearney, D. D. Kaufman, and M. L. Beall. 1964. Enzymic dehalogenation of 2,2-dichloropropionate. *Weed Soc. Am. Abstr.* Feb. 10–13, p. 13; *Biochem. Biophys. Res. Commun.* **14**:29–33.

131. P. C. Kearney, C. I. Harris, D. D. Kaufman, and T. J. Sheets. 1965. Behavior and fate of chlorinated aliphatic acids in soils. *Adv. Pest Control Res.* **6**:1–30.

132. P. C. Kearney, D. D. Kaufman, D. W. von Endt, and F. S. Guardia. 1969. TCA metabolism by soil microorganisms. *J. Agric. Food Chem.* **17**:581–583.

133. P. C. Kearney, R. J. Smith, J. R. Plimmer, and F. S. Guardia. 1970. Propanil and TCAB residues in rice soils. *Weed Sci.* **18**:464.

134. C. D. Kesner and S. K. Ries. 1967. Diphenamid metabolism in plants. *Science* **155**:210–211.

135. A. Kiss. 1966. Herbicides and soil algae. *Novenyyedolen* **2**:217–254. (*Weed Abstr.* **18**:976).

136. G. C. Klingman and F. M. Ashton. 1975. *Weed Science: Principles and Practice.* Wiley-Interscience. New York. 431 pp.

137. L. Klyuchnikov and A. M. Petrova. 1960. The effects of frequent use of herbicides on the soil microflora. *Mikrobiologiya* **29**:238–241. (*Soils Fertiliz.* **23**:1663).

138. E. Knuesli, D. Berrer, G. Dupuis and H. Esser. 1969. Triazines. In *Degradation of Herbicides.* Eds. P. C. Kearney and D. D. Kaufman. Dekker. pp. 51–78.

139. D. E. Kratochvil. 1950. Effect of several herbicides on soil microorganisms. *N. Centr. Weed Contr. Conf., Proc.* **7**:102–103.

140. D. E. Kratochvil. 1951. Determination of the effect of several herbicides on soil microorganisms. *Weeds* **1**:25–31.

141. T. L. Laanio, P. C. Kearney, and D. D. Kaufman. 1973. Microbial transformations of dinitramine. *Pestic. Biochem. Physiol.* **3**:271–277.

142. W. L. Latshaw and J. W. Zahnley. 1927. Experiments with sodium chlorate and other chemicals as herbicides for field bindweed. *J. Agric. Res.* **35**:757–767.

143. H. Lees and J. H. Quastel. 1945. Bacteriostatic effects of potassium chlorate on soil nitrification. *Nature (Lond.)* **155**:276–278.

144. G. Lenhard. 1959. The effects of 2,4-D on certain physiological aspects of soil microorganisms. *S. African J. Agric. Sci.* **2**:487–497.

145. H. A. B. Linke and R. Bartha. 1970. Transformation products of the herbicide propanil in soil. *Bacteriol Proc. 1970,* Abstr. A59, p. 9.

146. C. Loeppky and B. G. Tweedy. 1969. Effects of selected herbicides upon growth of soil algae. *Weed Sci.* **17**:110–113.

147. M. A. Loos. 1969. Phenoxyalkanoic acids. In *Degradation of Herbicides.* Eds. P. C. Kearney and D. D. Kaufman. Dekker. pp. 1–50.

148. M. A. Loos, J. M. Bollag, and M. Alexander. 1967. Phenoxyacetate herbicide detoxication by bacterial enzymes. *J. Agric. Food Chem.* **15**:858:860.

149. A. N. Macgregor. 1963. The decomposition of dichloropropionate by soil microorganisms. *J. Gen. Microbiol.* **30**:497–501.

150. I. C. MacRae and M. Alexander. 1965. Microbial degradation of herbicides in soil. *J. Agric. Food Chem.* **13**:72–76.

151. L. A. Magee and A. R. Colmer. 1955. Effect of some herbicides on respiration of *Azotobacter. Appl. Microbiol.* **3**:288–292.

152. B. C. McBride and R. S. Wolfe. 1971. Biosynthesis of dimethyl arsine by *Methanobacterium. Biochemistry* **10**:4312–4317.

153. G. W. McClure. 1972. Accelerated degration of phenylcarbamates in soil by a mixed suspension of IPC-adapted microorganisms. *J. Environ. Qual.* **1**:177–180.

154. G. W. McClure. 1974. Degradation of aniline herbicides by propham-adapted organisms. *Weed Sci.* **22**:323–329.

155. L. L. McCormick and A. E. Hiltbold. 1966. Microbiological decomposition of atrazine and diuron in soil. *Weed Sci.* **14**:77–82.

156. R. W. Meikle. 1972. Decomposition: qualitative relationships. In *Organic Chemicals in the Soil Environment.* Eds. C. A. I. Goring and J. W. Hamaker. Dekker. pp. 145–251.

157. R. W. Meikle, C. R. Youngson, R. T. Hedlund, C. A. I Goring, and W. W. Addington.

1974. Decomposition of picloram by soil microorganisms: A proposed reaction sequence. *Weed Sci.* **22**:263–268.

158. C. M. Menzie. 1969. *Metabolism of Pesticides.* U.S. Fish Wildl. Serv., Special Scient. Rep. Wildlife No. 127; *An Update.* 1974. Ibid. No. 184.

159. M. Mickovski and O. Verona. 1967. Decomposition of triazine herbicides by some soil fungi. *Agric. Ital. (Pisa)* **67**:67–76. (*Biochem. Abstr.* **67**:107564X).

160. D. S. Murray, W. L. Rieck, and J. Q. Lynd. 1969. Microbial degeneration of five substituted urea herbicides. *Weed Sci.* **17**:52–55.

161. K. N. Namideo. 1972. Degradation of paraquat dichloride. *Indian J. Exp. Biol.* **10**:133–135.

162. National Academy of Sciences. 1968. *Weed Control.* (Vol. 2 of *Plant and Animal Pest Control.*). Nat. Acad. Sci. (Washington) Publ. 1597. 471 pp.

163. V. K. Nayyar, N. S. Randhawa, and S. L. Chopra. 1970. Effect of simazine on nitrification and microbial populations in a sandy loam. *Indian J. Agric. Sci.* **40**:445–451.

164. A. S. Newman and C. E. Downing. 1958. Herbicides and the soil. *J. Agric. Food Chem.* **6**:352–353.

165. L. G. Nickell and A. R. English. 1953. Effect of maleic hydrazide on soil bacteria and other microorganisms. *Weeds* **2**:190–196.

166. P. E. Nilsson. 1951. The action of chlorate on some microbial phenomena in the soil. *Kungl. Lantbrukshogskolans Annal.* **18**:60–73.

167. R. J. Otten, J. E. Dawson, and M. M. Schreiber. 1957. The effects of several herbicides on nitrification in soil. *N.E. Weed Contr. Conf., Proc.* **11**:120–127.

168. D. A. Pate and H. H. Funderburk. 1965. Absorption, translocation, and metabolism of dichlobenil. *Southern Weed Contr. Conf., Proc.* **18**:605–606.

169. M. G. Payne and J. L. Fults. 1947. Some effects of 2,4-D, DDT and Colorado 9 on root nodulation in the common bean. *Am. Soc. Agron. J.* **39**:52–55.

170. D. Pimentel. 1971. *Ecological Effects of Pesticides on Non-target Species.* Office Sci. Technol., Exec. Office President, Gov't. Printing Office, Washington, D.C. 220 pp.

171. J. R. Plimmer, P. C. Kearney, and H. Chisaka. 1970. Transformations of prometryne. *Weed Sci. Soc. Am. Abstr. 1970,* No. 167, p. 87.

172. J. Pochon, P. Tardieux, and M. Charpentier. 1960. Recherches sur les interactions entre les aminotriazines herbicides et la microflora tellurique. *C. R. Acad. Sci.* **250**:1555–1556.

173. G. W. Probst and J. B. Tepe. 1969. Trifluralin and related compounds. In *Degradation of Herbicides.* Eds. P. C. Kearney and D. D. Kaufman. Dekker. pp. 255–282.

174. J. H. Quastel and P. G. Scholefield. 1953. Urethanes and soil nitrification. *Appl. Microbiol.* **1**:282–287.

175. A. A. Rakhimov and V. F. Rybina. 1963. Effect of some herbicides on soil microflora. *Uzbeksk. Biol. Zhur.* **7**:74–76 (*Biol. Abstr.* **45**, 83599).

176. E. H. Rapoport and G. Cangioli. 1963. Herbicides and the soil fauna. *Pedobiologia* **2**:235–238.

177. L. T. Richardson. 1970. Effects of atrazine on growth response of soil fungi. *Can. J. Plant Sci.* **50**:594–596.

178. W. F. Ritter, H. P. Johnson, W. G. Lovely, and M. Molnau. 1974. Atrazine, propachlor and diazinon residues on small agricultural watersheds. *Environ. Sci. Technol.* **8**:38–42.

179. E. L. Robinson. 1975. Arsenic in soil with five annual applications of MSMA. *Weed Sci.* **23**:341–343.

180. R. Rodriguez-Kabana, E. A. Curl, and H. H. Funderburk. 1968. Effect of atrazine on growth activity of *Sclerotium rolfsii* and *Trichoderma viride* in soil. *Can. J. Microbiol.* **14**:1283–1288.

181. D. A. van Schreven, D. J. Lindenbergh, and A. Koridon. 1970. Effects of several herbicides on bacterial populations and activity in soil. *Plant Soil* **33**:513–532.

182. M. Schroeder. 1970. The microbial decomposition of substituted urea herbicides. *Weed Sci. Soc. Am. Abstr.* No. 157. pp. 81–82. For identity of microorganisms, see Ph.D. Dissertation, *Mikrobieller Abbau von Harnstoffherbiziden.* Institute of Pathology, Christian-Albert University, Kiel, W. Germany: 1969.

183. N. E. Sharabi and L. M. Bordeleau. 1969. Biochemical decomposition of the herbicide Karsil and related compounds. *Appl. Microbiol.* **18**:369–375.

184. T. J. Sheets and A. S. Crafts. 1957. The phytotoxicity of four phenylurea herbicides in soils. *Weeds* **5**:93–101.

185. H. C. Sikka. 1972. Metabolism of endothall by aquatic microorganisms. Abstracts 164th ACS Mtg., Pesticides. In C. M. Menzie, *U.S. Fish Wildl. Serv., Spec. Scient. Rep.* No. 184. p. 193.

186. H. D. Skipper and V. V. Volk. 1972. Biological and chemical degradation of atrazine in three Oregon soils. *Weed Sci.* **20**:344–347.

187. H. D. Skipper, C. M. Gilmour, and W. R. Furtick. 1967. Microbial versus chemical degradation of atrazine in soils. *Soil Sci. Soc. Am. Proc.* **31**:653–656.

188. J. J. Skujins. 1967. Enzymes in soil. In *Soil Biochemistry.* Eds. A. D. McLaren and G. H. Peterson. Dekker. pp. 371–414.

189. R. A. Slepecky and J. V. Beck. 1950. The effect of 2,4-D on nitrification in soil *Soc. Am. Bacteriol., Proc.* **50**:17–18.

190. D. T. Smith and W. F. Meggitt. 1970. Persistence and degradation of pyrazon in soil. *Weed Sci.* **18**:260–264.

191. H. R. Smith, V. T. Dawson, and M. E. Wenzel. 1945. The effect of certain herbicides on soil micro-organisms. *Proc. Soil Sci. Soc. Am.* **10**:197–201.

192. J. W. Smith and T. J. Sheets. 1967. Persistence and decomposition of dichlobenil in soil. *Weed Soc. Am. Abstr. 1967 Mtg.* p. 76.

193. J. Sobieszczanski. 1969. The effect of herbicides on growth and morphology of some species of bacteria. *Acta Microbiol. Polon. Ser. B.* **1**:99–104.

194. P. Sprankle, W. F. Meggitt, and D. Penner. 1975. Adsorption, mobility, and microbial degradation of glyphosate in the soil. *Weed Sci.* **23**:229–234.

195. C. Stapp and R. Freter. 1952. Effect of 2,4-D in soils: reaction of soil bacteria to the growth substance. *Phytopath. Z.* **19**:20–33.

196. T. I. Steenson and N. Walker. 1956. Observations on the bacterial oxidation of chlorophenoxy-acetic acids. *Plant and Soil* **8**:17–32.

197. T. I. Steenson and N. Walker. 1957. The pathway of breakdown of 2,4-D and MCPA by bacteria. *J. Gen. Microbiol.* **16**:146–155.

198. T. I. Steenson and N. Walker. 1958. Adaptive patterns in the bacterial oxidation of 2,4-D and MCPA. *J. Gen. Microbiol.* **18**:692–697.

199. K. Steinbrenner, F. Naglitsch, and I. Schlicht. 1960. Die Einfluss der Herbizide Simazin auf die Bodenmikroorganismen und die Bodenfauna. *Albrecht Thaer Arch.* **4**:611–631.

200. C. R. Swanson. 1969. The benzoic and acid herbicides. In *Degradation of Herbicides*. Eds. P. C. Kearney and D. D. Kaufman. Dekker. pp. 299–320.

201. M. S. Tewfik and W. C. Evans. 1966. The metabolism of DNOC by soil microorganisms. *Biochem. J.* **99**:31–2P.

202. B. J. Thiegs. 1962. Microbial decomposition of herbicides. *Down to Earth* **18**:(2):7–10.

203. J. M. Tiedje and M. Alexander. 1969. Enzymatic cleavage of the ether bond of 2,4-dichlorophenoxy-acetate. *J. Agric. Food Chem.* **17**:1080–1084.

204. J. M. Tiedje, J. M. Duxbury, M. Alexander, and J. E. Dawson. 1969. 2,4-D metabolism: pathway of degradation of chlorocatechols by *Arthrobacter* sp. *J. Agric. Food Chem.* **17**:1021–1026.

205. C. M. Tu and W. B. Bollen. 1968. Effect of paraquat on microbial activity in soils. *Weed Res.* **8**:28–37.

206. C. M. Tu and W. B. Bollen. 1968. Interaction between paraquat and microbes in soils. *Weed Res.* **8**:38–45.

207. C. M. Tu and W. B. Bollen. 1969. Effect of Tordon herbicides on microbial activities in three Willamette valley soils. *Down to Earth* **25**(2):15–17.

208. B. G. Tweedy, C. Loeppky, and J. A. Ross. 1970. Metobromuron: Acetylation of the aniline moiety as a detoxification mechanism. *Science* **168**:482–483.

209. E. Valaskova. 1968. Die Empfindlichkeit von Bodenpilzen gegenuber Herbiziden. *Pflanzenschutz Berichte (Wien)* **38**:135–146.

210. A. Verloop and W. B. Nimmo. 1970. Metabolism of dichlobenil in sandy soil. *Weed Res.* **10**:65–70.

211. D. W. Von Endt, P. C. Kearney, and D. D. Kaufman. 1968. Degradation of monosodium methanearsonic acid by soil microorganisms. *J. Agric. Food Chem.* **16**:17–20.

212. N. Walker and T. I. Steenson. 1957. The persistence in soil of bacteria adapted to the decomposition of the hormone herbicides. *Rep. Rothamsted Exp. Stn.* 1957, p. 270; see also 1955 p. 66, 1957 p. 81, 1958 p. 70.

213. P. Wallnofer. 1969. The decomposition of urea herbicides by *Bacillus sphaericus* isolated from soil. *Weed Res.* **9**:333–339.

214. J. R. Warren, F. Graham, and G. Gale. 1951. Dominance of an actinomycete in the soil microflora after 2,4-D treatment of plants. *Phytopathology* **41**:1037–1039.

215. D. M. Webley, R. B. Duff, and V. C. Farmer. 1958. The influence of chemical structure on beta-oxidation by soil Nocardias. *J. Gen. Microbiol.* **18**:733–746.

216. J. S. Whiteside and M. Alexander. 1960. Measurement of microbiological effects of herbicides. *Weeds* **8**:204–213.

217. V. Wilkinson. 1969. Ecological effects of diquat. *Nature* **224**:618–619.

218. V. Wilkinson and R. L. Lucas. 1969. Inflence of herbicides on the competitive ability of fungi to colonise plant tissue. *New Phytol.* **68**:701–708.

219. C. L. Winely and C. L. San Clemente. 1968. Inhibition by certain pesticides of the nitrite oxidation of *Nitrobacter agilis*. *Bact. Proc. A.* 63, 11.

220. A. W. Winston and P. M. Ritty. 1961. What happens to phenoxy herbicides when applied to a watershed area. *N.E. Weed Control Conf., Proc.* **15**:396–400.

221. D. C. Wolf and J. P. Martin. 1975. Microbial decomposition of ring-^{14}C atrazine, cyanuric acid, and 2-chloro-4,6-diamino-*s*-triazine. *J. Environ. Qual.* **4**:134–139.

222. E. A. Woolson and P. C. Kearney. 1973. Persistence and reactions of ^{14}C-cacodylic acid in soils. *Environ. Sci. Technol.* **7**:47–50.

223. E. A. Woolson, J. H. Axley, and P. C. Kearney. 1971. The chemistry and phytotoxicity of arsenic in soils. *Soil Sci. Soc. Am. Proc.* **35**:938–943.

224. A. D. Worsham and J. Giddens. 1957. Some effects of 2,2-dichloropropionic acid on soil microorganisms. *Weeds* **5**:316–320.

225. W. A. Worth and A. M. McCabe. 1948. Differential effects of 2,4-D on aerobic, anaerobic, and facultative anaerobic organisms. *Science* **108**:16–18.

226. K. A. Wright and R. B. Cain. 1972. Microbial metabolism of pyridinium compounds. *Biochem. J.* **128**:543–599, 561–569.

227. C. R. Youngson, C. A. I. Goring, R. W. Meikle, H. H. Scott, and J. D. Griffith. 1967. Factors influencing the decomposition of Tordon herbicide in soils. *Down to Earth* **23**(2):2–11.

14

HERBICIDES IN WATER

A. HERBICIDES IN DRAINAGE WATER

The commonly used herbicides are, with the exception of trifluralin, much less toxic than insecticides to fish and aquatic arthropods (Table 14.1). Moreover, trout are not notably more susceptible than other species such as bluegill sunfish (Table 14.2). In general, those herbicides that leach are rapidly degraded, and those that are slowly degraded do not leach. However, the greatest losses from soil to drainage water occur after heavy rains, the runoff carrying herbicides not only in solution but also adsorbed onto suspended soil particles.

Phenoxys. When sampled in 1968, rivers in the western United States yielded evidence of the presence of small amounts of these herbicides in the following instances of 240 samples[59]:

	2,4-D	Silvex	2,4,5-T
No. of Samples Positive	36	10	24
Maximum Content, ppb	0.35	0.21	0.07
River with Maximum Content	James R., S.D.	Humboldt R., Nev.	Green R., Utah

The absolute maximum reported was 18 ppb of 2,4-D found in an irrigation return flow at Moses Lake in the interior of Washington State.[40] The amount of surface runoff in rainstorms following 2,4-D application was found to be 13% for the PGBE ester formulation and 4% for the amine salt when applied at 2.2 lb/a to fallow land with a 6% slope.[6] The maximum persistence of 2,4-D found in natural waters aerobically incubated in the laboratory was 120 days.[2] However, wells located around an industrial waste lagoon near Denver, Colorado came to contain enough 2,4-D during a period in the mid-1950s to cause crop damage when the water was used for irrigation.[71] For drinking water, although the standard treatments in filtration plants reduce unduly high surges down to acceptable concentrations, they are inadequate to eliminate persistent low level contaminations; for this, activated charcoal is required to remove pollutants such as 2,4-D.[15]

2,4,5-T applied to forestland is soon no longer found in runoff water, the herbicide usually disappearing from the runoff before the oil carrier; it is considerably more evanescent than picloram which persists in runoff for 1 yr after the application, and fenuron or bromacil which are still found in streams 2 yr after being applied.[73]

TCDD. Some commercial formulations of 2,4,5-T have contained the contaminant TCDD (2,3,7,8-tetrachlorodibenzo-*p*-dioxin) which is highly

Table 14.1. Toxicity of some common herbicides to fish and aquatic arthropods: 24-hr LC$_{50}$ levels in ppm (Pimentel, 1971).

	Rainbow Trout	Bluegill Sunfish	*Pteronarcys** Stoneflies	*Daphnia*[†] Waterfleas
2,4,5-T	1.3	1.4[‡]	--	1.5
Simazine	5	130	50	--
Atrazine	12.6	--	16	3.6
Monuron	--	1.8	--	106
Diuron	4.3[§]	12	3.6	1.4
Picloram	150	--	120	450
Trifluralin	0.21	0.10	13	0.24
Propham	--	32	--	--
Chlorpropham	--	10	--	10
Dalapon	340	480	100	11

* Sanders and Cope 1968   [§]48 hrs.   [†] D. magna or D. pulex

[‡] Butoxyethanol ester

toxic, teratogenic, extremely persistent, and liable to accumulate in food chains; there are also indications that 2,4,5-T itself has teratogenic potential at high doses.[69] Although 2,4,5-T has not been found in streams in the fall following a spring spraying of forestlands, it is possible that small amounts of TCDD could be continuously released, since its half-life in soil is more than 1 yr.[52] At a concentration of only 0.2 ppb, TCDD has been found to reduce the reproductive success of *Physa* snails and *Paranais* oligochaetes, while young coho salmon exposed to 0.1 ppb for 48 hr take up enough TCDD to kill them 10–80 days later.[63] Although about half of the 2,4,5-T batches tested before 1972 contained TCDD at contaminations ranging up to 10 ppm, current production practices have ensured that they never exceed 0.1 ppm.[113]

Picloram. Applied at the unusually heavy rate of 9 lb/a to a watershed area, picloram reached the stream draining it in concentrations up to 0.4 ppm[67]; but it is scarcely toxic to aquatic fauna (Table 14.1). Picloram, which leaches to considerable depths, has very occasionally reached aquifers in Texas. In forest spraying in Oregon, picloram was the most refractory to breakdown by red-alder forest-floor material, being only one-third degraded in 133 days; 2,4-D and amitrole were the least persistent of the forest herbicides. In the streams of this state, herbicide concentrations after spraying seldom exceeded 0.1 ppm, and 1 ppm has never been found in 7 yr of testing.[72] The very extensive spraying of 2,4-D, 2,4,5-T and picloram on tropical forest and mangrove swamps in South Vietnam did not result in a reduction of the reported catch of freshwater and saltwater fish, nor of shellfish (molluscs, cuttlefish, shrimp, and crabs), during the period 1965–1967.[97] In the sprayed mangrove swamps, there was a reduction in the number and variety of plankton species and large fish, but the fish larvae and eggs showed an increased abundance, with no constriction in the variety of species.[69]

Triazines. Atrazine has been found present in surface waters in Iowa, where it is extensively employed on the corn crop; the concentrations are of the order of 2 ppb in June and 0.2 ppb in August, and the average content of Des Moines drinking water in the summer of 1974 was approximately 0.5 ppb.[78] Applications at 3 lb/a to silt loam cornfields in Iowa with a 10–15% slope put 3 ppm atrazine in the runoff water, and 4.5 ppm in the suspended soil particles, after the first rain.[79] In experiments on a sandy loam with a 6% slope, a 0.5-inch rain removed 2%, a 2.4-inch rain removed 7%, of atrazine applied at this dosage.[107] The total loss of atrazine for the season from the Iowa fields was 15%; Pennsylvania cornfields with a similar slope treated at 2 lb/a lost 2.5% over the posttreatment season, putting average

Table 14.2. **Toxicity of herbicides to rainbow trout and bluegill sunfish: 48-hr LC_{50} figures in ppm.**

	Trout	Sunfish		Trout	Sunfish
Trifluralin	0.098[*]	0.019[a]	Simazine	56[d]	>100[b]
Vernolate	6.2[g]	9.2[b]	2,4,5-T acid	1.3[d]	0.5[d]
Diuron	4.3[h]	74[d]	2,4-D (PGBEE)	0.96[h]	0.90[a]
Diquat	12.3[h]	91[e]	2,4-D (BEE)	2.1[h]	3.7[d]
Dicamba	35[d]	40[b]	2,4-D (Dimethylamine)	250[*f]	166[c]
Dalapon	>500[*f]	115[d]	Silvex (PGBEE)	0.65[h]	16.6[a]

[a] Cope 1966   [b] Cope 1965   [c] Hughes and Davis 1963   [d] Bohmont 1967

[e] Surber and Pickering 1962   [f] Alabaster 1969   [g] WSSA 1967   [h] FWPCA 1968

[*] 24-hr LC_{50}

concentrations of 10 ppb atrazine into the runoff water and 120 ppb onto the suspended soil particles.[37]

The highest concentrations found in runoff water in Texas experiments were 40 ppb atrazine, to be compared to 50 ppb with propachlor and 40 ppb with trifluralin.[3] In Iowa, the 15% seasonal runoff loss for atrazine compares with 2.5% for propachlor and a mere 0.1% for the insecticide diazinon. In Hawaii with its light volcanic soils, the atrazine reaching the runoff sediments is slowly degraded by bacteria to the deethylated isopropylamino metabolite, which is much more biostable than the hydroxyatrazine produced by chemical degradation in soil.[35]

Prometryn, the triazine employed on cotton, when tested at 2.5 lb/a in Oklahoma on a sandy loam plot with a 1% slope, lost only 0.4% to runoff and was never washed to depths below 5 cm; the first good rain (artificially applied) gave 140 ppb in the runoff water, while the suspended particles contained 1.9 ppm, enough to constitute 10% of the prometryn washed away.[5]

Trifluralin. Although it is one of the most toxic herbicides to aquatic organisms (Table 14.1), trifluralin is so strongly absorbed to soil particles that runoff water and sediments have no lethal effect.[67] When applied at 1 lb/a to cottonfields in North Carolina, 83% of the herbicide runoff was on the suspended particles, and less than 1% was thus removed during the season; although the highest concentration in the runoff was 24 ppb, the maximum found in a small pond in the drainage area was only 1.6 ppb.[88] Applied to a silty clay loam in Louisiana with a grade of only 0.2%, trifluralin and diuron showed the least runoff (Table 14.3), while fenac and linuron showed quite high concentrations in the drainage water but very modest losses over the season.[109] Applied at 2 lb/a to clay loam in Texas with an 8% slope, picloram, 2,4,5-T, and dicamba yielded 2.2, 3.3, and 4.8

Table 14.3. **Runoff of herbicides from plots treated in the spring, 1971–1973: silty clay loam with a 0.2% grade (Willis et al., 1975).**

	Diuron	Linuron	Fenac	Trifluralin
Crop	Cotton	Soybean	Sugarcane	Cotton & Soybean
Dosage, kg/ha	0.84	2.24	3.36	1.40
Max. Concentration in Runoff, ppb	Trace	124	310	1.9
Rainfall before Max. Conc'n., cm	3.3	1.6	7.4	7.4
Max. Seasonal Loss to Runoff, in % of that applied	0.12	0.30	2.90	0.05

ppm, respectively, in the runoff water from a 1.3-cm artificial rain 1 day after the application. As much as 5.5% of the dicamba applied was lost during the season. Plots with 3% slope yielded half as much herbicide in the runoff as those with 8% slope, and plots treated at 1 lb/a yielded two-thirds as much herbicide runoff as those treated at 2 lb/a.[96]

DCPA. Application of the chlorinated compound DCPA (Dacthal) as a preemergent herbicide to the onion and cabbage fields of the lower Rio Grande Valley, Texas, resulted in concentrations in the Arroyo Colorado (which drained the area), amounting to 0.05 ppb in the summer and rising to a maximum of 1.6 ppb in February. Residues of DCPA in fish reached a maximum in March, being 2.8 ppm in menhaden in the Arroyo and 0.5 ppm in spotted sea trout at its outlet in the Laguna Madre.[62] However it is nontoxic to fish at 500 ppm in the water, and to mallards at 5000 ppm in the diet.[106] The fungicide HCB, a contaminant of commercial DCPA that can be percutaneously absorbed by spraymen[12] and can cause porphyria, was not found as residues in the Arroyo Colorado. **Amitrole** sprayed on 100 acres of clear-cut forestland at 2 lb/a to control *Rubus spectabilis* could no longer be detected 2 miles downstream on the sixth day after the spraying.[60] Two other experiments found amitrole to have become undetectable after 2 and 3 days, respectively.[67]

PCP. Of other herbicides used on field crops, PCP applied as its Na salt is the only one which leaches more readily and is more toxic than 2,4,5-T (Table 14.1) not only to trout but to a number of Japanese fish,[75] while sublethal levels depress the growth rate of *Poecilia*.[22] The temporary introduction of sodium pentachlorophenate as a herbicide in ricefields in Japan has been suspected as a frequent cause of mass deaths of fish and shellfish (Hashimoto, 1969[65]), particularly after heavy rains draining out to the sea coast.[67] PCP is the most dangerous among the herbicides that might drain into waste stabilization ponds of municipal sewage systems, in which the right population of oxygen-producing algae is necessary to counterbalance putrefying bacteria, since it is completely toxic to *Chlorella* at 7.5 ppb. Maximum contaminations of other herbicides which are safe for these algae are measured in parts per million, being 5 for 2,4-D, 10 for DNOC, 25 for 2,4,5-T, and 500 for TCA.[47]

B. EFFECTS OF HERBICIDES IN THE AQUATIC ENVIRONMENT

The excessive growth of aquatic plants in impounded or increasingly eutrophic water-bodies has demanded the use of herbicides for the management of fish and estuarine sea-food invertebrates. Aquatic weed control, directed principally against water hyacinth (*Eichhornia*), water fern (*Salvinia*), and alligator weed (*Alternanthera*), has become necessary for fish management over more than 1 million acres of ponds in the southern United States.[28] Elimination of the European water milfoil (*Myriophyllum*) and of eelgrass (*Zostera*) is necessary along the Atlantic coast to ensure a full harvest of oysters, clams and crabs. Overgrowths of *Elodea*, water fern (*Salvinia*), water lettuce (*Pistia*), duckweed (*Lemna*), and algae frequently require control and elimination. The side effects of herbicides on aquatic fauna have been reviewed by Mullison (1970) and by Frank (1972).

Relative Persistence of Aquatic Herbicides

For algae, copper sulfate applied at 0.5 ppm Cu controls most species, the threshold of damage to fingerlings of the most sensitive fish (trout) being 0.75 ppm.[51,68] Although the water concentrations of Cu do not disappear as fast as those of diquat (Table 14.4), Cu resembles diquat in being rapidly adsorbed onto sediments. Acrolein, which is the herbicide of choice for clearing irrigation ditches at 0.5 ppm concentration, is quickly effective and rapidly lost by volatilization, leaving the water safe for crops. Of the her-

Table 14.4. Disappearance of aquatic herbicides from the treated water: days after the application in parentheses (Frank, 1972).

	ppm applied	Highest ppm found	Final ppm remaining	Usual Application Rates
2,4-D, dimethylamine	1.5	0.14 (1)	0.004 (41)	1.5-2.0 ppm
Copper (sulfate)	0.5	0.42 (1)	0.19 (3)	0.1-2.0 ppm
Diquat	0.62	0.49 (1)	0.001 (8)	0.25-1.5 ppm
Endothall	1.2	0.79 (4)	0.001 (36)	0.5-4.0 ppm
Silvex, PGBE* ester	2.9	1.6 (7)	0.02 (182)	1.5-2.0 ppm
Fenac, granular	1.0	0.71 (8)	0.07 (160)	15-20 1b/a
Dichlobenil, granular	0.58	0.32 (36)	0.004 (160)	10-15 1b/a
2,4-D ester†, granular	1.33	0.067 (18)	0.001 (36)	20-40 1b/a‡

* Propylene glycol butyl ether

† Butoxyethanol ester

‡ Equivalent to 1.8 - 3.6 ppm

bicides commonly applied against submerged weeds (Table 14.4), endothall and diquat are short-lasting, while fenac, dichlobenil, and silvex are comparatively long-lasting, and give more complete control.[28] Dichlobenil is less persistent than fenac, but more persistent than 2,4-D.[76]

2,4-D is of intermediate persistence, and is widely applied as 2,4-D amine at 2–4 lb/a acid equivalent per acre in repeated treatments against the large floating weeds, and as 2,4-D ester in granular form at 20–40 lb/a acid equivalent for submerged weeds. 2,4-D and silvex reach their maximum water concentrations 1–2 wk after application, being taken up and then released from the vegetation. Residues of 2,4-D disappear from the sediment faster than from the water,[87] and the biodegradation of this herbicide follows first-order kinetics.[40] The breakdown of 2,4-D, which is oxidative and microbial, is suppressed where there is a deficiency of dissolved oxygen, no more than two-thirds of the acetic acid moiety being degraded over a 6-month period.[76]

The bipyridyliums paraquat and diquat persist in water for only 1–4 wk, whereas fenac and amitrole persist for at least 200 days.[36] Paraquat is more persistent than diquat in the bottom mud, 1–8 ppm being found there 4 yr after an application at 0.3 lb/a. The contrast between these bipyridylium compounds and the other aquatic herbicides is shown by the following residue determinations (ppm) made on pond hydrosoils[29]:

	Paraquat	Diquat	2,4-D	Dichlobenil	Fenac
Dosage Rate Applied	1.14	0.62	1.33	0.4	1.1
Posttreatment Day 1	3.56	0.47	4.96	6.41	1.31
Posttreatment Day 56	20.51	36.80	0.10	0.50	0.39

Effects on Fish and Amphibia

If used in ponds, acrolein is toxic to fish such as suckers and creek chub,[51] the 24-hr LC_{50} for bluegills being 80 ppb.[67] For safety to fish, dalapon is the herbicide of choice, the LC_{50} for brown trout being 400 ppm.[67] Dichlobenil and fenac when employed as aquatic herbicides have not been known to kill fish; brook trout were unaffected by fenac applied at 15 lb/a,[51] and fry of bass and sunfish tolerated 20–25 ppm of either herbicide.[42] Dichlobenil applied at the normal 0.5 ppm rate did not affect spawning,[67] but in ponds treated at 10 ppm there was some mortality among red-ear sunfish, although the survivors put on more weight.[19]

Endothall. This proved nontoxic to fry of pond fish (Table 14.5) and four successive applications at 1 ppm did not affect populations of brook trout, bluegills or smallmouth bass.[51] The LC_{50} values of disodium endothall for

Table 14.5. **Survival of fish fry in concentrations of aquatic herbicides: number of days that they survived, or reached the termination (T) of the exposure at 8 days (Hiltibran, 1967).**

	Conc'n ppm	Bluegill Sunfish	Green Sunfish	Lake Chub-sucker	Smallmouth Bass
2,4-D, dimethylamine	25	T	T	T	T
Arsenite (sodium)	8	T	T	-	T
Diquat (cation)	1.3	3*	-	2	4
Endothall	25	T	T	T	T
Silvex, PGBE ester	2.4	2	4	T	4[†]
Fenac	20	5	-	T	-
Dichlobenil, granular	25	T[‡]	T	T	T
2,4-D ester[§], granular	1.0	2	5	5	5

* 2.5 ppm for bluegills   [†] 1 ppm for bass   [‡] 10 ppm for bluegills   [§] PGBE ester

eight species of Missouri fish were all between 95 and 210 ppm,[100] and adult bluegills could tolerate continuous exposure to 100 ppm of the free acid.[55] Two strongly herbicidal salts of endothall were unfortunately toxic to fish, namely the dimethylamine (Table 14.5) and the cocoamine.[75]

Phenoxys. *2,4-D* as the free acid is relatively nontoxic to fish, 100 ppm causing 10% mortality in bluegills after 7 days, and 13% mortality in large-mouth bass after 14 days.[53] An extensive series of tests[49] revealed that the amine salts (with the exception of the cocoamine) were less toxic to bluegills than the esters, the 24-hr LC_{50} figures (ppm) for bluegills being as follows:

Alkanolamine salt	450	Iso-octyl (IO) ester	3.6
Dimethylamine salt	390	Butoxyethanol (BE) ester	2.1
Cocoamine salt	8	PGBE ester	2.1

Applied to estuaries at 30 lb/a acid equivalent, the BE ester caused about 10% mortality to menhaden,[7] while the IO ester killed nearly 50% of the pumpkinseed sunfish present.[77] The PGBE ester is relatively toxic to the fry of pond fish at 1 ppm acid equivalent (Table 14.5). The minimum concentration of 2,4-D (from the BE ester) that does not harm the growth and reproduction of fathead minnows, as derived from an exposure of parents and offspring for 10 months in flowing water, was 0.3 ppm; this is roughly one-twentieth of the observed 96-hr LC[50] figure.[64] Bluegills planted in Oklahoma ponds treated with the PGBE ester at 10 ppm 2,4-D sustained some mortality; although spawning was delayed by 2 wk, the ultimate production of fry was up to normal. While there was no mortality in ponds treated at 5 ppm, lesions did develop in liver, brain, and vascular system at this concentration, becoming more frequent at 10 ppm.[20] Continuous exposure of the red-ear sunfish to the PGBE ester at 3 ppm 2,4-D or more induced lesions to appear in liver and testis.[16]

The bluegills planted in the 2,4-D treated ponds put on more weight, those at 10 ppm attaining twice the length and three times the weight of those in untreated ponds. This effect, which is proportional to the dosage applied, is evidently due to the weed control making for greater accessibility of food-chain prey, supplemented by a diminution in the competition between the bluegills.[20]

Silvex in the free acid form is no more toxic to fish than dichlobenil, or fenac (Table 14.6). Applied at 20 ppm from its sodium or potassium salt, silvex did not kill the fry of pond fish,[42] and repeated applications at 2–3 ppm in Ontario ponds had no observable effect on largemouth bass, nor on rainbow or brook trout.[51] Unfortunately the PGBE ester and other esters are quite toxic[7b]; for example the LC[50] of the PGBE ester for the fathead minnow is 1.2 ppm as compared to 150 ppm with the sodium salt.[66] By contrast, application of the K salt of silvex to two ponds at 1.5–2 ppm

Table 14.6. Toxicity of aquatic herbicides to fish and aquatic arthropods: 24-hr LC[50] figures in ppm.

	Rainbow Trout[1]	Bluegill Sunfish[1,2]	*Pteronarcys* Stoneflies[3]	*Gammarus lacustris*[4]	*Daphnia magna*[5]
Sodium arsenite	60	44	140	---	6.5
Diquat (dibromide)	90	91	---	---	7.1
Endothall (acid)	1.2*	428	---	2.0	46
Silvex (acid)	21.9	14.5	5.2	---	100[†]
Fenac (Na salt)	7.5	19	220	22	>100
Dichlobenil	22	20	42	16	9.8
2,4-D (butyl ester)	250	1.3	8.5	2.1[‡]	>100[§]

* dimethylamine   † K salt   ‡ PGBE ester   § acid

[1] Pimentel 1971   [2] Hughes and Davis 1962   [3] Sanders and Cope 1968

[4] Sanders 1969   [5] Crosby and Tucker 1966

resulted in the sunfish population in the following 2-yr period showing a lower ratio of weight to length than in untreated ponds.[46] Applications of the PGBE ester of silvex to ponds at 8 lb/a caused fish mortalities in about half of the cases, the species which suffered being gizzard shad, smallmouth bass, and bluegills.[108] It was with silvex formulations that it was shown that hatched fry were considerably more susceptible to herbicides than the fish embryos in the eggs.[42] When the PGBE ester is applied to ponds at 9 kg/ha, the ester linkage is all hydrolyzed in 3 days, the silvex disappears from the water in 3 wk, and is no longer found in the sediment after 5 wk.[4]

Dimethylamine salt formulations of *2,4,5-T* were less toxic to bluegills than the esters,[49] as shown by the following 24-hr LC_{50} figures (ppm):

| Dimethylamine salt | 144 | PGBE ester | 17 |
| Iso-octyl (IO) ester | 28 | Butoxyethanol (BE) ester | 1.4 |

Bipyridyliums.   Diquat, which kills aquatic vegetation very rapidly, did not kill bluegills when applied to Ontario ponds at 3 ppm,[51] and it controlled *Elodea* without killing brook trout or sticklebacks when applied to Wisconsin ponds at 1 ppm.[41] However, it proved to be toxic to fry of bluegills and bass at 2.5 ppm (Table 14.5). Paraquat, with similar physical and biological properties, also failed to kill rainbow trout or other pond fish when applied at 1 ppm (House et al., 1967[75]), but an application at 0.5 ppm where the decomposition of the aquatic weeds reduced the oxygen saturation of the water down to only 4% resulted in kills of trout but not of perch.[65] Treatment of a Colorado pond with paraquat at 1.14 ppm killed the bluegills but not the channel catfish.[26]

Urea Herbicides.   Diuron kills aquatic vegetation almost as rapidly as diquat, but is much more persistent; the LC_{50} levels for fish range between 5 and 25 ppm.[75] Its use in spawning ponds at concentrations up to 0.4 ppm caused a slight reduction in the growth rate of carp but not of pike.[67] Bluegills, considered one of the more tolerant species, showed some fingerling mortality in ponds treated with diuron at 3 ppm; the not infrequent appearance of lesions and haemorrhages in their gill lamellae was probably connected with the pronounced oxygen deficiency in some of the treated ponds.[61] Monuron and fenuron are less toxic to bluegills and other freshwater fish than diuron or neburon.[102]

Triazine Herbicides.   Simazine and atrazine act extremely slowly on aquatic plants, the phytotoxic symptoms not appearing until 2-6 wk after the application.[101] When the fry of pond fish were exposed to simazine or atrazine at 10 ppm, the bluegills and chubsucker survived; the smallmouth bass and some green sunfish were killed, but the mortality could be avoided by using granules.[42] Long-term exposure to atrazine at only 0.7 ppm for several months induced no mortality among bluegills, but they lost their equilibrium at 0.2 ppm; at this concentration, spawning was reduced in fathead minnows, and brook trout became lethargic and their fry grew more slowly.[57] Applications at 2 ppm of simazine, which is less toxic to fish than atrazine but more toxic than diuron, have resulted in the death of some adult red-ear sunfish, and a generally reduced fish production in the treated ponds.[67]

Other Nitrogenous Herbicides. Amitrole and dalapon, not applied directly to water but frequently used to control vegetation along the banks of irrigation ditches, proved harmless to the fry of all four pond species tested at 50 ppm concentration. Propanil is unsuitable for use in ricefields, since the first metabolite dichloroaniline (DCA) is deleterious to fish. The thiocarbamate herbicides which are substituted suffer from their propensity to decrease leaf-cuticle formation; thus when EPTC is used on field crops, it requires the addition of a safener such as R-25788, the mixture being sold as Eradicane. Another unexpected consequence in ricefields came from the use of the organophosphorus nematicide Nemacur; when applied against the ring nematode *Criconemoides* in Louisiana, it stimulated the growth of sedges which are wild hosts for the nematode.[44]

Residues. Organic herbicides do not usually accumulate in fish sufficiently to cause a problem with residues, and in any case they are quite rapidly lost. The half-lives in the fish were found to be less than 1 wk for 2,4-D, less than 2 wk for dichlobenil, and less than 3 wk for diquat and endothall.[56] In the bluegills and bullheads in Tennessee Valley Authority reservoirs treated with 2,4-D from the BE ester at 40–100 lb/a, the maximum residue found was 0.15 ppm, contrasting with residues ranging up to 1.0 ppm in the mussel *Elliptio*.[90] A similar level of 2,4-D residue occurred in the pigtoe mussel *Pleurobema* after reservoir treatment with the dimethylamine salt at this dosage; while the levels in sunfish, bullheads, and bass were less than 0.1 ppm, those in the gizzard shad reached a maximum of 0.34 ppm.[111] After 84 days in pools treated with 2 ppm of this herbicide from its dimethylamine salt, bluegills and channel catfish contained about 1 ppm of 2,4-D-derived residues in their muscle tissue, but only a trace of 2,4-D itself.[87] Fish in water treated with silvex at 10 ppm had lost all the residue taken up 3 days after the treatment (Sutton et al., 1969[28]).

Data on fish residues, available for paraquat in the intestinal contents of channel catfish,[31] for dichlobenil in bluegills,[99] and for simazine in green sunfish,[80] show the same contrast, viz.:

Paraquat	2 ppm	5 months after pond treated at 1 ppm
Dichlobenil	0.15 ppm	2 months after pond treated at 20 ppm
Simazine	zero	1 wk after 3-wk exposure to 3 ppm

Sodium Arsenite. This is a cheap and safe material for combatting submerged vegetation, although it should be injected below the surface of the water if there is a danger of cattle drinking it. Arsenite at 8 ppm was tolerated by the fry of the pond fish (Table 14.5), and ponds in Ontario have been treated with 5 ppm without killing rainbow or brook trout.[51] The residues of arsenic from the organic arsenicals MSMA, DSMA, or cacodylic acid which have been found in fish (*Gambusia*) do not exceed the water concentrations by more than 2.5 times, although in *Daphnia* and algae they may exceed it by as much as 250 times.[50] The continuous use of sodium arsenite over a period of 10 yr along the shores of an inland lake gave maximum concentrations in yellow perch 4–5 yr old that did not exceed 0.07 ppm in their flesh.[58]

Amphibia. When herbicides used in water were tested for their toxicity to tadpoles of the western chorus frog and Fowler's toad, the following 96-hr LC_{50} values (in ppm) were found:

Pseudacris triseriata		*Bufo woodhousi fowleri*	
Silvex (BE ester)	10	Endothall	1.2
Paraquat	28	Molinate	14
2,4-D amine	100	Paraquat	26

The most toxic herbicide tested was trifluralin, with an LC_{50} of 0.10 ppm for the toad tadpoles.[83] 2,4-D free acid inhibits the development of *Rana temporaria* eggs at 5000 ppm, while at 500 ppm it causes malformations of the tail of the developing tadpole.[54]

Effects on Invertebrates

The herbicides applied against aquatic weeds are no more toxic to aquatic insects and crustacea than to fish, the LC_{50} levels (Table 14.6) being higher than the maximum concentrations found in the water. Faunal changes after the application are referable to the destruction of the vegetation rather than to direct toxic effects.

With aquatic herbicide applications in general, the bottom fauna are at first reduced, and then the release of plant nutrients allows the recovery of those species most tolerant of plant decay. The phytoplankton often recovers from the herbicidal impact within 2 wk.[28] The anaerobiosis resulting from the killed vegetation may affect certain fish-food organisms, but the total forage biomass available, although reduced, does not normally limit the fish production. Actual net increases in aquatic invertebrates have resulted from applications of paraquat, diquat, and dichlobenil.[8]

Endothall. In a Missouri farm pond inhabited by three species of sunfish, two species of bullheads and at least three species of minnows, the effect of endothall at 6 ppm was, in killing back *Potamogeton, Elodea,* and *Naias flexilis,* to eliminate the habitat of *Physa* and *Heliosoma* snails, thus depriving red-ear sunfish of their preferred food.[100] The organic detritus thus produced encourages the growth of oligochaete worms (Table 14.7), a preferred food of bluegills. Subsequently the flora recovers, led by the species most tolerant to the herbicide, for example, in Missouri ponds the muskweed *Chara vulgaris* followed by *Elodea,* and accompanied by the algae *Cladophora, Pithophora,* and *Spirogyra,* which had not been completely controlled.[100]

Bipyridyliums. In Britain, paraquat and diquat applications proved harmless to aquatic invertebrates, except at one site where severe oxygen deple-

Table 14.7. Effect of endothall (Na salt) at 5 ppm on the bottom fauna of a Missouri farm pond: numbers per square foot before and after (Walker, 1963).

	Untreated	1 week	6-8 wks	11-14 mths
Oligochaete worms	164.8	99.4	140.8	245.3
Musculium clams	27.4	33.8	7.8	0.5
Physa snails	2.2	0.2	0.0	0.0
Tendipedid midges	3.8	3.0	9.9	3.7
Chaoborine midges	3.2	3.7	6.1	20.8
Caddisfly larvae	3.7	5.0	4.9	0.0
Odonata nymphs	2.7	5.4	2.6	0.2
Mayfly nymphs	1.9	0.8	0.3	0.8
Water beetles	2.1	0.2	0.1	10.0
Total Biomass g/ft^2	2.10	1.28	1.30	1.59

tion due to the dead weeds caused some mortality among dragonfly nymphs, caddisfly larvae, and the aquatic beetles, and even in clams, leeches, and aquatic snails.[70] In a Wisconsin pond treated with diquat at 1 ppm to control *Elodea,* all of the bottom-feeding invertebrates were maintaining their abundance 1 wk after the application; the only species that suffered were those that depended on *Elodea* for their habitat, namely the aquatic snails (*Physa, Heliosoma,* and *Gyraulis*) and the amphipod *Hyalella azteca.*[41] Treatment of two Florida ponds with 0.5 ppm diquat resulted in an increase in the benthic fauna, oligochaetes in one and *Chaoborus* larvae in the other.[94] With paraquat and diquat, the masses of killed vegetation have led to high populations of Tendipedid midges, as observed in Britain[70] and New Zealand.[27]

When applied at 1 ppm, diquat or paraquat induce a 50% reduction in phytoplankton in the first 2 days, but they allow a rapid recovery as their residues disappear.[94] In a New Zealand lake so treated a massive growth of *Nitella* replaced the *Largarosiphon* weed eliminated.[27] In a Welsh reservoir the elimination of *Potamogeton* and *Myriophyllum* by 1 ppm diquat resulted in the macrophytic alga *Chara globularis* occupying two-thirds of its area 6 wk later.[11]

Triazines.   When simazine was applied to a Missouri farm pond[101] the slow killing characteristic of *s*-triazines as aquatic herbicides resulted in a fair survival of snails and clams, and no increases in oligochaetes and water beetles; however the simazine applications, and those with dichlobenil likewise, were followed by a considerable increase in Chaoborine phantom-midge larvae. Bass spawning ponds treated with simazine at 2 ppm, which eliminated overgrowths of net algae without harming the fry, had the effect of reducing other phytoplankton blooms and producing a high population of zooplankton a month after the application.[91]

Atrazine is more toxic to aquatic invertebrates than simazine, applications to ponds at 0.5–2.0 ppm reducing clams by 75%, and oligochaetes, caddisfly larvae, and mayfly nymphs by at least 50%; by contrast, aquatic snails, damselfly nymphs, and water beetles increased by at least 100% (Walker, 1962[75]). Although the 48-hr LC_{50} of atrazine by static test is 6.9 ppm for *Daphnia magna* and 5.7 ppm for *Gammarus fasciatus,* in continuous-flow tests a concentration of 0.25 ppm over three generations reduces the progeny production by *Daphnia,* and 0.14 ppm maintained until the F_1 reduces the developmental rate of *Gammarus.*[57] For this reason the maximum acceptable toxicant concentration (MATC) is set between 0.14 and 0.25 ppm for *Daphnia,* and between 0.06 and 0.14 ppm for *Gammarus*; these tolerances are still well above the 2 ppb atrazine concentrations found in runoff water after the height of the corn herbicide applications.[78]

Phenoxys. 2,4-D, applied at 45 kg/ha either in acid or ester formulation on granules dispersed over the ice of an inland lake, controlled water milfoil in the following spring without causing any distinguishable reductions in the numbers or biomass of bottom invertebrates.[45] Applied to reservoirs to give a concentration of 1 ppb, 2,4-D had little effect on bottom organisms such as *Hemagenia* mayflies.[90] But at 1 ppm it reduced the numbers of bottom-feeding and open-water invertebrates by 50% or more (Walker, 1962[75]). Even at 100 ppm, it does not affect *Philodina* rotifers, nor *Cladophora* algae.[33] Of the formulations of 2,4-D, the PGBE ester poses the most hazard to small crustacea (Table 14.8).

Silvex applied at 1.5 gal/a of the formulation (Kuron) for control of white water lily in a New York pond caused a 2-wk reduction in plankton and no reduction in the benthic invertebrates.[74] Applied as the K salt, silvex at water concentrations of 3–5 ppm in a Missouri farm pond gave rise to increased populations of midge larvae and oligochaetes due to the vegetation kill, with a subsequent surge of *Chaoborus* larvae, followed by an increase in Libellulid dragonfly nymphs 1 yr later (Table 14.9). Populations of the Tabanid *Chrysops* were rapidly decreased but partially recovered, while the numbers of snails, leeches, and damselflies remained relatively unchanged.[38] The PGBE ester of silvex is the most toxic formulation for freshwater crustacea (Table 14.8), but when it is applied to ponds the reductions in crustacea, and of rotifers, are temporary.[108] The potassium salt of silvex caused no such reductions in plankton organisms, nor did 2,4-D acid, fenac, dichlobenil, or endothall (Pierce, 1958–1969[67]).

Other Organic Herbicides. Of the herbicides not employed as aquatic herbicides (Table 14.10), trifluralin is the only one with an LC_{50} to cladocerans less than 1 ppm. Fenac, which is applied to ponds at 1–3 ppm, had no effect on the invertebrates, nor indeed on fish and amphibia.[67] Applied successfully at 2 ppm to control rooted plants such as *Potamogeton,* sedges, and rushes in draw-down areas in lakes, fenac did not reduce the numbers of the zooplankton crustacea, and so did not promote algal blooms.[89] Diuron, monuron, and DNBP are more toxic than fenac, while propham, EPTC, propanil, dalapon, and MCPA are even less toxic than fenac to *Daphnia* or *Gammarus.*[23,84,85] To the stonefly *Pteronarcys californica,* diuron and trifluralin are about as toxic as silvex, and picloram, paraquat and dalapon about as nontoxic as fenac.[86] Diuron and neburon applied at 3 ppm are directly toxic to bottom invertebrates such as immature mayflies, dragonflies, and midges, but these insects attain even greater numbers 3 wk after the treatment.[61,102] Dichlobenil and diquat are roughly similar in their effect on aquatic insects though not on crustacea, the 96-hr LC_{50} figures

Table 14.8. Toxicity of phenoxy herbicide formulations to freshwater crustacea: 48-hr LC_{50} figures in ppm (Sanders, 1970).

	Waterfleas	Seed-shrimp	Scud	Sowbug	Glass-shrimp	Crayfish
2,4-D acid	>100	--	3.2	--	--	--
dimethylamine salt	4.0	8.0	>100	>100	>100	>100
BE ester*	5.6	1.8	5.9	3.2	1.4	>100
PGBE ester†	0.10	0.32	2.6	2.2	2.7	>100
Silvex BE ester	2.1	4.9	0.74	40	8.0	60
PGBE ester	0.18	0.20	1.0	0.50	3.2	>100

* Butoxyethanol ester

† Propylene glycol butyl ether ester(s)

Table 14.9. Effect of silvex (K salt) at 3–5 ppm in a Missouri farm pond: percentage frequencies in 216 samples from treated plots taken regularly throughout the year after treatment (Harp and Campbell, 1964).

	Untreated	Treated		Untreated	Treated
Oligochaeta	82	161	Hemiptera		
			Belostoma	0	1
Hirudinea			Buenoa	1	3
Erpobdella punctata	12	22	Gerris	1	1
Helobdella & Placobdella	28	49	Mesovelia mulsanti	5	2
Gastropoda			Trichoptera		
Helisoma trivolvis	23	43	Agrypna vestita	4	13
Physa gyrina & sayii	11	22	Phryganea	1	0
Pelecypoda			Lepidoptera (Nymphula)	7	9
Musculium securis	0	15			
Sphaerium rhomboideum	0	2	Coleoptera		
			Celina	24	6
Ephemeroptera			Hydrovatus	0	1
Ameletus	1	8	Haliplus	5	4
Caenis	18	34	Peltodytes	12	12
Odonata			Diptera		
Anomalagrion hastatum	8	7	Ceratopogonidae spp.	3	4
Argia	5	2	Chaoborus	27	73
Ischnura	28	50	Chrysops	71	50
Anax junius	0	3	Tendipedidae spp.	93	185
Libellulidae spp.	52	104	Prionocera	8	4
Megaloptera (Chauliodes)	0	2			

Table 14.10. Toxicity of herbicides to freshwater crustacea: 48-hr LC_{50} figures in ppm (Sanders, 1970).

	Waterflea[a]	Seed-shrimp[b]	Scud[c]	Sowbug[d]	Glass-shrimp[e]	Crayfish[f]
Trifluralin	0.56	0.25	1.8	2.0	1.2	50
Molinate	0.60	0.18	0.39	0.40	1.0	5.6
Simazine	1.0	3.2	>100	>100	>100	>100
Vernolate	1.1	0.24	20	5.6	1.9	24
Dichlobenil	10	7.8	18	34	9.0	22
Diphenamid	56	50	>100	>100	58	>100

[a] Daphnia magna   [b] Cypridopsis vidua   [c] Gammarus fasciatus   [d] Asellus brevicaudus

[e] Palaemonetes kadakiensis   [f] Onconectes nais

(ppm) being as follows[110]:

		Diquat	Dichlobenil
Hyalella azteca	(Amphipod)	0.05	8.5
Callibaetis sp.	(Mayfly)	16.4	10.3
Limnephilus sp.	(Caddisfly)	33.0	13.0
Enallagma sp.	(Damselfly)	>100	20.7
Libellula sp.	(Dragonfly)	>100	>100

Diphenamid is considerably less toxic to freshwater crustacea than dichlobenil (Table 14.10), while dicamba resembles fenac in being nontoxic to all of them at 100 ppm.[84]

When tested in a model ecosystem, trifluralin was the only herbicide investigated which accumulated in the aquatic fauna, the biomagnification as compared with water being 930 in *Gambusia* fish and 17,700 in the snail *Physa*.[81] 2,4-D was represented by only small amounts of unknown metabolites, two in *Physa* and one in *Gambusia*. Alachlor and propachlor each left an unknown metabolite in the snail. In the crab *Uca*, dicamba left a conjugated metabolite, pyrazon left one unknown metabolite and pyrazon itself, and cyanazine (Bladex) left three unknown metabolites. Again with the benzothiodiazine bentazon the crab was the only species to show appreciable residues, namely bentazon itself as well as N-isopropylanthranilamide and anthranilic acid.[10]

Sodium Arsenite. Applied at an operational dosage of 4 ppm, this inorganic herbicide caused drastic reductions in copepods, cladocerans, and rotifers[21]; whereas there are notable differences in susceptibility between species of microcrustacea, this reduction is shown in the net decrease of O_2 consumption by the microfauna of treated ponds (Ball, 1966[67]). At lower concentrations sodium arsenite has the effect of increasing the number of benthic invertebrates and thus increasing the growth rate of bluegills in the ponds.[34] Cacodylic acid is one of the least toxic herbicides for the glass-shrimp *Gammarus lacustris*.[82] Arsenicals are bioaccumulated in aquatic species, but not biomagnified in their food chains. MSMA and As are concentrated about 5 times in crustacea, 30 times in algae, and 130 times in *Gambusia* fish, while cacodylic acid shows a higher bioaccumulation ratio.[112]

Estuarine Fauna. In estuaries infested with water milfoil, the 2,4-D esters are particularly effective, whereas endothall and silvex are not; moreover, these 2,4-D formulations do not kill plants such as *Potamogeton, Vallis-*

neria and *Ceratophyllum*.[7] The most effective formulation, the BE esters applied at 30 lb 2,4-D per acre, did not affect the Eastern oyster, the softshell clam and the blue crab in Chesapeake Bay, except in stagnant shallows where the oxygen deficiency resulting from the dead milfoil led to a high bacterial production of H_2S.[77] In Virginia tidal creeks, the anaerobiosis thus produced was the evident cause of a population reduction in the mollusc *Macoma baltica* and the amphipod shrimp *Leptocheirus plumosus*.[39] 2,4-D treatment with the BE ester did not harm the pigtoe mussel *Pleurobema cordatum*; even at 100 lb/a.[14] At this dosage it could also control the eelgrass *Zostera marina,* which is detrimental to oyster farming, without visibly affecting oysters and the blue clam *Mytilus edulis.* Oyster eggs and larvae are more sensitive, and whereas endothall and 2,4-D acid are practically nontoxic (Table 14.11), the BE ester of 2,4-D kills the larvae at 1 ppm[24]; fortunately the concentrations in practice do not exceed 0.1 ppm.[95]

Silvex (as the PGBE ester) is more toxic than the 2,4-D ester to eggs and larvae. The silvex ester, unlike 2,4-D acid, decreases shell growth in adult oysters, 1 ppm reducing it by nearly 25%.[13,14] The PGBE ester of 2,4,5-T was even more toxic. Whereas diquat, paraquat, and atrazine were without effect on oystershell growth, monuron, diuron, and neburon were inhibitory in ascending order. While endothall is nontoxic to the immature stages of the hard clam, some of the urea herbicides pose a hazard.[25]

Whereas 2,4-D ester caused negligible mortality of the brown shrimp *Penaeus aztecus* at 2 ppm in 2 days, the silvex ester at 0.24 ppm caused 50% mortality. The urea and bipyridylium herbicides mentioned previously, with the exception of neburon, were without effect on brown or white shrimp at 1 ppm. Neburon, diuron and the polyglycol butyl esters of silvex and 2,4,5-T were also toxic to estuarine fish such as spot and mullet. Of all the herbicides tested, acrolein was the most toxic, not only to fish but also to oysters and *Penaeus* shrimps.[13,14]

Effects on Phytoplankton

Tests of candidate herbicides for use against water milfoil in estuaries,[13,14] made in salt water on phytoplankton collected from the Florida coast of the Gulf of Mexico (Table 14.12), showed that the 2,4-D and 2,4,5-T acids were harmless, as also were picloram, dalapon, and the thiocarbamate herbicides tested. But the esters of these phenoxy compounds, and also silvex and the bipyridylium herbicides, were deleterious. At 1 ppm the urea herbicides, which are Hill-reaction inhibitors, caused 40–90% inhibition of their photosynthetic activity.[13] Further work on *Protococcus, Chlorella* and *Dunaliella* has also shown that diuron was the most growth-inhibiting of the

Table 14.11. Toxicity of herbicides to immature stages of oysters and clams: LC$_{50}$ figures in ppm (Davis and Hidu, 1969).

Eastern Oyster (Crassostrea virginica)			Hard Clam (Mercenaria mercenaria)		
	Eggs, 48-hr	Larvae, 14-day		Eggs, 48-hr	Larvae, 12-day
Silvex	5.9	0.71	Neburon	<2.4	<2.4
2,4-D Ester	8.0	0.74	Diuron	2.53	>5
MCPA	15.6	31.3	Monuron	>5	>5
2,4-D Salt	20.4	64	Fenuron	>10	>5
Endothall	28.2	48	Endothall	51	12.5
Amitrole	734	255			

Table 14.12.   Effect of herbicides at 1 ppm on estuarine phytoplankton: percent reduction in carbon fixation during 4 hr exposure (Butler, 1965).

Monuron	94	2,4-D, PGBE ester	44	Dalapon-Na	0
Neburon	90	Fenuron	41	2,4-D acid	0
2,4,5-T ester	89	DCPA	37	2,4-D dimethylamine	0
Diuron	87	Pebulate	24	2,4,5-T acid	0
Silvex*	78	2,4-D, BE ester	16	MCPA amine	0
Paraquat	53	Molinate	9	EPTC	0
Diquat	45	Picloram	0	Vernolate	0

* PGBE ester

urea herbicides and fenuron the least.[98] At 0.1 ppm, diuron reduced the biomass of the freshwater alga *Scenedesmus quadricaudata* by 99%, as compared to a 12% reduction with 2,4-D at this concentration.[92] Tests performed on marine unicellular algae (Table 14.13) show that the triazines, also Hill-reaction inhibitors, as well as the urea herbicides and trifluralin, inhibit their growth at concentrations below 1 ppm, whereas the phenoxys and the other herbicides would not cause inhibition at concentrations at least 10 times greater.[103]

Little is known of the effect of herbicides on aquatic bacteria, among which *Escherichia coli* is important; the much greater dilution of the organisms and the herbicide in water as compared to that in soil is partly counterbalanced by the tendency of the microorganisms to adsorb and concentrate those pesticides that are least water soluble.[105] The accumulation of dead vegetation resulting from a paraquat treatment does induce a many-fold increase in the heterotrophic bacteria in the water and mud, and a fivefold increase in bacteria resistant to paraquat, but the numbers return to normal about 2 wk later.[30]

Populations of phytoplankton are often stimulated as a result of the destruction of the macrophytes by the herbicide. Elimination of *Potamogeton* and *Chara* from a Florida pond by means of 1 ppm dichlobenil was soon followed by a phytoplankton bloom, consisting of *Gonyaulax, Oscillatoria* and other species, that lasted for a month; almost simultaneously a rich zooplankton of *Diaptomus* copepods and *Keratella* rotifers, along with ostracods and cladocerans, developed and persisted for the same period.[104] Phytoplankton blooms and subsequent zooplankton increases also followed applications of simazine or atrazine, although the response was considerably slower. The elimination of *Largarosiphon* from a shallow bay of a New Zealand lake by means of 0.5 ppm diquat was followed several weeks later by blooms of *Anabaena* and *Staurastrum* algae and green flagellates.[27] Blooms have not resulted from the application of sodium arsenite, silvex, or fenac.

Table 14.13. Effect of herbicides on marine unicellular algae: concentrations (ppm) at which growth over 10 days is 50% inhibited (Walsh, 1972).

Herbicide (free acid)	*Chlorococcum* sp.	*Dunaliella tertiolecta*	*Isochrysis galbana*	*Phaeodactylium tricornutum*
Diuron	0.01	0.02	0.01	0.01
Neburon	0.03	0.04	0.03	0.03
Monuron	0.10	0.15	0.13	0.10
Fenuron	0.75	1.5	0.75	0.75
Ametryn	0.01	0.04	0.01	0.02
Atrazine	0.10	0.30	0.10	0.20
Simazine	2.0	5.0	0.50	0.50
Prometone	0.50	1.5	1.0	0.25
2,4-D	50	75	50	50
2,4,5-T	100	125	50	50
Silvex	25	25	5.0	5.0
Trifluralin	2.5	5.0	2.5	2.5
Dalapon	50	100	20	25
Chloramben	50	50	15	25
Endothall	50	50	25	15
Dichlobenil	60	60	60	25
Paraquat*	50	20	5.0	10
Diquat[†]	200	30	15	15

* Dichloride   [†]Dibromide

REFERENCES CITED

1. J. S. Alabaster. 1969. Survival of fish in 164 pesticides, wetting agents, and miscellaneous substances. *Int. Pest Control* 11(2):29–35.

2. O. M. Aly and S. D. Faust. 1964. Studies on the fate of 2,4-D and ester derivatives in natural surface water. *J. Agric. Food Chem.* 12:541–546.

3. J. A. Axe, A. C. Mathers, and A. F. Wiese. 1969. Disappearance of atrazine, propazine, and trifluralin from soil and water. *Southern Weed Conf., Proc.* 22:367.

4. G. W. Bailey, A. D. Thruston, J. D. Pope, and D. R. Cochrane. 1970. The degradation kinetics of an ester of silvex and the persistence of silvex in water and sediments. *Weed Sci.* 18:413–416.

5. F. L. Baldwin, P. W. Santelmann, and J. M. Davidson. 1975. Prometryn movement across and through the soil. *Weed Sci.* 23:285–288.

6. A. P. Barnett, E. W. Hauser, A. W. White, and J. H. Holladay. 1967. Loss of 2,4-D in washoff from cultivated fallow land. *Weeds* 15:133–137.

7. G. F. Beaven, C. K. Rawls, and G. E. Beckett. 1962. Field observations upon estuarine animals exposed to 2,4-D. *N.E. Weed Control Conf., Proc.* 16:449–458.

8. E. Blok. 1969. Experiments with herbicides in fishing waters 1964–1967. *Weed Abstr.* 18:117, No. 746.

9. B. L. Bohmont. 1967. Toxicity of herbicides to livestock, fish, honeybees, and wildlife. *20th Western Weed Control Conf. Proc.* 21:25–27.

10. G. M. Booth, C. C. Yu, and D. J. Hansen. 1973. Fate, metabolism, and toxicity of Bentazon in a model ecosystem. *J. Environ. Quality* 2:408–411.

11. M. P. Brooker and R. W. Edwards. 1973. Effects of the herbicide paraquat on the ecology of a reservoir. *Freshwater Biol.* 3:157–175.

12. J. E. Burns, F. M. Miller, E. D. Gomes, and R. A. Albert. 1974. Hexachlorobenzene exposure from contaminated DCPA in vegetable spraymen. *Arch. Environ. Health* 29:192–194.

13. P. A. Butler. 1963. Commercial fisheries investigations. In *Pesticide Wildlife Studies 1962.* U.S. Fish Wildl. Serv. Circ. 167. pp. 11–25.

14. P. A. Butler. 1965. Effects of herbicides on estuarine fauna. *Southern Weed Control Conf., Proc.* 18:576–580.

15. G. Chesters and J. G. Konrad. 1971. Effects of pesticide usage on water quality. *J. Environ. Quality* 1:14–17.

16. O. B. Cope. 1964. Sport fisheries investigations. In *Pesticide Wildlife Studies 1963.* U.S. Fish Wildl. Serv. Circ. 199.

17. O. B. Cope. 1965. Sport fisheries investigations. In *Pesticide Wildlife Studies.* U.S. Fish Wildl. Serv. Circ. 226. pp. 51–63.

18. O. B. Cope. 1966. Contamination of the freshwater ecosystem by pesticides. *J. Appl Ecol.* 3(Suppl.):33–44.

19. O. B. Cope, J. P. McCraren, and L. Eller. 1969. Effects of dichlobenil in two fishpond environments. *Weed Sci.* 17:158–165.

20. O. B. Cope, E. M. Wood, and G. H. Wallen. 1970. Some chronic effects of 2,4-D on the bluegill (*Lepomis macrochirus*). *Trans. Am. Fish. Soc.* 99:1–12.

21. B. C. Cowell. 1965. The effects of sodium arsenite and silvex on the plankton populations in farm ponds. *Trans. Am. Fish. Soc.* **94:**371–377.

22. C. A. Crandall and C. J. Goodnight. 1962. Effects of sublethal concentrations of several toxicants on growth of the common guppy. *Limnol. Oceanogr.* **7:**233–239.

23. D. G. Crosby and R. K. Tucker. 1966. Toxicity of aquatic herbicides to *Daphnia magna. Science* **154:**289–291.

24. H. C. Davis. 1961. Effects of some pesticides on eggs and larvae of oysters and clams. *Comm. Fish. Rev.* **23**(12):8–23.

25. H. C. Davis and H. Hidu. 1969. Effects of pesticides on embryonic development of clams and oysters and on survival and growth of the larvae. *Fishery Bull. (U.S. Fish. Wildl. Serv.)* **67:**393–404.

26. R. D. Earnest. 1971. The effect of paraquat on fish in a Colorado farm pond. *Progr. Fish-Cult.* **33:**27–31.

27. G. R. Fish. 1966. Some effects of the destruction of aquatic weeds in Lake Rotoiti, New Zealand. *Weed Res.* **6:**350–358.

28. P. A. Frank. 1972. Herbicidal residues in aquatic environments. In *Fate of Organic Pesticides in the Aquatic Environment.* Adv. Chem. Ser. 111. pp. 135–148.

29. P. A. Frank and R. D. Comes. 1967. Herbicidal residues in pond water and hydrosoils. *Weeds* **15:**210–213.

30. J. C. Fry, M. P. Brooker, and P. L. Thomas. 1973. Changes in the microbial populations of a reservoir treated with paraquat. *Water Res.* **7:**395–406.

31. H. H. Funderburk. 1969. Paraquat and diquat. In *Degradation of Herbicides.* Eds. P. C. Kearney and D. D. Kaufman. Dekker. pp. 282–298.

32. FWPCA. 1968. *Water Quality Criteria.* Rep. Nat. Tech. Adv. Comm. to Sec. Interior. Fed. Water Poll. Contr. Admin., U.S. Dept. Interior 234 pp.

33. S. D. Gerking. 1948. Destruction of aquatic plants by 2,4-D. *J. Wildl. Manage.* **12:**221–227.

34. P. A. Gilderhus. 1966. Some effects of sublethal concentrations of sodium arsenite on bluegills and the aquatic environment. *Trans. Am. Fish. Soc.* **95:**289–296.

35. K. P. Goswami and R. E. Green. 1971. Microbial degradation of the herbicide atrazine and its 2-hydroxy analog in submerged soils. *Environ. Sci. Technol.* **5:**426–429.

36. A. R. Grzenda, H. P. Nicholson, and W. S. Cox. 1966. Persistence of four herbicides in pond water. *J. Am. Water Works Assoc.* **58:**326–332.

37. J. K. Hall, M. Pawlus, and E. R. Higgins. 1972. Losses of atrazine in runoff water and soil sediment. *J. Environ. Quality* **1:**172–176.

38. G. L. Harp and R. S. Campbell. 1964. Effects of the herbicide silvex on benthos of a farm pond. *J. Wildl. Manage.* **28:**308–317.

39. D. Haven. 1963. Mass treatment with 2,4-D on milfoil in tidal creeks in Virginia. *Southern Weed Conf., Proc.* **16:**345–350.

40. R. B. Hemmett and S. D. Faust. 1968. Biodegradation kinetics of 2,4-dichlorophenoxyacetic acid by aquatic microorganisms. *Residue Rev.* **29:**191–207.

41. W. Hilsenhoff. 1966. Effect of diquat on aquatic insects and related animals. *J. Econ. Entomol.* **59:**1520–1521.

42. R. C. Hiltibran. 1967. Effects of some herbicides on fertilized fish eggs and fry. *Trans. Am. Fish. Soc.* **96:**414–416.

43. E. Hindin, D. S. May, and G. H. Dunstan. 1964. Collection and analysis of synthetic organic pesticides from surface and ground water. *Residue Rev.* 7:130–156.

44. J. P. Hollis. 1972. Nematicide-weeds interaction in rice fields. *Plant Dis. Rptr.* 56:420–424.

45. F. N. Hooper. 1958. The effect of applications of pelleted 2,4-D upon the bottom fauna of Kent Lake, Michigan. *North Centr. Weed Control Conf., Proc.* 15:41.

46. A. Houser. 1963. Loss in weight of sunfish following aquatic vegetation control using the herbicide silvex. *Proc. Oklahoma Acad. Sci.* 43:232–237.

47. J. C. Huang and E. F. Gloyna. 1968. Effect of organic compounds on photosynthetic oxidation. *J. Water Res.* 2:347–366.

48. J. S. Hughes and J. T. Davis. 1962. Comparative toxicity to bluegill sunfish of granular and liquid herbicides. *Conf. S.E. Assoc. Game Fish Comm., Proc.* 16:319–323.

49. J. S. Hughes and J. T. Davis. 1963. Variations in the toxicity to bluegill sunfish of phenoxy herbicides. *Weeds* 11:50–53.

50. A. R. Isensee, P. C. Kearney, E. A. Woolson, G. E. Jones, and V. P. Williams. 1973. Distribution of alkyl arsenicals in model ecosystems. *Environ. Sci. Technol.* 7:841–845.

51. M. G. Johnson. 1965. Control of aquatic plants in farm ponds in Ontario. *U.S. Fish Wildl. Serv., Progr. Fish-Cult.* 27:23–30.

52. P. C. Kearney, E. A. Woolson, and C. P. Ellington. 1972. Persistence and metabolism of chlordioxins in soils. *Environ. Sci. Technol.* 6:1017–1019.

53. J. E. King and W. T. Penfound. 1946. Effects of two of the new formagenic herbicides on bream and large-mouth bass. *Ecology* 27:372–374.

54. J. Lhoste and P. Roth. 1946. Sur l'action des solutions aqueeuses de 2,4-D sur l'evolution des oeufs de *Rana temporaria. Compt. Rend. Soc. Biol.* 140:272–273.

55. H. L. Lindaberry. 1961. Considerations regarding the use of aquathol in potable watersheds. *N.E. Weed Contr. Conf., Proc.* 15:481–484.

56. K. J. Macek. 1969. Biological magnification of pesticide residues in food chains. *Environ. Health Series, Oregon State Univ.,* No. 1, pp. 17–21.

57. K. J. Macek, K. S. Buxton, S. Sauter, S. Gnilka, and J. W. Dean. 1976. *Chronic Toxicity of Atrazine to Selected Aquatic Invertebrates and Fishes.* Environmental Protection Agency, Washington. Processed Publ. EPA-600/3-76-047. 50 pp.

58. K. M. Mackenthun. 1960. Some limnological investigations on the long-term use of sodium arsenite as an aquatic herbicide. *Proc. N. Central Weed Control Conf.* 17:30–31.

59. D. B. Manigold and J. A. Schulze. 1969. Pesticides in selected western streams—a progress report. *Pestic. Monit. J.* 3:124–135.

60. R. B. Marston, D. W. Schultz, T. Shiroyama, and L. V. Snyder. 1968. Amitrole concentrations in creek waters downstream from an aerially sprayed watershed sub-basin. *Pestic. Monit. J.* 2:123–128.

61. J. P. McCraren, O. B. Cope, and L. Eller. 1969. Some chronic effects of diuron on bluegills. *Weed Sci.* 17:497–504.

62. F. M. Miller and E. D. Gomes. 1974. Detection of DCPA residues in environmental samples. *Pestic. Monit. J.* 8:53–58.

63. R. A. Miller, L. A. Norris, and C. L. Hawkes. 1973. Toxicity of TCDD in aquatic organisms. *Environ. Health Perspect. (1973).* pp. 177–186.

64. D. I. Mount and C. E. Stephan. 1967. A method for establishing acceptable toxicant levels for fish. *Trans. Am. Fish. Soc.* **96**:185–193.

65. R. C. Muirhead-Thomson. 1971. *Pesticides and Freshwater Fauna.* Academic. pp. 67–70.

66. W. R. Mullison. 1966. Some toxicological aspects of silvex. *Southern Weed Conf., Proc.* **19**:420–435.

67. W. R. Mullison. 1970. Effects of herbicides on water and its inhabitants. *Weed Sci.* **18**:738–750.

68. National Academy of Sciences. 1972. *Water Quality Criteria.* Environ. Studies Bd. and Nat. Acad. Engin. U.S. Gov't. Printing Office, Washington, D.C. 594 pp.

69. National Academy of Sciences. 1974. Defoliation in South Vietnam. *Nat. Acad. Sci. News Report* **24**(3–4):1, 4–11.

70. J. F. Newman and J. M. Way. 1966. Some ecological observations in the use of paraquat and diquat as aquatic herbicides. *Br. Weed Contr. Conf. Proc. 1966.* pp. 582–585.

71. H. P. Nicholson. 1969. Occurrence and significance of pesticide residues in water. *J. Washington Acad. Sci.* **59**:77–85.

72. L. A. Norris. 1971. Chemical brush control: Assessing the hazard. *J. Forestry* **69**:715–720.

73. J. H. Patric. 1971. Herbicides and water quality in American forestry. *Proc. Northeast. Weed Sci. Soc.* **25**:365–375.

74. M. Pierce. 1960. Progress report of the effect of Kuron on the biota of Long Pond, Dutchess County. *N.E. Weed Contr. Conf. Proc.* **14**:472–475 (also **12**:338–343).

75. D. Pimentel. 1971. *Ecological Effects of Pesticides on Non-target Species.* Office Sci. Technol., Exec. Office President, Gov't. Printing Office, Washington, D.C. 220 pp.

76. H. B. Pionke and G. Chesters. 1973. Pesticide–sediment–water interactions. *J. Environ. Quality* **2**:29–45.

77. C. K. Rawls. 1965. Field tests of herbicide toxicity to certain estuarine animals. *Chesapeake Sci.* **6**:150–151.

78. J. J. Richard et al. 1975. Analysis of various Iowa waters for selected pesticides: Atrazine, DDE, and dieldrin, 1974. *Pestic. Monit. J.* **9**:117–123.

79. W. F. Ritter, H. P. Johnson, W. G. Lovely, and M. Molnau. 1974. Atrazine, propachlor, and diazinon residues on small agricultural watersheds. *Environ. Sci. Technol.* **8**:38–42.

80. C. A. Rodgers. 1970. Uptake and elimination of simazine by green sunfish. *Weed Sci.* **18**:134–136.

81. J. R. Sanborn. 1974. *The Fate of Select Pesticides in the Aquatic Environment.* U.S. Environ. Prot. Agency, Ecol. Res. Series EPA-660/3-74-025. 83 pp.

82. H. O. Sanders. 1969. *Toxicity of Pesticides to the Crustacean* Gammarus lacustris. Tech. Paper 25, Bur. Sport. Fish. Wildl. U.S. Dept. Interior. 18 pp.

83. H. O. Sanders. 1970. Pesticide toxicities to tadpoles of the western chorus frog and Fowler's toad. *Copeia* **1970**(2):246–251.

84. H. O. Sanders. 1970. Toxicities of some herbicides to six species of freshwater crustaceans. *J. Water Poll. Contr. Fed.* **42**:1544–1550.

85. H. O. Sanders and O. B. Cope. 1966. Toxicities of several pesticides to two species of cladocerans. *Trans. Am. Fish. Soc.* **95**:165–169.

86.   H. O. Sanders and O. B. Cope. 1968. The relative toxicity of several pesticides to naiads of three species of stoneflies. *Limnol. Oceanogr.* **13**:112–117.

87.   D. P. Schultz. 1973. Dynamics of a salt of 2,4-D in fish, water, and hydrosol. *J. Agric. Food Chem.* **21**:187–192.

88.   T. J. Sheets, J. R. Bradley, and M. D. Jackson. 1973. Movement of trifluralin in surface water. *Proc. S. Weed Sci. Soc.* **26**:376. See also Water Resour. Inst., N. Carolina State Univ., *Rep. No. 60.*

89.   R. L. Simpson and D. Pimentel. 1972. Ecological effects of the aquatic herbicide fenac on small ponds. *Search* (Cornell Univ. Agric. Exp. Stn.) **2**(10):1–59.

90.   G. E. Smith and B. G. Isom. 1967. Investigation of effect of large-scale applications of 2,4-D on aquatic fauna and water quality. *Pestic. Monit. J.* **1**(3):16–21.

91.   J. R. Snow. 1963. Simazine as an algicide for bass ponds. *Progr. Fish-Cult.* **25**:34–36.

92.   L. Stadnyk, R. S. Campbell, and B. T. Johnson. 1971. Pesticide effect on growth and ^{14}C assimilation in a freshwater alga. *Bull. Environ. Contam. Toxicol.* **6**:1–8.

93.   E. W. Surber and Q. H. Pickering. 1962. Acute toxicity of endothall, diquat, dalapon, and silvex to fish. *Progr. Fish-Cult.* **24**:161–171.

94.   W. M. Tatum and R. D. Blackburn. 1962. Preliminary study of the effects of diquat on the natural bottom fauna and plankton in two subtropical ponds. *Proc. S.E. Assoc. Game Fish Comm.* **16**:301–307.

95.   M. L. J. Thomas and J. R. Duffy. 1968. Butoxyethanol ester of 2,4-D in the control of eelgrass and its effects on oysters and other benthos. *N.E. Weed Contr. Conf., Proc.* **22**:186–194.

96.   D. W. Trichell, H. L. Morton, and M. G. Merkle. 1968. Loss of herbicides in runoff water. *Weed Sci.* **16**:447–449.

97.   F. H. Tschirley. 1969. The ecological consequences of the defoliation program in Vietnam. *Science* **163**:779–786.

98.   R. Ukeles. 1962. Growth of pure cultures of marine phytoplankton in the presence of toxicants. *Appl. Microbiol.* **10**:532–537.

99.   C. van Valin. 1966. Persistence of 2,6-dichlorobenzonitrile aquatic environments. In *Organic Pesticides in the Environment.* Adv. Chem. Ser. 60. pp. 271–279.

100. C. R. Walker. 1963. Endothall derivatives as aquatic herbicides in fishery habitats. *Weeds* **11**:226–232.

101. C. R. Walker. 1964. Simazine and other *s*-triazine compounds as aquatic herbicides in fish habitats. *Weeds* **12**:134–139.

102. C. R. Walker. 1965. Diuron, fenuron, monuron, neburon, and TCA mixtures as aquatic herbicides in fish habitats. *Weeds* **13**:297–301.

103. G. E. Walsh. 1972. Effects of herbicides on photosynthesis and growth of marine unicellular algae. *Hyacinth Control J.* **10**:45–48.

104. G. E. Walsh, C. W. Miller, and P. T. Heitmuller. 1971. Uptake and effects of dichlobenil in a small pond. *Bull. Environ. Contam. Toxicol.* **6**:279–288.

105. G. W. Ware and C. C. Roan. 1970. Interaction of pesticides with aquatic microorganisms and plankton. *Residue Rev.* **33**:15–45.

106. Weed Science Society of America. 1970. *Herbicide Handbook.* 2nd edition. Dept. of Agronomy, Univ. of Illinois, Urbana. 368 pp.

107. A. W. White, A. P. Barnett, B. G. Wright, and J. H. Holladay. 1967. Atrazine losses from fallow land caused by runoff and erosion. *Environ. Sci. Technol.* **1**:740–744.

108. E. W. Whitney. 1968. The toxicity of silvex to aquatic fauna. *Weed Soc. Am. Abstr.* **1968**:63.

109. G. H. Willis, R. L. Rogers, and E. M. Southwick. 1975. Losses of diuron, linuron, fenac, and trifluralin in surface drainage water. *J. Environ. Qual.* **4**:399–402.

110. D. C. Wilson and C. E. Bond. 1969. The effects of the herbicides diquat and dichlobenil (Casoron) on pond invertebrates. I. Acute toxicity. *Trans. Am. Fish. Soc.* **98**:438–443.

111. T. A. Wojtalik, T. F. Hall, and L. O. Hill. 1971. Monitoring ecological conditions associated with wide-scale applications of DMA and 2,4-D to aquatic environments. *Pestic. Monit. J.* **4**:184–203.

112. E. A. Woolson. 1975. Bioaccumulation of arsenicals. In *Arsenical Pesticides.* Am. Chem. Soc. Symposium Series 7. pp. 97–107.

113. E. A. Woolson, R. F. Thomas, and P. D. J. Ensor. 1972. Survey of polychlorodibenzo-*p*-dioxin content in selected pesticides. *J. Agric. Food Chem.* **20**:351–359.

114. WSSA. 1967. *Herbicide Handbook.* Weed Science Society of America. Humphrey Press, Geneva, N.Y. 293 pp.

15

FUNGICIDES AND THE SOIL MICROFLORA

A. PERSISTENCE OF FUNGICIDES IN SOILS

A number of simple organic compounds are applied to greenhouse soil and seedbeds to control root-rot and damping-off fungi, as well as nematodes and soil insects. These include methyl bromide and chloropicrin, which are injected into the soil at about 5 ml per square foot. Dichloropropene (Telone) and dichloropropane–dichloropropene mixture (D-D) are applied to fields primarily as nematicides at 20–40 gal/a, while EDB is applied primarily as a fungicide at 1–4 gal/a. Allyl alcohol is used in soil drenches as a herbicide as well as a fungicide. More recent soil fumigants include metham (Vapam, a dithiocarbamate) and dazomet (Mylone, a thiodiazine). In addition, many of the dithiocarbamates (and also PCNB, Dexon, and even captan) have been applied in drenches to seedbeds and market-gardens, besides being used along with the quinones as surface treatment for seeds planted in fields. Benomyl, like other benzimidazole compounds, acts as a systemic fungicide; moreover, it has given indications of either repelling[66] or killing[59] certain soil nematodes.

Volatile Fungicides

The fumigant fungicides quickly dissipate from soils, as may be shown by bioassaying the treated soil with *Pythium*. Applications of allyl alcohol, like formalin, have lost their fungitoxicity by the second day treatment. Metham and dazomet have an effective life of 2 days,[17] their loss of activity being independent of microorganisms.[20] Both fungicides give rise to the toxicant methyl isothiocyanate.[70,71] Dazomet also degrades to formaldehyde and dimethylamine, and some N,N'-dimethyl urea is produced as a result of resynthesis.[20] Metham may combine in the soil either with its hydrolysis product methylamine to form dimethylthiourea (DMTU), or with methyl isothiocyanate to produce small amounts of dimethyl-thiuram disulfide (DMTD) (Fig. 15.1), and these two toxic residual contaminants may be left behind in the metham-treated soil.[93]

Nonvolatile Fungicides

The less volatile synthetic organic fungicides have half-lives exceeding 2 wk, as bioassays with *Pythium* and *Rhizoctonia* have shown, as follows[17]:

	Concn, ppm	Half-life, days		Concn, ppm	Half-life, days
Nabam	100	16	Ziram	1800	45
Chloranil	2000	17	Captan	250	70
Thiram	160	38	Zineb	800	77

While the period of appreciable fungicidal activity in soil is 60 days for thiram and at least 65 days for captan,[80] it considerably exceeds 8 months for PCNB (pentachloronitrobenzene). Dexon, a soil fungicide specific for *Pythium*, was still persisting at high concentrations more than 1 yr after application.[2]

Ferbam, ziram, and thiram produce the dimethyldithiocarbamate ion (DMDC) in fungi and yeasts[51] (Fig. 15.2). In alkali ferbam yields the DMDC ion, and in acid below pH 5 it yields dimethylamine and CS_2.[77] Soil microorganisms form amino-acid adducts with DMDC.[85] The soil fungi *Penicillium notatum, Glomerella cingulata* and *Fusarium roseum* produce CS_2 from thiram.[86] Thiram may exchange radicals with SH compounds

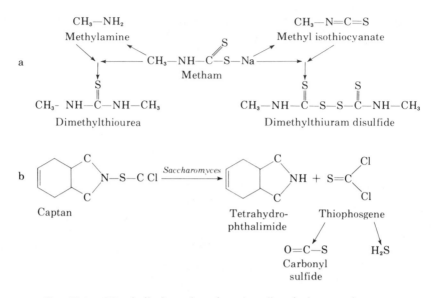

Fig. 15.1. Metabolic fate of metham in soil and of captan in yeast.

in cells, as demonstrated with *p*-nitrothiophenol.[76] Among the ethylene-bisdithiocarbamate fungicides, nabam, maneb, and zineb break down in water to yield sulfur and ETM (ethylene thiuram monosulfide) and ETU (ethylenethiourea).[96] When aerated in aqueous solution, they yield several degradation products (Fig. 15.2) including ETM, EDI (ethylene-diisothiocyanate), ETU, ethylenediamine, and sulfur.[24] Ethylenethiourea (ETU) is a known carcinogen, but in most soils it is rapidly broken down into ethyleneurea and thence to CO_2.[41] Carbonyl sulfide is also released as a vapor above nonsterile soil treated with nabam.[67] Ferbam and nabam lose their biological activity in soil much faster than captan does, and this loss is unaffected by sterilization.

Captan produces thiophosgene on contact with yeast and fungal tissues[52]; this is also active fungicidally, and gives rise to di(thiophosgene). In addition, tetrahydrophthalimide, carbonyl sulfide and H_2S may be produced by the subcellular components of *Saccharomyces pastorianus* from captan, and from folpet[84] (Fig. 15.1). The half-life of captan in moist silt loam can be as low as 3.5 days, as compared to 1 day for dichlone and 12 hr for anilazine; when the soil was dry, the half-lives were 2.5 days for anilazine and more than 50 days for captan.[9]

Dichlone, which like captan is applied to the foliage of apple trees, gave rise to soil residues in New York orchards which never exceeded 6 ppm nor penetrated more than 2 inches, while metiram (a mixture of two zinc di-

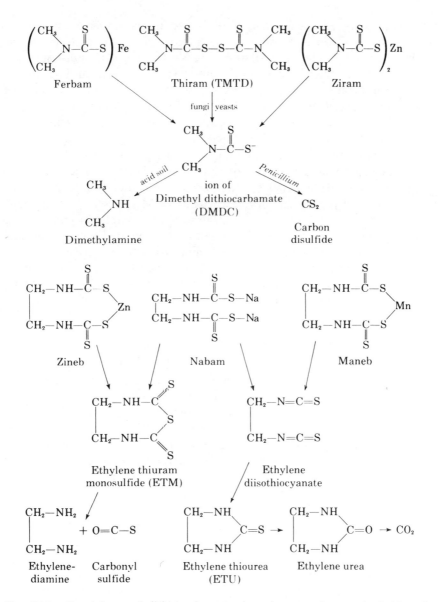

Fig. 15.2. Breakdown of dithiocarbamates by microorganisms and of thiocarbamates in water (Menzie, 1969, 1974).

thiocarbamates) left an average of 0.25 ppm in the upper 2 inches with small amounts down to 6 inches; neither fungicide could be found in the following spring.[48] When the persistent PCNB is applied to moist soil, it is slowly reduced to PCA (pentachloroaniline), which is stable but less biologically active; in submerged soil the PCA produced has already exceeded the PCNB remaining within 10 days of the application.[45] Among eight species of fungi, seven species of *Streptomyces,* and one *Nocardia* which were investigated, all of them degraded PCNB to PCA.[11]

The systemic fungicide benomyl applied to irrigated cotton fields at 100 mg/sq ft (ca. 10 lb/a) gave a soil concentration of 28 ppm which declined to 4 ppm in 18 wk.[35] In aqueous solution, benomyl is quite rapidly hydrolyzed to produce methyl-2-benzimidazole carbamate (MBC), which is probably the actual toxicant for fungi.[14] In soil, benomyl is completely converted to MBC in a few hours; this metabolite is further degraded to nonfungicidal metabolites, and four species of bacteria and two of fungi were isolated that can effect this degradation.[34] Carboxin is degraded to its sulfoxide (not to aniline), the conversion in soil being complete in 2 wk, and can be further oxidized to the sulfone in hydrosoils.[13]

Mercury Fungicides

Since mercuric chloride, applied on potato seed-pieces to protect them from *Rhizoctonia,* gradually lost its fungicidal activity, Booer (1944) concluded that the H_2S produced by the soil bacterium *Vibrio desulfuricans* must have sequestered the Hg as mercuric sulfide, although he could not detect it. A quarter-century later, Jensen and Jernelov (1969) demonstrated that hydrosoils from lakes or aquaria converted mercuric chloride to dimethylmercury, the activity being due to the microorganisms in it. Although mercuric sulfide can be produced under anaerobic conditions, the microorganisms in aquatic sediments can produce the methylmercury ion and dimethylmercury (Fig. 15.3) from phenylmercury, alkoxy-alkylmercury and even metallic Hg under anaerobic as well as aerobic conditions.[39] An anaerobic bacterium was isolated from sediment which could transfer methyl radicals from methylcobalamine to the mercuric Hg ion.[103] Vice versa, a pseudomonad was found that could remove the alkyl groups from salts of methylmercury, ethylmercury, and phenylmercury, producing metallic Hg and methane, ethane, and benzene, respectively.[28] Thus mercury may be metabolically cycled between methylated and demethylated forms by various soil and hydrosoil microorganisms.[41] On the other hand, microorganisms from Wisconsin lake sediments produced diphenylmercury as a major metabolite, methylmercury not appearing among the other metabolites.[58]

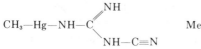

CH₃—Hg—NH—C⟨ NH / NH—C≡N   Methylmercury dicyandiamide
                                (MMD)

CH₃—Hg⁺   Hg₂
Methylmercury ion   Metallic mercury

CH₃—Hg—CH₃
Dimethylmercury

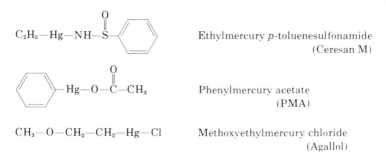

C₂H₅—Hg—NH—S(=O)⟨ ⟩   Ethylmercury *p*-toluenesulfonamide
                          (Ceresan M)

⟨ ⟩—Hg—O—C(=O)—CH₃   Phenylmercury acetate
                         (PMA)

CH₃—O—CH₂—CH₂—Hg—Cl   Methoxyethylmercury chloride
                         (Agallol)

Fig. 15.3. Interrelationship between organomercury fungicides, methyl mercury, and metallic mercury.

It was also demonstrated[42] that soil, provided it had not been sterilized, degraded PMA (phenylmercury acetate) to elemental Hg vapor which could be collected and weighed, the loss due to volatilization from the soil amounting to 10–15% in 4 wk. Subsequently a bacterium with the characteristics of a *Pseudomonas* was isolated from soil which indeed produced Hg vapor from PMA.[92] Whereas EMA (ethylmercury acetate) was lost at about the same rate by volatilization both as elemental Hg and the unchanged organomercurial, MMD (methylmercurydicyandiamide) was volatilized from the soil as the unchanged organic mercurial at about half this rate.[42]

MMD was also quite rapidly sorbed in the soil and there degraded by various species of *Bacillus,* while hydroxymercurichlorophenol (Semesan) was similarly broken down by the fungi *Aspergillus niger* and *Penicillium notatum.*[88] Semesan disappeared from the soil much more rapidly than the dithiocarbamates and captan, the loss depending on the activity of microorganisms.[69]

The use of mercurial-treated seed introduces about 1 gram of Hg per acre to the soil, which is no more than that added by rain or snow contaminated by industrial sources of mercury.[27] Nevertheless, since mercurial fungicides with alkyl groups are converted to the highly toxic and persistent methyl-mercury in soil, these were first replaced by alkoxy mercurials, and later all mercurials were suspended in the United States. The bulk of the environ-mental contamination derives from the aerial transport of mercury released in the burning of coal, and from water transport of the mercury used as catalysts in industry. Thus Lake Michigan sediments had come by the 1970s to contain 0.2 ppm methylmercury, and inflowing streams such as the Grand River contained 10 ppm of this contaminant in their sediments.[43]

Recent countermeasures to overcome the residue problem with mercurials and the resistance problem with the benzimidazoles have been the development of certain organophosphorus compounds as fungicides, particularly for powdery mildews[26]; examples are triamiphos (Wepsyn) and Dowco 199 (diethyl phthalimidophosphorothioate).

B.   EFFECTS OF FUNGICIDES ON THE SOIL FLORA

The materials developed as fungicides have been chosen for their lack of phytotoxicity to higher plants. Even the copper fungicides, which accumu-late in the soil year after year, had no effect on the growth of apple trees and grass cover in orchards of eastern England.[65] However, frequent spray-ing with Bordeaux mixture resulting in soil accumulations reaching 300 ppm Cu has adversely affected vegetable crops in Florida, while Cu fertil-izers were considered partly responsible for an iron-deficiency chlorosis of citrus seedlings.[57] The use of zinc oxysulfate as a bactericide on peach foliage in South Carolina has caused an excess of Zn in the soil sufficient to reduce the growth of the peach trees, and sulfur fungicides in Nova Scotia can make the soil so acid that beans are partially and carrots totally sup-pressed.[53] Nevertheless, the principal floral side effects of fungicides are among the fungi themselves.

Changes in Species Composition

The fumigant compounds applied to soil control nematodes as well as pathogenic fungi, such as methyl bromide, chloropicrin, and D-D mixture, have the effect of increasing the general bacterial counts.[18,44,61] This effect may derive from the removal of competition, and the utilization of the dead fungi or the fungicide as food for the bacteria.[56] The actinomycetes are even more tolerant of fungicides than the bacteria.[98] Among the nonvolatile fun-gicides, the dithiocarbamates such as nabam and thiram stimulate bacteria and actinomycetes, and captan stimulates bacteria, while the volatile dithiocarbamate metham (Vapam) stimulates the actinomycetes.[47] The persistent fungicide PCNB neither stimulates nor inhibits bacteria and actinomycetes even at 200 ppm in the soil.[45] The systemic fungicide benomyl causes no change in bacterial populations, and reduces the numbers of actinomycetes as well as fungi.[83]

Soon after the initial surge of bacteria and actinomycetes, the microflora in treated soil establishes a biological equilibrium largely restricted to a few tolerant species of fungi. *Pythium* and other Phycomycetes are the most susceptible, the order of tolerance among the pathogenic root-rot or damping-off fungi being:

Metham: *Pythium < Phytophthora < Rhizoctonia < Fusarium*
            < Verticillium
D-D mixture: *Pythium & Phytophthora < Fusarium < Verticillium*
Thiram: *Rhizoctonia < Pythium < Phytophthora*
Captan: *Pythium < Rhizoctonia < Phytophthora* and
             Sclerotium

Dexon is specific for *Pythium* but ineffective against *Rhizoctonia* which decomposes it; PCNB is remarkable in controlling *Rhizoctonia* and other root pathogens but not *Pythium*. EDB is also nontoxic to *Rhizoctonia*, while *Verticillium* is the most resistant pathogen to methyl bromide.[47] Of the useful nonpathogens in the soil, the chitinolytic fungi are the most sensitive to fungicides, and no other species can perform their function.[19]

The fungi surviving to become dominant in the new ecosystem are species in the genera *Aspergillus, Penicillium, Mucor, Pyrenochaeta,* and *Trichoderma.* Often *Trichoderma viride* becomes predominant, mainly because it has the property of suppressing other fungi. The following species are encouraged by the following fungicides[3]:

D-D mixture: *Trichoderma viride, Gliocladium penicilloides,*
             Pyrenochaeta sp.
Metham   *Penicillium restrictum, Pyrenochaeta* sp.
CS$_2$ *Trichoderma viride, Fusarium solani, Penicillium luteum*
Ferbam and thiram: *Penicillium* spp., *Trichoderma* spp.
Chloropicrin and allyl alcohol: *Trichoderma viride*

Captan applied to tree-nursery soil at 125 lb/a, a dosage sufficient to inhibit nitrification temporarily, encouraged the growth of *Fusarium* and *Penicillium* as well as *Trichoderma.*[1] PCNB fails to control the pathogenic *Fusarium oxysporum* because this fungus absorbs less of the fungicide than susceptible species such as *Rhizoctonia solani*; moreover it converts more of it to pentachloroaniline (PCA), and in addition produces pentachloro-thioanisole (PCTA) which the susceptible fungus cannot.[72]

The replacement of the native fungal flora by *T. viride* is the most pronounced after soil treatments with allyl alcohol,[75,105] a strong fungicide which is utilized by this resistant species as food.[37] The effectiveness of carbon disulfide to control *Armillaria mellea* is mainly due to this mild fungicide favoring *T. viride*, which then acts as a biological control agent.[5,29] This simplified ecosystem, with a low fungal population dominated by *T. viride*, may persist in citrus soils for 3 yr after the initial D-D treatments which induced it.[55] It is of interest that *T. viride* can reconstitute the turf fungicide chloroneb from its metabolite dichloromethoxyphenol by restoring the second methoxy group.[100]

A change in the constellation of fungal diseases of plants was particularly noted when benomyl was applied to turf or soil. This systemic fungicide, while extremely toxic to the Deuteromycetes among the Fungi Imperfecti, in particular to the pathogens *Verticillium, Fusarium, Botrytis,* and *Rhizoctonia,* and the nonpathogens *Trichoderma, Penicillium,* and *Aspergillus,* is considerably less toxic to most Phycomycetes, as well as bacteria, actinomycetes, and yeasts.[78] Applications of benomyl to turf in New South Wales gave rise to foci of an undescribed Basidiomycete species, which made its appearance only in areas treated with this compound.[87] The use of benomyl soil drenches on carnations in greenhouses has led to severe infections with the leaf-spot fungus *Alternaria dianthi* in Massachusetts and Florida.[54] Cowpeas in plots treated with benomyl and other benzimidazole fungicides were most severely attacked by the wet stem rot fungus *Pythium aphanidermatum,* a situation considered to be caused by the fungicide suppressing antagonists and competitors.[101]

Effects on Nitrifying Organisms

Among the bacteria, the plant pathogens survive soil treatments with fungicides, but the nitrifying bacteria *Nitrosomonas* and *Nitrobacter* are sensitive to them. A decrease in nitrification was observed in soils fumigated with CS_2 as early as 1895, and was a characteristic side effect when chloropicrin and D-D mixture were applied to control nematodes in Hawaiian pineapple fields.[47] Operational concentrations of methyl bromide,[102] EDB, and metham,[46] and the nonvolatile fungicides zineb, maneb, and nabam,[25] all inhibit nitrification. The particular effect of maneb, as also of Dyrene, is to suppress the NH_4-oxidizing *Nitrosomonas* but not the NO_2-oxidizing

Nitrobacter.[21] Other nitrification suppressors include dicloran (2,6-dichloro-4-nitroaniline), but not PCNB quintozene or TCNB tecnazene.[10]

Meanwhile the ammonifiers have not been inhibited,[91,98] and ammonium ions accumulate to the point where they can cause serious root injury.[3] It should be noted that while mild fungicides such as EDB may actually stimulate ammonification, potent fungicides such as chloropicrin may suppress it.[47] D-D mixture applied to control the nematode *Pratylenchus penetrans* tends to induce higher NH_4 and lower NO_3 levels in the soil, although there are no consistent microfloral changes.[23] The decrease in nitrate production may reduce yields in certain crops such as tobacco,[68] and require rectification by nitrate fertilizers.

At the other end of the scale, denitrification is inhibited by ziram, zineb, and ferbam[64]; applications of maneb can cause the accumulation of nitrite for this reason, and so can nabam, whereas captan inhibits denitrification without the accumulation of nitrite.[6]

Nitrification has recovered its normal level within 1–2 months after the application of methyl bromide or chloropicrin,[102] D-D mixture, or EDB,[46] but with heavy applications of D-D or EDB at 50 gal/a the recovery may take 3 months.[91] The less volatile fungicides such as nabam and dazomet inhibit nitrification for about 2 months.[12]

Some species of the nitrogen-fixing *Rhizobium* bacteria may be sensitive to certain of the fungicides, which may thus depress root nodulation of leguminous crops. Even thiram and dichlone when used as seed treatments may be quite toxic to *Rhizobium* populations.[81]

Effects on Algae

Both nabam and captan proved to be inhibitory to freshwater algae at 1 ppm concentration.[50] When tested on estuarine algae in salt water, nabam at this concentration completely inhibited *Dunaliella, Phaeodactylum,* and *Monochrysis,* while it took 10 ppm to inhibit *Chlorella* and *Proloccus.* These two species were completely inhibited by ethylmercury phosphate, a constituent of Lignasan, at only 6 ppb[94]; this fungicide is the only mercurial not suspended in the United States, being used for injecting elms to control *Ceratostomella ulmi.*

Development of Fungicide Resistance in Microorganisms

The development of resistance among susceptible target species of fungi has been reviewed by Georgopoulos and Zaracovitis (1967) and more recently by Lacy and Vargas (1976). Resistance first appeared as a problem in the citrus blue mold *Penicillium digitatum*, a pest of stored lemons, which was combatted by fumigation with biphenyl or by treatments with its close relative sodium orthophenylphenate. This biphenyl resistance was presaged by the appearance of resistant colonies in Israel in 1940; it became a real problem in California packing-houses by 1960, some 5 yr after the introduction of the phenylphenate salt, when resistant strains were found in about one-half of the samples taken.[22,32] Biphenyl resistance has been found in laboratory isolates of two other species of *Penicillium* and also in six more species of fungi.

In the field (Table 15.1), populations of the stem-rot fungus *Rhizoctonia solani* in Louisiana cotton lands have come to differ widely in their susceptibility to the soil fungicide PCNB, which for that reason was giving inconsistent results; isolates from treated fields tended to be more tolerant than those from untreated fields, and the most resistant strains were found in Natchitoches parish,[82] while laboratory colonies resistant to pentachloronitrobenzene were isolated from *Sclerotium rolfsii* in Greece and *Botrytis allii* and *B. cinerea* in England. The potato fungus *Fusarium caeruleum* (*solani*) was found to throw mutants for TCNB resistance on being cultured in tetrachloronitrobenzene in the laboratory.[60] A strain of the cherry rot fungus *Rhizopus arrhizus* strongly resistant to dicloran was discovered among field collections from the San Joaquin Valley of California.[97] The failure of seed treatments of hexachlorobenzene to control common bunt of wheat was becoming frequent in certain parts of Australia by the season of 1964–1965, and it was concluded that the populations of *Tilletia foetida* in three districts of Victoria were resistant to HCB.[49]

In addition to these resistances in the substituted benzenes, a tolerance to dodine (*n*-dodecylguanidine acetate) developed in the apple scab fungus *Venturia inaequalis* in orchards along the Lake Ontario and Lake Geneva shores of New York State, becoming evident in 1965 after it had been used for 6 yr.[89] In Michigan, the dodine tolerance became evident in 1974, after 10 yr of dodine sprays.[40] A resistance to the triazine fungicide anilazine (Dyrene) developed in the dollar-spot fungus *Sclerotinia homoeocarpa* on the turf grass at Urbana, Illinois after 5 yr of its use.[73] It did not develop on a plot treated with actidione and thiram at Urbana, but it did develop on golf greens treated with cadmium succinate and thiram in Pennsylvania.[15]

Resistance of fungi to copper fungicides has not become a problem despite their many years of use, the only exception recorded being in apple orchards in the mountain area of Georgia, United States. Here the black-rot fungus *Physalospora obtusa* was often not controlled by Bordeaux mixture. Two series of tests of conidial samples from orchards in various countries, exposed to a standard lethal deposit, showed the following

Table 15.1. Some instances of fungicide resistance developed in the field.

Biphenyl	Penicillium digitatum	California 1960
	(blue mold of citrus)	
PCNB	Rhizoctonia solani	Louisiana 1960
	(stem mold of cotton)	
Hexachlorobenzene	Tilletia foetida	Australia 1964-65
	(common bunt of wheat)	
Dodine	Venturia inaequalis	New York 1965
	(apple scab)	
Anilazine	Sclerotinia homoeocarpa	Illinois 1970
	(dollar spot of turf)	
Bordeaux Mixture	Physalospora obtusa	Georgia 1952
	(black rot of apple)	
Organic Mercury	Pyrenophora avenae	United Kingdom 1967
	(chrysanthemum rust)	New Zealand 1968
Oxycarboxin	Puccinia horiana	Japan 1974
	(chrysanthomum rust)	
Benomyl	Botrytis cinerea	Netherlands 1970
	(soft rot of cyclamen)	
	Venturia inaequalis	Australia 1970-71
	(apple scab)	Michigan 1974
	Erisyphe & Sclerotinia	Michigan 1973
	(powdery mildews of turf)	
Thiabendazole	Penicillium digitatum	California 1971

percentage germination[90]:

County	Number of years sprayed	Percentage Germination	
		January	February
Gilmer (sprayed)	30	33.2	35.0
Gilmer (unsprayed)	0	1.6	0.8
Habersham	25	3.0	3.6
Union	10	1.2	8.6
Fannin	8	59.8	65.0
White and Rabun	0	12.4	0.6

It is evident that resistance was present in two of the four orchard areas that had been treated for many years, and absent in the two areas that had not been treated. Copper resistance may be induced in a number of fungi cultured on media containing Cu^{2+} ions[4] even when, as in the case of *Piricularia oryzae,* the parent stock had been proved to be uniformly susceptible.[104] The same result may be obtained with single-spore isolates of *Penicillium notatum,*[79] and the induced Cu resistance usually extends to Hg, glyodin, and captan. It appears that mutations occurring during the period of culturing on copper provide the genetic basis of these induced resistances, but that they may be supplemented by temporary phenotypic adaptations which are lost when exposure to copper is discontinued.

Resistance of fungi to mercurial seed-treatments became evident in 1968, when oat seed treated with ethylmercury salts (Ceresan) were no longer freed from the leaf-spot fungus *Pyrenophora avenae* in parts of Scotland and Northern Ireland.[74] The same situation was reported from New Zealand, and out of 26 conidial isolates from seed oats in 13 places in the United Kingdom, 11 of them proved resistant to PMA (phenylmercury acetate). The conidia subsequently obtained from the PMA-resistant isolates gave rise to both R and S colonies, while those obtained from the S gave only S colonies, although they could acquire a transitory tolerance on being grown on PMA-treated media.[31] Mercury-resistance could be induced by continuous culturing of *Penicillium notatum* on Hg-treated media.[79]

Among the new systemic fungicides, resistance has recently appeared to the oxathiin compound oxycarboxin in the chrysanthemum rust fungus *Puccinia horiana* in parts of Japan, where spores from treated fields showed a tenfold resistance sufficient to negate control.[36] Carboxin-resistant isolates have been obtained in laboratory cultures of *Ustilago maydis* and *U. hordei.*[49]

The first report of benomyl resistance came from the Netherlands, where

Botrytis cinerea causing soft-rot in the petioles of nursery-grown cyclamens came to resist control with this benzimidazole compound exactly 1 yr after the first of four successive applications[7]; the resistance was at least 2000 times the normal level, and was stable in benomyl-free culture. In the apple scab fungus *Venturia inaequalis* benomyl resistance first appeared in 1970–1971, causing control failures in the orchards of South Australia; there was cross-resistance to thiophanate-methyl.[99] In Michigan orchards, apple-scab resistance to benomyl developed in 3 yr, as compared to the 10 yr required for development of dodine tolerance.[40] Subsequently, benomyl-resistant strains of powdery mildew and other fungi in the genera *Cercospora, Sphaerotheca, Oidopsis, Septoria* and *Penicillium* have appeared in parts of the United States, Britain, the Netherlands, Israel, and Japan.[49] The powdery mildew *Erisyphe graminis* infecting grass turf has developed resistant strains in benomyl-treated plots in Michigan[95] as has the dollar-spot fungus *Sclerotinia homoeocarpa*.[49] The strains were cross-resistant to thiophanate and thiophanate-methyl, which resemble benomyl in that the first metabolite MBC is the active fungicidal compound. There was cross-resistance to thiabendazole in *Sclerotinia,* but not in *Erisyphe.* Resistance to thiabendazole itself now occurs in a high percentage of the strains of *Penicillium digitatum* and *P. italicum* on lemons in California packing houses.[33]

The genetic basis for fungicide resistance has been investigated in laboratory isolates of fungus species that have not posed any field resistance problems, and the influence of monogenes, oligogenes, or polygenes has been characterized in the various investigations.[30] Resistance mechanisms have been sought in the case of oxathiin compounds, and have been attributed to a less sensitive target enzyme system (succinic oxidase) or to reduced absorption.[49] Benomyl resistant strains of *Aspergillus nidulans* have been found to have a characteristic modification in the tubulin of the spindle fibers formed in mitosis which resists the binding of the toxic metabolite MBC to it.[16] Mercury resistance in *Pyrenophora avenae* has been associated with a greater protective binding of Hg due to an increased level of anthraquinones.

REFERENCES CITED

1. V. P. Agnihotri. 1971. Persistence of captan and its effects on microflora, respiration, and nitrification of a forest nursery soil. *Can. J. Microbiol.* **17**:377–383.
2. R. Alconero and D. J. Hagedorn. 1968. The persistence of Dexon in soil and its effects on soil microflora. *Phytopathology* **58**:34–40.
3. M. Alexander. 1969. Microbial degradation and biological effects of pesticides in soil.

In *Soil Biology: Reviews of Research*. UNESCO, Paris (Place de Fontenoy, 75 Paris 7e). pp. 209–240.

4. J. Ashida. 1965. Adaptation of fungi to metal toxicants. *Annu. Rev. Phytopathol.* **3**:153–174.

5. D. E. Bliss. 1951. The destruction of *Armillaria mellea* in citrus soils. *Phytopathology* **41**:665–683.

6. J. M. Bollag and N. M. Henninger. 1976. Influence of pesticides on denitrification in soil and with an isolated bacterium. *J. Environ. Qual.* **5**:15–18.

7. G. J. Bollen and G. Scholten. 1971. Acquired resistance to benomyl and some other systemic fungicides in a strain of *Botrytis cinerea* in cyclamen. *Netherlands J. Plant Pathol.* **77**:83–90.

8. J. R. Booer. 1944. The behaviour of mercury compounds in soil. *Ann. Appl. Biol.* **31**:340–359.

9. H. P. Burchfield. 1959. Comparative stabilities of Dyrene, dichlone, and captan in a silt loam soil. *Contrib. Boyce Thompson Inst.* **20**:205–215.

10. J. C. Caseley and F. E. Broadbent. 1968. The effect of five fungicides on soil respiration and some nitrogen transformations. *Bull. Environ. Contam. Toxicol.* **3**:58–64.

11. C. I. Chacko, J. L. Lockwood, and M. J. Zabik. 1966. Chlorinated hydrocarbon pesticides: Degradation by microbes. *Science* **154**:893–895.

12. P. Chandra and W. B. Bollen. 1961. Effects of nabam and mylone on nitrification, soil respiration, and microbial numbers. *Soil Sci.* **92**:387–393.

13. W. T. Chin, G. M. Stone, and A. E. Smith. 1970. Degradation of carboxin in water and soil. *J. Agric. Food Chem.* **18**:731–732.

14. G. P. Clemons and H. D. Sisler. 1969. Formation of a fungitoxic derivative from benomyl. *Phytopathology* **59**:705–706.

15. H. Cole, B. Taylor, and J. Duich. 1968. Evidence of differing tolerances to fungicides among isolates of *Sclerotinia homoeocarpa*. *Phytopathology* **58**:683–686.

16. L. C. Davidse. 1975. Antimitotic activity of methyl benzimidazol-2-yl carbamate in fungi and its binding to cellular protein. In *Microtubules and Microtubule Inhibitors*. Eds. M. Borgers and M. de Brabander. North Holland Publ. Co., Amsterdam. pp. 483–495.

17. K. H. Domsch. 1958. Die Wirkung von Bodenfungiziden. II. Wirkungsdauer. *Z. Pflanzenkrankh. Pflanzenschutz* **65**:651–657.

18. K. H. Domsch. 1959. Die Wirkung von Bodenfungiziden. III. Quantitative Veranderung der Bodenflora. *Z. Pflanzenkrankh.* **66**:17–26.

19. K. H. Domsch. 1970. Effects of fungicides on microbial populations in soil. In *Pesticides in the Soil*. Internat. Symposium Michigan State Univ. pp. 42–46.

20. N. Drescher and S. Otto. 1968. Uber den Abbau von Dazomet in Boden. *Residue Rev.* **23**:49–54.

21. H. D. Dubey and R. L. Rodriguez. 1970. Effect of Dyrene and maneb on nitrification and ammonification. *Soil Sci. Soc. Am. Proc.* **34**:435–439.

22. R. Duran and S. M. Norman. 1961. Differential sensitivity to biphenyl among strains of *Penicillium digitatum*. *Plant Disease Rptr.* **45**:475–480.

23. J. M. Elliott, C. F. Marks, and C. M. Tu. 1974. Effects of the nematicides DD and Mocap on soil nitrogen, soil microflora and *Pratylenchus penetrans*. *Can. J. Plant Sci.* **54**:801–809.

24. R. Engst and W. Schnaak. 1970. Untersuchungen zum Metabolismus der fungiziden Athylen-bis-dithiocarbamaten Maneb und Zineb. III. *Z. Lebens-Unter. u-Forsch.* **143**:99–103.

25. C. F. Eno. 1957. *Effect of Soil Applications of Carbamate Fungicides on the Soil Microflora.* Fla. Agric. Exp. Stn., Rep. 142.

26. C. Fest and K. J. Schmidt. 1973. *The Chemistry of Organophosphorus Pesticides.* Springer-Verlag, Berlin. 339 pp.

27. N. Fimreite. 1970. Mercury uses in Canada and their possible hazards as sources of mercury contamination. *Environ. Poll.* **1**:119–131.

28. F. Furukawa, T. Suzuki, and K. Tonomura. 1969. Decomposition of organic mercurial compounds by mercury resistant bacteria. *Agric. Biol. Chem.* **33**:128–130.

29. S. D. Garrett. 1957. Effect of a soil microflora selected by CS_2 fumigation on survival of *Armillaria mellea. Can. J. Microbiol.* **3**:135–149.

30. S. G. Georgopoulos and C. Zaracovitis. 1967. Tolerance of fungi to organic fungicides. *Annu. Rev. Phytopathol.* **5**:109–130.

31. W. Greenaway and J. W. Cowan. 1970. The stability of mercury resistance in *Pyrenophora avenae. Trans. Brit. Mycol. Soc.* **54**:127–138.

32. P. R. Harding. 1962. Differential sensitivity to sodium orthophenylphenate by biphenyl-sensitive and biphenyl-resistant strains of *Penicillium digitatum. Plant Disease Rptr.* **46**:100–104.

33. P. R. Harding. 1972. Differential sensitivity to thiabendazole by strains of *Penicillium italicum* and *P. digitatum. Plant Disease Rptr.* **56**:256–259.

34. A. Helweg. 1973. Persistence of benomyl in different soil types and microbial break-down of the fungicide in soil and agar culture. *Tijdsskr. Planteavl.* **77**:232–243.

35. R. B. Hine, D. L. Johnson, and C. J. Wenger. 1969. The persistence of two benzimida-zole fungicides in soil and their fungistatic activity against *Phymatotrichum omnivorum. Phytopathology.* **59**:798–801.

36. W. Iida. 1975. On the tolerance of plant pathogenic fungi and bacteria to fungicides in Japan. *Japan. Pestic. Information* **23**:13–16.

37. H. L. Jensen. 1959. Allyl alcohol as a nutrient for microorganisms. *Nature* **183**:903.

38. S. Jensen and A. Jernelov. 1969. Biological methylation of mercury in aquatic organisms. *Nature* **223**:753–754.

39. A. Jernelov. 1969. Conversion of mercury compounds. In *Chemical Fallout.* Eds. M. W. Miller and G. G. Berg. Charles C Thomas. pp. 68–74.

40. A. L. Jones and R. J. Walker. 1976. Tolerance of *Venturia inaequalis* to dodine and benzimidazole fungicides in Michigan. *Plant Disease Rptr.* **60**:40–44.

41. D. D. Kaufman. 1974. Degradation of pesticides by soil microorganisms. In *Pesticides in Soil and Water.* Ed. W. D. Guenzi. Soil Sci. Soc. Am., Madison, Wis. pp. 133–202.

42. Y. Kimura and V. L. Miller. 1964. The degradation of organomercury fungicides in soil. *J. Agric. Food Chem.* **12**:253–257.

43. D. H. Klein. 1973. *Mercury in the Environment.* Environ. Prot. Technol. Series (U.S. Environ. Protect. Agency) EPA-660/2.73.008, 23 pp.

44. H. W. Klemmer. 1957. Response of bacterial, fungal, and nematode populations of Hawaiian soils to fumigation. *Proc. Soc. Am. Bacteriol.* **57**:12.

45. W. H. Ko and J. D. Farley. 1969. Conversion of PCNB to pentachloroaniline in soil and the effect of these compounds on soil microorganisms. *Phytopathology* **59**:64–67.

46. H. Koike. 1961. The effects of fumigants on nitrate production in soil. *Proc. Soil Sci. Soc. Am.* **25**:204–206.

47. W. A. Kreutzer. 1963. Selective toxicity of chemicals to soil microorganisms. *Annu. Rev. Phytopathol.* **1**:101–126.

48. R. J. Kuhr, A. C. Davis, and J. B. Bourke. 1974. Dissipation of Guthion, Sevin, Polyram, Phygon, and Systox from apple orchard soil. *Bull. Environ. Contam. Toxicol.* **11**:224–230.

49. M. L. Lacy and J. M. Vargas. 1977. Resistance of fungi to fungicides. In *Chemie der Pflanzenschutz und Schadlingsbekampfungsmittel.* Ed. R. Wegler. Springer-Verlag, Berlin. Vol. IV. pp. 239–256.

50. N. Lazaroff. 1967. Algal response to pesticide pollutants. *Bacteriol. Proc. 1967,* No. G149. p. 48.

51. R. A. Ludwig and G. D. Thorn. 1960. Chemistry and mode of action of dithiocarbamate fungicides. *Adv. Pest. Control Res.* **3**:219–252.

52. R. J. Lukens. 1963. Thiophosgene split from captan by yeast. *Phytopathology* **53**:881.

53. A. W. MacPhee, D. Chisholm, and C. R. McEachern. 1960. The persistence of certain pesticides in soil and their effect on crop yields. *Can. J. Soil Sci.* **40**:59–62.

54. W. J. Manning and P. M. Papia. 1972. Benomyl soil treatments and natural occurrence of *Alternaria* leaf spot on carnation. *Plant Disease Rptr.* **56**:9–11.

55. J. P. Martin, W. C. Baines, and J. O. Ervin. 1957. Influence of soil fumigation for citrus replants on the fungus population of the soil. *Proc. Soil Sci. Soc. Am.* **21**:163–166.

56. J. P. Martin and J. O. Ervin. 1952. Effect of fumigation on soil organisms. *Calif. Citrogr.* **38**:6.

57. J. P. Martin and P. F. Pratt. 1958. Fumigants, fungicides, and the soil. *J. Agric. Food Chem.* **6**:345–348.

58. F. Matsumura, Y. Gotoh, and G. M. Boush. 1971. Phenyl mercuric acetate: Metabolic conversion by microorganisms. *Science* **173**:49–51.

59. J. M. McGuire and M. J. Goode. 1970. Effect of benomyl on *Xiphinema americanum* and tobacco ringspot virus infection. *Phytopathology* **60**:1150–1151.

60. R. E. McKee. 1951. Mutations appearing in *Fusarium caeruleum* cultures treated with tetrachloronitrobenzene. *Nature* **162**:611.

61. C. D. McKeen. 1954. Methyl bromide as a soil fumigant for controlling soil-borne pathogens in vegetable seed beds. *Can. J. Bot.* **32**:101–115.

62. C. M. Menzie. 1969. *Metabolism of Pesticides.* Bureau Sport Fish Wildl., U.S. Dept. Interior, Special Scientific Report, Wildlife No. 127. p. 190.

63. C. M. Menzie. 1974. *Metabolism of Pesticides: An Update.* Bureau Sport Fish Wildl., U.S. Dept. Interior, Special Scientific Report, Wildlife No. 184. p. 181.

64. S. Mitsui, I. Watanabe, M. Hohma, and S. Honda. 1964. The effect of pesticides on denitrification in paddy soil. *Soil Sci. Plant Nutr.* **10**:15–23.

65. K. Mellanby. 1962. *Pesticides and Pollution.* Collins, London. pp. 99–102.

66. P. M. Miller. 1969. Benomyl and thiabendazole suppress root invasion by larvae of *Heterodera tabacum. Phytopathology* **58**:1040–1041.

67. W. Moje, D. E. Munnecke, and L. T. Richardson. 1964. Carbonyl sulphide, a volatile fungitoxicant from nabam in soil. *Nature* **202**:831–832.

68. H. D. Morris and J. E. Giddens. 1963. Responses of several crops to ammonium and nitrite forms of nitrogen as influenced by soil fumigation. *Agron. J.* **55**:372–374.

69. D. E. Munnecke. 1958. The persistence of nonvolatile diffusible fungicides in soil. *Phytopathology* **48**:581–585.

70. D. E. Munnecke, K. H. Domsch, and J. W. Eckert. 1962. Fungicidal activity of air passed through columns of soil treated with fungicides. *Phytopathology* **52**:1298–1306.

71. D. E. Munnecke and J. P. Martin. 1974. Release of methylisothiocyanate from soils treated with Mylone. *Phytopathology* **54**:941–945.

72. T. Nakanishi and H. Oku. 1969. Metabolism and accumulation of PCNB by phytopathogenic fungi in relation to selective toxicity. *Phytopathology* **59**:1761–1762.

73. J. F. Nicholson, W. A. Meyer, J. U. Sinclair, and J. D. Butler. 1971. Turf isolates of *Sclerotinia homoeocarpa* tolerant to Dyrene. *Phytopath. Z.* **72**:169–172.

74. K. M. Old. 1968. Mercury-tolerant *Pyrenophora avenae* in seed oats. *Trans. Br. Mycol. Soc.* **51**:525–534.

75. A. J. Overman and D. S. Burgis. 1956. Allyl alcohol as a soil fungicide. *Phytopathology* **46**:523–535.

76. R. G. Owens. 1969. Metabolism of fungicides and related compounds. *Ann. N.Y. Acad. Sci.* **160**:114–143.

77. R. G. Owens and J. H. Rubinstein. 1964. Chemistry of the fungicidal action of thiram and ferbam. *Contrib. Boyce Thompson Inst.* **22**:241–258.

78. J. F. Parr. 1974. Effects of pesticides on microorganisms in soil and water. In *Pesticides in Soil and Water*. Ed. W. D. Guenzi. Weed Sci. Soc. Am., Madison, Wis. pp. 315–340.

79. A. D. Partridge and A. E. Rich. 1962. Induced tolerance to fungicides in three species of fungi. *Phytopathology* **52**:1000–1004.

80. D. Pimentel. 1971. *Ecological Effects of Pesticides on Non-Target Organisms*. Office Science Technol., Exec. Office President, Gov't. Printing Off., Washington. 220 pp.

81. M. Ruhloff and J. C. Burton. 1951. Compatibility of rhizobia with seed protectants. *Soil Sci.* **72**:283–290.

82. M. N. Shatla and J. B. Sinclair. 1963. Tolerance to pentachloronitrobenzene among cotton isolates of *Rhizoctonia solani*. *Phytopathology* **53**:1407–1411.

83. M. R. Siegel. 1975. Benomyl-soil microbial interactions. *Phytopathology* **65**:219–220.

84. M. R. Siegel and H. D. Sisler. 1968. Fate of the moieties of folpet in *Saccharomyces cerevisiae*. *Phytopathology* **58**:1123–1133.

85. A. K. Sijpesteijn, J. Kaslander, and G. J. M. van der Kerk. 1962. On the conversion of sodium DMDC into its α-aminobutyric acid derivative by microorganisms. *Biochim. Biophys. Acta* **62**:587–589.

86. H. D. Sisler and C. E. Cox. 1951. Release of carbon disulfide from tetramethylthuiram disulfide by fungi. *Phytopathology* **41**:465.

87. A. M. Smith, B. A. Stynes, and K. J. Moore. 1970. Benomyl stimulates growth of a basidiomycete on turf. *Plant Disease Rptr.* **54**:774–775.

88. W. C. Spanis, D. E. Munnecke, and R. A. Solberg. 1962. Biological breakdown of two organic mercurial fungicides. *Phytopathology* **52**:455–462.

89. M. Szkolnik and J. D. Gilpatrick. 1969. Apparent resistance of *Venturia inaequalis* to dodine in New York apple orchards. *Plant Disease Rptr.* **53**:861–864.

90. J. Taylor. 1953. The effect of continued use of certain fungicides on *Physalospora obtusa*. *Phytopathology* **43**:268–270.

91. G. D. Thornton. 1952. Some effects of D-D, EDB, and chloropicrin on microbiological action in several Florida soils. *Soil Crop Sci. Soc. Fla., Proc.* **12**:68–71.

92. K. Tonomura, K. Maeda, F. Futai, T. Nakagami, and M. Yamada. 1968. Stimulative vaporization of PMA by mercury resistant bacteria. *Nature* **217**:644–646.

93. N. J. Turner and M. E. Corden. 1963. Decomposition of sodium N-methyldithiocarbamate in soil. *Phytopathology* **53**:1388–1394.

94. R. Ukeles. 1962. Growth of pure cultures of marine phytoplankton in the presence of toxicants. *Appl. Microbiol.* **10**:532–537.

95. J. M. Vargas. 1973. A benzimidazole resistant strain of *Erisyphe graminis. Phytopathology* **63**:1366–1368.

96. J. W. Vonk and A. K. Sijpesteijn. 1970. Studies on the fate of ethylene-bisdithiocarbamate fungicides and their decomposition products. *Ann. Appl. Biol.* **65**:489–496.

97. D. J. Weber and J. M. Ogawa. 1965. The mode of action of 2,6-dichloro-4-nitroaniline in *Rhizopus arrhizus. Phytopathology* **55**:159–165.

98. R. N. Wensley. 1953. Microbiological studies of the action of some selected soil fumigants. *Can. J. Bot.* **31**:277–308.

99. T. Wicks. 1974. Tolerance of the apple scab fungus to benzimidazole fungicides. *Plant Disease Rptr.* **58**:886–889.

100. M. V. Wiese and J. M. Vargas. 1973. Interconversion of chloroneb and 2,5-dichloro-4-methoxyphenol by soil microorganisms. *Pest. Biochem. Physiol.* **3**:214–222.

101. R. J. Williams and A. Ayanaba. 1975. Increased incidence of *Pythium* stem rot in cowpeas treated with benomyl and related fungicides. *Phytopathology* **65**:217–218.

102. J. P. Winfree and R. S. Cox. 1958. Comparative effects of fumigation with chloropicrin and methyl bromide on mineralization of nitrogen. *Plant Disease Rptr.* **42**:807–810.

103. J. M. Wood, F. S. Kennedy, and G. C. Rosen. 1968. Synthesis of methyl mercury compounds by extracts of a methanogenic bacterium. *Nature* **220**:173.

104. Y. Yamasaki, H. Niizeki, S. Tsuchiya, and T. V. Suwa. 1964. Studies of drug-resistance of the rice-blast fungus *Piricularia oryzae. Bull. Nat. Inst. Agric. Sci. (Japan) D* **11**:1–110.

105. M. Yatazawa, D. J. Persidsky, and S. A. Wilde. 1960. Effect of allyl alcohol on micropopulation of prairie soils. *Proc. Soil Sci. Soc. Am.* **24**:313–316.

16

FUNGICIDES: EFFECTS ON INVERTEBRATES AND VERTEBRATES

A. EFFECTS ON INVERTEBRATE FAUNA

Arthropod Predators and Parasites in Orchards

The fungicides at first applied to pome and citrus orchards were materials such as sulfur and the dinitro compounds which also had an insecticidal effect on orchard insects and mites, many of which are useful natural enemies rather than pests. The present wide array of fungicides available demands that their potential side effects on predators and parasites, particu-

larly those liberated for biological control, be tested systematically following the methods pioneered by workers in California and Nova Scotia.

Past Experience and Tests.   The use of sulfur in dust form to control apple
scab and apple mildew in Nova Scotia orchards led to outbreaks of the eye-
spotted bud moth *Spilonota ocellana* which reached a peak in 1927. When
dusts were replaced by sprays, wettable sulfur still suppressed the predators
Haplothrips faurei, Leptothrips mali, Anystis agilis, and *Anthocoris musculus.*[62] Flotation sulfur in particular so reduced the predator *Hemisar-
coptes malus* and the parasite *Aphytis mytilaspidis* that outbreaks of the
oystershell scale *Lepidosaphes ulmi* followed in the 1930s; this was
eventually remedied by the substitution of ferbam and the copper fungicides.[38,40] The replacement of wettable sulfur by glyodin allowed the
parasite *Agathis laticinctus* to steadily increase its numbers.[62]

Extensive studies on the populations of predators and parasites were
made by experimentally spraying plots in these Nova Scotia orchard sites
with fungicides.[41] The results (Table 16.1) show the deleterious effects of
wettable sulfur and lime–sulfur, as compared to copper, mercurial, and
organic fungicides. Dichlone was the only deleterious organic fungicide,
reducing populations of *Typhlodromus* and *Stethorus*. Among the mite
predators, the Phytoseiids were the most susceptible and *Anystis* the least,
and among the hymenopterous parasites *Agathis* was less susceptible than
Aphytis. The several Mirid species investigated were all as unaffected as the
two species tabulated.

For the fungicides employed in the citrus orchards of California, the
laboratory experiments of Bartlett (1963) performed with deposits simulat-
ing those applied in the field (Table 16.2) show that dinocap, sulfur and
lime–sulfur are toxic to the hymenopterous parasites, but not to the Coccinellids. The sulfur formulations and dinocap are also harmful to
Amblyseius, but not to the Chrysopid predator, to which Bordeaux mixture
is deleterious.[7] Mancozeb (a zinc–manganese ethylene bisdithiocarbamate)
applied to pear orchards at 9 kg/ha offers a means of controlling the pear
psylla without harming the Chrysopid, Mirid, or Anthocorid predators of
pest mites and insects.[42]

Hymenopterous Parasites.   The effect of fungicides has been investigated
on *Trichogramma cacoeciae,* an egg parasite commonly reared in large
numbers for release as a biological control agent. When the adults were
confined to residues,[65] there was no fungicide that did not cause 100% kill
after 56 hr of exposure. When compounds were compared for the rapidity

Table 16.1. Effect of fungicides on predators and parasites in apple orchards (Lord, 1949; MacPhee and Sanford, 1954, 1956, 1961).

		Bordeaux	Fixed Cu	Sulfur	Lime-S	Ferbam	Dichlone	Captan	PMA
P	Typhlodromus pyri	+0	0	++	++	++	+	0	0
A	Anystis agilis	0	0	+	-	0	0	0	0
T	Haplothrips faurei	0	0	+	+	0	+	0	0
T	Leptothrips mali	-	-	+	-	0	0	0	0
M	Diaphnidia pellucida	0	0	0	-	0	-	-	-
M	Deraeocoris fasciolus	0	-	0	-	-	0	0	0
An	Anthocoris musculus	-	-	+0	-	0	0	0	-
C	Stethorus punctum	0	-	+	-	0	+	-	0
B	Agathis laticinctus	0	-	0	-	0	-	-	-
Ap	Aphytis mytilaspidis	0	0	+-	++	0	0	0	0

0 no effect;   +0 possibly a reduction;   + partial reduction;   ++ practical elimination;
- not determined.

P Phytoseiid;   A Anystid;   T Thysanopteran;   M Mirid;   An Anthocorid;
C Coccinellid;   B Braconid;   Ap Aphelinid.

Table 16.2. Effect of fungicides on insect parasites and predators from citrus orchards (Bartlett 1963, 1968).

		Bordeaux	Sulfur	Lime-S	Ferbam	Captan	Zineb	Dinocap
A	Aphytis lingnanensis	O	H	H	O-L	O-L	O-L	H
E	Metaphycus luteolus	O-L	M	M	O	L	O	M-H
E	Metaphycus helvolus	O	M-H	M	O	O	O	M
E	Leptomastix dactylopii	O	M	M	O	O	O-L	H
Co	Cryptolaemus montrouzieri	O-L	O-L	O-L	O-L	O	O	O
Co	Hippodamia convergens	O	O	O	O	O	O	O
Co	Rodolia cardinalis	O	O-L	O	O	O	O-L	O
Co	Lindoros lophanthae	O-L	L	O-L	O	O-L	O	O
Co	Stethorus picipes	O-L	O	O	O	O	O	L
Ch	Chrysopa carnea	L-M	L	O	O	O	O	L
P	Amblyseius hibissci	O-L	M-H	H	O	O	O-L	L-M

O no toxicity; L low toxicity; M medium toxicity; H high toxicity

A Aphelinid wasp; E Encyrtid wasp; Co Coccinellid lady beetle;

Ch Chrysopid lacewing; P Phytoseiid mite

with which they killed, they could be separated into three groups as follows:

Slow kill	Intermediate*	Rapid Kill†
Captan	Dichlofluanid	Dinocap
Dodine	Metiram	Mancozeb
Thiram	Triamiphos	Binapacryl
Zineb	Dithianon	
Oxythioquinox	* ⅓ dead	† ⅔ dead
	after 10 hr	after 4 hr

When the criterion of the side effect was the extent to which the parasitization of *Sitotroga* eggs was inhibited, the compounds separated into the following three categories:

Harmless	Suppressive	Inhibitory
Captan	Mancozeb	Maneb
Folpet	Dinocap	Sulfur
Captafol	Propineb	Pyrazophos
Thiophanate	Zineb	Dinobuton
Cu oxychloride	Thiram	Oxythioquinox

confirming the harmlessness of captan, but not that of oxythioquinox, to *T. cacocciae.*[29] At normal deposit densities, however, oxythioquinox, binapacryl, propineb, and zineb had no effect on the mortality or fecundity of *Aphytis holoxanthus,* parasite of the Florida red scale in Israel.[53] To the Pteromalid *Mormoniella vitripennis,* dinocap, thiram, thioquinox, and oxythioquinox were as harmless as captan even at 10 times the normal concentration, although the last two were toxic at 100 times the normal concentration; binapacryl, an acaricide used against powdery mildew, was harmless only at the normal concentration.[3] Field studies showed that the fungicide–acaricides dinocap and oxythioquinox, as well as the mildew fungicide triamiphos, were just as dangerous to *Aphelinus mali* as wettable sulfur, while thiram, zineb, and even captan could cause some kills.[68] But the general experience in Germany[56] and the Netherlands[9] is that captan is harmless to all predator and parasite populations in the orchard. Ferbam, which in German apple orchards had reduced the incidence of the egg parasite *Trichogramma* by 75 percent,[59] proved as nontoxic as captan and zineb to the parasites of California citrus orchards.

Predaceous Mites. Lime–sulfur and dinocap were fungicides that favored infestations of the European red spider mite *Panonychus ulmi* in England by reducing the Phytoseiid *Typhlodromus pyri* and the Mirid predator *Blepharidopterus angulatus*.[46] In the Netherlands too, the dinocap applied to control apple mildew was found to reduce the population of the predaceous Phytoseiid *Typhlodromus tiliae* by 83% in the first week rising to 98% by the third week after spraying, the reductions in the pest Tetranychid *Panonychus ulmi* being 92% in the first week and 97% in the second week, after which they started to recover their numbers.[67] In Michigan, dinocap caused considerable mortality to *Amblyseius fallacis*.[17] It was also highly toxic to the predaceous mite *Zetzellia mali,* and to the Stigmaeid mite *Agistemus fleschneri* predaceous on the European red spider mite.[16] The proprietary fungicide Dikar (a mixture of dinocap with two dithiocarbamates) naturally had the same effect on *Agistemus*.[47]

Ferbam, zineb, and dichlone were deleterious to the *Typhlodromus* predaceous mites in American orchards, although not as bad as elemental sulfur, while glyodin and captan had little or no harmful effect.[15] Deposits of folpet, maneb, and captan, but not thioquinox, were harmless to the phytoseiid mites *Amblyseius fallacis* and *Phytoseiulus persimilis*.[57] In European orchards, triamiphos was harmless to *Typhlodromus tiliae,* as were captan and thiram.[67] Residues of dodine, or of captan or dithianon, on apple-leaf discs were harmless to *Amblyseius fallacis*.[17] Dodine also had no effect on the Coccinellid *Stethorus punctum,* nor on the predator *Agistemus fleschneri*.[16] Benomyl, in contrast to captan, was very deleterious to the *Agistemus* predator of *Panonychus ulmi,* as shown by the following mite densities per leaf resulting from the annual treatments[47]:

		Agistemus	*Panonychus*
Benomyl	2 oz/100 gal, 5 times	0.03	1.40
Captan	20 oz/100 gal, 5 times	0.44	13.80
Control	Untreated	0.40	0.41

Soil Invertebrates

The effect of fungicides on soil arthropods was investigated in West Virginia orchards, where it was found that the effect of a year's sprays of micronized sulfur, which amounted to 250 lb/a, was not to reduce the numbers of soil arthropods; ferbam, aggregating 75 lb/a in a season, actually increased the numbers by five times.[25] In Argentina[50] it was found that Nipagin (methyl *p*-hydroxybenzoate) and Omadine (a pyridinethione salt) were without effect. However, captan and zineb at the very high dosages of 5000 and 10,000 kg/ha, respectively, caused 60 and 70% reductions; with captan, the marked decrease in *Isotoma thermophila* was accompanied by an increase in *Proisotoma minuta.*

Fumigant fungicides such as D-D mixture and methyl bromide, which are also nematicides, are extremely lethal to populations of springtails and mites in the treated soil; the Collembola have not yet recovered 6 months after the treatments, while the recolonization of mites takes up to 2 yr, and even then it is restricted to a few species such as *Alliphis halleri, Geolaelaps aculeifer,* and *Pergamasius* spp.[21] The soil fumigant dazomet (Mylone), a fungicide and nematicide, when applied at 325 lb/a strongly reduced the populations of Collembola but not of insects in general, and considerably reduced Enchytraeid worms and the Oribatid but not the predaceous Gamasine mites.[19] These fumigants, including metham and chloropicrin, are all nematicidal and proved to be also toxic to earthworms and millipedes in the soil.[20]

Although copper fungicides have a small effect on predators and parasites, their use over a number of years in English orchards has put as much as 2000 ppm into the surface layer of the soil, which results in the disappearance of earthworms.[43] Since *Lumbricus terrestris,* normally present in biomasses up to 1 ton/a, pulls into the soil about 90% of each annual leaf fall (amounting to about 0.5 tons dry weight), there is an accumulation of leaf litter in these treated orchards, providing a source of ascospores of the apple scab to infect the apple foliage in the following spring.[51] In two orchards with a long history of heavy spraying with copper fungicides the only earthworms to be found were a few *L. castaneus,* among the matted and undecayed surface leaf litter.[52] With captan, on the other hand, a soil contamination of 500 lb/a was without effect on *Lumbricus* although it killed some *Helodrilus* earthworms (DeVries, 1962[49]).

The systemic fungicide benomyl, being a carbamate and a cholinesterase inhibitor, is toxic to earthworms. Orchards at Long Ashton, U.K., sprayed with benomyl at 0.25 lb/a seven times in the season, came to have no *Lumbricus terrestris* in the soil, and only a few *Allolobophora.* Where the benomyl deposits are insufficient to kill, they prevent the contaminated leaves from being eaten by the earthworms. Thiophanate-methyl, MBC, and thiabendazole are also toxic, but thiabendazole is not repellent.[61] Pastures in southwestern Ontario sprayed with benomyl at 7 lb/a showed a 95% reduction in numbers and 91% reduction in biomass of the earthworm population, which consisted of *Allolobophora caliginosa* and some *L. terrestris.*[63] Benomyl and thiabendazole when applied to tomato and tobacco had the effect of reducing the entry of the nematode *Heterodera tabacum* into their roots, probably because these benzimidazoles are repellent to the nematode larvae.[45]

B. HAZARDS OF FUNGICIDES TO VERTEBRATES

Mercurial Seed Dressings

The organic mercurials used in seed treatments, at the rate of about 20 ppm by weight, for protection against root-rot fungi have been suspected of reducing populations of seed-eating birds. Bobwhite quail could be killed with grain treated with ethylmercury phosphate after 13–20 days' feeding, although the quail preferred untreated grain (Springer, 1957[49]). When pheasants were fed newly sprouted corn seeds treated with fungicides, taking up to 30 seeds per day for 13 days, ethylmercury phosphate (Semesan Jr.) and mercuric phenyl cyanide (Barbak) caused kidney lesions that were much milder and more temporary than those caused by thiram[37]; since there was no mortality in excess of the controls, nor any reduction in egg or chick production, the authors concluded that the mercurials were not responsible for reductions in pheasant populations. Domestic pigeons force-fed with grain treated with dieldrin or with ethylmercury chloride (Ceresan) were all killed after 7 days on the insecticide but survived without symptoms 6 wk on the mercurial, showing that it was dieldrin that was responsible for the extensive deaths in the spring of 1956 of wood pigeons and pheasants in British croplands sown to treated seed.[14]

However in Sweden, where the principal seed fungicide employed was MMD (methylmercury dicyandiamide), deaths of seed-eating birds were associated with the use of such methylmercury seed treatments. Pheasants fed on MMD-treated grain daily died after 29–61 days, with liver residues of 30–130 ppm, while a reduced egg-hatch was also observed.[11] Residue determinations on the livers of seed-eating birds (gallinaceous, corvine and passerine) and predaceous hawks and owls in Sweden revealed that they exceeded the cited thresholds (ppm) in the following percentages of instances:

Threshold (ppm)	>1	>2	>5	>10	<20
Seed eaters found dead	70	48	30	20	13
Seed eaters shot	66	41	19	12	4
Predators examined	85	62	36	18	11

It will be noted that the residues are higher in those seed eaters found dead than those randomly collected. Among the predaceous birds, of which 26 out of the 412 examined were found dead or dying, the residues were highest in the white-tailed eagle, peregrine falcon, and kestrel, species that had been disappearing in Sweden. Mercury residues were also high in mammalian predators such as the single specimens of red fox, marten, and polecats which were found dying.[11] Consequently the use of MMD and ethylmercury halides as seed dressings was restricted in 1965 and banned in 1966.

Assays made on wood pigeons in southern Sweden before and after the ban[69] found the liver concentrations of Hg to exceed the thresholds cited in the following percentages of birds assayed:

	>2 ppm	>5 ppm
1964	46	30
1966	6	Nil

thus proving that these alkylmercuries rather than general environmental Hg contamination had been responsible for the residues in wildlife. Among the predators, the increase in Hg residues on the introduction of the alkyl-mercuries in the 1940s, and their abrupt fall on their suspension in 1966, was proved by determinations made on the feathers of the goshawk.[35] In Britain, where the principal mercurial seed-dressing was phenylmercury urea (Leytosan), deaths of seed-eating birds were rare.[43]

Indeed, phenylmercury and methoxyethyl mercury were found to be de-graded in the bodies of warm-blooded vertebrates to inorganic Hg as well as methylmercury. On the other hand, methyl- and ethylmercury are eliminated very slowly, with half-lives exceeding 2 months.[8] Feeding experi-ments in which the hydroxides of these organic mercuries were given to chickens to the total amount of 222 mg Hg over 139 days resulted in the following residues in muscle[36] (the liver residues being some 20 times greater):

	Phenyl-mercury	Methoxyethyl-mercury	Methyl-mercury
ppm Hg	0.23	0.39	1.25
Percent as Methylmercury	87	59	94

On the basis of these results, alkoxymethyl mercuries were allowed to be substituted in Sweden, although in France some mortality of wildlife has been ascribed to such seed-dressings.[11] In the United States, methoxyethyl mercuries such as the acetate (MEMA) and the chloride (Agallol) remained available, but phenylmercury urea (Agrox) was discontinued, and sub-sequently all mercurials were suspended in 1975.

On the Canadian prairies in 1968 and 1969, where the only environmental source of mercury was treated seed, the average liver Hg residues were 3 ppm in ring-necked pheasants, 1–2 ppm in grey partridge and horned larks, about 300 ppm in three species of owls, and approximately 1 ppm in Swainson's hawk and the prairie falcon; pheasant eggs with more than 0.5 ppm Hg, coming from birds with over 3 ppm liver residues, showed a reduced hatch.[23] In the United States in 1969, mallard wings contained on the average 1.3 ppm on the Atlantic flyway, 0.35 ppm on the Mississippi flyway, and 0.15 on the Central and California flyways.[30] Bald eagle eggs from Florida to Alaska contained 0.2–0.3 ppm Hg,[71] well below the level of 1.3 ppm found to be the threshold for reduced hatchability in Sweden.[11]

The deleterious effect of methylmercuries on reproductive success was confirmed in mallards, where MMD at 3 ppm Hg in the diet reduced egg-laying and increased embryonic and duckling mortality.[32] MMD on treated seed posed a field hazard to the reproductive success of pheasants, while PMA (phenylmercury acetate) did not.[2] Methylmercury in the diet of breeding females did not reduce eggshell thickness in mallards, pheasants, ringdoves, or American sparrow hawks,[48] although it did so in coturnix quails and was a minor factor contributing to the eggshell thinning in brown pelicans.[10] The seed protectant ethylmercury *p*-toluenesulfonamide (Ceresan M) did not cause eggshell thinning in mallard,[27] pheasants,[58] or coturnix quail[26]; but a reduction in egg laying and hatch rate, besides occasional mortality, has occurred among pheasants feeding on grain treated with this mercurial.[58]

Mercury in Food Chains

Mercurials used as foliar fungicides can be a source of general environmental contamination, and so can the phenylmercury acetate (PMA) used as an anti-slime fungicide in the pulp and paper industry. Even larger sources of mercury pollution are the burning of coal and the use of mercury electrodes in the chlorine–alkali industry. Whatever the source, the mercury finds its way into water; in Sweden the freshwater fish such as perch and pike came to contain 0.5–3.5 ppm Hg in their axial muscle,[35,70] and fish-eating birds such as the white-tailed eagle accumulated 300–400 ppm Hg in their pectoral muscle.[34] In North America, food-chain accumulation reached its peak in a 175-ppm residue in the liver of a great blue heron in Lake St. Clair on the United States–Canadian boundary, where mercury from a chloralkali plant had caused extensive contamination of water and fish.[18] At least two bald eagles appear to have been killed by mercury, and secondary poisoning from contaminated prey has been experimentally proved in young hawks.[60]

The consumption of wheat seed treated with mercurial fungicides has caused epidemics of paralysis and death among human populations, notably in Iraq. The consumption of fish and shellfish contaminated with mercury wastes from chloralkali plants caused an epidemic of paralysis among human populations at Minamata, Japan in the years following 1956.[24,28] Here the Hg residues in the fish ranged up to 50 ppm, whereas those resulting from the general pollution of British as well as Swedish fisheries were usually between 1 and 2 ppm. The intensity of concentration from water to fish is characteristically 1000–2500 times.[33] In Lake Cayuga, New York, the residues in lake trout increased from 0.2 ppm for 1-yr fish up to 0.5 ppm in 6-yr fish, the proportion of methylmercury salts increasing in the process.[5] The input and fate of mercury in the biotic environment has been reviewed by Vostal (1972).

Other Fungicides

Although Bordeaux mixture has the additional advantage of repelling rabbits in vegetable gardens,[49] copper sulfate fungicides as applied to field crops in France have poisoned sheep and chickens, the continuous intake of copper causing a serious jaundice.[4] Of the soil fumigants, chloropicrin is very toxic but the vapor is self-warning, and methyl bromide is dangerous because it is toxic and not self-warning, while EDB is equally toxic and its vapor is persistent.

Virtually all the fungicides used as foliar treatments are safe for man and animals (Table 16.3) and for game birds (Table 16.4). One exception is Dexon, employed for soil treatment against *Pythium* root fungi. The seed protectant dichlone (Phygon) is particularly toxic to fish (LC_{50} 0.1 ppm for bluegills) and freshwater crustacea.[55] Even captan, and more so its relatives folpet and captafol, are sufficiently toxic to zebrafish larvae that the speed at which the fish's heartbeat is stopped at 1 ppm concentration can be used as a bioassay for these compounds.[1] In model ecosystem studies, captan did not accumulate in *Gambusia* mosquitofish, although small amounts of an unknown metabolite were present.[54]

Table 16.3. Acute toxicity of fungicides to the laboratory rat; LD_{50} figures in mg/kg (Pimentel 1971; *Farm Chemicals Handbook*, 1974).

Dexon	64	Dichlone	1300	Dicloran	>10000
Nabam	395	Ziram	1400	Benomyl	>10000
Thiram	780	Thiabendazole	3100	PCNB	>12000
Dinocap	980	Zineb	5200	Thiophanate	>15000
Dodine	1000	Captan	9000	Ferbam	>17000

Table 16.4. Oral toxicity of fungicides to game birds: acute LD_{50}* and chronic LC_{50}† figures.

	Mallard		Pheasant		Coturnix	
	mg/kg	ppm	mg/kg	ppm	mg/kg	ppm
Captan	--	>5000	--	>5000	--	>5000
Folpet	>2000	--	--	--	--	--
Nabam	>2560	--	707	--	2120	--
Zineb	>2000	--	>2000	--	--	--
Thiram	>2800	--	673	--	--	--
Dichlone	>2000	--	--	--	--	--
Dyrene	>2000	--	--	--	--	--
Cadmium succinate	--	1325	--	--	--	2700
Bordeaux mixture	>2000	>2000	--	--	--	--
Panogen‡	56	--	55	--	--	--
Ceresan M	> 173§	--	28§	--	51§	--

* Single oral dose in capsules, 2-week observation period

   (Tucker and Crabtree 1970)

† Treated feed for 5 days, clean feed for 3 days

   (Heath et al. 1972, Pimentel 1971)

‡ Cyano(methylmercuri)guanidine

§ Data for 7.7% formulation calculated to pure ethylmercurytoluene–

   sulfonamide

Hexachlorobenzene, employed as a protectant against bunt fungi of wheat, led to at least 348 cases of cutaneous porphyria when HCB-treated seed was consumed in Turkey 25 yr ago.[13] Contamination of cattle from commercial manufacturing sources in Louisiana in 1972 led to HCB residues in the human population, but they were insufficient to cause cutaneous porphyria.[12] HCB can be a hazard to seed-eating birds, since low concentrations reduce egg hatchability, and the ingestion of treated seeds in quantity has killed birds in Europe.[60] In a model terrestrial–aquatic ecosystem, HCB showed a lower bio-accumulation and a higher biodegradability than any of the chlorinated insecticides, accumulating even more in *Physa* snails than in *Gambusia* mosquitofish.[44]

REFERENCES CITED

1. Z. H. Abedi and D. E. Turton. 1968. Note on the response of zebra fish larvae to folpet and difolatan. *J. Assoc. Off. Anal. Chem.* **51**:1108–1109.

2. W. J. Adams and H. H. Prince. 1972. Survival and reproduction of ring-necked pheasants consuming two mercurial fungicides. In *Environmental Mercury Contamination*. Eds. R. Hartung and B. D. Dinman. Ann Arbor Science Publishers, Inc. pp. 306–317.

3. G. W. Ankersmit, J. T. Locher, H. H. W. Velthuis, and K. W. B. Swart. 1962. Effect of insecticides, acaricides, and fungicides on *Mormoniella vitripennis*. *Entomophaga* **4**:251–255.

4. O. Antoine. 1966. Les antiparasitaires et les animaux domestiques; risques pour les poissons, les abeilles et la vie sauvage. *Parasitica (Gembloux)* **22**:107–114.

5. C. A. Bache, W. H. Gutenmann, and D. J. Lisk. 1971. Residues of total mercury and methylmercuric salts in lake trout as a function of age. *Science* **172**:951–952.

6. B. R. Bartlett. 1963. The contact toxicity of some pesticide residues to hymenopterous parasites and coccinellid predators. *J. Econ. Entomol.* **56**:694–698.

7. B. R. Bartlett. 1968. Outbreaks of two-spotted spider mites and cotton aphids following pesticide treatment. I. Pest stimulation vs. natural enemy destruction as the cause of outbreaks. *J. Econ. Entomol.* **61**:297–303.

8. M. H. Berlin et al. 1969. Maximum allowable concentrations of mercury compounds. *Arch. Environ. Health* **19**:891–905.

9. A. F. H. Besemer. 1964. The available data on the effect of spray chemicals on useful arthropods in orchards. *Entomophaga* **9**:263–269.

10. L. J. Blus, R. G. Heath, C. D. Gish. A. A. Belisle, and R. M. Prouty. 1971. Egg-shell thinning in the brown pelican. *BioScience* **21**:1213–1215.

11. K. Borg, H. Wanntorp, K. Erne, and E. Hanko. 1969. Alkyl mercury poisoning in terrestrial Swedish wildlife. *Viltrevy, Swedish Wildlife* **6**:302–376.

12. J. E. Burns and F. M. Miller. 1975. Hexachlorobenzene contamination: Its effects on a Louisiana population. *Arch. Environ. Health* **30**:44–48.

13. C. Carm and G. Nigogasyan. 1963. Acquired toxic porphyria tarda due to HCB. *J. Amer. Med. Assoc.* **183**:88–91.

14. R. B. A. Carnaghan and J. D. Blaxland. 1957. The toxic effect of certain seed-dressings on wild and game birds. *Vet. Rec.* **66**:324–325.

15. D. W. Clancy and H. J. McAlister. 1958. Effects of spray practices on apple mites and their predators in West Virginia. *Proc. 10th Int. Congr. Entomol.* **4**:597–601.

16. B. A. Croft. 1975. *Integrated Control of Apple Mites*. Mich. State Univ. Extension Bull. E-825. 12 pp.

17. B. A. Croft and E. E. Nelson. 1972. Toxicity of apple orchard pesticides to Michigan populations of *Amblyseius fallacis*. *Environ. Entomol.* **1**:576–579.

18. E. H. Dustman, L. F. Stickel and J. B. Elder. 1972. Mercury in wild animals, Lake St.

Clair, 1970. In *Environmental Mercury Contamination*. Eds. R. Hartung and B. D. Dinman. Ann Arbor Science Publishers, Inc. pp. 46–52.

19. C. A. Edwards and J. R. Lofty. 1971. Nematicides and the soil fauna. *Proc. 6th Brit. Insectic. Fungic. Conf.* **1**:158–166.

20. C. A. Edwards and J. R. Lofty. 1971. *The Biology of Earthworms*. Chapman and Hall, London. 283 pp.

21. C. A. Edwards and A. R. Thompson. 1973. Pesticides and the soil fauna. *Residue Rev.* **45**:1–79.

22. *Farm Chemicals Handbook*. 1974. Meister Publishing Co., 37841 Euclid Ave., Willoughby, Ohio 44094. 466 pp.

23. N. Fimreite, R. W. Fyfe, and J. A. Keith. 1970. Mercury contamination of Canadian prairie seed-eaters and their avian predators. *Can. Field-Nat.* **84**:269–276.

24. L. J. Goldwater. 1972. Human toxicology of mercury. In *Environmental Toxicology of Pesticides*. Eds. F. Matsumura, G. M. Boush, and T. Misato. Academic. pp. 165–175.

25. E. Gould and E. O. Hamstead. 1951. The toxicity of cumulative spray residues in soil. *J. Econ. Entomol.* **44**:713–717.

26. M. A. Haegele and R. K. Tucker. 1974. Effects of 15 common environmental pollutants on eggshell thickness in mallards and coturnix. *Bull. Environ. Contam. Toxicol.* **11**:98–102.

27. M. A. Haegele, R. K. Tucker, and R. H. Hudson. 1974. Effects of dietary mercury and lead on eggshell thickness in mallards. *Bull. Environ. Contam. Toxicol.* **11**:5–11.

28. M. Harada. 1975. Minamata disease: A medical report. In *Minamata*. Eds. W. E. Smith and A. M. Smith. Holt, Reinhart, and Winston. pp. 180–192.

29. S. A. Hassan. 1974. Eine Methodes zur Prufung der Einwirkung von Pflanzenschutzmitteln auf Eiparasiten der Gattung *Trichogramma;* Ergebnisse einer Versuchsreihe mit Fungiziden. *Z. Angew. Entomol.* **76**:120–124.

30. R. G. Heath and S. A. Hill. 1974. Nationwide organochlorine and mercury residues in wings of adult mallards and black ducks during the 1969–70 hunting season. *Pestic. Monit. J.* **7**:153–164.

31. R. G. Heath, J. W. Spann, E. F. Hill, and J. F. Kreitzer. 1972. *Comparative Dietary Toxicity of Pesticides to Birds*. U.S. Bur. Sports Fish. Wildl., Special Scientific Report, Wildlife No. 152. 57 pp.

32. G. Heinz. 1974. Effects of low dietary levels of methyl mercury on mallard reproduction. *Bull. Environ. Contam. Toxicol.* **11**:386–392.

33. A. V. Holden. 1973. Mercury in fish and shellfish: A review. *J. Food Technol.* **8**:1–25.

34. S. Jensen, A. G. Johnels, M. Olsson, and T. Westermark. 1972. The avifauna of Sweden as indicators of environmental contamination with mercury and chlorinated hydrocarbons. *Proc. 15th Internat. Ornithol. Congr.* pp. 455–465.

35. A. G. Johnels and T. Westermark. 1969. Mercury contamination of the environment in Sweden. In *Chemical Fallout*. Eds. M. W. Miller and G. G. Berg. Charles C Thomas. pp. 221–241.

36. A. Kiwimae, A. Swensson, U. Ulfvarson, and G. Westoo. 1969. Methylmercury compounds in eggs from hens after oral administration of mercury compounds. *J. Agric. Food Chem.* **17**:1014–1016.

37. D. L. Leedy and C. R. Cole. 1950. The effects on pheasants of corn treated with various fungicides. *J. Wildl. Manage.* **14**:218–225.

38. F. T. Lord. 1947. The influence of spray programs on the fauna of apple orchards in Nova Scotia. II. Oystershell scale. *Can. Entomol.* **79**:196–209.

39. F. T. Lord. 1949. Ibid. III. Mites and their predators. *Can. Entomol.* **81**:202–214, 217–230.

40. A. W. MacPhee and C. R. MacLellan. 1971. Ecology of apple orchard fauna and development of integrated pest control in Nova Scotia. *Proc. Tall Timbers Conf. Ecol. Animal Control Habitat Mgr.* No. 3, pp. 197–208.

41. A. W. MacPhee and K. H. Sanford. 1954, 1956, 1961. The influence of spray programs on the fauna of apple orchards in Nova Scotia. VII and 1st and 2nd Suppl. *Can. Entomol.* **86**:128–135; **88**:631–639; **93**:671–673.

42. R. D. McMullen and C. Jong. 1971. Dithiocarbamate fungicides for control of pear psylla. *J. Econ. Entomol.* **64**:1266–1270.

43. K. Mellanby. 1967. *Pesticides and Pollution.* Collins, London. pp. 99–102.

44. R. L. Metcalf, I. P. Kapoor, P. Y. Lu, C. K. Schuth, and P. Sherman. 1973. Model ecosystem studies of the environmental fate of six organochlorine pesticides. *Environ. Health Perspect., Exp. Issue* **4**:35–44.

45. P. M. Miller. 1969. Suppression by benomyl and thiobendazole of root invasion by *Heterodera tabacum. Plants Dis. Rptr.* **53**:963–966.

46. R. C. Muir. 1964. The influence of certain fungicides and insecticides on *Panonychus ulmi* and its predators. *Rep. East Malling Res. Stn. 1964.* pp. 167–170.

47. E. E. Nelson, B. A. Croft, A. J. Howitt and A. L. Jones. 1973. Toxicity of apple orchard pesticides to *Agistemus fleschneri. Environ. Entomol.* **2**:219–222.

48. D. B. Peakall and J. L. Lincer. 1972. Methylmercury: Its effect on eggshell thickness. *Bull. Environ. Contam. Toxicol.* **8**:89–90.

49. D. Pimentel. 1971. *Ecological Effects of Pesticides on Non-target Organisms.* Office Science Technol , Exec. Off. Pres., Gov't. Printing Office, Washington. 220 pp.

50. E. H. Rapoport and L. Sanchez. 1968. Effect of organic fungicides on the soil microfauna. *Pedobiologia* **7**:317–322.

51. F. Raw. 1962. Studies of earthworm populations in orchards. I. Lead burial in apple orchards. *Ann. Appl. Biol.* **50**:389–404.

52. F. Raw and J. R. Lofty. 1960. Earthworm populations in orchards. *Rep. Rothamsted Exp. Stn., 1959.* pp. 134–135.

53. D. Rosen. 1967. Effect of commercial pesticides on the fecundity and survival of *Aphytis holoxanthus. Israel J. Agric. Res.* **17**:47–52.

54. J. R. Sanborn. 1974. *The Fate of Select Pesticides in the Aquatic Environment.* U.S. Environ. Prot. Agency, Ecol. Res. Series EPA-660/3-74-025. 83 pp.

55. H. O. Sanders. 1970. Toxicity of some herbicides to six species of freshwater crustaceans. *J. Water Poll. Contr. Fed.* **42**:1542–1550.

56. H. Schneider. 1958. Untersuchungen uber den Einfluss neuzeitlicher Insektizide und Fungizide auf die Blutlauszehrwespe (*Aphelinus mali*). *Z.Angew. Entomol.* **43**:173–196.

57. F. F. Smith, T. J. Henneberry, and A. L. Boswell. 1963. The pesticide tolerance of *Typhlodromus fallacis* and *Phytoseiulus persimilis. J. Econ. Entomol.* **56**:274–278.

58. J. W. Spann, R. G. Heath, J. F. Kreitzer, and L. N. Locke. 1972. Ethyl mercury *p*-toluenesulfonamide; lethal and reproductive effects on pheasants. *Science* **175**:328–331.

59. W. Stein. 1961. Der Einfluss verschiedener Schadlingsbekampfungsmittel auf Eiparasiten der Gattung *Trichogramma. Anz. Schadlingskunde* **34**:87–89.

60. W. H. Stickel. 1974. Effects on wildlife of new pesticides and other pollutants. *Proc. 53rd Ann. Conf. Western Assoc. State Game Fish Comm.* pp. 484–491.

61. A. Stringer and M. A. Wright. 1973. The effect of benomyl and some related compounds on *Lumbricus terrestris* and other earthworms. *Pestic. Sci.* **4**:165–170.

62. H. T. Stultz. 1955. The influence of spray programs on the fauna of apple orchards in Nova Scotia. VIII. *Can. Entomol.* **87**:79–85.

63. A. D. Tomlin and F. L. Gore. 1974. Effects of insecticides and a fungicide on the numbers and biomass of earthworms in pasture. *Bull. Environ. Contam. Toxicol.* **12**:487–492.

64. R. K. Tucker and D. G. Crabtree. 1970. *Handbook of Toxicity of Pesticides to Wildlife.* U.S. Dept. Interior, Fish and Wildl. Ser., Resources Publ. No. 84, 131 pp.

65. H. Ulrich. 1968. Versuche uber die Empfindlichkeit von *Trichogramma* gegenuber Fungiziden. *Anz. Schadlingsk.* **41**:101–106.

66. J. Vostal. 1972. Transport and transformation of mercury in nature. In *Mercury in the Environment.* Eds. L. Friberg and J. Vostal. CRC Press, Cleveland, Ohio. pp. 15–27.

67. M. van de Vrie. 1962. The influence of spray chemicals on predatory and phytophagous mites on apple trees in the Netherlands. *Entomophaga* **7**:243–250.

68. M. van de Vrie. 1967. The effect of some pesticides on the predatory bugs *Anthocorus* and *Orius* and the woolly aphid parasite *Aphelinus mali. Entomophaga, Mem. Hors. Ser.* **3**:95–101.

69. H. Wanntorp, K. Borg, E. Hanko and K. Erne. 1967. Mercury residues in wood pigeons in 1964 and 1966. *Nord. Vet. Med.* **19**:474–477.

70. G. Westoo. 1969. Methylmercury compounds in animal foods. In *Chemical Fallout.* Eds. M. W. Miller and G. G. Berg. Charles C Thomas, Springfield, Illinois. pp. 75–93.

71. S. N. Wiemeyer, B. M. Mulhern, F. J. Ligas, R. J. Hensel, J. E. Mathisen, F. C. Robards, and S. Postupalsky. 1972. Residues of organochlorine pesticides, chlorinated biphenyls, and mercury in bald eagle eggs and changes in shell thickness. *Pestic. Monit. J.* **6**:50–55.

AUTHOR INDEX

Note: The numbers in parentheses are reference numbers and show that an author's work is referred to although his name is not mentioned in the text. Numbers in *italics* indicate the pages on which the full references appear.

Abbott, D. C., 299(2), 302(1), *309*
Abdellatif, M. A., 68(1), *88*
Abedi, Z. H., *463*(1)
Abou-Donia, M. B., 223(1), *240*
Adam, Y., 117(124), *132*
Adams, J. B., 350(1), 350(2), *358*, 374(67), *393*
Adams, L., 172(1), *194*, 204, *213*, 219, 220, *240*
Adams, R. E., 87(2), *88*
Adams, R. F., 237(118), *246*
Adams, W. J., 460(2), *463*
Addington, W. W., 381(157), *397*
Addison, R. F., 299(3), 300(3), 301(3), *309*
Adelson, B. J., 109(117), 110(117), *132*
Adkisson, P. L., 22(43), *26*, 50(1), 50(75), 51(75), *57, 61*
Agnihotri, V. P., 439(1), *445*
Agthe, C., 19(1), *24*
Ahmed, M. K., 50(2), *57*, 118(1), *126*
Ahmed, W., 148(135), *167*, 187(153), *202*
Ainardi, V. R., 256(73), *268*
Akamine, E. K., 368(1), *390*
Alabaster, J. S., 185(2), *194*, 405(1), *426*
Albert, R. A., 407(12), *426*
Albert, T. F., 222(3), *240*
Albert, W. B., 94(2), *126*
Albone, E. S., 100(3), *126*, 275(4), 276(4), *309*
Alconero, R., 433(2), *445*
Aldrich, J. A., 172(67), *197*
Aldrich, J. W., 212(37), *215*
Aleeva, L. V., 207(2), *213*

Alexander, M., 99(4, 43, 135), 100(73, 135), 109(68), *127, 128, 130, 133*, 275(152), *317*, 325(1), 330(1), 332(37), *339, 341*, 363(2, 91, 129, 150), 364(2, 94), 365(24, 148), 369(24, 52, 98), 369(203, 204), 372(91), 382(42), 385(216), *390, 391, 392, 393, 395, 396, 397, 400*, 439(3), 441(3), *445*
Allen, L. D., 366(51), *393*
Allison, D. B., 181(2), 181(4), *194*
Aly, O. M., 369(3), *390*, 403(2), *426*
Ames, P. L., 250(1), *264*
Anas, R. E., (5), *310*
Andersen, R. N., 327(40), *341*
Anderson, C. A., *et al*, 107(5), *127*, 238(4), *240*
Anderson, D. W., 250(40), 250(83), 251(4), 252(3), 253(3), 257(83), 258(83), 258(3), 259(83), 260(2), 261(3), *264, 266, 268*
Anderson, G. R., 364(4), 384(4), 385(4), *390*
Anderson, H. L., 67(64), 76(64), 78(64), *90*
Anderson, J., 157(83), *164*
Anderson, J. L., 352(4), *358*
Anderson, J. M., 179(5), 179(131), 180(6), 180(7), 180(83, 138), 181(138), *194, 198, 200, 201*, 282(156), *317*
Anderson, J. P. E., 100(6), *127*, 377(5), *390*
Anderson, L. D., 55(3), 56(3), *57*, 57(3), 351(3), *358*
Anderson, R. B., 182(8), *194*

467

SUBJECT INDEX

Abalone, red, 306
Abate, *see* Temephos
Accipiter nisus, 262
Acheta, 107
Achromobacter, 99, 364, 366, 386
   degradation of herbicides, 368, 369, 373, 376
Acris crepitans, 210, 211
Acris gryllus, 211
Acrolein, 408, 409, 422
Actidione, 442
Actinomycetes stimulated by pesticides, 124, 439
Aculus schlechtendali, 45, 46
Acute versus chronic toxicity to birds, 239, 240
Adaptation of soil bacteria, 368, 369
Adenostoma fasciculatum, 355
Adsorption, of herbicides, 335–338
   of organochlorines, 273–275
Aechmophorus occidentalis, 285, 286
Aedes nigromaculis, 21, 22, 150
Aedes sollicitans, 159
Aerobacter aerogenes, 366, 373
   on DDT in hydrosoil, 276, 280, 281
   on insecticides in soil, 99, 100, 102, 120, 122
Agathis laticinctus, 452, 453
Agistemus fleschneri, 45, 456
Agonoderus lecontei, 72
Agonum dorsale, 73, 74
Agonum retractum, 35
Agoseris, 322
Agriolimax reticulatus, 85
Agriotes sputator, 351
Agrobacterium, 364, 366, 376
Agrobacterium tumefaciens, 99, 281, 373
Agrostis tenuis, 346

Alachlor, 16, 421
   fate in soil, 330, 333, 337, 338
Alamine, 11, 151
Alcaligenes, 364
Aldicarb, 120, 121, 239
   on beneficial insects, 50, 65, 70
   on earthworms and nematodes, 84, 87
Aldicarb alcohol sulfone, 120, 121
Aldicarb amide sulfone, 120
Aldicarb sulfone, 120, 121
Aldicarb sulfoxide, 120, 121
Aldrin, 4, 11–14, 16, 17, 19, 20, 22, 54, 102
   on amphibia and snakes, 210, 211, 288
   on aquatic invertebrates, 138, 141–146, 150, 154, 156
   in atmosphere, 299
   effect on algae, 157, 281
   effect on birds, 220, 228, 229, 231, 235
   effect on earthworms, 71, 77, 79–81
   effect on fish, 169, 173–176, 178, 179, 193, 289
   epoxidation in soil, 101
   fate in water, 275, 276
   persistence in soil, 95, 96, 101, 106
   residues in mice, 208, 288
   in runoff water, 272, 273, 293
   on slugs and nematodes, 85–87
   on soil arthropods, 65, 67, 69, 71, 74–76
   on soil bacteria and pathogens, 123, 124
Aldrin diol, 102, 276
Alewife, 250, 286, 297
Alfalfa ecosystem, 31, 55, 57
Algae, 276–278, 408, 441
   effect of herbicides, 384, 390, 422, 424
   effect of insecticides, 155, 157, 159
Allethrin, 4, 139, 145, 146
Alliphis halleri, 457

DATE DUE

GAYLORD			PRINTED IN U.S.A.